Facility Location

Springer
*Berlin
Heidelberg
New York
Barcelona
Hong Kong
London
Milan
Paris
Tokyo*

Zvi Drezner · Horst W. Hamacher
(Editors)

Facility Location
Applications and Theory

With 60 Figures and 20 Tables

Springer

Professor Zvi Drezner
California State University Fullerton
College of Business and Economics
CA 92834 Fullerton
USA
zdrezner@fullerton.edu

Professor Dr. Horst W. Hamacher
University of Kaiserslautern
Department of Mathematics
Box 3049
67653 Kaiserslautern
Germany
hamacher@mathematik.uni-kl.de

ISBN 3-540-42172-6 Springer-Verlag Berlin Heidelberg New York

Library of Congress Cataloging-in-Publication Data applied for
Die Deutsche Bibliothek – CIP-Einheitsaufnahme
Facility location: applications and theory; with 20 tables / Zvi Drezner; Horst W. Hamacher (ed.). – Berlin; Heidelberg; New York; Barcelona; Hong Kong; London; Milan; Paris; Tokyo: Springer, 2002
 ISBN 3-540-42172-6

This work is subject to copyright. All rights are reserved, whether the whole or part of the material is concerned, specifically the rights of translation, reprinting, reuse of illustrations, recitation, broadcasting, reproduction on microfilm or in any other way, and storage in data banks. Duplication of this publication or parts thereof is permitted only under the provisions of the German Copyright Law of September 9, 1965, in its current version, and permission for use must always be obtained from Springer-Verlag. Violations are liable for prosecution under the German Copyright Law.

Springer-Verlag Berlin Heidelberg New York
a member of BertelsmannSpringer Science+Business Media GmbH

http://www.springer.de

© Springer-Verlag Berlin · Heidelberg 2002
Printed in Germany

The use of general descriptive names, registered names, trademarks, etc. in this publication does not imply, even in the absence of a specific statement, that such names are exempt from the relevant protective laws and regulations and therefore free for general use.

Hardcover-Design: Erich Kirchner, Heidelberg

SPIN 10841482   42/2202-5 4 3 2 1 0 – Printed on acid-free paper

Preface

Wouldn't it be nice to have a book on location analysis which

- contains most of the main stream topics,
- could be used for teaching purposes,
- could be used for self study or reference,
- would motivate the material not only by reviewing the theoretical results, but also by giving some of the practical success stories using location theory as model for problems in industry and management science?

As the reader may guess, this question is a rhetoric one – sure it would be nice to have such a book, and many who want to learn about the subject or teach corresponding courses in university departments of mathematics, management science, computer science, geography, civil engineering, etc. would be likely users of such a book. Practitioners who want to find out, whether one of their problems can be solved with one of the location tools would love to get a quick overview of what is available in this area. Researchers could use the book to check whether there is any new literature on a location subject they would want to work on.

This book is an attempt to satisfy this wish list. We were lucky enough to find many of the most prominent researchers and practitioners in the world, who contributed 14 chapters to a book in which the largest part of the location area is covered. A most up-to-date collection of references is provided at the end of each chapter.

In the first chapter **"The Weber Problem"**, Zvi Drezner, Kathrin Klamroth, Anita Schöbel, and George O. Wesolowsky introduce the classic of location theory, the Weber problem in the plane. Many historical notes show, how location theory has evolved over time, and that the idea of this book – to show the power of models from location theory to solve real-world problems – is well-founded by the positive experience over the last century. The authors discuss the single facility Weber problem with Euclidean distance in some details and introduce the reader to modifications of this basic model by changing the distance to ℓ_p-norms or gauges, and by finding more than one new facility.

In situations, where certain regions are not allowed to be used for siting new facilities, or may not even be used for trespassing more advanced models are required. These situations occur obviously very often, if geographic characteristics have to be taken into account in regional planning problem, machines or conveyer belts have to be considered in production environments, etc. The authors introduce restricted and barrier location problems as corresponding models.

The last topic which is discussed in this chapter is devoted to line locations or locations of objects. Planning of railway lines of highways and plant layout are examples where the algorithms sketched by the authors can be used.

A point is covered by a facility when its distance to the facility is within some threshold radius. Obviously, many locational questions in practice deal with this problem, in particular in the social area, where "good" covering may mean the difference between life and death (for instance in a good covering of neighborhoods by health services, see Chapter 4). Frank Plastria reviews in Chapter 2 many classical, but also very recent **"Continuous Covering Location Problems"**. When locating a useful and desirable facility one may seek full covering, i.e. a covering of all points of a given set using a smallest possible radius, or maximal covering, when one wants to cover as many points as possible within a fixed radius. When locating a necessary, but undesirable facility, one will rather look for empty or minimal covering, i.e. try to cover none or the smallest number of possible points within the largest possible or given radius. Frank describes in Chapter 2 applications of such questions occurring within a given geographical region, their models and corresponding solution methods. In the plane, distance is traditionally measured the Euclidean way, and covering problems then amount to placing circular disks. This seems easy, since it uses traditional geometric objects only. But, from an algorithmic point of view, various difficulties need to be overcome using a number of ideas and constructs from computational geometry.

As in Chapter 1, need arises to use other distances than the Euclidean one to model real-world problems. While some theoretical properties nicely carry over, solution methods must be adapted to the new geometric objects involved. Further topics which are discussed by Frank include other environments than the plane, the need to locate several facilities simultaneously, and line and dimensional facilities.

The third chapter by John Current, Mark Daskin and David Schilling presents several classic **"Discrete Network Location Models"**. Here, a "discrete" problem is one for which there are a fixed number of candidate facility sites and the locations of these are known with certainty. Consequently - in contrast to the first two chapters - "continuous" problems in which the facilities can be located anywhere in the feasible region are not addressed. A "network" problem is one in which the set of potential facility sites is located at the nodes of some underlying network. In general, it is assumed that demand also exists at some known set of nodes on the underlying network.

As basic models, the set covering, maximal covering, p-center, p-dispersion, p-median, fixed charge hub and maxisum problem are introduced. While these basic models assume direct delivery from a facility, many facilities such as solid waste collections substations and distribution centers provide collection and/or distribution functions in which demand is served by multiple drop off and/or pickup routes. In order to efficiently plan this services, location- routing models can be used. Obviously, the design of the network is in all models of crucial importance. Corresponding models are reviewed. In all models, multiple criteria, time dependency and stochasticity can be important to mirror reality as well as possible. Several classes solution techniques are presented to solve all of these problems. The chapter concludes with a discussion of recent directions in modifying and extending these classic problems.

The chapter **"Location Problems in the Public Sector"** written by Vladimir Marianov and Daniel Serra deals with the location of facilities or services in discrete space or networks, that are related to the public sector, such as emergency services (ambulances, fire stations, and police units), school systems and postal facilities. The main difference between this class of problems and the location of private facilities is in the objective function. Profit maximization and capture of larger market shares from competitors are the main criteria in private applications, while social cost minimization, universality of service, efficiency and equity are the goals in the public sector. Since these objectives are in general difficult to measure, they are frequently surrogated by the minimization of locational and operational costs needed for full coverage by the service, or the search for maximal coverage given an amount of available resources.

Vladimir and Daniel first focus on public facility location models that use some type of coverage criterion, with special emphasis on emergency services. The second section examines models based on the p-median problem and some of the issues faced by planners when implementing this formulation in real-world locational decisions. The last section examines new trends in public sector facility location modeling

In the fifth chapter **"Consumers in Competitive Location Models"** Tammy Drezner and H.A. Eiselt extend previous models in a decisive way. Instead of assuming that the customers are allocated by certain rules to the facilities, customers are in this type of models considered free to choose their facility. Assuming some knowledge about the customers attributes (disposable income, age, level of education, etc.), complete market and distance information and a "rational" behavior of all customers, each of them will evaluate potential facilities by a utility function. Consequently, they will choose the facility which has for them the highest utility.

After introducing the main elements of service-oriented competitive location models, Tammy and H.A. show, how the major decisions made by consumers to arrive at their purchasing decision are structured. The paper then demonstrates ways to incorporate facility features in the decision-making process. A discussion concerning the lack of rationality on the part of consumers as well as more complex behavioral patterns conclude the paper.

The p-median problem has been introduced in Chapter 3. Burcin Bozkaya, Jianjun Zhang, and Erhan Erkut show in the sixth chapter **"An Efficient Genetic Algorithm for the p-Median Problem"** how to tackle this difficult problem using a genetic algorithm (GA): GA is a modern heuristic method that has been applied to a number of combinatorial optimization problems with success. However, there are very few applications of GAs to location-allocation problems. In this chapter, the authors provide an overview of GA concepts, review related previous studies, and develop a GA considering three different crossover operators as well as mutation and invasion features. They report the results of their computational experiments with medium-sized random instances of the p-median problem. It turns out, that the GA produces solutions are of comparable quality to those produced by a well-known and widely used heuristic for this problem. The obvious conclusion is that GA has the potential to provide very good solutions for the p-median problem.

Because location problems with many demand points are difficult to solve, most large-scale location problems are solved by first grouping the demand points into representative aggregate points, and then locating servers with respect to these aggregate points. In Chapter 7 entitled **"Demand Point Aggregation for Location Problems"** Richard L. Francis, Timothy J. Lowe, and Arie Tamir study the resulting error. They provide a useful upper bound on this error which leads to algorithms to determine, where aggregate demand points should be placed with the objective of minimizing the value of this error bound. They give an example of one such algorithm when the objective of the original location problem is the p-median problem. They also report on some computational experimentation with the algorithm. Finally, they generalize their approach to obtain error bounds for an entire class of location problems.

The objective of Chapter 8 **"Location Software and Interface with GIS and Supply Chain Management"**, co-authored by Thorsten Bender, Holger Hennes, Jörg Kalcsics, M. Teresa Melo, and Stefan Nickel, is to bridge the gap between location theory and practice by focussing on the development of software capable of addressing the different needs of a wide group of users. For those interested in non-commercial applications (e.g. students and researchers), the library of location algorithms (LoLA) can be of considerable assistance. LoLA contains a collection of efficient algorithms for solving planar, network and discrete facility location problems. These algorithms are available as callable library and therefore can be incorporated into other applications. To address the specific needs of users working with large amounts of demographic data (e.g. urban planners), LoLA was linked to a geographical information system (GIS), thus taking advantage of the additional functionality that such a tool offers. Finally, there is a wide group of practitioners who need to solve large problems and require special purpose software with a good data interface. For these users, a commercial location software tool in the area of supply chain management (SCM) was developed. The tool is embedded in the Advanced Planner and Optimizer SCM software developed by SAP AG, Germany.

The interrelation between **"Telecommunication and Location"** is the subject of Chapter 9. The authors, Eric Gourdin, Martine Labbé, and Hande Yaman review models for telecommunication network design where location problems are involved. The key observation here is, that telecommunication networks are naturally structured in a multi-layer hierarchical architecture.

At a lower level of such an architecture, collectors are used to collect the traffic and to sent it to a higher level. The network design is then done in an iterative way by deciding on the number and location of concentrators, and the assignment of the terminals to these concentrators, the design of the access network and then the design of the backbone network. The location of the collectors is therefore of crucial importance for the network design problem.

The authors consider three classes of models, namely uncapacitated, capacitated and dynamic ones. For each class, they discuss the core problem, its generalizations and the solution methods in the literature.

In the chapter **"Reserve Design and Facility Siting"**, Charles ReVelle and Justin C. Williams compare five classes of discrete siting models as to form and solution method against parallel parcel selection models for species protection. The location set covering model is compared to the species set covering problem, and the maximal covering location problem is viewed alongside the maximal covering species problem. As well, location covering models seeking backup and redundant coverage are seen to have distinct analogs in parcel selection models. In addition, the maximum expected covering problem has a mirror image in a model that seeks the maximum expected number of species protected through appropriate parcel choice. Finally, the maximum availability location problem, which maximizes calls covered with alpha reliability given a resource or facilities limit, has parallels in a parcel selection model which seeks to maximize the number of species protected with the stated reliability.

Oded Berman and Dmitry Krass consider in Chapter 11 **"Facility Location Problems with Stochastic Demands and Congestion"** a location problem under two sources of uncertainty: the exact timing of customer demands, and the potential congestion at the service facilities. These problems, referred to as "Location Problems with Stochastic demands and Congestion" (LPSDC), arise in location of emergency service facilities, as well as certain non-emergency (e.g., retail or health service) facilities. The goal of this chapter is to review the main approaches for LPSDC models on networks, identify the assumption structures implicit in these approaches, and discuss potential avenues for future work in this area.

After formulating a general LPSDC model, Oded and Dmitry point out the inherent complexities of the model related to describing the stochastic behavior of the underlying system. They provide a classification of the available approaches into cover-type and median-type models, with a further sub-division into fixed and mobile service models. Cover-type models attempt to assure specified standards of coverage (e.g., how many minutes a call for service is allowed to wait in the queue) through the use of service level constraints. By using strong assumptions about the underlying system, these constraints can be linearized, leading to solvable integer programming formulations. However, the effect of the assumptions on the quality of the underlying solutions is not clear – in particular, there is often no guarantee that the resulting solutions in fact satisfies the service level constraints. Median-type models attempt to minimize average response time to calls for service (they also contain a penalty term for calls that are rejected by the system). These models generally contain a much more accurate representation of the stochastics of the underlying system than the cover-type models. As a result, the formulation is often not available in closed form, leading t o a variety of location-allocation heuristic approaches. Some important questions for future research are outlined, and strengths and weaknesses of different approaches are discussed.

Chapter 12 deals with **"Hub Location Problems"**. The importance of this topic is, in particular, motivated by the fact, that hub networks play a very important role in modern transportation and telecommunication systems. In transportation, this includes air passenger and freight travel, express shipments (e.g. overnight delivery), large truck systems, postal operations and rapid transit systems. In the telecommunication area, computer communication, telephone networks, video conferences, distributed computer processing, etc. all rely on a smart installation of hubs. Rather than serving every origin-destination demand with a direct link, a hub network provides service via a smaller set of links between origins/destinations and hubs, and between pairs of hubs. The authors concentrate on new discrete or network hub location models in which location of the hubs is a key decision. Research that addresses hub network design, without determining hub locations is not considered. Since there have been previous survey articles on hub location, emphasis is on new work in this area.

Oliver Karch, Hartmut Noltemeier and Thomas Wahl consider in Chapter 13 the interrelation between **"Location and Robotics"**. An autonomous mobile robot, that has to navigate in its environment must be able to answer the following three questions: "Where am I?", "Where am I going?", and "How should I get there?", which are summarized under notion of *localizing a robot.* The authors give an overview of the localization problem and then consider the sensors with which the robots are equipped, the different types of localization (relative and absolute), and the corresponding localization methods. Variants of the localization problem, where some restrictive assumptions about the robot and its environment allow a formulation as a pure geometric problem are introduced. The latter problem is solved using methods from Computational Geometry. Due to the restrictive assumption, the resulting solution is, in general not directly applicable in practice, where the data is noisy. Therefore, the authors introduce distance functions of model the resemblance between the noisy sensor data and structures of the original method.

The last chapter is written by Franz Rendl, who discusses various aspects of **"The Quadratic Assignment Problem"** (QAP). This problem can be used to model various real- life problems including the design of a university campus or a hospital, the layout of a factory, wiring problems, the assignment of letters to typewriters, ranking or archaeological data, etc. On the other hand, notoriously difficult optimization problems such as the traveling salesman problem, the graph partition problem, or the max clique problem are special cases of QAP. Franz gives several equivalent formulations of QAP, describes various ways of relaxations, which can be used to solve it either exactly of by heuristics. He also discusses complexity issues and reports on computational experience with some of the methods described in his chapter.

May 2001

Zvi Drezner
Horst W. Hamacher

Contents

1 The Weber Problem 1
Zvi Drezner, Kathrin Klamroth, Anita Schöbel, George O. Wesolowsky
1.1 Introduction 1
1.2 History and Literature Review 3
1.3 Solution Procedures 6
1.4 Properties of the Weber Problem 11
1.5 Other Distance Measures 13
1.6 Multiple Facilities 16
1.7 Restricted Weber Problems 18
1.8 Line Location and Dimensional Facilities 20
1.9 Extensions 23
1.10 Epilogue 24
References 24

2 Continuous Covering Location Problems 37
Frank Plastria
2.1 Introduction 37
2.2 Full Covering 45
2.3 Maximal Covering 56
2.4 Empty Covering 60
2.5 Minimal Covering 63
2.6 Push-Pull Covering Models 64
2.7 Positioning Models 65
2.8 Multiple Facility Covering Location Models 65
2.9 Extensive Facility Covering Location Models 69
References 72

3 Discrete Network Location Models 81
John Current, Mark Daskin, David Schilling
3.1 Introduction 81
3.2 Basic Facility Location Models 82
3.3 Location-Routing Models 95
3.4 Facility Location-Network Design Models 96
3.5 Multiobjective Models 96
3.6 Dynamic Location Models 98
3.7 Stochastic Location Models 98
3.8 Solution Approaches for Location Models 101
3.9 Conclusions 107
References 108

4 Location Problems in the Public Sector ... 119
Vladimir Marianov, Daniel Serra
- 4.1 Introduction ... 119
- 4.2 Covering Models in the Public Sector ... 120
- 4.3 p-Median Models in Public Facility Location ... 132
- 4.4 Conclusions ... 142
- References ... 143

5 Consumers in Competitive Location Models ... 151
Tammy Drezner, H. A. Eiselt
- 5.1 Introduction ... 151
- 5.2 Incorporating Facilities Features ... 155
- 5.3 The Lack of Rationality ... 160
- 5.4 More Complex Customer Behavior ... 164
- 5.5 Conclusions ... 169
- References ... 169

6 An Efficient Genetic Algorithm for the p-Median Problem 179
Burçin Bozkaya, Jianjun Zhang, Erhan Erkut
- 6.1 Introduction ... 179
- 6.2 The p-Median Problem ... 180
- 6.3 Genetic Algorithms ... 182
- 6.4 Review of the Relevant Literature ... 184
- 6.5 The Proposed Genetic Algorithm ... 186
- 6.6 Computational Study ... 192
- 6.7 Conclusions ... 202
- References ... 204

7 Demand Point Aggregation for Location Models ... 207
Richard L. Francis, Timothy J. Lowe, Arie Tamir
- 7.1 Introduction ... 207
- 7.2 The Aggregation Problem ... 208
- 7.3 Aggregation Error ... 210
- 7.4 Guidelines for Aggregation ... 214
- 7.5 An Aggregation Algorithm ... 215
- 7.6 Computational Experience ... 219
- 7.7 Error Bound Generalizations ... 222
- 7.8 Summary ... 229
- References ... 230

8 Location Software and Interface with GIS and Supply Chain Management ... 233
Thorsten Bender, Holger Hennes, Jörg Kalcsics, M. Teresa Melo, Stefan Nickel
- 8.1 Introduction ... 233
- 8.2 LoLA– Library of Location Algorithms ... 235

8.3	LoLA goes GIS	249
8.4	Supply Chain Management	255
8.5	Outlook	271
	References	272

9 Telecommunication and Location ... 275
Eric Gourdin, Martine Labbé, Hande Yaman

9.1	Introduction	275
9.2	Uncapacitated Models	278
9.3	Capacitated Concentrator Location Problem	288
9.4	Capacitated Models	299
9.5	Dynamic Models	301
	References	302

10 Reserve Design and Facility Siting ... 307
Charles ReVelle, Justin C. Williams

10.1	Introduction	307
10.2	Set Covering Problems	308
10.3	Maximal Covering Problems	311
10.4	Redundant/Backup Coverage Problems	313
10.5	Chance Constrained Covering Models	316
10.6	Expected Covering Models	323
10.7	Conclusion	326
	References	326

11 Facility Location Problems with Stochastic Demands and Congestion ... 329
Oded Berman, Dmitry Krass

11.1	Introduction	329
11.2	Coverage Problems with Stochastic Demand and Congestion	339
11.3	Problems with Median-Type Objective: The Stochastic Queue Model	356
11.4	Conclusions and Open Problems	368
	References	369

12 Hub Location Problems ... 373
James F. Campbell, Andreas T. Ernst, Mohan Krishnamoorthy

12.1	Introduction	373
12.2	Background	374
12.3	Recent Trends	381
12.4	Models and Taxonomy	383
12.5	Applications	388
12.6	Solving Hub Location Problems	393
12.7	Conclusions	400
	References	402

13 Location and Robotics ... 409
Oliver Karch, Hartmut Noltemeier, Thomas Wahl

13.1 Introduction ... 409
13.2 Related Problems ... 410
13.3 A Short Overview of the Localization Problem ... 410
13.4 Solving the Geometric Problem ... 413
13.5 A Sharper Bound for $|\mathcal{EC}|$ ... 417
13.6 Problems in Realistic Scenarios ... 422
13.7 Adaptation to Practice ... 424
13.8 Suitable Distances for $d(\mathcal{S}, V^*)$ and $D(V_1^*, V_2^*)$ ... 426
13.9 Our Implementation RoLoPro ... 432
13.10 Experimental Tests ... 435
13.11 Possible Enhancements to the Algorithms ... 436
References ... 436

14 The Quadratic Assignment Problem ... 439
Franz Rendl

14.1 Introductory Example ... 439
14.2 Equivalent Formulations of QAP ... 440
14.3 Applications ... 442
14.4 Computational Complexity of QAP ... 443
14.5 Relaxations of QAP ... 443
14.6 Heuristics ... 453
14.7 Computational Experience ... 454
14.8 Bibliographical Notes ... 455
References ... 455

1 The Weber Problem*

Zvi Drezner[1], Kathrin Klamroth[2], Anita Schöbel[3], and George O. Wesolowsky[4]

[1] College of Business and Economics, California State University, Fullerton, CA 92834. e-mail: zdrezner@fullerton.edu
[2] Department of Computer Science and Mathematics, University of Applied Sciences, Dresden, Germany. e-mail: klamroth@math.ku.dk
[3] Department of Mathematical Sciences, University of Kaiserslautern, Germany. e-mail: schoebel@mathematik.uni-kl.de
[4] Faculty of Business, McMaster University, Hamilton Ont. L8S-4M4, Canada. e-mail: wesolows@mcmail.cis.McMaster.CA

1.1 Introduction

The Weber problem discussed in this chapter has a long and convoluted history. Many players, from many fields of study, stepped on its stage, and some of them stumbled. The problem seems disarmingly simple, but is so rich in possibilities and traps that it has generated an enormous literature dating back to the seventeenth century, and continues to do so. Many of the people writing on this problem and its variations have had a basic difficulty: what to call it. As can be seen by perusing the references, some of the many names that have been used are: the Fermat problem, the generalized Fermat problem, the Fermat-Torricelli problem, the Steiner problem, the generalized Steiner problem, the Steiner-Weber problem, the Weber problem, the generalized Weber problem, the Fermat-Weber problem, the one median problem, the median center problem, the minisum problem, the minimum aggregate travel point problem, the bivariate median problem, and the spatial median problem.

The main object of this chapter is not a comprehensive history but rather an attempt to put into perspective the efforts of many people in different disciplines who struggled with various versions of this problem, often unaware that others had gone before them. Rather than being a drawback, the parallelism of these many efforts is not only a tribute to the enduring importance of this problem in several fields but has also resulted in a great variety of clever and inventive methods. The problem has given rise to an extraordinary number of generalizations, extensions and modifications. It would literally require volumes to do them justice; space permits only a brief and somewhat arbitrarily selected summary. Reviews of the Weber problem can be found in Wesolowsky (1993), Love et al. (1988) and Francis et al. (1992).

---

* Part of this chapter is based on the paper by Wesolowsky (1993). All excerpts from that paper are included with permission from Elsevier Science.

Facility Location: Applications and Theory.
Edited by Z. Drezner and H.W. Hamacher
© 2002 Springer-Verlag, ISBN 3-540-42172-6

1.1.1 Definition of the Weber Problem

We are to find the "minisum" point (x^*, y^*) which minimizes the sum of weighted Euclidean distances from itself to n fixed points with co-ordinates (a_i, b_i). The weights which are associated with the fixed points are denoted by w_i. One simple (and simplistic) scenario for the problem is that we wish to locate a warehouse and that the weights w_i are the costs per unit distance of shipping the requirements to customers located at the fixed points (a_i, b_i); (x^*, y^*) is then the warehouse location that minimizes the transportation cost. One can also view (x^*, y^*) as a two-dimensional generalization of the simple (one-dimensional) median of n weighted values, and hence the name "spatial median".

The problem can be stated as:

$$\min_{x,y} \left\{ W(x,y) = \sum_{i=1}^{n} w_i d_i(x,y) \right\} \tag{1.1}$$

where $d_i(x,y) = \sqrt{(x-a_i)^2 + (y-b_i)^2}$ is the Euclidean distance between (x, y) and (a_i, b_i).

1.1.2 The Dual Formulation

A different approach to finding the location of the minisum or spatial median point (x^*, y^*) is to solve a problem that is dual to (1.1). For a discussion see Scott et al. (1995). Consider the programming problem:

$$\max_{U,V} \left\{ D(U,V) = -\sum_{i=1}^{n}(a_i u_i + b_i v_i) \right\} \tag{1.2}$$

subject to:

$$\sum_{i=1}^{n} u_i = 0$$

$$\sum_{i=1}^{n} v_i = 0$$

$$\sqrt{u_i^2 + v_i^2} \leq w_i$$

The maximum value of D in (1.2) is equal to the minimum value of W in (1.1). At optimality, the vectors $(u_i, v_i)^T$ "point" from the fixed points (a_i, b_i) to the minisum point (x^*, y^*), and thus any two such vectors can be used to solve for its location. A geometrical solution to the weighted three-point Weber problem and a historical review of the dual problem is found in Martini (1996).

1.2 History and Literature Review

The following is a brief history of the problem of finding the spatial median, (the minisum Euclidean distance point). W. Kuhn (1967) provided an excellent historical sketch. One of his sources was an article by M. Zacharias (1913). Other sources providing details of early solutions are Pottage (1983), Honsberger (1973), and Dörrie (1965).

Who actually first proposed the problem, or in what form, will probably never be known. It is usual to credit Pierre de Fermat (1601-1665) with proposing a basic form of the spatial median problem by issuing the challenge (Kuhn, 1967) "let he who does not approve of my method attempt the solution of the following problem: given three points in the plane, find a fourth point such that the sum of its distances to the three given points is a minimum". It is also usual to credit the Italian mathematician and student of Galileo, Evangelista Torricelli (1608-1647) with the solution (for details see Section 1.3.1 below). However, as usual, the history of the problem is a bit murky. Other mathematicians also worked on the problem at the time. Pottage (1983) mentions treatments by Cavalieri, Viviani and Roberval. Torricelli himself had several methods for this problem; see Honsberger (1973), and Dörrie (1965).

Not everyone credits Fermat with originating the problem. Melzak (1983) says "This problem was proposed and solved by the Italian mathematician Battista Cavalieri (1598-1647); then it was proposed by the French mathematician Pierre Fermat (1601- 1655) and solved again by the Italian scientist Evarista Torricelli (1608-1647)." Zacharias (1913), on the other hand, credits Torricelli with both posing and solving the problem; hence the name "Torricelli point" (see Section 1.3.1). Other geometrical solutions and "rediscoveries" continued into the twentieth century (Honsberger, 1973).

In 1647, Cavalieri's "Exerciones Geometricae" showed that the three lines joining the Torricelli point to the vertices form angles of $120°$ with each other. Another geometrical method of finding this "unweighted median" point was given by Simpson (Thomas Simpson (1710-1761)) in the "Doctrine and Application of Fluxions" (London 1750). (For details see Section 1.3.2 below.) Smith (1923) calls him "that strange mathematical genius". Simpson also suggested, as an exercise, generalizing the problem to include different weights.

The dual problem (see for example, Scott et al., 1995) also has early origins. The use of Simpson lines is already an implicit use of the dual. The Ladies Diary or Woman's Almanack (1755) contains the problem: "In the three sides of an equiangular field stand three trees, at the distances of 10, 12, and 16 chains from one another. To find the content of the field, it being the greatest the data will admit of?". This geometrical problem was stated in a more academic manner in Annales de Mathematiques Pures et Appliques,Vol. I (1810-11), on page 384, as "given any triangle, circumscribe the largest possible equilateral triangle about it."

It should be noted that for $n = 3$, finding the spatial median is equivalent to finding the shortest network (tree) spanning three points (see Winter, 1985). This latter problem has been popularized by Courant and Robbins (1941) as the Steiner (Jacob Steiner (1796-1863)) problem. However, as Kuhn (1967) says, "Although this very gifted geometer (Steiner) of the 19th century can be counted among the dozens of mathematicians who have written on the subject, he does not seem to have contributed anything new, either to its formulation or its solution."

In the twentieth century, the problem passed to those who claimed there was a use for it. Alfred Weber (1909) used a weighted three point version of the problem to depict industrial location minimizing transport cost; the three fixed points were two sources of materials with different weights and a weighted market location ("place of consumption") respectively. A mathematical appendix to his book, written by Georg Pick, gives a geometrical construction procedure. Pick refers to "an old apparatus which was invented by Varignon (see Section 1.3.4)", and uses the mechanical analogue for explanation as well as suggesting it for the solution of problems with $n \geq 3$. Tellier (1972) also obtained an explicit solution using trigonometry to find the optimum location, as well as discussing the conditions under which one of the points is optimum.

Very shortly after the English translation of Weber's book appeared, Eels (1930) published an article about the "unfortunate error" that "had repeatedly occurred in various publications of the United States Census Bureau" and "from this source ... spread to various books on sociology and population problems, and seems to have remained unchallenged for almost twenty years". Later authors like Schärlig (1973) have continued the lesson. It seems that the Census Bureau had been calculating the center of gravity for populations but attributing to it the property of the "point of minimum aggregate travel", which is the spatial median. Eels expounded on this error at considerable length, and then offered a solution of his own to the three point problem; however these points were restricted to form an isosceles triangle.

Eels' article unleashed a volume of correspondence so large that the editor, Ross (1930), had to abandon the "usual procedure to publish (the correspondence) in full". Professor Corrado Gini of the University of Rome and President of the Central Institute of Statistics of the Kingdom of Italy wrote referring to his article with Galvani (1929). He said that the problem had been fully discussed in his article and the Census errors noted. Also, contrary to Eels, he pointed out that "the designation of 'median point' should properly be given to the point of minimum aggregate travel, and not to the intersection of two orthogonal lines, which is not invariant with respect to the system of co-ordinates." This variation in definitions of median still continues. Hall (1988) says that "in two dimensions, the median minimizes the average rectangular distance and the mean minimizes the average squared distance". Professor Gini would have been upset.

The correspondence is also notable in that a Professor E.B. Wilson resolved the spatial median problem with $n = 3$ and references were given to yet more such re-discoveries. Even more intriguing was the mention of a Mr. Douglas E. Scates of Cincinnati Ohio, who "with his associates seem to be making progress toward establishing a working model for a general population". Scates (1933) did publish a method (without associates) in Metron; it used essentially trial and error. Other approximation (as opposed to iteration) methods were developed at the time.

It was Endre Vaszonyi Weiszfeld (1936) (a Hungarian mathematician, who wrote in French in a Japanese journal), now known as Andrew Vaszonyi, who provided a practical method for finding the spatial median or the Euclidean minisum point for large n and unequal weights (see Section 1.3.5 for details). This method is the iteration procedure (see equation (1.4)) which is the trick of partially separating out the point (x, y) from the extremum equations and using the result in an iterative way to improve the solution. His method, uniquely suited to the computer age, lay dormant and unknown until a series of rediscoveries in the late fifties and early sixties.

This method was first rediscovered by Miehle (1958), who was dealing with a more complex problem of link length minimization arising out of the Steiner problem. Miehle also has interesting photos of analogue machines. The most complete treatment was by Kuhn and Kuenne (1962), who gave the necessary and sufficient conditions for the optimum to be at a fixed point. A more recent paper on this topic is Juel and Love (1986). The same procedure was again proposed by Cooper (1963), who used it as part of his algorithm for location-allocation, and who, like Miehle, borrowed the basic idea from numerical analysis.

Ostresh (1978a, 1978b) defined a slightly different step when an iteration falls on a fixed point. Katz (1974) showed that local convergence is generally linear. An extensive discussion of convergence was given by Morris (1981) and Brimberg (1989). An accelerated Weiszfeld algorithm was proposed by Drezner (1992, 1996). There have been many additional studies of the Weiszfeld algorithm when modified to apply to variations of the spatial median problem.

In the late sixties, computer programs for the optimization of non-linear functions started to be readily available in great abundance. Furthermore, even though its derivatives do not exist at the fixed points, $W(x, y)$ is convex (Love, 1967). However, the Weiszfeld iteration procedure is simple, elegant, and improves the solution at each step; it continued to be worked on.

Applications of the spatial median problem have been both of the conjectural and the actual kind. Unfortunately, the problem is an integral part of more complex problems in many fields and therefore its applications are often "buried" and are not easily found by researchers from outside the field. Compounding this difficulty is the fact that there is a virtually perverse lack of consistency in naming the problem. We already discussed its incarnation as the "minimum of aggregate travel" among demographers and geographers, and its appearance in the economic theory of industrial location. Riveline (1967) reported the use of a version of the Varignon Frame as part of the analysis of optimum gallery location in French coal mines. Burstall, Leaver and Sussams (1962) used a simple Varignon Frame as an aid to the location of factories in London. It can be used in cluster analysis (Cooper, 1973) and in related statistical techniques. Overton (1983) gives several references for the problem's use in the physical application of discretizing minimal surfaces. Ostresh (1977) writes "it has application to the siting of steel mills and schools, houses of ill repute and hospitals." It is not clear if Ostresh was familiar with case studies on all of these applications.

1.3 Solution Procedures

1.3.1 Torricelli Point

A solution to the unweighted $n = 3$ problem attributed to Torricelli is as follows. The three points are joined by lines to make a triangle. Equilateral triangles are constructed on the sides with the vertices pointing outward. The three circles through the vertices of the equilateral triangles intersect in the spatial median point, which is labeled the "Torricelli Point" (sometimes called "Fermat point") in the example in Figure 1.1.

Fig. 1.1. Geometrical constructions on a problem with equal weights

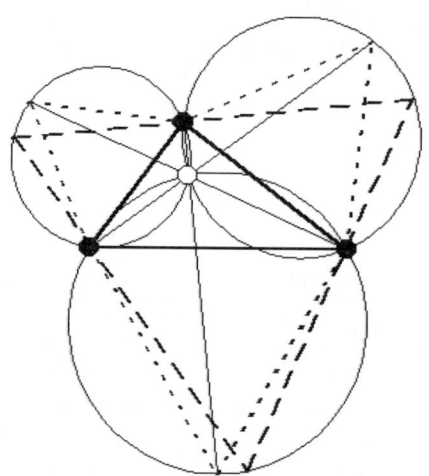

● Fixed Point
○ Torricelli Point

1.3.2 The Simpson Lines

Simpson suggested that the outside vertices of the equilateral triangles proposed by Torricelli are used by joining these vertices to the opposite fixed points. These are known as Simpson Lines. The Simpson lines intersect at the minisum (Torricelli) point. See the example in Figure 1.1. Simpson lines are not to be confused with "Simson lines", named after R. Simson (1687-1768), who, to make matters worse, was not, according to Coxeter (1969), their discoverer.

Both the solutions by Torricelli and Simpson apply when the triangle formed by the fixed points has no angle greater than $120°$; if an angle were greater than $120°$ they would give a point outside the triangle. Such a point can not be optimal. In this case, the fixed point associated with the angle which is $\geq 120°$ is the optimum location.

1.3.3 The Dual Problem

The dual problem is to find the largest equilateral triangle circumscribing a given triangle. The solution was given in Volume II (1811-12) by Rochat et al. (1811): "Thus the largest equilateral triangle circumscribing a given triangle has sides perpendicular to the lines joining the vertices of the given triangle to the point such that the sum of the distances to these vertices is a minimum ... One can conclude that the altitude of the largest equilateral triangle that can be circumscribed about a given triangle is equal to the sum of distances from the vertices of the given triangle to the point at which the sum of distances is a minimum." This problem is therefore a version of the dual for three unweighted fixed points (for this history of the dual, we are indebted to Kuhn (1976)). The equilateral triangle at optimality is given by the heavy-dotted triangle in Figure 1.1. It should be mentioned that this equilateral triangle was already incorporated into the constructions of Torricelli (Honsberger, 1973).

The formal statement of the dual dates back to just after the rediscoveries of the Weiszfeld procedure. Witzgall (1964) and Kuhn (1967) independently stated the problem in essentially the form in (1.2). White (1976) provided a Varignon Frame interpretation of the dual. Guccione and Gillen (1991) wrote a note on an economic interpretation of the dual wherein a transportation authority maximizes revenue and distances become rectilinear. Many other papers on the dual have appeared in the context of various generalized versions of our problem.

1.3.4 The Varignon Frame

Varignon proposed a mechanical analogue device which has actually been used in practice. The device yields valuable insights into the problem. Figure 1.2 shows a diagram of the device, which is called a Varignon Frame. A board is drilled with n holes corresponding to the co-ordinates of the n fixed points; n strings are tied together in a knot at one end, the loose ends are passed, one each, through the holes, and are attached to physical weights below the board which have the same magnitudes as the constants w_i. If the device were not subject to the ills of the physical world and there were no friction, the strings were infinitely thin, the holes infinitely small, and so on, then the final position of the knot would be at the optimum point (x^*, y^*).

Fig. 1.2. The Varignon Frame

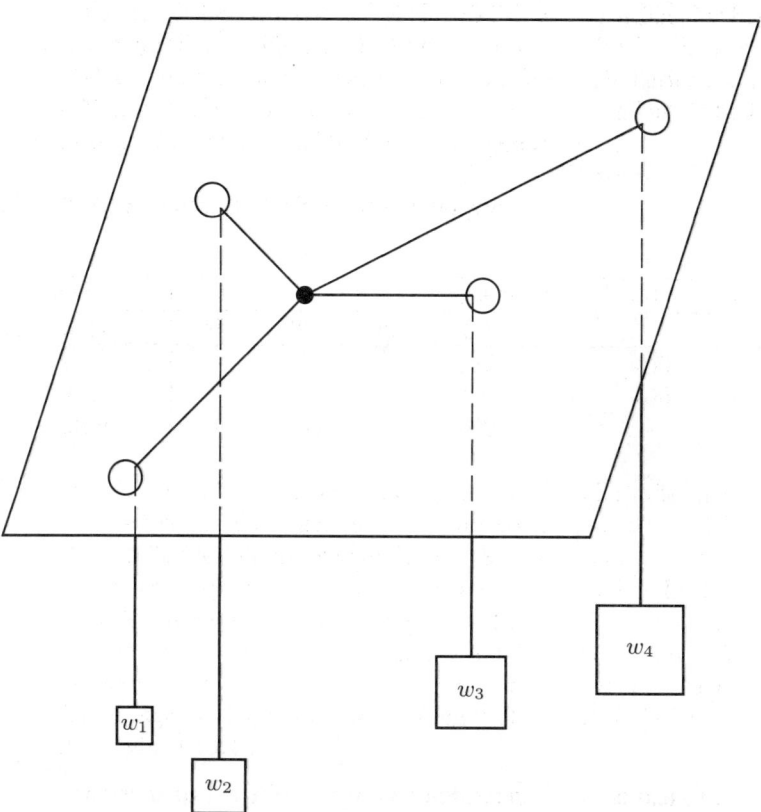

Why is the optimum point at the knot? Consider the force component, in the x direction, exerted by a single weight w_i on the knot in Figure 1.2; the knot is at co-ordinates (x, y). This force component in the x direction will be $\frac{w_i(x-a_i)}{d_i(x,y)}$ and will be to the left (have a negative sign) if $a_i > x$. It is evident that the sum of such components from all the weights is equal to the first expression in (1.3). Similarly, the sum of components in the negative y

direction is the second expression in (1.3). The resultant force vector exerted on the knot by all the strings is therefore zero at the same point (x^*, y^*) where conditions (1.3) are satisfied.

Note that if any one of the holes on the Varignon Frame is moved "out" on the line of its string with the knot, the optimum point is not affected; this is analogous to the property of the simple median that the points which do not coincide with the median can be "stretched out" or interchanged on any one side at a time without affecting the median.

The Varignon Frame provides an interpretation of the dual variables at (u^*, v^*): if the solution is not at a fixed point i, the vector $(u_i, v_i)^T$ is the negative of the force vector exerted on the knot by weight w_i.

1.3.5 The Weiszfeld Algorithm

The simplest and most commonly used technique to solve the Weber problem is called the "Weiszfeld procedure" (Weiszfeld, 1936).

If we differentiate and set the partial derivatives equal to zero to obtain the first order conditions for optimality we have:

$$\frac{\partial W(x,y)}{\partial x} = \sum_{i=1}^{n} \frac{w_i(x-a_i)}{d_i(x,y)} = 0$$

$$\frac{\partial W(x,y)}{\partial y} = \sum_{i=1}^{n} \frac{w_i(y-b_i)}{d_i(x,y)} = 0$$

(1.3)

It can be shown that $W(x,y)$ is convex, so that (1.3) define a minimum. However, it is immediately evident that these derivatives do not exist when (x,y) coincides with fixed point i, because then $d_i(x,y) = 0$. Equations (1.3) can not, in general, be solved explicitly for (x,y) if $n > 3$.

We can extract x from the first equation in (1.3) if we ignore its presence in $d_i(x,y)$ and we can do the same for y from the second equation in (1.3). The result can be formulated as an iterative procedure if we consider the extracted (x,y) to be a new iteration $(k+1)$ and the (x,y) trapped in the distance term to be the old iteration (k). To be specific:

$$\left(x^{(k+1)}, y^{(k+1)}\right) = \left(\frac{\sum_{i=1}^{n} \frac{w_i a_i}{d_i(x^{(k)},y^{(k)})}}{\sum_{i=1}^{n} \frac{w_i}{d_i(x^{(k)},y^{(k)})}}, \frac{\sum_{i=1}^{n} \frac{w_i b_i}{d_i(x^{(k)},y^{(k)})}}{\sum_{i=1}^{n} \frac{w_i}{d_i(x^{(k)},y^{(k)})}}\right)$$

(1.4)

In other words, once we have a point $(x^{(k)}, y^{(k)})$, we can obtain the next, and hopefully better, point by substituting in the right hand side of (1.4). The trick of partially isolating the solution variables for the purpose of obtaining an iterative solution method is well known in modern numerical analysis and belongs to the class of procedures known as one-point iteration methods (Dahlquist and Björck, 1974, Chapter 6.) The term "one point" arises because only the current point is used to determine the succeeding one.

The Weiszfeld procedure has its quirks. Kuhn (1973) showed that it will fail if the iteration falls on a fixed point; also see Chandrasekaran and Tamir (1989). The following simple problem with five points gives the Weiszfeld algorithm a very hard time. Four points, each with a weight of 1, are placed at the corners of a square of side 1 centered at (0,0), and a fifth point (which is the optimal solution) with a weight of 4 is placed at (100,0). Applying the Weiszfeld algorithm on this problem starting at (50,0) reaches (90.44,0) after 100,000 iterations, (97.447,0) after 200,000 iterations, and (99.999887,0) (not even an accuracy of 10^{-4}) after a million iterations.

1.3.6 Other Iterative Methods

Austin (1959) started with a general n but with equal weights in the "point of minimum aggregate distance" tradition. He obtained the iteration equations but used them in a different way. He noted that for any starting point, they showed a centroid solution with weights inversely proportional to the distances. He then suggested that they be regarded as points projected from the starting point onto a circle of arbitrary radius around the starting point. The co-ordinates on the circle could be read graphically and their centroid point could be found. The next iteration would be halfway between this centroid point and the previous point. Austin briefly suggested a generalization to arbitrary weights (which was slightly incorrect). Seymour (1970) compared this method computationally with the Weiszfeld algorithm. However, it can be readily shown that this is a simple gradient method with fixed step size. While the Weiszfeld procedure is also gradient descent (Cooper and Katz, 1981), it has a variable step size of $1/\sum \frac{w_i}{d_i(x^{(k)},y^{(k)})}$.

Porter (1963) proposed a rather interesting method for solving for the point of minimum aggregate travel (with equal weights); he stated that the point lay between the centroid and a line perpendicular to the bisector separating the distribution into equal halves. Court (1964) showed that this method was incorrect. In his reply to Court, Porter (1964) wrote "Arnold Court has chased the point of minimum travel back into hiding, where it lies convulsed in helpless laughter at our inability to pin it down". However, with a little help from a computer, it could now indeed be pinned down.

There have also been numerous other iteration schemes proposed. Convergence of the Weiszfeld procedure is known to be slow in the vicinity of fixed points. To use standard nonlinear minimization techniques, one could eliminate the problem with derivatives at the fixed points by using a hyperbolic approximation; an example is $d_i^H(x,y) = \sqrt{(x-a_i)^2 + (y-b_i)^2 + \epsilon}$ where ϵ is very small (see Wesolowsky and Love (1972), Eyster, White and Wierwille

(1973)). Drezner (1986) proposed an interpretation to the value of ϵ. If the point has an area, then the average distance between the facility to a demand point resembles $d_i^H(x, y)$ with ϵ being proportional to the area of the demand point. All orders of derivatives are now always continuous. This approximation can be used with standard unconstrained optimization packages but can also be easily adapted to the Weiszfeld procedure. Unfortunately, using such an approximation may get the objective function close to optimum, but the actual point found may not be close to the true optimizing point in cases where the cost function is very "flat".

Many methods specifically adapted to the median problem or its generalizations have also been proposed. Vergin and Rodgers (1967) used gradient methods and Love (1969) applied convex programming to a problem in three dimensions. Over the years, still other iterative methods were proposed. These include: Seymour and Weindling (1975), Harris (1976), El-Shaieb (1978), Cooper and Katz (1981), Overton (1983) (for equal weights) and Rosen and Xue (1991). An algebraic programming method was given by Chandrasekaran and Tamir (1990).

1.3.7 Using a Lower Bound

Although the convergence of the solution may be "slow" in some problem configurations, the potential improvement in the sum of weighted distances in the succeeding iterations may be quite small. A method of dealing with this that was suggested by Love and Yeong (1981), is to have a continuously updated lower bound on $W(x, y)$ during iteration and stop the iterations once the difference between the value of the objective function and the lower bound is smaller than a given tolerance.

The basic idea of Love and Yeong (1981) is that the convex hull and the current value of the gradient determine an upper bound on the improvement that can be made. In effect, the upper bound on the possible improvement is the magnitude of the gradient times the distance from the current point to the farthest point in the convex hull.

Further works on such bounds include Elzinga and Hearn (1983), Juel (1984), and Drezner (1984). Wendell and Peterson (1984) derived a lower bound from the dual.

1.4 Properties of the Weber Problem

1.4.1 Optimality of a Fixed Point

What happens when one of the weights is larger than the sum of all the others? The knot in the Varignon frame will disappear down the hole of that weight. Is this the condition that the optimum location (x^*, y^*) be at a fixed point? Does one weight have to "overpower" all the others? No, all that is necessary is that over a hole, the net force exerted by the other weights is less than or equal to the weight on the string through that hole; this will guarantee that that hole is the optimum location even when the weight associated with that hole is very small. To see this, consider a board with only two holes, each with an equal weight. The knot will languish anywhere between them. If we now drill a new hole anywhere on a line between the two old ones, and attach a very small weight to the string through it, the new string will pull the knot to a position over its hole (but not down the hole). Since derivatives of $W(x, y)$ do not exist at the fixed points, we should expect strange behavior of the knot at these locations. The "hole" conditions, which are both necessary and sufficient for the optimum to occur at (a_r, b_r), are:

$$\left(\sum_{i \neq r} \frac{w_i(a_r - a_i)}{d_i(a_r, b_r)}\right)^2 + \left(\sum_{i \neq r} \frac{w_i(b_r - b_i)}{d_i(a_r, b_r)}\right)^2 \leq w_r^2 \qquad (1.5)$$

1.4.2 How Likely is the Optimal Location on a Fixed Point?

Drezner and Simchi-Levi (1992) investigated the likelihood of the solution to be at a fixed point. Suppose n points are randomly generated in a disc. What is the probability that the optimal solution is one of the fixed points? Intuitively, one would think that this probability increases with n because the disc becomes denser and denser with fixed points and there is "less room" for location on non-fixed points. Drezner and Simchi-Levi (1992) analyzed the case where all weights are equal. They found that the probability that the optimal solution is on a fixed point is approximately $\frac{1}{n}$. That means that with 10 fixed points the probability is about 10%, but with 1,000 fixed points the probability is only 0.1% (or there is a 99.9% probability that the optimal location is *not* on a fixed point). Another interesting result in that paper concerns the difference between the optimal value of the objective function and the best value of the objective function on a fixed point. They showed that the best value of the objective function among all fixed points is expected to be $1 + \frac{3}{n}$ times the optimal value. That means that when solving a problem with 1,000 fixed points, selecting the fixed point with the lowest value of the objective function is expected to be only 0.3% higher than the optimum.

One might conclude that for a problem with 1,000 fixed points one can evaluate the value of the objective function at the fixed points, pick the best one, and expected to be only 0.3% over the optimum. This naive approach is not a good one and illustrates the efficiency of the Weiszfeld algorithm. One iteration of the Weiszfeld algorithm requires about the same computing time as one evaluation of the value of the objective function. Therefore, this naive approach will require about the same time as 1,000 Weiszfeld iterations. However, for most problems of that size, the number of Weiszfeld iterations is only in the single digit.

Kuhn (1967) wrote "as for the statements of Courant and Robbins that the generalization of the problem to more than three points is a sterile generalization". (Actually, Courant and Robbins (1941) did not use the word "sterile". What they really said in reference to the four point unweighted problem is "This problem, which was also treated by Steiner, does not lead to interesting results. It is one of the superficial generalizations not infrequently found in mathematical literature"). Indeed, "sterile" seems to be a grossly inappropriate word given the number of generalizations, variations, modifications, extensions and downright mutations that the problem has given birth to. It would be a taxonomist's nightmare to attempt a consistent categorization, and in any case there is not the space available in this chapter. We will, however, attempt a brief sampler and refer the reader to recent literature reviews such as the survey of representative location problems by Brandeau and Chiu (1989) and the references in Love, Morris, and Wesolowsky (1988).

1.5 Other Distance Measures

1.5.1 Minimizing the Sum of Squared Euclidean Distances

In this case the problem is

$$\min_{x,y} \left\{ C(x,y) = \sum_{i=1}^{n} w_i d_i^2(x,y) \right\} \qquad (1.6)$$

$C(x,y)$ is separable into a sum of two components, one containing only x, and one containing only y. By simple calculus we can show (White, 1971) that the optimum point $(x^\bullet, y^\bullet)$ is given by:

$$(x^\bullet, y^\bullet) = \left(\frac{\sum_{i=1}^{n} w_i a_i}{\sum_{i=1}^{n} w_i}, \frac{\sum_{i=1}^{n} w_i b_i}{\sum_{i=1}^{n} w_i} \right) \qquad (1.7)$$

This point is known, of course, as the centroid or center of gravity. To give this point, our Varignon analogue machine must be modified somewhat. We must untie the knot joining the strings, and then tie a knot on each individual string large enough so that it can not pass through the hole. If we were fortunate enough to have a weightless board, the board would now balance on a needle point placed underneath at $(x^\bullet, y^\bullet)$. To put it another way, the centroid is a point such that if we draw any line through it, the weight×distance components projected onto the line would sum to the same value on either side of the point.

1.5.2 Minimizing the Sum of Rectilinear Distances

Rectilinear, rectangular, or Manhattan distances are distances that are often used to approximate travel in a grid. In this case the distance between the facility and a demand point i is given by $d_i^R(x, y) = |x - a_i| + |y - b_i|$. Minimizing the sum of weighted distances to find the optimum location (x°, y°) now becomes:

$$\min_{x,y} \left\{ R(x, y) = \sum_{i=1}^{n} w_i d_i^R(x, y) \right\} \tag{1.8}$$

This problem is easy to solve because the objective function is separable, as was the one in the preceding problem. For example, to find x° we minimize $\sum w_i |x - a_i|$. As can easily be verified by examining the piecewise linear derivative of $\sum w_i |x - a_i|$, this is done by finding the median of the weighted a_i's, or, in other words, the midpoint of the weights w_i arranged along the x-axis at their corresponding locations a_i. This median may be some value a_i, or a range that includes two adjacent values. To summarize, (x°, y°) is found by finding the median of the weighted a_i values and the median of the weighted b_i values respectively. This solution is sometimes called the co-ordinatewise median (Rousseeuw and Leroy, 1987).

1.5.3 p-Norm Distances

A generalization of Euclidean distance is the ℓ_p distance, which is given by

$$\ell_{pi} = \sqrt[p]{|x - a_i|^p + |y - b_i|^p},$$

for the demand point i and a facility (x, y). It can depict a wide variety of distance measures; the Euclidean distance is the special case $p = 2$ and rectilinear distance is $p = 1$. Love and Morris (1979) showed that ℓ_p distances can be used to approximate road distances. Weber problems with respect to p-norms have been studied, for example, in Brimberg (1989) and in Brimberg and Love (1995). For solving these location problems Weiszfelds's algorithm (see Section 1.3.5) can be adapted; and this is mostly done by using a hyperbolic approximation. Linear convergence of this method could be shown for $1 \leq p \leq 2$, while for $p > 2$ counterexamples for convergence have been found, see Morris (1981), Brimberg and Love (1992), and Brimberg and Love (1993). Note that the "hole" conditions of the Varignon frame for the Weber problem with Euclidean distance (see equation (1.5)) can also be generalized to p-norm distances, see Juel and Love (1981); hence optimal solutions at demand points can easily be found.

Recently, Ortega, Mesa and Sánchez (2000) proposed an iterative method for solving p-norm location problems with $1 < p < 2$. Their approach is based on an approximation of round norms by block norms which they use to develop an iterative linear programming approach.

1.5.4 Block Norms and Polyhedral Gauges

If the unit ball of a norm γ is a polyhedral set, the norm is called a block norm and it is called a polyhedral gauge if symmetry is not required any more. Examples for block norms are the rectilinear norm, or the Chebychef norm, both having polyhedral unit balls with four vertices. The halflines starting at the origin and passing through the vertices of the polyhedral gauge are called fundamental directions. Such distances can be used, for example, to approximate road networks in a planar setting. Location problems with polyhedral gauges can be formulated as linear programs, see Ward and Wendell (1985) and are therefore easily solvable. If we draw the fundamental directions starting at each of the demand points we get a partition of the plane into polyhedral cells, and the objective function is linear on each cell. In particular, Durier and Michelot (1985) showed that there always exists an optimal solution at a cell vertex.

1.5.5 Other Distance Metrics

In addition to the basic generalizations mentioned above, many other types of distances have been investigated. Examples are:

- central metrics (Perreur and Thisse, 1974),
- distance functions based on altered norms (Love and Morris (1979) and Love, Truscott and Walker (1985)),
- weighted one-infinity norms (Ward and Wendell, 1980),
- mixed norms (Planchart and Hurter (1975) and Hansen, Perreur and Thisse (1980)),
- block and round norms (Thisse, Ward and Wendell, 1984),
- mixed gauges (Durier and Michelot, 1985),
- asymmetric distances (Hodgson, Wong and Honsaker, 1987).
- weighted sums of order p (Brimberg and Love (1995), Üster (1999))

Note that there are many techniques to solve locations problems with arbitrary gauges, among them for example a grid-approximation technique, see Carrizosa and Puerto (1995), or a primal-dual approach, see Michelot (1993). Hansen et al. (1985) developed a geometrical Branch-and-Bound algorithm, which was improved and applied by Plastria (1992). In a new approach, Carrizosa et al. (2000) use polyhedral gauges to approximate any other gauge in a planar location problem, for example, by using the sandwich approximation algorithm of Burkard, Hamacher and Rote (1991).

The space in which location can take place has also been generalized. Love (1969) extended the problem to three dimensions. Also, the earth's surface is approximately planar only on a small scale. Drezner and Wesolowsky (1978), Aly, Kay and Litwhiler (1979), and Drezner (1985) are among those using spherical distances. Wesolowsky (1983) and Plastria (1995) give a review of spherical location problems. More general spaces have also been employed, for example Eckhardt (1980) used Banach spaces.

1.6 Multiple Facilities

One of the earliest and most straight-forward generalizations was to add more new facilities. This was done in two basically different ways. One was to simply add additional facilities, with given interactions between themselves and the demand points. The other one is to assign each demand point one facility for service (location-allocation models).

1.6.1 The Multifacility Model

If we have m new facilities and the weight between facility j and demand point i is w_{ij} and between facility j and s is v_{js}, then we have:

$$\min_{(x_j,y_j)j=1,\ldots,m} \left\{ \sum_{i=1}^{n}\sum_{j=1}^{m} w_{ij}\sqrt{(x_j - a_i)^2 + (y_j - b_i)^2} \right. \\ \left. + \sum_{j=1}^{m-1}\sum_{s=j+1}^{m} v_{js}\sqrt{(x_j - x_s)^2 + (y_j - y_s)^2} \right\} \quad (1.9)$$

The facilities thus have predetermined shipments to the demand points and to each other. Miehle (1958), a "co-rediscoverer" of the Weiszfeld procedure, considered a problem of this type.

The multifacility Weber problem is a convex optimization problem with a nondifferentiable objective function since two objects (demand points or facilities) may coincide, i.e., have the same location.

For block norm distances and polyhedral gauges, (1.9) can be formulated as a linear programming problem (Ward and Wendell, 1985). However, the dominating set result (compare Section 1.5.4) transfers to the multifacility case only for polyhedral gauges with at most 4 fundamental directions (see Hansen, Perreur and Thisse (1980) and Michelot (1987) for a counterexample for general gauges). Efficient algorithms are available particularly for the case of rectilinear distances, see, for example, Dax (1986).

For general distance metrics, hyperbolic approximations of the objective function can be used to avoid nondifferentiability (Wesolowsky and Love (1972), Eyster, White and Wierwille (1973)), or primal-dual methods may be applied (Idrissi, Lefebvre and Michelot, 1989). Many authors developed solution methods for Euclidean distances (see, for example, Calamai and Charalambous (1980), Rado (1988) and Xue, Rosen and Pardalos (1996) who recently gave a polynomial time dual algorithm). Interior point and related methods were succesfully applied by Andersen (1996) and for gauge distances by Fliege (2000) who also proved that his method has polynomial complexity.

Other solution concepts are based on coincidence conditions to identify points of nondifferentiablility (see Fliege (1997) for a very general discussion), or use special treatment for points of coincidence (Overton, 1983).

1.6.2 Location-Allocation Models

The other way to incorporate more than one new facility is to have a location-allocation model. Cooper (1963), also a co-rediscoverer of the Weiszfeld procedure, considered a location-allocation problem. Here the demand points have fixed total demands which are to be optimally allocated to the facilities. An example of a simple location-allocation problem is:

$$\min_{\substack{(x_j,y_j)j=1,\ldots,m \\ w_{ij};i=1,\ldots,n;j=1,\ldots,m}} \left\{ \sum_{i=1}^{n} \sum_{j=1}^{m} w_{ij} \sqrt{(x_j - a_i)^2 + (y_j - b_i)^2} \right\} \quad (1.10)$$

subject to:

$$\sum_{j=1}^{m} w_{ij} = w_i \text{ for } i = 1, \ldots, n$$

$$w_{ij} \geq 0 \text{ for } i = 1, \ldots, n; j = 1, \ldots, m$$

This problem is particularly difficult because the objective function is neither convex nor concave (Cooper, 1967) and may have a large number of local minima. Eilon, Watson-Gandy and Christofides (1971) described an example problem with 50 demand points and 5 facilities for which 61 local minima were found, the worst of which had a deviation of 40.9% from the best. The problem can also be interpreted as an enumeration of the Voronoi partitions of the customer set, and it was proven to be NP-hard in Megiddo and Supowit (1984).

There have been many attempts at both heuristic and exact solutions. A well-known heuristic approach is the sequential location-allocation procedure (Cooper, 1964). The method alternates between a location and an allocation phase until no further improvement is made. Other heuristics include local search methods (see, for example, Love and Juel (1982), Brimberg and Mladenovic (1996b)), modifications of the objective function (Chen, 1983), methods based on clustering (see, for example, Sullivan and Peters (1980), Moreno, Rodrigez and Jimenez (1990)), a projection method (Bongartz, Calamai and Conn, 1994), Tabu search (Brimberg and Mladenovic, 1996a) and a p-Median plus Weber heuristic (Hansen, Mladenovic and Taillard, 1998). A comparative study of selected heuristics can be found in Brimberg et al. (2000).

Exact solution methods were for a long time restricted to relatively small problem sizes. Branch and Bound algorithms were developed in Kuenne and Soland (1972), Ostresh (1973), Ostresh (1975), Drezner (1984) and Rosing (1992), among others. Love and Morris (1975) concentrated on rectilinear distances, and Love (1976) applied dynamic programming to problems where all demand points are located on a line. Brimberg and Love (1998) later generalized this approach to certain classes of planar problems. More recently, the application of global optimization techniques has increased the size of problem instances that can be solved exactly. Examples are a d.-c. programming method for the two facility case (Chen et al., 1998) and a column generation approach (Krau, 1997). Approximation schemes for the problem were developed by Lin and Vitter (1992a and 1992b) and by Arora, Raghavan and Rao (1998), who gave an ε-approximation scheme for the Euclidean location-allocation problem.

An exhaustive overview of applications of location-allocation problems is given in Hodgson, Rosing and Shmulevitz (1993). For a more detailed overview about multifacility location problems and location-allocation problems, we refer to the survey of Plastria (1995).

1.7 Restricted Weber Problems

Since the Weber problem is based on a geometrical representation of location problems, many authors tried to incorporate geographical reality into the model. Typical examples are the consideration of forbidden regions where the placement of facilities is forbidden, and barrier regions where additionally traveling is prohibited. Forbidden regions can be used to model, for example, state parks or other protected areas, densely populated neighborhoods or regions where the geographic characteristics such as the land slope forbid the construction of the desired facility. Similarly, barrier regions can be used to model military areas, mountain ranges, lakes, big rivers, or, on a smaller scale, machines and conveyor belts in an industrial plant. Applications of restricted location problems mentioned in the literature involve location and routing of robots (Lozano-Perez and Wesley, 1978), circuit board design (LaPaugh (1980), Foulds and Hamacher (1993)), the exclusion of regions for the placement of unwanted (obnoxious) facilities (Karkazis (1988), Erkut and Neumann (1989), Brimberg, Love and Wesolowsky (1993)), facility layout (Francis et al., 1992) and robotic assembly work place planning (Francis et al., 1994). Further applications can be found in Hamacher and Nickel (1995) and Plastria (1995), among others.

1.7.1 Weber Problems with Forbidden Regions

A forbidden region is generally modeled by a closed (not necessarily connected) set R. Consequently, the feasible region for the facility is given by $\mathcal{F} = \mathbb{R}^2 \setminus \text{int}(R)$ and the Weber problem with forbidden regions can be formulated as

$$\min_{(x,y) \in \mathcal{F}} \left\{ F(x,y) = \sum_{i=1}^{n} w_i d_i(x,y) \right\} \tag{1.11}$$

We assume in the following that the optimal solution of the corresponding unconstrained problem is not feasible in the restricted case since otherwise the problem could be solved by the methods discussed in the previous sections.

A central result for the Weber problem with forbidden regions is the boundary theorem: If R is a connected set and if d is any norm distance, then the restricted Weber problem with the forbidden region R has an optimal solution on the boundary of R.

The boundary theorem was proven for polyhedral forbidden regions and l_p distances in Aneja and Parlar (1994) and under more general assumptions in Hamacher and Nickel (1995). It can also be interpreted as a visibility property as developed in Hansen, Peeters and Thisse (1982). From the algorithmic point of view, the boundary theorem justifies a search algorithm on the boundary of R as formulated in Hansen, Peeters and Thisse (1982), Francis et al. (1992) and Aneja and Parlar (1994).

For polyhedral gauges and convex forbidden regions more efficient solution methods based on a discretization result presented in Hamacher and Nickel (1994) can be applied. In this case the intersections of the boundary of R with the fundamental directions of the polyhedral gauge originating from the demand points determine a finite dominating set that contains at least one optimal solution of the problem. This discretization result can be extended to so-called bumpy sets and also to the case of attraction and repulsion (Nickel and Dudenhöffer, 1996). Algorithms for multi-facility problems and multicriteria problems with forbidden regions can be found, for example, in Nickel (1997) and Fliege and Nickel (2000).

A different approach based on a geometrical Branch and Bound algorithm was suggested by Hansen et al. (1985) and Plastria (1992). The big square small square method splits the feasible region into squares, and either rejects a square or further subdivides it based on the evaluation of lower bounds.

1.7.2 Weber Problems with Barriers

Barrier regions impose even stronger restrictions on the Weber problem since not only the location of facilities is restricted to a feasible region $\mathcal{F}$ but also traveling is prohibited in the interior of the barriers, resulting in non-convex distance functions. If a norm distance d is given to model unconstrained distances, the barrier distance $d_i^B(x,y)$ between the facility and a demand point i is defined as the length of a shortest path connecting the two points and not intersecting the interior of a barrier. The resulting Weber problem with barriers is given by

$$\min_{(x,y)\in\mathcal{F}} \left\{ B(x,y) = \sum_{i=1}^n w_i d_i^B(x,y) \right\} \quad (1.12)$$

Barriers were first introduced to location modeling by Katz and Cooper (1981). The authors consider a Weber problem with the Euclidean metric and with one circular barrier. A heuristic algorithm is suggested that is based on the sequential unconstrained minimization technique (SUMT) for nonlinear programming problems.

Most of the work on location problems with barriers concentrates on special barrier shapes or special distance functions. Assuming that all barrier sets are polyhedra allows for the construction of a visibility graph of the demand points and the extreme points of the barrier polyhedra. Two nodes u, v of this graph are connected by an edge of length $d(u, v)$ if the corresponding points in the plane have distance $d^B(u, v) = d(u, v)$.

The visibility graph was used by Aneja and Parlar (1994), Butt (1994) and Butt and Cavalier (1996) for the evaluation of the objective function value at solution points in the context of heuristic and iterative algorithms. Klamroth (2001a) and Klamroth (2001b) showed that an optimal solution of the nonconvex barrier problem can be found by solving a finite and, in the case of line barriers, polynomial number of related unconstrained subproblems, implying exact as well as heuristic solution methods.

From the point of view of special distance functions, rectilinear and, more general, block norm distances played a central role for the development of discretization based solution procedures. Larson and Sadiq (1983) identified an easily determined finite dominating set for rectilinear distances. This result was later generalized by Batta, Ghose and Palekar (1989) who also included forbidden regions into the model, and by Savaş, Batta and Nagi (2001) who located finite size facilities acting as barriers themselves. Similar discretization results were developed by Hamacher and Klamroth (2000) for general block norm distances. The computational efficiency of these methods was significantly improved by Segars (2000) who showed that the consideration of a much smaller dominating set is sufficient to solve the problem.

A different approach to handle the nonconvexity of the objective function can be seen in the application of global optimization methods (see, for example, Hansen, Jaumard and Tuy (1995)). Krau (1997) generalized the big square small square method (see Section 1.7.1) to handle polyhedral barrier sets as well as forbidden regions. On the other hand, Fliege (1997) suggested to model the physical barriers by suitable barrier functions (in the sense of nonlinear optimization), an approach which may yield good approximations for a broad class of location problems with barriers including multifacility problems.

1.8 Line Location and Dimensional Facilities

Another natural extension of the classical Weber problem is to locate not a point, but a dimensional facility, such as a line, a line segment, a path, a square, or a circle. Such models are motivated by applications, for example, in railway or highway planning, or in plant layout. They are also strongly related to the field of robust statistics (finding regression lines, L_1-fit of points) and, especially in higher dimensions, to computational geometry (transversal theory, set width problems).

Finding a least-distance line was first modeled as a location problem by Wesolowsky (1975) and has drawn much attention and generated many publications since this time.

Given a set of demand points in the plane, the line location problem is to find a straight line l which is as close as possible to the given points, i.e.,

$$\min_{\text{lines } l} \left\{ L(l) = \sum_{i=1}^{n} w_i \gamma_i(l) \right\} \tag{1.13}$$

where $\gamma_i(l)$ is the closest distance between (a_i, b_i) and any point on the line. As distance measures, mainly norms have been considered. In particular, a median line minimizes the sum of (weighted) distances to the given point set.

To prove results for line location problems, the following approach is useful: Consider the vertical distance (which is frequently used in statistics) between a point (a, b) and a non-vertical line $l_{s,c}$ with slope s and intercept c, given as

$$d_{ver}((a, b), l_{s,c}) = |as - b + c|.$$

Note that d_{ver} is a convex function, yielding that the line location problem is piecewise linear and convex and can therefore be solved easily, see Megiddo and Tamir (1983). The following results have been transferred from line location problems with the vertical distance to all norm distances, by a method developed in Schöbel (1998).

- All median lines are pseudo-halving, i.e., the sum of weights of the demand points in each of the open halfspaces, separated by the line l is smaller or equal than the sum of all weights.
- There exists a median line passing through at least two of the demand points.

This has first been shown for the case that γ is the Euclidean distance by Wesolowsky (1975), independently also by Morris and Norback (1980), Megiddo and Tamir (1983), and Korneenko and Martini (1993), the latter in the more general context of locating a hyperplane in n dimensional space. The generalization to all norm distances has been developed in Schöbel (1998) and Schöbel (1999a). If the distance between the given points and the line is measured by a smooth norm, it even can be shown that all optimal lines contain at least two of the given points, see Martini and Schöbel (1998b).

Note that for gauges, slightly weaker results have been derived by Plastria and Carrizosa (2001).

Using these results, a polynomial-time enumeration of all lines passing through two of the given demand points is possible. However, many suggestions for specialized algorithms have been developed. For the Euclidean distance Lee and Ching (1985) proposed an $O(M^2)$ time approach which is the best known so far, while for the rectilinear distance an optimal linear-time approach was suggested in Zemel (1984). With respect to block norms the problem can be solved in time $O(GM)$ where G is the number of fundamental directions of the block norm, see Schöbel (1996).

Line location problems can further be specialized and extended in numerous ways. A recent overview has been given by Díaz-Báñez et al. (2001). Some possible extensions, all minimizing $L(l)$ — many of them not completely solved so far — are the following.

- Most of the mentioned results can be transferred to locating hyperplanes in normed spaces, see Martini and Schöbel (1998a), or Plastria and Carrizosa (2001).
- The NP-hardness of the location of more than one line has been shown by Megiddo and Tamir (1982), but only for minimizing the maximum distance to the given facilities we are aware of further results.
- Also for the location of line segments, only few results have been derived in the case of minimizing the sum of distances. The problem has first been mentioned by McKinnon and Barber (1972) in the context of designing a transportation network, and a bicriterial model and some first results have been given by Schöbel (1999a).
- Results for the location of half-lines have been derived by Morris and Norback (1983).
- Restricted line location problems are important from a practical point of view. They have been considered in Morris and Norback (1983), in Robert (1991), and in a more general context in Schöbel (1999b).
- A much more general approach is to consider (polygonal) sets instead of demand points and to approximate them by a line, a problem closely related to finding stabbing lines in computational geometry. Numerous results for this problem, in particular with the Euclidean distance, have been obtained by Robert and Toussaint (1994).
- Also non-linear facilities have been considered. Examples are the location of a circle line in Laporte et al. (1994), the location of a circle in the out-of-roundness problem (see, for example, Yeralan and Ventura (1988), Drezner et al. (2001)) and the location of rectlilinear paths in Drezner et al. (1985).

1.9 Extensions

As already mentioned, there are numerous other extensions and generalizations of the Weber problem, and we cannot report on all of them in this chapter. But at least some of them — both well established and recent approaches — should be briefly mentioned.

Many generalizations are based on modifications of the objective function in the framework of Weber objectives. As an example, transportation costs were made more general and required to be simply increasing and continuous functions of distances by Hansen et al. (1985). Tellier and Polanski (1989) and Drezner and Wesolowsky (1990) removed the restriction that weights must be non-negative. If all weights are negative, the facility is "pushed away" from all the demand points and hence is called an obnoxious facility. This leads to a new objective function that maximizes the sum of distances to the demand points. A recent overview of obnoxious facility location has been given by Lozano and Mesa (2000).

Since in the process of locating a new facility usually several decision makers with different, sometimes even conflicting objectives are involved, multicriteria location models are another natural extension of the classical, single criterion models. The goal often is to determine efficient solutions, i.e. locations which can only be improved in one of the criteria by accepting higher costs in at least one other criterion. Starting from point-objective models where each demand point has its own associated objective function (see, for example, Wendell, Hurter and Lowe (1977)), other multicriteria location models have been suggested (see Current, Min and Schilling (1990) for an overview). This includes multicriteria Weber problems with two or more minisum objective functions as discussed, for example, in Hamacher and Nickel (1996) and Puerto and Fernández (1998).

A very general approach to model objective functions in location problems is the concept of ordered Weber problems originally developed by Puerto and Fernández (2000). Ordered Weber problems include most of the models and extensions mentioned above as special cases. They can be solved efficiently for the case of polyhedral gauges, see Nickel et al. (1999), or Rodríguez-Chía et al. (2000).

In other generalizations, the Weber problem was treated as stochastic or as dynamic optimization problem. For example, demand points need not be fixed, or even be points. Cooper (1974) described the locations of the demand points by probability distributions. The demand points can also be represented by areas with some associated density function. This approach may be used to model a set of demand points that is too large to be represented by individual points, or to model a probability distribution of demand over an area (Love (1972) and Drezner and Wesolowsky (1980)). An example where the weights of the demand points are stochastic was treated by Aly and White (1978).

When the problem is viewed as a cost minimizing location problem, it is static in the sense that the parameters are assumed to be constant over the decision horizon. When there are anticipated changes such as those in weight or locations of demand points and if re-locations of the facility are possible, the problem becomes "dynamic" rather than static. Erlenkotter (1981) summarized different approaches to dynamic location.

1.10 Epilogue

The Weber problem is the cornerstone of locational analysis. It is the first location problem ever posed, and gave rise to numerous extensions and models. It is the root of the tree spanning location models. Every model discussed in this book can be traced back to the Weber problem.

The Weber problem was extended by considering other environments (such as networks, or the globe) leading to a variety of distance measures. Objectives considered in many models include the minisum (the original Weber objective) as well as the minimax, maximin (used for modeling obnoxious facility location where being close is a detriment), competitive (when attempting to attract as much demand as possible from competing facilities) and composites of these measures. Many models assume stochastic or dynamic rather than deterministic demand. Barriers to travel or constraints are considered in many models as well. Extensions to the location of multiple facilities abound. Most of these models (location-allocation or by other terminologies p-median or p-center models) assume that demand is serviced by the closest facility, leading to difficult combinatorial problems of simultaneously allocating demand to facilities and locating the facilities (location-allocation models).

All these extensions are summarized in the LOLA classification scheme (see Chapter 8). Each model is classified by five characteristics describing its uniqueness in the vast ocean of location models.

References

Aly, A. and J. White (1978) "Probabilistic Formulations of the Multifacility Weber Problem," *Naval Research Logistics Quarterly*, 25, 531-547.

Aly, A., D. Kay and D. Litwhiler Jr. (1979) "Location Dominance on Spherical Surfaces," *Operations Research*, 27, 972-981.

Andersen, K.D. (1996) "An Efficient Newton Barrier Method for Minimizing a Sum of Euclidean Norms," *SIAM Journal of Optimization*, 6, 74-95.

Aneja, Y.P. and M. Parlar (1994) "Algorithms for Weber Facility Location in the Presence of Forbidden Regions and/or Barriers to Travel," *Transportation Science*, 28, 70-76.

Arora, S., Raghavan, P. and S. Rao (1998) "Approximation Schemes for Euclidean k-Medians and Related Problems," *Proceedings of the 30th ACM STOC*, 106-113.

Austin, T., Jr. (1959) "An Approximation to the Point of Minimum Aggregate Distance," *Metron*, 1, 10-21.

Batta, R., A. Ghose and U.S. Palekar (1989) "Locating Facilities on the Manhatten Metric with Arbitrarily Shaped Barriers and Convex Forbidden Regions," *Transportation Science*, 23, 26-36.

Bongartz, I., P.H. Calamai and A.R. Conn (1994) "A Projection Method for l_p Norm Location-Allocation Problems," *Mathematical Programming*, 66, 283-312.

Brandeau, L.B. and S.S. Chiu (1989) "An Overview of Representative Problems in Location Theory," *Management Science*, 35, 645-674.

Brimberg, J. (1989) "Properties of Distance Functions and Minisum Location Models," Ph.D. Thesis, McMaster University.

Brimberg, J., P. Hansen, N. Mladenovic and E.D. Taillard (2000) "Improvements and Comparison of Heuristics for Solving the Uncapacitated Multisource Weber Problem," *Operations Research*, 48, 444-460.

Brimberg, J. and R.F. Love (1992) "Local Convergence in a Generalized Fermat-Weber Problem," *Annals of Operations Research*, 40, 33-66.

Brimberg, J. and R.F. Love (1993) "Global Convergence of a Generalized Iterative Procedure for the Minisum Location Problem with l_p Distances," *Operations Research*, 41, 1153-1163.

Brimberg, J. and R.F. Love (1995) "Properties of Ordinary and Weighted Sums of Order p," *RAIRO-Operations Research*, 29(1), 59-72.

Brimberg, J. and R.F. Love (1998) "Solving a Class of Two-Dimensional Uncapacitated Location-Allocation Problems by Dynamic Programming," *Operations Research*, 46, 702-709.

Brimberg, J., R.F. Love and G.O. Wesolowsky (1993) "The Minisum Problem with Infeasible Regions," *Studies in Locational Analysis*, 4, 29-33.

Brimberg, J. and N. Mladenovic (1996a) "Solving the Continuous Location-Allocation Problem with Tabu Search," *Studies in Locational Analysis*, 8, 23-32.

Brimberg, J. and N. Mladenovic (1996b) "A Variable Neighborhood Algorithm for Solving the Continuous Location-Allocation Problem," *Studies in Locational Analysis*, 10, 1-12.

Burkard, E.R., H.W. Hamacher and G. Rote (1991) "Sandwich Approximation of Univariate Convex Functions with an Application to Separable Convex Programming," *Naval Research Logistics*, 38, 911-924.

Burstall, R., R. Leaver and J. Sussams (1962) "Evaluation of Transport Costs for Alternative Factory Sites - A Case Study," *Operational Research Quarterly*, 13, 345-354.

Butt, S.E. (1994) "Facility Location in the Presence of Forbidden Regions," Ph.D. Thesis, Department of Industrial and Management Systems Engineering, Pennsylvania State University.

Butt, S.E. and T.M. Cavalier (1996) "An Efficient Algorithm for Facility Location in the Presence of Forbidden Regions," *European Journal of Operational Research*, 90, 56-70.

Calamai, P. and C. Charalambous (1980) "Solving Multifacility Location Problems Involving Euclidean Distances," *Naval Research Logistics Quarterly*, 27, 609-620.

Carrizosa, E., H.W. Hamacher, R. Klein and S. Nickel (2000) "Solving Nonconvex Planar Location Problems by Finite Dominating Sets," *Berichte des ITWM*, No. 18, Kaiserslautern.

Carrizosa, E. and J. Puerto (1995) "A discretizing algorithm for location problems," *European Journal of Operational Research* 80, 166-174.

Chandrasekaran, R. and A. Tamir (1989) "Open Questions Concerning Weiszfeld's Algorithm for the Fermat-Weber Location Problem," *Mathematical Programming*, 44, 293-295.

Chandrasekaran, R. and A. Tamir (1990) "Algebraic Optimization: The Fermat-Weber Location Problem," *Mathematical Programming*, 46, 219-224.

Chen, P.-C., P.Hansen, B. Jaumard and H. Tuy (1998) "Solution of the Multisource Weber and Conditional Weber Problems by D.-C. Programming," *Operations Research*, 46, 548-562.

Chen, R. (1983) "Solution of Minisum and Minimax Location-Allocation Problems with Euclidean Distances," *Naval Research Logistics Quarterly*, 30, 449-459.

Cooper, L. (1963) "Location-Allocation Problems," *Operations Research*, 11, 331-343.

Cooper, L. (1964) "Heuristic Methods for Location-Allocation Problems," *SIAM Reviews*, 6, 37-53.

Cooper, L. (1967) "Solutions of Generalized Locational Equilibrium Models," *Journal of Regional Science*, 7, 1-18.

Cooper, L. (1973) "N-Dimensional Location Models: An Application to Cluster Analysis," *Journal of Regional Science*, 13, 41-54.

Cooper, L. (1974) "A Random Locational Equilibrium Problem," *Journal of Regional Science*, 14, 47-54.

Cooper, L. and I. Katz (1981) "The Weber Problem Revisited," *Computers and Mathematics with Applications*, 7, 225-234.

Courant, R. and H. Robbins (1941) *What is Mathematics?*, Oxford University Press.

Court, A. (1964) "The Elusive Point of Minimum Travel," *Annals of the Association of American Geographers*, 54, 400-403.

Coxeter, H.S.M. (1969) *Introduction to Geometry*, Wiley.

Current, J., H. Min and D. Schilling (1990) "Multiobjective Analysis of Facility Location Decisions," *European Journal of Operational Research*, 49, 295-307.

Dahlquist G. and A. Bjorck (1974) *Numerical Methods*, prentice-Hall Series in Automatic Computation, Prentice-hall, Englewood Cliffs, NJ.

Dax, A. (1986) "A Note on Optimality Conditions for the Euclidean Multifacility Location Problem," *Mathematical Programming*, 36, 72-80.

Díaz-Báñez, J.M., J.A. Mesa and A. Schöbel (2001) "Continuous Location of Dimensional Structures," *Technical Report in Wirtschaftsmathematik, University of Kaiserslautern*.

Dörrie, H. (1965) *100 Great Problems of Elementary Mathematics, their History and Solution*, (translated by David Antin). Dover.

Drezner, Z. (1984) "The Two-Center and Two Median Problems," *Transportation Science*, 18, 351-361.

Drezner, Z. (1985) "A Solution to the Weber Location Problem on the Sphere," *Journal of the Operational Research Society*, 36, 333-334.

Drezner Z. (1986) "Location of Regional Facilities," *Naval Research Logistics Quarterly*, 33, 523-529.

Drezner Z. (1992) "A Note on the Weber Location Problem," *Annals of Operations Research*, 40, 153-161.

Drezner Z. (1996) "A Note on Accelerating the Weiszfeld Procedure," *Location Science*, 3, 275-279.

Drezner Z. and D. Simchi-Levi (1992) "Asymptotic Behavior of the Weber Location Problem on the Plane," *Annals of Operations Research*, 40, 163-172.

Drezner, Z., G. Steiner and G.O. Wesolowsky (1985) "One Facility Location with Rectilinear Tour Distances," *Naval Research Logistics Quarterly*, 32, 391-405.

Drezner, Z., S. Steiner and G.O. Wesolowsky (2001) "On the Circle Closest to a Set of Points," *Computers and operations Research*, forthcoming.

Drezner Z. and G.O. Wesolowsky (1978) "Facility Location on a Sphere," *Journal of the Operational Research Society*, 29, 997-1004.

Drezner, Z. and G.O. Wesolowsky (1980) "Optimal Location of a Facility Relative to Area Demands," *Naval Research Logistics Quarterly*, 27, 199-206.

Drezner, Z. and G.O. Wesolowsky (1990) "The Weber Problem on the Plane with some Negative Weights," *INFOR*, 29, 87-89.

Durier, R. and C. Michelot (1985) "Geometrical Properties of the Fermat-Weber Problem," *European Journal of Operational Research*, 20, 332-343.

Eckhardt, U. (1980) "Weber's Problem and Weiszfeld's Algorithm in General Spaces," *Mathematical Programming*, 18, 186-196.

Eels, W.C. (1930) "A Mistaken Conception of the Center of Population," *Journal of the American Statistical Association*, XXV, 33-40.

Eilon, S., C.D.T. Watson-Gandy and N. Christofides (1971) *Distribution Management: Mathematical Modelling and Practical Analysis*, Hafner, New York.

El-Shaieb, A. (1978) "The Single Source Weber Problem - Survey and Extensions," *Journal of Operational Research*, 29, 469-476.

Elzinga J., and D.W. Hearn (1983) "On Stopping Rules for Facilities Location Algorithms," *IIE Transactions*, 15, 81-83.

Erkut, E. and S. Neumann (1989) "Analytical Methods for Location of Undesirable Facilities," *European Journal of Operational Research*, 40, 275-291.

Erlenkotter, D. (1981) "A Comparative Study of Approaches to Dynamic Location Problems," *European Journal of Operational Research*, 6, 133-143.

Eyster, J., J. White and W. Wierwille (1973) "On Solving Multifacility Location Problems Using a Hyperboloid Approximation Procedure," *AIIE Transactions*, 5, 1-6.

Fliege, J. (1997) *Effiziente Dimensionsreduktion in Multilokationsproblemen*, Shaker Verlag, Aachen.

Fliege, J. (2000) "Solving Convex Location Problems with Gauges in Polynomial Time," *Studies in Locational Analysis*, 14, 153-171.

Fliege, J. and S. Nickel (2000) "An Interior Point Method for Multifacility Location Problems with Forbidden Regions," *Studies in Locational Analysis*, 14, 23-46.

Foulds, L.R. and H.W. Hamacher (1993) "Optimal Bin Location in Printed Circuit Board Assembly," *European Journal of Operational Research*, 66, 279-290.

Francis, R.L., H.W. Hamacher, C.-Y. Lee and S. Yeralan (1994) "On Automating Robotic Assembly Workplace Planning," *Transactions of the Institute of Industrial Engineers*, 11E, 47-59.

Francis, R.L., F. Leon, L.F. McGinnis and J.A. White (1992) *Facility Layout and Location: An Analytical Approach*, Prentice-Hall, New York, 2nd edition.

Gini, C. and L. Galvani (1929) "Di talune estensioni dei concetti di media ai caratteri qualitativi," *Metron*, 8.

Guccione, A. and W. Gillen (1991) "An Economic Interpretation of Kuhn's Dual for the Steiner-Weber Problem: A Note," *Journal of Regional Science*, 31, 93-95.

Hall, R.W. (1988) "Median, Mean, and Optimum as Facility Locations," *Journal of Regional Science*, 28, 65-81.

Hamacher, H.W. and K. Klamroth (2000) "Planar Location Problems with Barriers under Polyhedral Gauges," *Annals of Operations Research*, 96, 191-208.

Hamacher, H.W. and S. Nickel (1994) "Combinatorial Algorithms for Some 1-Facility Median Problems in the Plane," *European Journal of Operational Research*, 79, 340-351.

Hamacher, H.W. and S. Nickel (1995) "Restricted Planar Location Problems and Applications," *Naval Research Logistics*, 42, 967-992.

Hamacher, H.W. and S. Nickel (1996) "Multicriteria Planar Location Problems," *European Journal of Operational Research*, 94, 66-86.

Hansen, P., B. Jaumard and H. Tuy (1995) "Global Optimization in Location," in *Facility Location*, Z. Drezner (ed.), Springer Series in Operations Research, 43-68.

Hansen, P., N. Mladenovic and E. Taillard (1998) "Heuristic Solution of the Multi-source Weber Problem as a p-Median Problem," *Operations Research Letters*, 22, 55-62.

Hansen, P., D. Peeters and J.-F. Thisse (1982) "An Algorithm for a Constrained Weber Problem," *Management Science*, 28, 1285-1290.

Hansen, P., D. Peeters, D. Richard and J.-F. Thisse (1985) "The Minisum and Minimax Location Problems Revisited," *Operations Research*, 33, 1251-1265.

Hansen, P., D. Peeters and J.-F. Thisse (1982) "An Algorithm for a Constrained Weber Problem," *Management Science*, 28, 1285-1295.

Hansen, P., J. Perreur and J.-F. Thisse (1980) "Location Theory, Dominance and Convexity: Some Further Results," *Operations Research*, 28, 1241-1250.

Harris, B. (1976) "Speeding Up Iterative Algorithms - The Generalized Weber Problem." *Journal of Regional Science*, 16, 411-413.

Hodgson, M.J., K.E. Rosing and F. Shmulevitz (1993) "A Review of Location-Allocation Applications Literature," *Studies in Locational Analysis*, 5, 3-29.

Hodgson, M.J., R.T. Wong and J. Honsaker (1987) "The p-Centroid Problem on an Inclined Plane," *Operations Research*, 35, 221-233.

Honsberger, R. (1973) *Mathematical Gems from Elementary Combinatorics, Number Theory, and Geometry I. The Dolciani Mathematical Expositions*, published by the Mathematical Association of America.

Idrissi, H.F., O. Lefebvre and C. Michelot (1989) "Duality for Constrained Multifacility Location Problems with Mixed Norms and Applications," *Annals of Operations Research*, 18, 71-92.

Juel, H. (1984) "On a Rational Stopping Rule for Facilities Location Algorithms," *Naval Research Logistics Quarterly*, 31, 9-11.

Juel, H. and R. Love (1981) "Fixed Point Optimality Criteria for the Weber Problem with Arbitrary Norms," *Journal of the Operational Research Society*, 32, 891-897.

Juel, H. and R. Love (1986) "A Geometrical Interpretation of the Existing Facility Solution Condition for the Weber Problem," *Journal of the Operationol Research Society*, 37, 1129-1131.

Karkazis, J. (1988) "The General Unweighted Problem of Locating Obnoxious Facilities on the Plane," *Belgian Journal of Operations Research, Statistics and Computer Science*, 28, 3-49.

Katz, I. (1974) "Local Convergence in Fermat's Problem," *Mathematical Programming*, 6, 89-104.

Katz, I. and L. Cooper (1981) "Facility Location in the Presence of Forbidden Regions, I: Formulation and the Case of Euclidean Distance With One Forbidden Circle," *European Journal of Operational Research*, 6, 166-173.

McKinnon, R.D. and G.M. Barber (1972) "A New Approach to Network Generation and Map Representation: The Linear Case of the Location-Allocation Problem," *Geographical Analysis*, 4, 156-168.

Klamroth, K. (2001a) "Planar Weber Location Problems with Line Barriers," *Optimization*, to appear.

Klamroth, K. (2001b) "A Reduction Result for Location Problems with Polyhedral Barriers," *European Journal of Operational Research*, 130, 486-497.

Korneenko, N.M. and H. Martini (1993) "Hyperplane Approximation and Related Topics." In *New Trends in Discrete and Computational Geometry*, János Pach (ed.), Springer-Verlag, 135-162.

Krau, S. (1997) "Extensions du Problème de Weber," Ph.D. Thesis, Département de Mathématiques et de Génie Industriel, Université de Montréal.

Kuenne, R.E. and R.M. Soland (1972) "Exact and Approximate Solutions to the Multisource Weber Problem," *Mathematical Programming*, 3, 193-209.

Kuhn, H.W. (1967) "On a Pair of Dual Nonlinear Programs," *Nonlinear Programming*, J. Abadie (ed.), New Holland.

Kuhn, H.W. (1973) "A Note on Fermat's Problem," *Mathematical Programming*, 4, 98-107.

Kuhn, H.W. (1976) "Nonlinear Programming: A Historical View," *Nonlinear Programming*, American Mathematical Society.

Kuhn, H. and R. Kuenne (1962) "An Efficient Algorithm for the Numerical Solution of the Generalized Weber Problem in Spatial Economics," *Journal of Regional Science*, 4, 21-34.

Laporte, G., J.A. Mesa and F. Ortega (1994) "Assessing Topological Configuration for Rapid Transit Networks," *Studies in Locational Analysis*, 7, 105-121.

Larson, R.C. and G. Sadiq (1983) "Facility Locations with the Manhattan Metric in the Presence of Barriers to Travel," *Operations Research*, 31, 652-669.

Lee, D.T. and Y.T. Ching (1985) "The Power of Geometric Duality Revisited," *Information Processing Letters*, 21, 117-122.

Lin, J.-H. and J.S. Vitter (1992a) "ε-Approximation with Minimum Packing Constraint Violation," *Proceedings of the 24th ACM STOC*, 771-782.

Lin, J.-H. and J.S. Vitter (1992b) "Approximation Algorithms for the Geometric Median Problem," *Information Processing Letters*, 44, 148-162.

Love, R.F. (1967) "A Note on the Convexity of Siting Depots," *The International Journal of Production Research*, 6, 153-154.

Love, R.F. (1969) "Locating Facilities in Three-Dimensional Space By Convex Programming," *Naval Research Logistics Quarterly*, 16, 503-516.

Love, R.F. (1972) "A Computational Procedure for Optimally Locating a Facility with Respect to Several Rectangular Regions," *Journal of Regional Science*, 12, 233-242.

Love, R.F. (1976) "One-Dimensional Facility Location-Allocation Using Dynamic Programming," *Management Science*, 22, 614-617.

Love, R.F. and H. Juel (1982) "Properties and Solution Methods for Large Location-Allocation Problems," *Journal of the Operations Research Society*, 33, 443-452.

Love, R.F. and J.G. Morris (1975) "A Computation Procedure for the Exact Solution of Location-Allocation Problems with Rectangular Distances," *Naval Research Logistics*, 22, 441-453.

Love, R.F. and J.G. Morris (1979) "Mathematical Models of Road Travel Distances," *Management Science*, 25, 130-139.

Love, R.F., J.G. Morris and G.O. Wesolowsky (1988) *Facilities Location: Models & Methods*, North-Holland.

Love, R.F., W.G. Truscott and J. Walker (1985) "Terminal Location Problem: A Case Study Supporting the Status Quo," *Journal of the Operational Research Society*, 36, 131-136.

Love, R. and W. Yeong (1981) "A Stopping Rule for Facilities Location Algorithms," *AIIE Transactions*, 13, 357-362.

Lozano, A.J. and J.A. Mesa (2000) "Location of Facilities with Undesirable Effects and Inverse Location Problems: A Classification," *Studies in Locational Analysis*, 14, 253-291

Lozano-Perez, T. and M. Wesley (1978) "An Algorithm for Planning Collision Free Paths Among Polyhedral Obstacles," *Communications of the ACM*, 22, 560-570.

Martini H. (1996) "A Geometric Generalization of the Vecten-Fasbender Duality," *Studies in Locational Analysis*, 10, 53-65.

Martini, H. and A. Schöbel (1998a) "Median Hyperplanes in Normed Spaces – a Survey," *Discrete Applied Mathematics*, 89, 181-195.

Martini, H. and A. Schöbel (1998b) "A Characterization of Smooth Norms," *Geometriae Dedicata*, 77, 173-183.

Megiddo, N. and K.J. Supowit (1984) "On the Complexity of Some Common Geometric Location Problems," *SIAM Journal on Computing*, 13, 182-196.

Megiddo, N. and A. Tamir (1982) "On the Complexity of Locating Linear Facilities in the Plane," *Operations Research Letters*, 1, 194-197.

Megiddo, N. and A. Tamir (1983) "Finding Least-Distance Lines," *SIAM J. on Algebraic and Discrete Methods*, 4(2), 207-211.

Melzak, Z.A. (1983) *Invitation to Geometry*, Wiley.

Michelot, C. (1987) "Localization in Multifacility Location Theory," *European Journal of Operational Research*, 31, 177-184.

Michelot, C. (1993) "The Mathematics of Continuous Location," *Studies in Locational Analysis*, 5, 59-83.

Miehle, W. (1958) "Link-Length Minimization In Networks," *Operations Research*, 6, 232-243.

Moreno, J., C. Rodrigez and N. Jimenez (1990) "Heuristic Cluster Algorithm for Multiple Facility Location-Allocation Problem," *Operations Research*, 25, 97-107.

Morris, J.G. (1981) "Convergence of the Weiszfeld Algorithm for Weber Problems Using a Generalized "distance" Function", *Operations Research*, 29, 37-48.

Morris, J.G. and J.P. Norback (1980) "A Simple Approach to Linear Facility Location," *Transportation Science*, 14(1), 1-8.

Morris, J.G. and J.P. Norback (1983) "Linear Facility Location - Solving Extensions on the Basic Problems," *European Journal of Operational Research*, 12, 90-94.

Muralimohan, R. and A.J.G. Babu (1983) "Mathematical Modelling of the Weber Problem in the Presence of Convex Forbidden Regions," Tims/Orsa Joint National Meeting, Chicago.

Nickel, S. (1995) *Discretization of Planar Location Problems*, Shaker Verlag, Aachen.

Nickel, S. (1997) "Bicriteria and Restricted 2-Facility Weber Problems," *Mathematical Methods of Operations Research*, 45, 167-195.

Nickel, S. and E.-M. Dudenhöffer (1996) "Weber's Problem with Attraction and Repulsion under Polyhedral Gauges," *Journal of Global Optimization*, 11, 409-432.

Nickel, S., Puerto, J., Rodríguez-Chía, A.M., and Weissler, A. (1999) "Multicriteria Ordered Weber Problems," *Technical Report in Wirtschaftsmathematik* 53, University of Kaiserslautern.

Ortega, F.A., J.A. Mesa and A.B. Sánchez (2000) "An Iterative Method for Solving the Weber Problem in R^2 with l_p Norms, $p \in (1,2)$, Based in Linear Programming," *Studies in Locational Analysis*, 14, 137-152.

Ostresh, L., Jr. (1973) "TWAIN – Exact Solutions to the Two Source Location-Allocation Problem." In *Computer Programs for Location-Allocation Problems*, G. Rushton, M.F. Goodchild and L.M. Ostresh, Jr. (ed.), Monograph Number 6, Department of Geography, University of Iowa, Iowa City, IA.

Ostresh, L., Jr. (1975) "An Efficient Algorithm for Solving the Two Center Location-Allocation Problem," *Journal of Regional Science*, 15, 209-216.

Ostresh, L., Jr. (1977) "The Multifacility Location Problem: Applications and Descent Theorems," *Journal of Regional Science*, 17, 409-419.

Ostresh, L., Jr. (1978a) "On the Convergence of a Class of Iterative Methods for Solving the Weber Location Problem," *Operations Research*, 26, 597-609.

Ostresh, L., Jr. (1978b) "Convergence and Descent in the Fermat Location Problem," *Transportation Science*, 12, 153-164.

Overton, M. (1983) "A Quadratically Convergent Method for Minimizing a Sum of Euclidean Norms," *Mathematical Programming*, 27, 34-63.

LaPaugh, A.S. (1980) "Algorithms for Integrated Circuit Layout: An Analytic Approach," Ph.D. Thesis, Massachusetts Institute of Technology.

Perreur, J. and J. Thisse (1974) "Central Metrics and Optimal Location," *Journal of Regional Science*, 14, 411-421.

Planchart, A. and A. Hurter Jr. (1975) "An Efficient Algorithm for the Solution of the Weber Problem With Mixed Norms," *SIAM Journal of Control*, 13, 650-655.

Plastria, F. (1992) "GBSSS: The Generalized Big Square Small Square Method for Planar Single Facility Location," *European Journal of Operational Research*, 62, 163-174.

Plastria, F. (1995) "Continuous Location Problems," Chapter 11 in *Facility Location - a Survey of Applications and Methods*, Z. Drezner (ed.), Springer.

Plastria, F. and E. Carrizosa (2001) "Gauge-Distances and Median Hyperplanes," *Journal of Optimisation Theory and Applications*, 110, to appear.

Porter, P.W. (1963) "What is the Point of Minimum Aggregate Travel," *Annals of the Association of American Geographers*, 53, 224-232.

Porter, P.W. (1964) "A Comment on the Elusive Point of Minimum Travel," *Annals of the Association of American Geographers*, 54, 406.

Pottage, J. (1983) *Geometrical Investigations*, Addison-Wesley.

Puerto, J. and F.R. Fernández (1998) "A Convergent Approximation Scheme for Efficient Sets of the Multi-Criteria Weber Location Problem," *Sociedad de Estadística e Investigación Operativa*, 6, 195-204.

Puerto J. and Fernández F.R. (2000) "Geometrical properties of the symmetric single facility location problem," *Journal of Nonlinear and Convex Analysis*, 1(3), 1-22.

Rado, F. (1988) "The Euclidean Multifacility Location Problem," *Operations Research*, 36, 485-492.

Riveline, C. (1967) "Optimum Transportation Network by Mechanical Analogy," 31'st National Meeting, Operations Research Society of America, New York.

Robert, J.M. (1991), "Linear Approximation and Line Transversals," Ph.D. Thesis, School of Computer Sciences, McGill University, Montreal.

Robert, J.M. and G.T. Toussaint (1994) "Linear Approximation of Simple Objects," *Computational Geometry*, 4, 27-52.

Rochat M., Vecten, Fauquier, and Pillate (1811) "Questions Resolues: Solutions des deux problemes proposes a la page 384 du premier volume des annales," *Ann. Math. Pures et Appl.*, 2(1811/1812), 88-96.

Rodríguez-Chía, A.M., Nickel S., Puerto J., and Fernández, F.R. (2000) "A flexible approach to location problems," *Mathematical Methods of Operations Research*, 51, 69-89.

Rosen, J.B. and G.L. Xue (1991) "Computational Comparison of two Algorithms for the Euclidean Single Facility Location Problem," *ORSA Journal on Computing*, 3, 207-212.

Rosing, K.E. (1992) "An Optimal Method for Solving the (Generalized) Multi-Weber Problem," *European Journal of Operational Research*, 58, 414-426.

Ross, F.A. (1930) "Editor's Note on the Center of Population and Point of Minimum Travel," *Journal of the American Statistical Association*, XXV, 447-452.

Rousseeuw, P.J. and A.M. Leroy (1987) *Robust Regression and Outlier Detection*, Wiley.

Savaş, S., R. Batta and R. Nagi (2001) "Finite-Size Facility Placement in the Presence of Barriers to Rectilinear Presence," Working Paper, Dept. of Industrial Engineering, University at Buffalo, Buffalo, NY, submitted to *Operations Research*.

Scates, D.E. (1933) "Locating the Median of the United States," *Metron*, 11, 49-65.

Schärlig, A. (1973) "About the Confusion between the Center of Gravity and Weber's Optimum," *Regional and Urban Economics*, 13, 371-382.

Schöbel, A. (1996) "Locating Least-Distant Lines with Block Norms," *Studies in Locational Analysis*, 10, 139-150.

Schöbel, A. (1998) "Locating Least Distant Lines in the Plane," *European Journal of Operational Research*, 106(1), 152-159.

Schöbel, A. (1999a) *Locating Lines and Hyperplanes – Theory and Algorithms*, Kluwer.

Schöbel, A. (1999b) "Solving Restricted Line Location Problems via a Dual Interpretation," *Discrete Applied Mathematics*, 93, 109-125.

Scott C.H., T.R. Jefferson, and S. Jorjani (1995) "Conjugate Duality in Facility Location," in *Facility Location*, Z. Drezner (ed.), Springer Series in Operations Research, 89-101.

Segars, R., Jr. (2000) "Location Problems with Barriers Using Rectilinear Distance," Ph.D. thesis, Department of Mathematical Sciences, Clemson University, Clemson, SC.

Seymour, D. (1970) "Note on Austin's "An Approximation to the Point of Minimum Aggregate Distance," *Metron*, 28, 412-421.

Seymour, D. and J. Weindling (1975) "An Iterative Curve Fitting Approach for Solving the Weber Problem in Spatial Economics," *Annals of Regional Science*, 9, 14-24.

Smith, D.E. (1923) *History of Mathematics Vol 1.*, Ginn and Company.

Sullivan, P.J. and N. Peters (1980) "A Flexible User Oriented Location-Allocation Algorithm," *Journal of Environmental Management*, 10, 181-193.

Tellier, L. (1972) "The Weber Problem: Solution and Interpretation," *Geographical Analysis*, 4, 215-233.

Tellier, L.-N. and B. Polanski (1989) "The Weber Problem: Frequency of Different Solution Types and Extension to Repulsive Forces and Dynamic Processes," *Journal of Regional Science*, 29, 387-405.

Thisse, J.-F., J. Ward and R. Wendell (1984) "Some Properties of Location Problems with Block and Round Norms," *Operations Research*, 32, 1309-1327.

Üster, H. (1999) "Weighted Sums of Order p and Minisum Location Models," Ph.D. Thesis, McMaster University, Canada.

Vergin, R. and J. Rogers (1967) "An Algorithm and Computational Procedure for Locating Economic Facilities," *Management Science*, 13, B240-254

Ward, J. and R. Wendell (1980) "A New Norm for Measuring Distance Which Yields Linear Location Problems," *Operations Research*, 28, 836-843.

Ward, J. and R. Wendell (1985) "Using Block Norms for Location Modelling," *Operations Research*, 33, 1074-1090.

Weber, A. (1909) *Über den Standort der Industrien, Tübingen*, (English translation by Friedrich, C. J. (1929). *Theory of the Location of Industries*, University of Chicago Press.

Weiszfeld, E. (1936) "Sur le Point Pour Lequel la Somme des Distances de n Points Donnes est Minimum," *The Tohoku Mathematical Journal*, 43, 355-386.

Wendell, R.E., A.P. Hurter and T.J. Lowe (1977) "Efficient Points in Location Problems," *AIIE Transactions*, 9, 238-246.

Wendell, R. and E. Peterson (1984) "A Dual Approach for Obtaining Lower Bounds to the Weber Problem," *Journal of Regional Science*, 24, 219-228.

"Location of the Median Line for Weighted Points," *Environment and Planning A*, 7, 163-170.

Wesolowsky G.O. (1983) "Location Problems on a Sphere," *Regional Science and Urban Economics*, 12, 495-508.

Wesolowsky, G.O. (1993) "The Weber Problem: History and Perspectives," *Location Science*, 1, 5-23.

Wesolowsky, G.O. and R.F. Love (1972) "A Nonlinear Approximation Method for Solving a Generalized Rectangular Distance Weber Problem," *Management Science*, 18, 656-663.

White, J.A. (1971) "A Quadratic Facility Location Problem," *American Institute of Industrial Engineers Transactions*, 3, 156-157.

White, D.J. (1976) "An Analogue Derivation of the Dual of the General Fermat Problem," *Management Science*, 23, 92-94.

Winter, P. (1985) "An Algorithm for the Steiner Problem in the Euclidean Plane," *Networks*, 15, 323-345.

Witzgall, C. (1964) "Optimal Location of a Central Facility: Mathematical Models and Concepts," *National Bureau of Standards Report 8388*, Gaithersberg, Maryland.

Xue, G.L., J.B. Rosen and P.M. Pardalos (1996) "A Polynomial Time Dual Algorithm for the Euclidean Multifacility Location Problem." *Operations Research Letters*, 18, 201-204.

Yeralan, S. and J.A. Ventura (1988) "Computerized Roundness Inspection," *International Journal of Production Research*, 26, 1921-1935.

Zacharias, M. (1913) "Elementargeometrie und elementare nicht-euklidische Geometrie in Synthetischer Behandlung," *Encyklopädie der mathematischen Wissenschaften (Geometrie)*, W. Fr. Meyer and H. Mohrmann (ed.), Leipzig, 1914-1931.

Zemel, E. (1984) "An $O(n)$ Algorithm for the Linear Multiple Choice Knapsack Problem and Related Problems," *Information Processing Letters*, 18, 123-128.

2 Continuous Covering Location Problems

Frank Plastria[1]

BEIF - Department of Management Informatics, GOLD - Research Group on Location and Distribution, Vrije Universiteit Brussel, Pleinlaan 2, B-1050 Brussels, Belgium. e-mail: Frank.Plastria@vub.ac.be

2.1 Introduction

A location problem arises whenever a question is raised like

- where are we going to put the thing(s) ?

The next two questions then immediately follow:

1. which places are available ?
2. on what basis do we choose ?

2.1.1 Locational Space

The answer to first question determines the locational space. We have a *continuous location problem* when this space is described by way of continuous variables, usually coordinates.

In most applications this space is either planar — just think of an integrated circuit, a piece of paper, your desktop, a shopfloor, a piece of land or a country (if not too large) — or on a sphere, when considering a really large region like a continent or even the whole earth; in these cases we need two coordinates to describe a position, and our locational space is two-dimensional. For positioning within a building, underwater or in the air it will also be necessary to take height or depth into account, so a third coordinate will be needed. Certain more theoretical frameworks may even call for more dimensions. Problems with one dimension also occur when we are locating on a line (which might be straight, curved and/or broken), such as a single stretch of highway, waterway or railway.

Continuous location problems also assume one cannot give an exhaustive list of all individual available places, as is the case in discrete location problems which are also discussed in another chapter. Here we deal with the somewhat more vague situation where we do not really know which sites are available, but rather that these are 'all over the place' and we want to find out where to look for good candidates. Thus, continuous location models can be considered as *site generating*, (Love et al., 1988), and will always have some geometrical flavour.

What we do have to take into account, however, is that in order to be eligible sites must come from some *feasible region(s)*. Hopefully this is described

---

Facility Location: Applications and Theory.
Edited by Z. Drezner and H.W. Hamacher
© 2002 Springer-Verlag, ISBN 3-540-42172-6

by constraints of not too complicated form, e.g. the rectangular shape of a printed circuit board. When the locational space is of geographic nature, the feasible region is almost always described as the insides of a series of polygonal region, the boundary of which is specified by a sequence of successive corner points, which are to be connected by straight line segments; such data are typical when using GIS (Geographic Information Systems). More complicated regions will and do arise when technical, economical or political rules compell the chosen sites not to be too close to certain sensitive places, leading to so *restricted* location problems (see (Hamacher and S.Nickel, 1995)) and/or perhaps to be within reach of some places, sometimes even the same as the sensitive ones (which evidently does not simplify the task). It may also happen that one should locate on a network, but taking into account continuous effects like pollution spread. In such a case the usual network models (see chapter 3) are insufficient and one needs the explicit description of the embedding of this network as part of the plane, e.g. under the form of straight line segments and/or curves.

Sometimes the model does not call for a feasible region and we then call the problem *unconstrained*. It must be clear, however, that such models are rather of theoretical interest and virtually all real world applications do include spatial constraints. Their presence invariably introduces new complexities into the model, as compared to their unconstrained counterparts, and have therefore often not been considered. If one is lucky, the unconstrained model, obtained by neglecting the spatial constraints actually present in the real world, may produce a solution that, by a mighty chance, is also feasible, and this will then evidently also solve the model including the constraints. As a rule this will not happen, and one will really have to take the constraints into account in the solution process. But the knowledge of such infeasible solutions to the unconstrained model is often quite helpful in solving the constrained problem. Thus some unconstrained models still have their uses as a first step in the solution process, and their study should not be neglected. In fact the largest part of the quite impressive and steadily growing literature on continuous location problems concerns unconstrained models.

2.1.2 Goal

Question 2 asks for the specification of some *goal*. In the kind of models we are dealing with in this book this goal is always to discover best choices, or at least choices that cannot (easily) be improved upon. Therefore our models are always stated as optimisation problems, which ask to (try to) minimise some cost, damage or discomfort and/or maximise some profit, quality or wellbeing.

In the easier problems this goal is stated as a single criterion, but in many (some say in all) applications there are several simultaneous criteria. While the first situation leads to classical, but by no means always simple problems of finding a minimum or maximum point of some function of the variables involved, in the second situation the actual aim is less clearly defined. Answers may be sought in different ways, and, what is perhaps worse, possibly leading to different answers. The most popular approach is to ask for determination of all *efficient* (also called *Pareto optimal* or *nondominated*) solutions, which are those solutions for which no other feasible solution exists which is simultaneously at least as good for all criteria and strictly better for at least one of the criteria. Unfortunately this set is usually quite hard to determine and often too large to be directly useful for a decision maker, since it does not point towards a precise solution.

It may be argued, however, that this last situation is perhaps not so bad at all. Indeed on the one hand, continuous location models are often used as first approximations to real world problems: even if you know that an adequate model for your problem should ideally be stated as discrete or on a network, such representations usually call for rather huge amounts of data which cannot always readily be obtained without great effort or costs. One may then use a less data-demanding continuous space description as a 'quick but dirty' method to find out which regions are the most promising ones to be studied with more precision. On the other hand, optimisation models as a rule call for mathematically precise formulations of variables, constraints and objectives, thus they only describe the quantitative aspects of the problem at hand, disregarding the frequently present qualitative aspects, evaluation of which should be left to the decision-makers. In addition, the data on which the 'hard' models are based are often only approximations, either because of unavailable data or uncertainty, e.g. due to the constantly changing environment they reflect. In such situations the optimisation model is to be considered as no more than a decision aid tool, the task of which is not to determine precise decisions (optimal solutions), but rather to suggest the better decisions by elimination of the outright bad ones. This may be obtained by a sensitivity analysis of the obtained solutions, which in location modelling takes the form of a region of nearly-optimal solutions, or containing optimal solutions to any situation that might arise.

2.1.3 Distance

Apart from local characteristics like cost of land which may differ from place to place, what always distinguishes one place from another place is its position relative to fixed points. And one of the most basic properties of relative position is the *distance*. Virtually all continuous location models are concerned with distances between points.

Distance is the mathematical description of the idea of proximity, and this may take many different forms depending on the application. For example, in a mountainous region it happens frequently that one can easily communicate verbally between two places across a chasm, whereas moving physically from one place to another may call for a large detour because of lack of wings.

Therefore knowledge of where two points are (by their coordinates, say) is insufficient to calculate their distance; one should also know what type of distance measure is to be considered. And it turns out that very many possible distance measures exist.

In a two or higher dimensional affine space environment, like a plane, the most familiar one is the *Euclidean* distance, which measures how far two points lie 'as the crow flies', by considering the length of the straight line segment joining two points. If the two points are given, e.g. in the plane, by their coordinates $X = (x, y)$ and $P = (a, b)$, relative to an orthonormal coordinate system (this means the axes are perpendicular and use the same scale), then the Euclidean distance is obtained as

$$d^{\text{euc}}(X, P) = \sqrt{(a-x)^2 + (b-y)^2} \qquad (2.1)$$

This is perhaps appropriate in the verbal communication distance example mentioned before, but certainly not for travel distance, except in the very unusual case where one can travel in all directions at the same ease, like sometimes in a desert.

Other, quite different distance measures arise when only some directions of travel are allowed. This happens in practice with some types of machinery. An automated drill for example is often moved over a drilling table by two motors, one in horizontal and one in vertical direction (disregarding a third motor which moves it up and down). For technical reasons it may be that the motors cannot work in parallel, and in this case the total time needed for moving the drill from one position to another position is the sum of the horizontal and the vertical movement times. In this case the appropriate distance measure will be the *rectangular* distance given by

$$d^{\text{rect}}(X, P) = |a - x| + |b - y| \qquad (2.2)$$

If, however, simultaneous movement of both motors is allowed, the drill may be moved in much shorter time from point X to point P: just start both motors together and stop each separately when the respective horizontal and vertical position of P is reached. The destination P will be reached when both motors have stopped, so the total travel time is the largest among horizontal and vertical travel times, and we have another distance measure: the *max* distance

$$d^{\text{max}}(X, P) = \max\{\,|a - x|\,,\,|b - y|\,\} \qquad (2.3)$$

Note that the path the drill actually has followed in this last case consists of two parts, a diagonally oriented part, while both motors worked together, and either a horizontal or vertical part, depending on which motor stopped first; also the actual speeds along these parts differ: the velocity along the diagonal is a factor $\sqrt{2}$ quicker than on the last part!

This situation may be broadly generalised by considering that only movements are allowed taken from a finite number of fixed directions and corresponding speeds. There will be many possible ways to move from one point to another, and one always chooses the one taking shortest time, which determines the distance. One so obtains a large class of distance measures called *block* or *polyhedral* distances. This name stems from the fact that any 'ball of radius r', i.e. the set of all points within distance r from an origin, is a (convex) polygon (or polyhedron in higher dimensions), which may be obtained as follows: always starting from the same origin, find in each allowed direction and corresponding speed the point reached in time r; finally take the convex hull of all these points (in the planar case this is the smallest convex polygon enclosing them).

For example, in a plane, the unit ball for the rectangular distance (the ball of radius 1 around the origin $(0,0)$) is the 'diamond' shaped tilted square with corner points $(1,0), (-1,0), (0,-1), (0,1)$, while for the max-distance it turns out to be the unit square with corners $(1,1), (1,-1), (-1,-1), (-1,1)$ (see figure 2.1). The fact that these two unit balls are different shows these distance measures really differ. Also, since the euclidean distance unit ball is a circle of radius 1, we see that in the plane euclidean distance is quite different from either rectangular or max distance, and, since a circle is not a polygon, that the euclidean distance is not a block distance.

Fig. 2.1. Unit balls for several distances in the plane

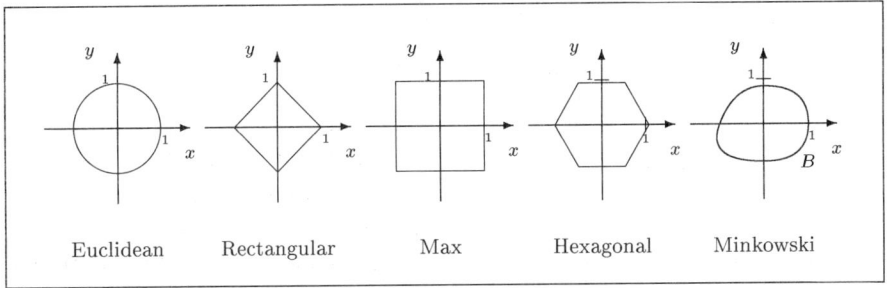

Euclidean   Rectangular   Max   Hexagonal   Minkowski

Apart from rectangular and max distances, more general block distances are usually not directly applicable in practice. They are very often used, however, to approximate other distance notions, and often lead to location problems that are easier to solve, as will become clear later on. For example, one very easily obtains a quite good approximate value, within a few percent, of the euclidean distance between two points on a map, using a hexagonal distance found by superposing a regular hexagonal grid on the map and counting the number of hexagonal cells that should be traversed. This little trick has become quite common with city-maps under the form of a loose transparent sheet with a preprinted hexagonal pattern.

Taking another huge leap towards general distances we arrive at the family of *Minkowski* distances, which consists of all possible distance notions which satisfy the celebrated triangle inquality (no side of a triangle is shorter than the sum of the other two) and are convex functions; these are obtained as follows. Consider any convex, closed and bounded set B which contains the origin, and assume this is the unit ball for some distance measure. This means that the set B consists exactly of all the points within distance 1 of the origin. In order to obtain the ball with radius r, say, we just blow up B by a factor r, keeping the origin fixed. Mathematically this is very easily done, by just considering all points $X = (a, b)$ lying in B (note that we take a two-dimensional example) and multiply the coordinates by r : (ra, rb). All these points form the blown up set rB. We now call the B-distance from the origin to some point P the lowest radius r for which P lies within rB. Finally, distances from some other point than the origin are obtained by translating first the origin to this point. This defines the Minkowski distance with unit ball B by the formula :

$$d^B(X, P) = d^B(0, P - X) = \min\{\ r > 0 \mid P - X \in rB\} \qquad (2.4)$$

In practice barriers to travel exist, like lakes, rivers or the chasm mentioned before, or speeds should be adapted to different terrains or to external influences like slopes, winds or currents. The influence of a steady current (if not too strong) is relatively easy to take into account, just by adding a linear perturbation term to the usual formula (2.1) (see (Plastria, 1992)), destroying of course the nice symmetry of euclidean distance: distance one way is not anymore equal to distance coming back! The other effects make the calculation of distance between just two points already a difficult problem and call for quite complex computational techniques. When speed is a general function of the position, one will have to resort to numerical approximation techniques (Fliege, 1997). When speed is constant on polygonal regions and their boundary segments, geometric approaches can be used (Herschberger and Suri, 1999; Mitchell and Papadimitriou, 1991). Handling a few barriers is now more or less well understood and used in location questions, see e.g. (Butt and Cavalier, 1996) and references therein. A somewhat simpler case of barriers arises when the space consists of a polygonal region and all movement must remain within this region (Guibas and Herschberger, 1989). This applies for travel on an island or on a lake, where one is allowed to travel in all directions at the same ease under provision to always stay on the island (lake); applications are also found in manufacturing with molds (Bose and Toussaint, 2000), in which also a totally different distance notion appears, *link* distance, giving the minimum number of turns needed along a broken line path within the region.

Real-world distances are usually still more complicated, and, as a rule, include lots of barriers and are not invariant with translation, as shown for

example in (Spiekerman and Wegener, 1994); see also the nice site at http://irpud.raumplanung.uni-dortmund.de/irpud/index_e.htm.
Therefore if we want a really close description of real world distances much more complex distance notions will have to be considered, which at present seem out of reach. An important application field of such generalised distances is to be found in robotics, with the many degrees of freedom in the movements that e.g. manipulator arms can produce, and the complicated barrier-filled environment within which they must function. Here we enter the realm of Riemann spaces, partial differential equations combined with quite hard combinatorics, in which continuous location theory has not really entered as yet. One notable exception is the quite simple analysis of the rest-position for a stationary crane (Rodriguez-Ramos and Francis, 1983).

Another easy situation is obtained when locational space is a sphere. Here the traditional distance notion corresponding to euclidean distance is the great circle distance, the properties of which are quite well understood. This is a quite adequate description of flight distances, although it disregards common disturbing effects like winds and/or Coriolis forces.

2.1.4 Objective(s)

In a typical location model we have a given set of fixed points, the *demand* points P_m ($m \in \mathcal{M}$), which is usually finite so $\mathcal{M} = \{1, \ldots, M\}$, although some models consider also continuously distributed demand.

The facilities for which we seek good sites $X_1, \ldots, X_N$ will in some way interact with the demand points, and possibly among themselves, and the effects of this interaction depend primarily on the distance between the interacting points. The objectives considered summarise all these effects into one global measure.

Some interactions are *pulls*, meaning that the general goal is improved when the interaction distance is decreased; in other words a pull means the closer the better. The typical example of a pull is service time, which increases with distance and should normally be minimised.

Other interactions *push* in the sense that it is desirable to have the interacting points as far apart as possible. This happens for instance with (potentially) dangerous installations, which ought to be installed as far as possible from sensitive places in order to minimise the all-over risk.

Many, if not most, applications, however, include both push and pull interactions, and therefore will call for a delicate balancing act between these two quite contradictory effects. The choice of an adequate objective is then not always obvious, mainly because the two effects are not immediately measured on comparable scales. In such cases the general goal can then perhaps better be expressed as consisting of two objectives (Carrizosa and Plastria, 1999).

2.1.5 Mathematical and Computational Tools

Since we want not only to come up with a 'solution' to an optimisation problem, but also obtain more information like how good this solution is, where other good, possibly better, solutions are to be found, what happens when data changes, etc. we need both analytical and computational tools.

Continuous locational questions are always of a very geometrical nature, so we naturally need *geometry*, both under its *analytical* form as under its *classical* planar or spatial form, particularly the recent *computational* geometry (cfr the introductory books (Preparata and Shamos, 1985) and (de Berg et al., 1997) and the excellent website with a wealth of information about publications, software and research at
http://compgeom.cs.uiuc.edu/~jeffe/compgeom/compgeom.html).

Location models also call for optimisation, so we also need the classical tools of *analysis*, particularly *convex analysis* (since Minkowski distances are always convex functions), and of course the computational methods of *linear programming*, *nonlinear programming*, in particular *convex optimisation* (e.g.(Hiriart-Urruty and Lemaréchal, 1993)), and *global optimisation* (see (Hansen et al., 1995)).

It turns out that continuous location models for a single facility may already be quite hard to solve, and so their multiple facility counterparts can be expected to be much harder still. This is probably the reason why multifacility location models in a continuous space are much less popular. But this is currently changing with the use of *metaheuristic* methods, which start to be applied to such rather hopelessly complex looking problems.

2.1.6 Scope

This chapter further consists of an introduction to several variants of some basic and directly applicable continuous location models for a single facility of the more geometrical kind. Models for the location of several point-facilities and of facilities with shape different from a point will be touched upon in two final sections. For a general overview of continuous location models we refer to (Plastria, 1995).

All the models we discuss here are based on some idea of *covering*. Such location models make use of an *action radius*, a threshold distance within which a demand point is considered to be covered. Among the many variant models based on covering we single out the following four basic ones.

In pull situations we either want *full covering* at smallest possible action radius, or, when the action radius is fixed we want *maximal covering*, i.e. to cover the largest demand possible.

Conversely, in push situations we either seek to place a facility with the largest possible action radius such that none of the demand points is covered, *empty covering*, or, when the action radius is again fixed, *minimal covering* to find a placement covering the least demand.

In order to keep things relatively simple, we will mainly discuss the two-dimensional case, so when not stated otherwise always assume our model is a planar one. Basic models are discussed in some detail, with particular emphasis on the first type, full covering, and this mainly for two reasons. On the one hand this model allows us to introduce many of the basic ideas also needed later on, and on the other hand since this model was most intensively studied, by far, and has a documented history of well over a century. Where appropriate we indicate to what extent the theory and methods generalise, and include pointers to the more recent results. More details on methods and extensions of these and other models are to be sought in the cited references, and in the other chapters of this book.

In their simplest planar versions with euclidean distances covering type models do not require much more than standard geometrical insight. Typically in these models one can prove that an optimal solution may always be found within a finite candidate set, which effectively reduces the problem to a discrete location problem. Actual construction, evaluation and (possibly implicit) enumeration of such a *finite dominating set* is then often achieved using techniques of computational geometry.

One of the central aims of computational geometry is to develop data-structures and techniques to handle them in order to solve questions of geometrical nature efficiently by computer. Efficiency is measured by computational effort, and expressed using 'big O' notation: if n is some measure of the quantity of data needed to state the problem, an algorithm has complexity $O(f(n))$ when the number of elementary operations it involves is bounded by $c.f(n)$, c being some constant independent of n. This notion of effort is not very precise, since several things need to be further clarified, such as what kind of data is used, what kind of arithmetic is used, what is an 'elementary' operation, how large is c, is this number of operations always the same for fixed n, or a worst case estimate, or rather an average behaviour ? In general this effort gives an indication of the behaviour for large datasets, although 'large' can still be very different from one case to another. For example a method A taking always exactly $56n$ steps is of $O(n)$, and thus considered as 'better' than any $O(n \log n)$ method B, even when this latter uses in the worst case only $4n \log_2 n$ comparable steps; A will, however, be able to beat B only when $\log_2 n > 14$, or $n > 2^{14} = 16384$, and this only in a worst case situation, which typically is of a rather exceptional nature. Very probably method B will remain better than A for most problem instances with much larger n.

2.2 Full Covering

2.2.1 Euclidean Distance

Where is a good place for locating an emitter, alarm siren or a facility from where some emergency service should be dispensed? Two typical properties of an emittor (siren) are first that electromagnetic (acoustic) waves propagate in straight lines uniformly in all directions, so euclidean distance is appropriate, and secondly that reception quality steadily decreases with distance, so we are in a pull situation if we want the received emissions to be as powerful as possible: the closer is the better.

If we know exactly all the places at which our emissions have to be received, and we want to keep the power of our emitter within bounds, we may use the following full covering model:

> Given the points P_m ($m \in \mathcal{M}$) in the plane we must find the circular ball with minimum radius covering them all; its centre is then the optimal site.

We may immediately observe the following useful property: since the circular ball we look for is always a convex set, and since it contains all the points P_m, it will also contain their convex hull. Now this convex hull is fully determined by its *extreme* points only, so we can delete all non-extreme points. This is very useful when $\mathcal{M}$ is infinite: when it consists of several regions all over which the emissions should be receivable, we may reduce $\mathcal{M}$ to the set of extreme points of the convex hull of all these regions together. In practice each region is probably given (or approximated) as a polygon, so we can consider only their cornerpoints, of which there are a finite number only. But also for finite sets $\mathcal{M}$ usually only a few points are extreme, so the property allows to strongly reduce the size of $\mathcal{M}$, as illustrated in figure 2.2.

Fig. 2.2. Extreme points (•) of a union of polygonal regions

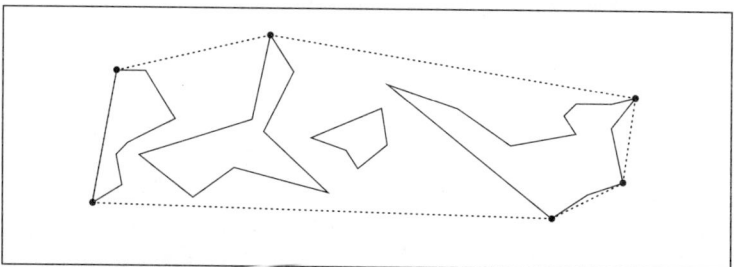

Unconstrained We will distinguish between a demand point *covered* by a circle and a demand point *on* the circle. In the first case the distance of the demand point to the centre is less than or equal to the radius. In the second case this distance is exactly equal to the radius.

A circle covering all demand points P_i is called a *covering circle* (or spanning circle). We wish to determine a covering circle of smallest radius or *minimal covering circle* (MCC). This simple question is documented to have been raised already in the middle of last century in a one-line article by (Sylvester, 1857), but is probably older still.

The following properties of the MCC are fundamental

1. At least two demand points P_m lie on the MCC.
2. If there are only two demand points P_j and P_k on the MCC, these form a diameter of the MCC.
3. If three or more demand points are on the MCC (which is then fully determined), three among them form an acute triangle
4. Any circle satisfying one of previous two properties and which covers all demand points is the MCC.

The proof relies on simple geometric properties of circles: in all cases illustrated in figure 2.3 the covering circle can be reduced after moving along the arrow(s). In fact this leads to a scheme for finding the MCC, independently developed by Pierce (according to (Sylvester, 1860)) and by (Chrystal, 1885). A variant of this scheme was more recently proposed by (Chakraborty and Chaudhuri, 1981) and shown to take $O(M^2)$ steps.

Fig. 2.3. Excluding circles as MCC

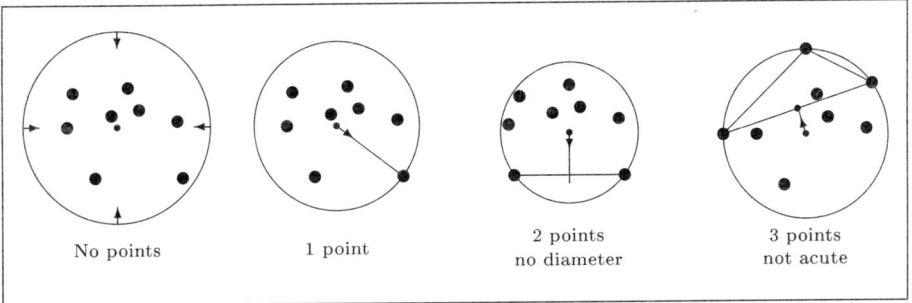

However probably the simplest efficient method to solve the full covering problem with euclidean distances in practice was developed by (Elzinga and Hearn, 1972a). It relies on elementary planar geometry and consists of constructing successively larger and larger circles passing either through two of the demand points forming a diameter or through three of the demand points forming an acute triangle, until a covering circle is found.

Fig. 2.4. Elzinga-Hearn : Dropping the fourth point

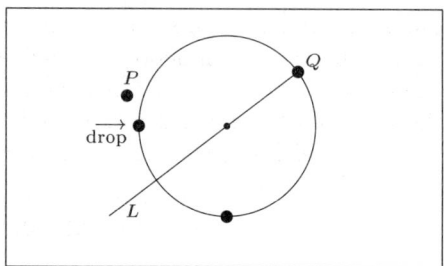

Elzinga-Hearn method

Initialise Pick any two demand points.
Handle two points.
   Let $\mathcal{C}$ be the circle defined by the two demand points.
    – If $\mathcal{C}$ covers all demand points, $\mathcal{C}$ is the MCC: Stop.
    – If $\mathcal{C}$ does not cover all demand points, then add any demand point outside $\mathcal{C}$ and proceed with these three demand points.
Handle three points
    – Check the shape of the triangle formed by the three demand points. If there is an obtuse angle, drop the demand point at that angle and proceed with the two remaining demand points.
    – Let $\mathcal{C}$ be the circle defined by the three demand points.
       – If $\mathcal{C}$ covers all demand points, $\mathcal{C}$ is the MCC: Stop.
       – Otherwise add any demand point, P say, outside $\mathcal{C}$.
        Drop one of the former three demand points chosen as follows (see figure 2.4):
            * Let Q be the triangle's farthest point from P.
            * Let L be the line through Q and $\mathcal{C}$'s centre, then the triangle's two other points lie on opposite sides of L:
            * Drop the one on P's side of L.
        Proceed with the three now remaining demand points.

This method always stops after a finite number of steps, since each successive constructed circle has a radius larger than the previous one, so no cycling can occur. Trials with randomly generated demand points turn out to need very few steps. And if the demand points to start with and the demand points added at each handling are well chosen this behaviour is improved. When choosing a demand point outside the circle $\mathcal{C}$ it seems best to take the one farthest outside $\mathcal{C}$, i.e. the demand point at largest distance from $\mathcal{C}$'s centre. (Drezner and Shelah, 1987), however, construct an example in which this method takes $O(M^2)$ steps for randomly chosen starting points. Starting with the two demand points which are most apart seems also to be an excellent choice. However it takes quite some work (naively $O(M^2)$) to find these points; it is easier to choose e.g. the leftmost demand point (lowest x-

coordinate) and the demand point farthest from it; note that in our example in figure 2.2 this choice will immediately give the final answer, as it would in the example in (Drezner and Shelah, 1987). It seems that the question of the worst case and/or average case behaviour of the method is in fact not precisely known, although many authors, e.g. (Preparata and Shamos, 1985) state an $O(M^2)$ worst case complexity.

(Maffioli and Righini, 1994) describe another variant of the method.

A slightly different view of the full covering problem is more appropriate for applications in emergency services (provided euclidean distance applies, e.g. when movement is done by air) and leads to another method based on computational geometry.

For any fixed centre, the smallest radius of a covering circle is given by the distance to the farthest demand point. Since we want to make this radius as small as possible, we are seeking that centre which minimises the maximal distance to any demand point. Therefore the full covering problem is often called a *minimax location* or *centre* problem, classified as $1/P = \mathbb{R}^2/w_m = 1/d^{\text{euc}}/\max$.

Fig. 2.5. Farthest point Voronoi diagram and minimum covering circle

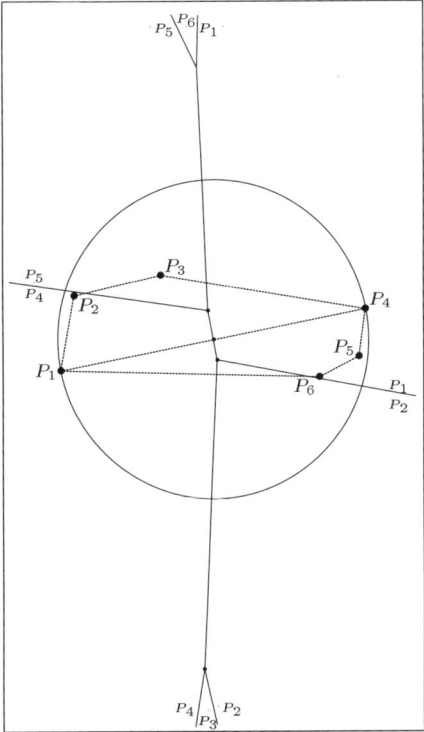

For any fixed centre some farthest demand point is always an extreme demand point (this holds in general for any Minkowski distance). So for each extreme demand point we may ask for all points of the plane for which this is a farthest one. This set turns out always to be a convex and unbounded region. All these regions meet along straight line segments, parts of the bisector between two extreme demand points, i.e. the line passing trough the midpoint of and orthogonal to the segment joining these points. They build up the *farthest point Voronoi diagram*, see figure 2.5. The line-segments common to two regions and nodes common to three regions always form a tree structure, from which it follows that for M extreme demand points, at most $M - 2$ such nodes and $2M - 3$ such segments appear (indeed, close up the diagram with a point at infinity at which all M unbounded segments arrive, yielding a graph on the sphere with f faces, v nodes and e edges, where $f = M$, $2e = M + 3(v - 1)$ and by Euler's equation $f - e + v = 2$).

Evidently, no inner point to any region can be the centre of the MCC: one can always move that centre closer to its farthest demand point, leading to a smaller covering circle. Therefore only nodes of the farthest point Voronoi diagram or points on its segments can be optimal. But for all centres on each such segment there are exactly two farthest points, and we know that this may only lead to a MCC in case this centre is exactly their midpoint. It follows that only the midpoint of a *diameter* of the set of demand points, i.e. two demand points which are farthest apart, can be of this type, of which there usually is only one!

Computational geometry shows how to construct the farthest point Voronoi diagram for M (not necessarily extreme) points in a number of steps which is $O(M \log M)$ (Aurenhammer, 1991; Okabe et al., 1992), so computation time increases only slightly quicker than M itself (it cannot be done quicker). Once the farthest point Voronoi diagram has been constructed, one has to check at most $M - 2$ nodes and at most one point on no more than $2M - 3$ segments, which is a quite quick job.

In fact the properties of the MCC directly imply that the optimal centre must also lie within the convex hull of the demand points. But a good part of the farthest point Voronoi diagram lies outside this convex hull, so too much work is done by constructing it fully.

It should be mentioned that (Megiddo, 1983a) and (Dyer, 1986) designed an analytical solution method of $O(M)$ worst case complexity, so theoretically more efficient. This is based on a prune and search technique, which identifies (in time linear in M) at least a fixed fraction (e.g. 1/16) of demand points which cannot lie on the circle, and iterating on the remaining demand points until only 2 or 3 points remain; the total work done will still be $O(M)$.

Numerical experiments with thousands of demand points (not necessarily extreme) seem to have indicated that the Elzinga Hearn method remains unbeaten for practice. However, yet another recent and quite simple method (Welzl, 1991) might be competitive for large problems. It was designed to have an *expected* $O(M)$ time complexity. It is based on adding one demand point after the other, using the simple observation that when the MCC is known for a set of demand points and a new demand point p is added then either p lies in this MCC and this circle remains the best, or p must lie on the new MCC. In that case it suffices to find the smallest circle MCC(p) passing through p and enclosing all other demand points. This is a simpler problem which can be handled similarly, by adding the points sequentially, keeping dynamically track of the MCC(p), where it often remains the same, and sometimes a new circle will have to be found, now passing through two known points, which again is a similar but easier problem based on the same arguments and the fact that three points uniquely determine the circle passing through them. If the demand points are handled in random order, it can be shown that the average running time is $O(M)$ (therefore randomising the list of demand points is the first step to be taken in theory).

Just finding the minimal covering circle may not be enough, but some sensitivity analysis may be called for. (Ohsawa and Imai, 1997) study what happens when a larger radius is allowed. They show how to use the farthest point Voronoi diagram to construct the set of all possible full covering disks with fixed radius and study the volume of this set as a function of the radius.

Constrained In presence of a feasible region S, the constrained euclidean distance minmax single facility location problem, i.e. finding the smallest covering circle with centre in S (CMCC), may be solved using the properties

- Either the CMCC is equal to the MCC, its centre lying in S.
- Otherwise the optimal site can only be

    either a point of S closest to some destination P_i, and then P_i lies on the CMCC

    or the point of intersection of the bisector of two destinations P_i and P_j with the boundary of S, closest to P_i and P_j, and then both P_i and P_j lie on the CMCC.

Therefore we may use the following method.

Constructing the CMCC

Step 1 Solve the unconstrained problem, by any method, yielding the MCC-circle $\mathcal{C}$. If $\mathcal{C}$'s centre is in S, stop: $\mathcal{C}$ is also the CMCC.

Step 2 Try out all points P_i in turn. For each one calculate all points X of S closest to P_i and for each consider the new circle $\mathcal{C}$ with centre X and radius $d^{\text{euc}}(X, P_i)$. If $\mathcal{C}$ is a covering circle: stop, $\mathcal{C}$ is the CMCC.

Step 3 Try out all pairs P_i, P_j. For each one construct the bisector of P_i, P_j. Let X be the point of intersection (if it exists) of this line with the boundary of S, closest to P_i and P_j and take the circle $\mathcal{C}$ centered at X and radius $d^{\text{euc}}(X, P_i) = d^{\text{euc}}(X, P_j)$. If $\mathcal{C}$ is a covering circle, store it as possible candidate CMCC.

Finally If not stopped during one of the steps 1, 2, the CMCC is the smallest among the candidate circles found in step 3.

This method calls for being able to find which points of S are closest to a given point, and also the intersections of a line (bisector) with S's boundary. In the common case when S is a union of polygons this is a simple task. In particular when S is convex, for any P_i the closest point in S is unique.

In fact the third step does too much work, and leads to a time complexity of $O(M^2)$. Of course not all pairs of demand points should be checked, but only those that can be the simultaneous farthest demand points of some point in the plane. In other words, only the pairs to which corresponds a segment of the farthest point Voronoi diagram should be checked. Therefore step 3 can better be replaced by computing as candidate centres only the intersection points between this diagram and S's boundary. This method brings us back to $O(M \log M)$ (Woeginger, 1998), which cannot be improved in general, even when S is the complement of a convex set. (Hurtado et al., 2000) show how construct the CMCC in linear time when S is convex using a prune and search technique as in the unconstrained case. It is not clear whether the method of (Welzl, 1991) can be extended to the presence of locational constraints.

2.2.2 Rectangular Distance

If we consider the unconstrained full covering problem, but with rectangular distances, things become much easier. The problem is now to find the smallest *diamond*, i.e. a square tilted over 45° covering all demand points. Such a diamond always has four sides, which we will call LT (Left-top), RT (Right-top), LB (Left-bottom) and RB (Right-bottom). Both LT and RB are upward sloping lines at 45° and both RT and LB are downward sloping lines at 45°.

Given any diamond covering the points P_i, we may push these sides down (for LT and RT) and up (for LB and RB) until they reach a first destination. Pushing further one would obtain a destination on the wrong side! In this way we obtain the smallest tilted rectangle containing all points P_i, and any diamond covering all P_i must contain this rectangle.

Fig. 2.6. Full covering with rectangular distance

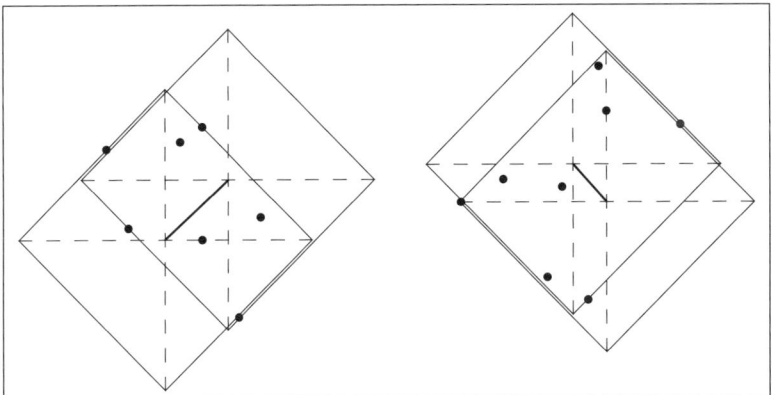

Usually this rectangle is not a square, and there exist many diamonds of smallest radius containing it, and the centre of any of these diamonds is always an optimal site. As shown in figure 2.6 the set of optimal sites is then a line segment parallel to those opposite sides of the rectangle which are farthest apart. As such this may be a line segment tilted at 45°, either upward or downward.

Exceptionally this rectangle is a square, in which case it is the smallest diamond covering all destinations, and its centre is then the unique optimal site.

We refrain here from detailing the analytic calculation of this simple geometric construction. It may be found in detail in (Francis et al., 1992), where one also finds how to construct the level curves of the objective function.

2.2.3 Other Distances

The case of max distance is exactly similar to the rectangular distance case, except that the unit ball is now a (untilted) square, making things even easier. It is easy to see that this construction directly extends to max distance in higher dimensions. This is not the case, however, for rectangular distance, since the simple relationship with max distance in the plane is then lost: e.g. 3-d max distance balls are cubes, whereas the rectangular distance balls are regular octahedrons. (Späth, 1978) completely solves the full covering problem with rectangular distance in 3-dimensional space.

(Plastria, 1994) indicates how the simple geometric construction can be extended to a general planar block distance; in a first step one has to construct a polygon having the same edge directions than the distance's unit ball, and tightly enclosing the demand points, and in a second step a smallest unit ball enclosing this polygon. This second step is harder in general: it consists of at most one shrinking step for each edge of the unit ball, where each shrinking resembles the first step. In total an $O(M)$ method.

Furthest point Voronoi diagrams can be extended to other distance measures in theory, but their construction is less evident and leads to several new difficulties (Klein, 1989). Edges will, however, not remain straight line segments and are usually hard to construct, while for block distances some edges may grow into regions of equal distance. See (Okabe et al., 1992) and the references therein for details. The principle of how to use these Voronoi diagrams to solve full covering problems remain unchanged, however, even in presence of locational constraints.

The algorithm of (Welzl, 1991) directly generalises to any Minkowski distance, under the condition that one is able to construct the solution for just 2 or just 3 demand points in the plane — in higher dimensions, n say, one must be able to solve up to $n+1$ points (Sharir and Welzl, 1992). Indeed, the most important property of the MCC carries over to these distances, thanks to the following nice application of the celebrated theorem of Helly in convex geometry, to minmax optimisation with quasiconvex functions (i.e. functions with convex lower level sets, see e.g. (Avriel et al., 1988)):

Theorem 1 *(Drezner, 1982; Carrizosa and Plastria, 1995b) A minimum of the maximum of a finite set of quasiconvex functions on $\mathbb{R}^n$ can always be obtained by considering no more than $n + 1$ of these.*

The trouble is, that one doesn't know which $n+1$ functions to choose, just like we didn't know in advance which 2 or 3 demand points would lie on the MCC. It is probably therefore that approaches proposed for dimensions higher than 2 (or 3) are rather general iterative methods of convex programming, see e.g. (Elzinga and Hearn, 1972b), (Drezner and Wesolowsky, 1980), (Jacobsen, 1981), (Plastria, 1987), (Pelegrín and Cánovas, 1998a). Arbitrary dimensional applications in approximation theory, under the name of Chebyshev problem, are discussed in (Beer and Pai, 1990).

(Bose and Toussaint, 2000) develop geometrical algorithms for the full covering problem for a simple polygon both for euclidean distance inside the polygon as for link distance. They look in particular at applications in molding, in which the sought site must lie on the polygon's boundary.

The full covering problem on the sphere using spherical distance has been much less studied. (Drezner and Wesolowsky, 1983b) describe a stepwise improvement method of unknown complexity, which can never be less than $O(M \log M)$, as stated by (Hurtado et al., 2000). However, when one can specify an open halfsphere in which the centre must lie, (Xue and Sun, 1995) and (Gómez et al., 1997) indicate how methods of linear worst case complexity can be constructed. See also (Patel, 1995),(Sarkar and Chaudhuri, 1996) and (Das et al., 1999).

2.2.4 Extensions

The most commonly studied extension of the full covering problem is to consider weighted euclidean distances in the plane : $1/P = \mathbb{R}^2/w_m \geq 0/d/\text{max}$. Purely geometric methods remain applicable thanks to the fact that bisectors, i.e. the sets of points at equal weighted distances of two given demand points are now circles. (Hearn and Vijay, 1982) extend both the oldest Chrystal-Peirce as well as Elzinga-Hearn's methods to the weighted case; the latter was also proposed by (Charalambous, 1982). Also Megiddo's $O(M)$ method can be adapted accordingly (Megiddo, 1983b), as does Welzl's method.

An interesting application of the distance weighted problem is described by (Zwick, 1994): two parts, a female one containing circular holes and a male one with corresponding circular pegs, should match in such a way that all pegs can simultaneously fit into their corresponding holes. In order to check this, coordinates of all holes P_i and pegs P'_i are measured, together with their corresponding radii r_i and r'_i ($\leq r_i$). We assume that the direction of both parts is fixed, and only translations are allowed. The parts will match if and only if a position X on the female part can be found for the male part's origin such that for all i we have $d^{\text{euc}}(P_i, X + P'_i) \leq r_i - r'_i$, or, equivalently, $\frac{1}{r_i - r'_i} d^{\text{euc}}(P_i - P'_i, X) \leq 1$. This can be checked by minimising the maximum of these expressions, and comparing this minimal value with 1. The parts will fit if and only if this minimal value is no more than 1, and then the difference with 1 gives a measure of slack in the fitting. Note that also allowing rotations between the parts considerably complicates matters, as developed further by (Zwick, 1994).

One may also consider on the one hand that the way distance is measured depends on the demand point, leading to mixed-distance minimax location problems, and on the other hand that perception may be a nonlinear (but increasing) function of distance which may differ between demand points, e.g. because of different reception power in the emittor example. This leads us to the generalised minimax location problem $1/P = \mathbb{R}^2/f_m \nearrow /d_m/\text{max}$ which may be written as

$$\min_{X \in S} \max_{m \in \mathcal{M}} f_m(d_m(X))$$

where the functions f_m are strictly increasing. The objective function of this problem is always quasiconvex and cannot have a local, nonglobal minimum; in many cases it even has a unique minimum (Pelegrín et al., 1985). For convex feasible sets S general optimisation techniques of nonlinear programming may be applied, and any local optimal solution found will be guaranteed to be a global optimum. In particular standard optimisation software, including the solver tool in an omnipresent tool like the spreadsheet Excel will do a good job in solving small scale cases of such problems. For large scale cases theorem 1 still applies, and a strategy like Welzl's algorithm might still be applied, although the steps will now involve solving a small scale version of the problem with $n+1$ (3 in the plane) demand points instead of constructing circles. Special purpose methods also have been developed (Plastria, 1987), (Frenk et al., 1996).

Another special type of constraint is a *restriction*: a region inside which it is forbidden to locate. When the unconstrained optimal solution is forbidden, it will be sufficient, because of quasiconvexity of the objective, to search only along the boundary of the connected component of the forbidden region which contains the unconstrained optimum. If this boundary is polygonal, we have on each segment a one-dimensional quasiconvex function to minimise, a quite easy task for nonlinear programming. For more general boundaries, one obtains by way of a parametric representation, a (usually multimodal) one-dimensional function to minimise, for which good techniques of global optimisation are available, like (Hansen et al., 1992) and (Blanquero and Carrizosa, 2000). When all distance measures are possibly different block distances direct geometric approaches can again be constructed (Nickel, 1998).

2.3 Maximal Covering

2.3.1 Fixed Radius

The situation becomes quite different when we consider the covering radius to be fixed, and the question is to cover the most demand possible. This kind of question naturally appears when the action radius cannot be chosen, and when it turns out to be too small for full coverage. Now a weight is associated with each demand point, usually indicating some population present at that point.

For the sake of simplicity we will only consider the planar euclidean distance case, and the question then becomes

– find that (centre X of a) circular disk of given radius r covering the largest possible total weight.

A demand point P_i is covered, and its weight counted, when it lies within the disk, or, equivalently (because euclidean distance is symmetric), when X lies within the disk of radius r centered at P_i. We can therefore draw all disks of radius r centered at some demand point, associating the corresponding weight to all points of the closed disks, and consider the planar diagram these form all together. The plane is cut into pieces of constant summed weight, as shown in figure 2.7. The small dots indicate demand points of weight 1, the larger ones are of weight 2. Choosing any site within some region will lead to the indicated coverage. Note that disks are considered to be closed, so any point on a disk's boundary is also considered to be covered by that disk. Therefore edge points in this diagram belong to the region formed by the intersection of all disks covering it. Figure 2.7 clearly illustrates some disturbing properties: there usually are many optimal solutions, in fact a

Fig. 2.7. Maximal covering problem (weights 1 and 2)

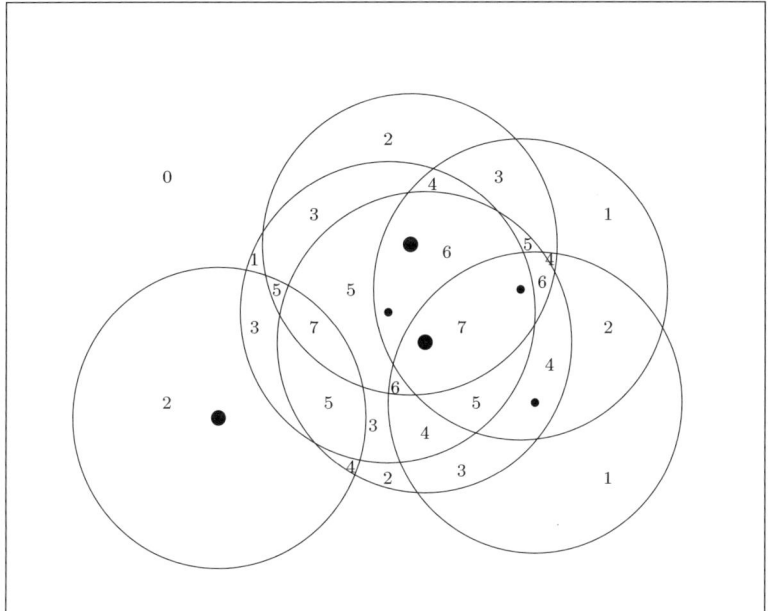

whole region of them, and this optimal region is not necessarily connected! This clearly shows that the objective function cannot be (quasi)convex and thus traditional iterative methods of nonlinear programming will not be able to find guaranteed global optimal solutions.

But it is evident that any nonzero coverage region, thus also the optimal one, either consists of one full disk or must have at least one 'corner' point, i.e. a point lying on two disk boundaries. Therefore it is sufficient to check only demand sites or such corner points and there are at most $M(M-1)$ of these latter because two circles have at most two intersection points. How this can be done in $O(M^2 \log M)$ time is described in (Drezner, 1981), and improved to $O(M^2)$ by (Chazelle and Lee, 1986).

When constraints are present one will also have to check points where disk boundaries intersect feasible region boundaries.

After having identified all optimal points, the full set of optimal sites can be reconstructed as intersections of corresponding disks with the feasible region. For a given fixed radius this set may consist of several pieces as shown in figure 2.7, yielding a region of sites all allowing the same maximal coverage.

2.3.2 Varying Radius

Usually the larger the radius is, the larger the costs will be, e.g. due to higher energy consumption, larger service costs, higher service times. One should then wonder how far this radius can be reduced without altering the coverage, which leads us to consider the question:

- given a desired coverage, what is the minimum radius that still allows to obtain this coverage, and where should the centre be placed?

This problem was studied by (Carrizosa and Plastria, 1998b) under the name of *minimal quantile* location problem .

Let us see intuitively what happens when the radius is reduced: the optimal regions decrease in size as intersections of shrinking disks. Some optimal regions may disappear, but as long as we have at least one optimal region left we may reduce the radius even further. It is only when no optimal region remains with nonempty interior that this process stops. Typically, for euclidean distance, this happens when either two intersecting disks touch, or when three of them have only one point in common: in other words when the optimal site either lies in the midpoint between two demand points, or at same distance of three of them forming an acute triangle, and the radius equals the distance to these demand points. Therefore only such sites have to be checked, of which there are only a finite number.

This can also be seen as follows: let us suppose we know an optimal solution to the minimal quantile problem. This means we know a disk which covers some subset A of demand points with total weight at least the required coverage, and no other disk with smaller radius can do the same. But that disk must then have smallest radius to cover A, so must be the MCC for the demand points in A. The properties of the MCC given in section 2.2.1 show that its centre may always be found among the midpoints of pairs of A and centres of acute triangles with vertices in A, always with corresponding radius. Since we do not know A in advance we should check all pairs and all adequate triplets of demand points.

This also shows that in presence of locational constraints we must add some more candidate optimal sites, being intersection points of bisectors of pairs of demand points with the feasible region's boundary, and extreme points of each connected part of the feasible region.

For each pair or triplet of demand points we construct their smallest enclosing disk and calculate the total weight of covered demand points. We retain only those disks with sufficient coverage and the retained disk with smallest radius gives our optimal minimal quantile solution.

What is even more interesting is that for any desired coverage the same set of candidate optimal sites and radius appears. Therefore a complete listing and checking of all these $O(M^3)$ candidate disks allows us to solve simultaneously all minimal quantile problems for all fixed coverages, as follows.

2 Continuous Covering Location Problems   59

Minimal quantile problem for all coverages

- We first sort all the candidate disks by decreasing coverage, and among ties (there will usually be many) retain only the one(s) with smallest radius.
- Secondly we run sequentially through the remaining list and delete any disk with larger radius than the previously retained disk: indeed this latter disk gives a higher coverage at lower radius, so is clearly to be preferred.
- Each disk that remains is now a minimal quantile solution for its own coverage, but also for any lower coverage up to the coverage obtained by the next disk which remains in the list.

But, as explained in the beginning of this section, each such disk is also a maximal covering solution for its own radius. And it remains so for all higher radii up to the radius of the previous remaining disk! In other words, we have at the same time obtained a full solution of the maximal covering problems for all fixed radii.

Therefore this gives us the complete trade off information between minimal radius and maximal coverage.

Fig. 2.8. Maximal covering with varying radius

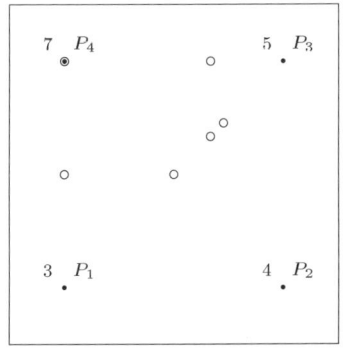
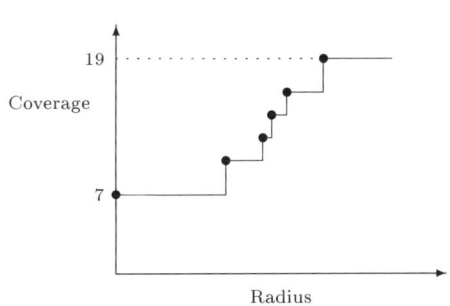

Optimal sites : o   Trade off curve

Figure 2.8 shows an example of such analysis in a slightly more complicated situation. There are four demand points with weights as indicated in the picture at the left; all distances are euclidean, but distance to P_3 counts double as compared to the distances to the other three demand points. No locational constraints are present. After processing all candidate disks, we remain with the six optimal sites shown at the left, yielding the trade-off curve shown at the right. This allows us to read off immediately the highest coverage achievable at given radius and inversely the lowest radius allowing to obtain a given coverage.

What we have done here is in fact to construct the set of efficient solutions to the bicriterion problem of simultaneously maximising the coverage and minimising the radius. This allows then to maximise any objective (even a nonlinear possibly discontinuous one) that is decreasing with the radius and increasing with the coverage, for example the profit one may expect from radio transmissions, when income from commercials increases with the public reached (coverage), while costs are due to the emission power necessary to obtain the action radius (often by a squared or even cubed distance law). The only solutions we have to check are the ones we just constructed, and we select the one giving the best objective value. Observe that in the example of figure 2.8 any linear objective (positive coefficient for coverage and negative one for radius) will always either lead to zero radius or to full coverage! For another somewhat more complex application in competitive location modelling, see (Plastria, 1997).

(Carrizosa and Plastria, 1998b) discusses (weighted) euclidean distances as outlined above, leading to an $O(M^4)$ method, announced to be reducible to $O(M^3 \log M)$, and also the case of (possibly mixed) block distances, leading to a $O(M^5)$ method. As shown by (Carrizosa and Plastria, 1995b) the basic principles of this analysis remain valid in much more general situations, not necessarily location problems.

2.4 Empty Covering

The question where to locate some undesirable facility has become quite crucial in many circumstances. The general reaction of the public is to want to live as far as possible from potentially dangerous sites, like storage of explosives, or annoying sites like soccer stadiums generating lots of noise and commotion. Still, such facilities are often necessary, so sites must be chosen for them.

The simplest goal is to ask for a site, within some feasible region, which is farthest possible from any sensitive places, like urban settlements. In figure 2.9 these points are shown as black dots and the feasible region as the inside of the two dotted lines.

2.4.1 Euclidean Distance

The distance notion appropriate in most cases is, in first approximation, euclidean. In other words we want to find where to position the largest possible disk, not covering any of a given number of fixed points. This can also be seen as maximising the minimum euclidean distance to given points, or by sign inversion $1/P = \mathbb{R}^2/w_m = -1/d^{\text{euc}}/\max$.

It is therefore useful to consider in the plane the question 'What is the closest demand point?', which yields the (closest point) Voronoi diagram as illustrated in figure 2.9. To every P_j corresponds one convex cell of all points closer to P_j than to any other, obtained as intersection of all halfplanes containing P_j with boundary a bisector with each of the other P_i. Note that the unbounded cells correspond exactly to the extreme demand points.

When choosing the centre of an empty disk in the plane the largest radius possible is found as the distance to the demand point in the centre's cell of this diagram. When the centre lies inside the cell we can always move the centre away from this demand point, thereby increasing the empty disk's radius. Any centre on a Voronoi edge (thus at equal distance to the two closest demand points) can be moved along the edge, away from the closest demand point's mid, also increasing the empty disk's radius. These movements are allowed provided the feasible regions allow so. It follows that the centre of a largest empty disk must either be a corner point of the feasible region, or at the intersection of a Voronoi edge with the feasible region's boundary, or be a Voronoi point, i.e. a node of the Voronoi diagram. This guides us to a finite candidate set, as depicted by circles in figure 2.9, together with the largest empty circle (here with just one demand point on its boundary).

Computational geometry techniques allow to construct the Voronoi diagram in $O(M \log M)$ time and to intersect this diagram with polygonal regions efficiently, see (Toussaint, 1983).

Fig. 2.9. Closest point Voronoi diagram and largest empty disk

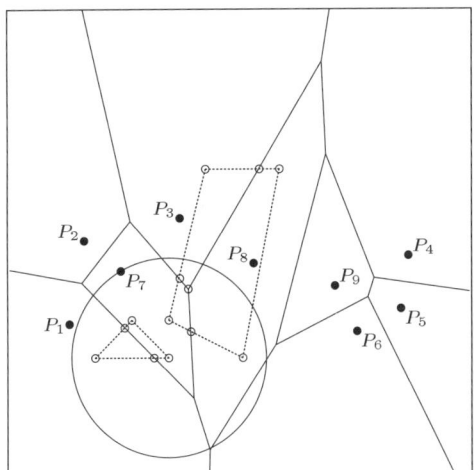

2.4.2 Variant Models

The empty covering model quite directly extends to differently weighted distances $1/P = \mathbb{R}^2/w_m < 0/d^{\text{euc}}/\max$. Most of the outlined geometric approach can be adapted (Melachrinoudis and Cullinane, 1986), in particular using multiplicatively weighted Voronoi diagrams (Aurenhammer and Edelsbrunner, 1984; Melachrinoudis and MacGregor Smith, 1995).

Extended versions of this problem assume the points P_i to have different sensitivities to the presence of the undesirable facility. This can be expressed in general by the minimisation of the maximum effect on all P_i, effects now being general decreasing functions of euclidean distance, classified as $1/P = \mathbb{R}^2/w_m : f_m \searrow/d^{\text{euc}}/\max$ (Hansen et al., 1981). These authors suggest the use of an interactive graphical method (extending the proposals of (Drezner and Wesolowsky, 1980; Melachrinoudis and Cullinane, 1986) for weighted distances), which can handle any kind of locational constraints: iteratively pick any feasible point and remove from the feasible region all disks around the demand points which would yield no better objective value. It is the picking of the feasible point that is most easily handled by using a graphical representation, let the pattern recognition power of the human eye do the job.

Automating this task is much harder, and requires special assumptions on the shape of the feasible region. For regions which are intersections of polygons with complements of circular disks the cutting plane technique described in (Carrizosa and Plastria, 1998a) may be applied in general, and, for simple weighted distances the extended geometric approach of (Melachrinoudis and Cullinane, 1985). This last approach by way of generalised Voronoi diagrams, however, may utterly fail for general decreasing functions: (Alonso et al., 1998) construct an example where the number of generalised Voronoi nodes to be checked would effectively be infinite.

When weighted distances are considered the question directly arises which weights are to be used, which is far from evident in applications. Therefore it is of interest to observe the jumpy behaviour of optimal sites when weights are modified (Erkut and Öncü, 1991) or the very disconnected full set of possible optimal solutions for any kind of weights (Carrizosa et al., 1995).

As indicated by (Drezner and Wesolowsky, 1983b), the spherical empty covering model is easily seen to be equivalent to a full covering model, replacing each P_i by its antipodal point.

Other distance functions have been considered, mainly the rectangular distance, see (Drezner and Wesolowsky, 1983a; Mehrez et al., 1986; Melachrinoudis, 1988; Appa and Giannikos, 1994), but the application potential of these models seems rather reduced, since it is not clear how a push effect might propagate by rectangular distance. Applications may be found in other fields, e.g. mapdesign in which a text of a given shape must fit as large as possible in between other features, see (Okabe and Suzuki, 1997) for pointers to such studies. (Orlowski, 1990) studies the related question of finding an empty rectangle of largest area.

In such applications, and in fact in most push situations, however, one should not protect a (small) finite set of points, but rather sensitive areas, usually represented as a collection of polygonal regions. In the empty covering vision, one should then find the largest disk with empty intersection with these areas, or maximise the minimum distance of the chosen site to the sensitive area. Such question have been studied in the plane by (Karkazis and Karagiorgis, 1986; Karkazis, 1988), — but more recent developments in the construction of Voronoi diagrams for areas can also be used (see (Okabe et al., 1992) and the references therein)— and on the sphere by (Hurtado et al., 2000). (Drezner and Wesolowsky, 1996) consider the case when the area to be protected is a network in the plane consisting of straight line segment edges, but allow edges and nodes to use differently weighted distances.

A different type of applications leading to very similar, but mainly three-dimensional models arises in the manufacturing industry when it is required to cut a largest possible piece of given shape out of a block of material. Such *design centering* problems call for the quite heavy machinery of global optimisation, see (Konno et al., 1997).

2.5 Minimal Covering

2.5.1 Fixed Radius

Just like in pull covering models, it is often the case that largest empty disks are just too small for practical purposes. It may then be necessary to accept coverage to some degree. Two directly related questions then directly arise:

Minimal covering
  For a given radius r what is the smallest possible coverage achievable?
Maximal radius
  For a given maximal level of coverage, what is the largest radius possible?

The first question was introduced and studied by (Drezner and Wesolowsky, 1994). It is very similar to the maximal covering problem: the covering function represented in figure 2.7 remains the same with the exception that, for reasons of mathematical well-posedness, one must consider now that the covering disks are open, so that points on their boundary are not covered! This does not change much and the problem can be solved by almost the same techniques. The solution method for minimal covering proposed by (Drezner and Wesolowsky, 1994) is adapted from (Drezner, 1981).

2.5.2 Varying Radius

(Plastria and Carrizosa, 1999) show how to solve simultaneously the minimal covering problem for all radii and the maximal radius problem for all coverages, by considering both objectives simultaneously as a bicriterion circular disk problem. They show that efficient disks must satisfy similar geometric properties as discussed for maximal covering models. Any efficient disk may contain in its boundary

- 1 demand point, and its centre is some corner point of S
- 2 demand points, and its centre lies on S's boundary
- 3 (or more) demand points forming an acute triangle

Therefore all efficient disks can be obtained by enumerating and evaluating (in $O(M^3 \log M)$ calculation time) all these $O(M^3)$ candidate disks. This yields the full trade-off information between maximal radius and minimal coverage.

2.6 Push-Pull Covering Models

As pointed out in the introduction, most applications cannot be described either by purely pull models nor by purely push models. Therefore it is necessary to take both types of effects simultaneously into account.

This can in fact be done in different ways, as discussed in (Carrizosa and Plastria, 1999). Either one can fix a bound on one objective and optimise the other, or one can directly search for all nondominated solutions for the bicriteria problem, i.e. all those solutions which cannot be improved for both objectives simultaneously.

The former approach is nothing else than considering locational constraints of a particular type. When one fixes a bound on the push effect with respect to some affected points, this may be seen as excluding a number of disks from the feasible region, leading to restricted full or maximal covering models, some cases of which were discussed in sections 2.2.4 and 2.3. If, on the other hand one fixes a bound on the pull effect on demand points, the feasible region becomes (part of) an intersection of disks, with as push objective empty or minimal covering models, as discussed in sections 2.4 and 2.5.

Applying the covering idea to bicriteria push-pull situations brings us to considering both minimax and maximin criteria. In other words we want to simultaneously be as close as possible to any, and hence the farthest among demand points, while be as far as possible from any, and hence from the closest among affected points. In general demand points and affected points are different, but these sets might overlap or even be the same! (Ohsawa, 2000) indicates how to construct the set of nondominated solution points by considering the closest point Voronoi diagram of affected points and farthest point Voronoi diagram of demand points. All such solutions are shown to lie on some edge of one of these diagrams or on the boundary of the feasible region. In case of a polygonal region this leads to a finite set of segments to inspect, and efficiently determining the nondominated parts of these is again well studied matter of computational geometry.

2.7 Positioning Models

An interesting field of application of covering-like models in higher dimensional spaces is found in marketing theory, as surveyed by (Schmalensee and Thisse, 1988), under the names of brand positioning or product positioning.

A particular type of products is characterised by n quantitative attributes. All existing brands are then represented by points $B_k \in \mathbb{R}^n$ ($k \in \mathcal{K} = \{1, \ldots, K\}$), the coordinates indicating their value for each attribute. Customer groups $i \in \mathcal{M}$, defined by similar tastes for the products, can also be represented by demand points $P_i \in \mathbb{R}^n$, describing their idea of 'ideal' product attributes. Each customer group is assumed to buy that brand which is closest to its ideal product in terms of the euclidean distance in $\mathbb{R}^n$.

Any company wanting to sell a new brand in this market faces the problem of choosing the attribute mix which will be bought by as many customer groups as possible. With the assumptions above, this product positioning problem calls for finding a point $X \in \mathbb{R}^n$ which is closer to as many as possible points P_i than any of the B_k. Furthermore, technical constraints restrict the available choices for X to some feasible region $S \subset \mathbb{R}^n$.

This problem has many similarities with maximal covering problems, discussed in section 2.3. For each customer group i we may consider its region of attraction $\mathcal{A}_i$: the ball centered at P_i and radius the distance to the closest B_k. Choosing any X within $\mathcal{A}_i$ will ensure that customer group i will opt for the new product. Hence we must try to find some X lying in as many of the balls $\mathcal{A}_i$, or corresponding to the highest weight, if customer groups are weighted by their size or importance.

Geometrically this is the same problem as maximal covering, except that the balls to consider are of different radius, and that the dimension may be quite high. (Crama et al., 1995; Crama and Ibaraki, 1997) have shown these problems to be NP-hard when the dimension n is part of the input. Since it is sufficient to check only intersection points of at most n balls, an adaptation of the search technique of (Drezner, 1981) achieves the task in $O(M^n \log M)$ time, which, although polynomial for fixed n, becomes rapidly prohibitive when n increases. Recently (Hansen et al., 1998) developed some new and quite efficient techniques to carry out this search more effectively, and were able to solve product positioning problems in $\mathbb{R}^{20}$ as large as $M = 500$, $K = 40$.

It must be noted that these product positioning problems are part of the family of competitive location models, reviewed in (Plastria, 2000) and in chapter 5 of this book.

2.8 Multiple Facility Covering Location Models

Models considering the simultaneous location of several facilities introduce the new question of allocation of demand points: which interaction effects between demand points and facilities, and possibly between the different facilities, have to be taken into account in the model. Several possibilities exist and lead to quite different models.

The simplest case is when the allocations are fully fixed, and it is customary to name such multifacility location models.

Most of the time, however, the exact allocations will be determined by the sites which will be chosen, so that the choice of allocations is part of the model. Such models are usually called location-allocation models. Clearly these will typically be harder than multifacility models. Indeed, each time an allocation is chosen, the solution of the location part of the model will involve solving a multifacility location model.

2.8.1 Multifacility Location Models

When only interactions between demand points and facilities are considered, and these are fully prespecified, each facility may be located separately from all others, and the model falls apart into single facility location models for each facility. Therefore multifacility models always assume some form of inter-facility interaction.

Pull models In pull type covering models this means one wants to minimise the maximum (weighted) distance among a number of given demand-point–facility pairs and inter-facility pairs, classified as $p/P/\bullet/d/\max$. This is always a convex optimisation problem and can typically be solved using convex optimisation techniques, see e.g. (Chatelon et al., 1982) and (Idrissi et al., 1989). For rectangular distances the model can be solved by linear programming (Morris, 1973) and/or as a network flow problem (Dearing and Francis, 1974).

Push models Push type models of this kind attempt to locate a given number of facilities so as to maximise the minimum (weighted) distance between all fixed-point-facility pairs and all inter-facility pairs. (Drezner and Wesolowsky, 1985) describe this type of model, but solve only a one-dimensional version of it, while (Brimberg and Mehrez, 1994) give a branch and bound scheme, but only applicable in case distances are rectangular.

Note that for these push models the allocations are only seemingly fixed. In fact one doesn't know in advance which facility is closest to any of the demand points, but the objective automatically makes the choice. Therefore the distinction between multifacility model and location-allocation model vanishes in this case!

2.8.2 p-Full Covering

p-centre The main multifacility version of the full covering problem is traditionally called the *p-centre problem*, classified as $p/P/\text{alloc}/d^{\text{euc}}/\text{max}$. It calls for covering all demand points by p balls of minimal equal radius.

That this problem is probably much harder than its single facility version can be inferred from its NP-hardness, which was independently shown by (Fowler et al., 1981), (Masuyama et al., 1981) and (Megiddo and Supowit, 1984) for euclidean and by (Hsu and Nemhauser, 1979) and (Megiddo and Supowit, 1984) for rectangular distances in the plane when the number p of balls is part of the input. However, for one-dimensional problems p-centre problems become 'easy': (Megiddo et al., 1981) and (Frederickson and Johnson, 1983) show these to be solvable in $O(M \log M)$. Strangely, as a consequence, when one considers again the planar case, but with all demand points lying on a fixed set of straight lines, the p-centre problem also remains polynomially solvable (Aupperle and Keil, 1989).

When one considers p as fixed, the problems can be solved in time polynomial in M, but exponential in p, typically $O(M^{3p})$, as shown by (Capoyleas et al., 1991), using (Drezner, 1984a). Indeed, let the optimal radius be r. Assuming we know an optimal solution, we can associate with each ball the set of demand points which are closest to the ball's centre, and this distance is never more than r. Therefore, taking for each of these p sets of demand points the smallest covering ball, these will all have radius at most r (with at least one equality, because of r's optimality). Enlarging these latter balls to radius r will yield a new optimal p-centre solution, with the additional property that each centre lies at the centre of a smallest covering ball of some set of demand points. The results of section 2.2.1 show that there are only $O(M^3)$ candidate points for such centres. So we can solve the p-centre problem by considering only these $O(M^3)$ points, each with a corresponding radius and covered set, and checking all choices of p out of these, yielding an overall $O(M^{3p})$ complexity. In order to avoid full enumeration, the $O(M^3)$ candidate sites may be considered as potential sites for (also hard) discrete problems of covering type (Ryzhkov, 1973),(Vijay, 1985) : given a radius R, find the minimum number of sites needed to cover all demand points within distance R, and search for the smallest R for which this number is (no more than) p. For small p and/or rectangular distances, more efficient direct methods may be devised, (Drezner, 1984b), (Drezner, 1987), (Ko et al., 1990).

The general problem remains, however, quite hard. Therefore many heuristics have been developed, either special purpose, such as interactive methods, (Ezra et al., 1994), direct methods (Chen, 1983), (Drezner, 1984a), (Dyer and Frieze, 1985), (Pelegrín, 1991), (Pelegrín and Cánovas, 1998b), or reduction of problem size by aggregation procedures (Rayco et al., 1997) and (Rayco et al., 1999), or, more recently, making use of metaheuristic ideas like simulated annealing (Maffioli and Righini, 1994). Other metaheuristics have been fruitfully used in the discrete version of this problem, or for its minisum cousin the p-median problem, like variable neighbourhood methods (Mladenović et al., 2000), genetic algorithms (Bozkaya et al., 2000), tabu search (Rolland et al., 1996), heuristic concentration (Rosing and ReVelle, 1997) and its variant gamma heuristic (Rosing et al., 1999), and might also be adapted to p-centre.

p-centre sum One of the properties of the p-centre model is that the objective value is determined by only one of the facilities, the one determining the maximum 'radius', while for the other facilities this radius is far from tight. This means that the optimal site of only one facility is well determined, while the others allow for a whole region of optimal sites.

This degeneracy may be decreased by using as an objective rather the sum of the covering radii associated with all facilities, leading to a new problem, called p-centre sum. In continuous environment the only study of this model is found in (Plastria, 1990).

2.8.3 p-Maximal Covering

The corresponding problem of covering the maximal possible weight of demand-points by way of a given number p of disks of given radius seems only to have been tackled by way of heuristics, see (Watson-Gandy, 1982) for euclidean distance and (Drezner, 1986) for rectangular distance.

2.8.4   Variants

p-dispersion Dispersion problems are pure push problems which only take into account facilities, so do not consider any demand points. The question is then to position a given number p of points as far as possible from each other within some feasible region. This is equivalent to packing p unit balls within a smallest possible copy of the feasible region. Such problems are classics in the area of recreational mathematics and have attracted the attention of many amateurs and professionals (Rogers, 1964), (Fejes Tóth, 1972). For the most complete and steadily updated source see the websites Packomania
http://hydra.nat.uni-magdeburg.de/packing/packing.html
and Erich's Packing Center at
http://www.stetson.edu/efriedma/packing.html.

At the moment of publication, the most recent results concerning packing circles in a circle were published by (Graham et al., 1998), and for circles within a square by (Nurmela and Östergård, 1999).

That this is also a serious matter with useful applications is shown e.g. by the derived cubature formulae for numerical integration on the sphere (Fliege and Maier, 1999).

The three-dimensional question 'what is the densest possible infinite packing of equal spheres?' has long been known as Kepler's problem: in 1611 Kepler asserted without proof that the greatest density packing of equal spheres is the so-called face-centered cubic packing, the classic way of piling cannonballs. In 1900, Hilbert made this conjecture of Kepler part of his famous list of unsolved mathematical problems. Many attempts to prove this conjecture have been shown to be incomplete, and the quest seems to go on. Two recent attempts, the first theoretical (Hsiang, 1993) and the second using global optimisation techniques (Hales, 1998) are not yet fully accepted, although a consensus seems to exist upon the probable correctness of Kepler's conjecture.

Continuous p-covering Continuous covering problems aim at covering all points of a continuous area by p balls of smallest radius. For $p = 1$ this is equivalent to covering all the extreme points of the area, as pointed out in section 2.2.1. But when $p > 1$ this equivalence is totally lost and the problems become much harder.

Again such questions are known as notorious in recreational mathematics, but have recently gained in interest for their immediate application potential in the area of global communication devices. Classical references are again (Rogers, 1964), (Fejes Tóth, 1972), and a few more recent contributions are found in (Rubinstein and Samorodnitsky, 1986), (Fejes Tóth et al., 1989) and (Suzuki and Drezner, 1996).

2.9 Extensive Facility Covering Location Models

We do not want to terminate this overview of location models of covering type without mentioning some of the work about extensive facilities. In all models we discussed so far, the facilities to be located were always point facilities. There exist many situations in which this assumption is invalid, and therefore attention should be payed to other types of facilities. This brings in many new questions, such as the shape of the facility and how covering is achieved, which in part depends again on the way distance is measured from it. Below we indicate a number of possibilities, classified by facility shape, without any attempt at completeness.

2.9.1 Line

Vertical distance The general problem of locating a line to approximate given two-dimensional points is best known from statistics as the problem of *linear regression*, where one of the coordinates is considered as the explained variable, and the other is explanatory. One must then determine an affine estimating function of the explanatory variable minimising some global measure of the deviations between observed (given by the explained variable) and estimated value (given by the estimating function applied to the explanatory variable's value). Depending upon the measure one obtains several types of regression lines: classical regression obtains from the sum of squared deviations (least squares), while Chebyshev (or ℓ_∞) regression uses the maximum absolute deviation.

This latter may be interpreted as a covering objective: what are the two parallel (nonvertical) lines enclosing all given points between them and with smallest vertical intercept difference ? The line lying in the middle between these two is then the Chebyshev regression line. As a consequence of theorem 1 at least three of the data-points lie on the union of upper and lower lines and not all on the same line, see e.g. (Appa and Smith, 1973), and (Schöbel, 1999) for more details on this 'vertical distance' situation.

In order to obtain regression estimators which are more robust against measurement errors, (Rousseeuw, 1984) suggested the use of the median absolute deviation as an objective to be minimised, in other words, what are the two parallel (nonvertical) lines enclosing half of the given points between them and with smallest vertical intercept difference? Recently several quite efficient techniques for computing the resulting *least median of squares* regression lines were developed, see (Edelsbrunner and Souvaine, 1990) and (Xu and Shiue, 1993).

A further generalisation is to consider other quantiles of the set of absolute (or squared) deviations between observed and estimated values. (Carrizosa and Plastria, 1995a) show how one may obtain such *least quantile of squares* regression lines for all quantiles simultaneously in $O(M^3)$ time

Closest point distance When the coordinates cannot readily be classified into explanatory and explained there is no distinguished 'vertical' direction, so any line direction is acceptable and the deviation measure becomes the closest point distance to the line. In case of euclidean distance, this means taking the orthogonal distance to the line. But other distance measures might be used.

The corresponding full covering model is to find the *centre* line, minimising the maximal distance between the datapoints and the line. For euclidean distance this can also be seen as the narrowest band (region between two parallel lines) which covers all datapoints, where width is measured orthogonally. The main result in this case, valid in fact for any symmetric Minkowski distance, says that one of the boundary-lines of some optimal band will contain an edge of the convex hull of the data-points, while the other boundary line contains a third data-point, see e.g. (Morris and Norback, 1983), which leads to a simple $O(M \log M)$ algorithm by running around this convex hull (note that it is only the construction of this convex hull which takes so much time, and, once known, finding the narrowest band takes $O(e)$ where e stands for the number of extreme points).

For ample information on this and other line and higher dimensional hyperplane location problems, see (Schöbel, 1999).

2.9.2 p Lines

(Megiddo and Tamir, 1982) have shown that detecting whether all datapoints lie on a set of p parallel lines is NP-hard (p being part of the input). Therefore the same holds for the full covering problem by p lines: if it were polynomially solvable, one could check if the minimal covering radius is zero, answering previous question in polynomial time.

2.9.3 Broken Line

(Díaz-Báñez and Mesa, 1998) construct efficient algorithms for finding a rectilinear line (only horizontal and vertical pieces permitted) with a given number of bends, minimising the maximum vertical distance to given datapoints. Both the case where left and right endpoints are arbitrary, and also when these endpoints are constrained to given vertical segments are considered.

2.9.4 Line Segment

Locating a line segment of given length is different from line location, since the closest point of the segment may be its endpoint, distance to which is then measured differently. Special purpose methods for the full covering problem for line segment location are developed by (Imai et al., 1992) and (Agarwal et al., 1993).

2.9.5 Circle

Finding a circle minimising the maximal (euclidean) distance from the datapoints is equivalent to full covering by a minimum width annulus (the circular band between two concentric circles). This is the *out-of-roundness* estimation problem from metrology, where one should test produced pieces for being sufficiently round, (ASME, 1972). This can also be seen as finding a central point minimising the difference between the distance to a farthest and to a nearest data-point.

(Rivlin, 1979) has characterised the local minima of this last function by the following property: there should be (at least) two furthest data-points and two closest data-points from the central point, and additionally these should circularly alternate. It follows that the central point should lie on the intersection between furthest and closest point Voronoi diagrams. It should be noted that the central point might also lie at infinity, e.g. when all datapoints lie on one line, in which case the optimal circle is a line, so has infinite radius. This allows to solve the problem in $O(M^2)$ time, since the number of such intersection points can be $O(M^2)$ (Roy and Zhang, 1992), a complexity which can at least be lowered to $O(M^{1.5+\epsilon})$ (Agarwal and Sharir, 1996).

Remarkably, when the order in which the data-points should be seen from the central point is given, as is normally the case for roundness measurements of an approximate circular piece, the problem becomes easier: there can be at most one local optimum yielding the correct order, as shown by (García-López et al., 1998), who describe an $O(M \log M)$ method for this case.

Finally (Gass et al., 1998) indicate that the variant problem in which the maximum difference between *squared* distance to data-point and *squared* radius is sought, can be solved by linear programming, and suggest that this yields approximate estimates of the out-of-roundness error sufficiently good for practice.

2.9.6 Grid

Locating a regular rectangular array of grid-points in such a way as to minimise the maximum distance of data-points to closest grid-points is a special case of the more general model of optimal gridpositioning studied in (Plastria, 1991) for which a general global optimisation algorithm was advocated. In the case of max-distance (Rayco et al., 1999) indicate how the problem can be solved in $O(M \log^2 M)$ by reduction to two 1-centre problems on a circular network.

Acknowledgements The author thanks Emilio Carrizosa for his many useful remarks, and Rafael Blanquero and Maria Paola Scaparra for several references.

References

Agarwal, P., Efrat, A., Sharir, M., and Toledo, S. (1993). Computing a segment center for planar point set. *Journal of Algorithms*, 15:314–323.

Agarwal, P. and Sharir, M. (1996). Efficient randomized algorithms for some geometric optimization algorithms. *Discrete & Computational Geometry*, 16:317–337.

Alonso, I., Carrizosa, E., and Conde, E. (1998). Maximin location: discretization not always works. *TOP*, 6:313–319.

Appa, G. and Giannikos, I. (1994). Is linear programming necessary for single facility location with maximin of rectilinear distance ? *Journal of the Operational Research Society*, 45:97–107.

Appa, G. and Smith, C. (1973). On l_1 and Chebyshev estimation. *Mathematical Programming*, 5:73–87.

ASME (1972). Measurement of Out-Of-Roundness. Technical Report ANSI B89.3.1-1972, American Society of Mechanical Engineers, NewYork.

Aupperle, L. and Keil, J. (1989). Polynomial algorithms for restricted euclidean p-centre problems. *Discrete Applied Mathematics*, 23:25–31.

Aurenhammer, F. (1991). Voronoi diagrams, a survey of a fundamental geometric data structure. *ACM Computing Surveys*, 23:345–405.

Aurenhammer, F. and Edelsbrunner, H. (1984). An optimal algorithm for constructing the weighted Voronoi diagram in the plane. *Pattern Recognition*, 17:251–257.

Avriel, M., Diewert, W., Schaible, S., and Zhang, I. (1988). *Generalized concavity*. Plenum, New York.

Beer, G. and Pai, D. (1990). On convergence of convex sets and relative Chebyshev centers. *Journal of Approximation Theory*, 62:147–169.

Blanquero, R. and Carrizosa, E. (2000). On covering methods for d.c. optimization. *Journal of Global Optimization*. forthcoming.

Bose, P. and Toussaint, G. (2000). Computing the constrained euclidean, geodesic and link centre of a simple polygon with applications. *Studies in Locational Analysis*, 15. forthcoming.

Bozkaya, B., Zhang, J., and Erkut, E. (2000). An efficient genetic algorithm for the p-median problem. this book.

Brimberg, J. and Mehrez, A. (1994). Multifacility location using a maximin criterion and rectangular distances. *Location Science*, 2:11–19.

Butt, S. and Cavalier, T. (1996). An efficient algorithm for facility location in the presence of forbidden regions. *European Journal of Operational Research*, 90:56–70.

Capoyleas, V., Rote, G., and Woeginger, G. (1991). Geometric clustering. *Journal of Algorithms*, 12:341–356.

Carrizosa, E., Muñoz, M., and Puerto, J. (1995). On the set of optimal points to the weighted maxmin problem. *Studies in Locational Analysis*, 7:21–33.

Carrizosa, E. and Plastria, F. (1995a). The determination of a 'least quantile of squares regression line' for all quantiles. *Computational Statistics & Data Analysis*, 20:467–479.

Carrizosa, E. and Plastria, F. (1995b). On minquantile and maxcovering optimisation. *Mathematical Programming*, 71:101–112.

Carrizosa, E. and Plastria, F. (1998a). Locating an undesirable facility by generalized cutting planes. *Mathematics of Operations Research*, 23:680–694.

Carrizosa, E. and Plastria, F. (1998b). Polynomial algorithms for parametric minquantile and maxcovering planar location problems with locational constraints. *TOP*, 6:179–194.

Carrizosa, E. and Plastria, F. (1999). Location of semi-obnoxious facilities. *Studies in Locational Analysis*, 12:1–27.

Chakraborty, R. and Chaudhuri, P. (1981). Note on geometrical solution for some minimax location problems. *Transportation Science*, 15:164–166.

Charalambous, C. (1982). Extension of the Elzinga-Hearn algorithm to the weighted case. *Operations Research*, 30:591–594.

Chatelon, J., Hearn, D., and Lowe, T. (1982). A subgradient algorithm for certain minimax and minisum problems - the constrained case. *SIAM Journal on Control and Optimization*, 20:455–469.

Chazelle, B. and Lee, D. (1986). On a circle placement problem. *Computing*, 36:1–16.

Chen, R. (1983). Solution of minisum and minimax location-allocation problems with euclidean distances. *Naval Research Logistics Quarterly*, 30:449–459.

Chrystal, G. (1885). On the problem to construct the minimum circle enclosing n given points in the plane. *Proceedings of the Edinburgh Mathematical Society*, 3:30–33.

Crama, Y., Hansen, P., and Jaumard, B. (1995). Complexity of product positioning and ball intersection problems. *Mathematics of Operations Research*, 20:885–894.

Crama, Y. and Ibaraki, T. (1997). Hitting or avoiding balls in euclidean space. *Annals of Operations Research*, 69:47–64.

Das, P., Chakraborty, R., and Chaudhuri, P. (1999). A polynomial time algorithm for a hemispherical minimax location problem. *Operations Research Letters*, 24:57–63.

de Berg, M., van Kreveld, M., Overmars, M., and Schwarzkopf, O. (1997). *Computational geometry. Algorithms and applications*. Springer-Verlag.

Dearing, P. and Francis, R. (1974). A network flow solution to a multifacility minimax location problem involving rectilinear distances. *Transportation Science*, 8:126–141.

Díaz-Báñez, J. and Mesa, J. (1998). Location of rectilinear center trajectories. *TOP*, 6:159–177.

Drezner, Z. (1981). On a modified one-center problem. *Management Science*, 27:848–851.

Drezner, Z. (1982). On minimax optimisation problems. *Mathematical Programming*, 22:227–230.

Drezner, Z. (1984a). The p-centre problem - heuristic and optimal algorithms. *Journal of the Operational Research Society*, 35:741–748.

Drezner, Z. (1984b). The planar two-center and two-median problems. *Transportation Science*, 18:351–361.

Drezner, Z. (1986). The p-cover problem. *European Journal of Operational Research*, 26:312–313.

Drezner, Z. (1987). On the rectangular p-center problem. *Naval Research Logistics*, 34:229–234.

Drezner, Z., editor (1995). *Facility location: a survey of applications and methods*. Springer-Verlag.

Drezner, Z. and Shelah, S. (1987). On the complexity of the Elzinga-Hearn algorithm for the 1-center problem. *Mathematics of Operations Research*, 12:255–261.

Drezner, Z. and Wesolowsky, G. (1980). Single facility l_p distance minimax location. *SIAM Journal on Algebraic and Discrete Methods*, 1:315–321.

Drezner, Z. and Wesolowsky, G. (1983a). Location of an obnoxious facility with rectangular distances. *Journal of Regional Science*, 23:241–248.

Drezner, Z. and Wesolowsky, G. (1983b). Minimax and maximin facility location problems on a sphere. *Naval Research Logistics Quarterly*, 30:305–312.

Drezner, Z. and Wesolowsky, G. (1985). Location of multiple obnoxious facilities. *Transportation Science*, 19:193–202.

Drezner, Z. and Wesolowsky, G. (1994). Finding the circle or rectangle containing the minimum weight of points. *Location Science*, 2:83–90.

Drezner, Z. and Wesolowsky, G. (1996). Obnoxious facility location in the interior of a planar network. *Journal of Regional Science*, 35:675–688.

Dyer, M. (1986). On a multidimensional search technique and its application to the euclidean one-center problem. *SIAM Journal on Computing*, 15:725–738.

Dyer, M. and Frieze, A. (1985). A simple heuristic for the p-centre problem. *Operations Research Letters*, 3:285–288.

Edelsbrunner, H. and Souvaine, D. (1990). Computing least median of squares regression lines and guided topological sweep. *Journal of the American Statistical Association*, 85:115–119.

Elzinga, D. and Hearn, D. (1972a). Geometric solutions for some minimax location problems. *Transportation Science*, 6:379–394.

Elzinga, D. and Hearn, D. (1972b). The minimum covering sphere problem. *Management Science*, 19:96–104.
Erkut, E. and Öncü, T. (1991). A parametric 1-maximin location problem. *Journal of the Operational Research Society*, 42:49–55.
Ezra, N., Handler, G., and Chen, R. (1994). Solving infinite p-center problems in euclidean space using an interactive graphical method. *Location Science*, 2:101–109.
Fejes Tóth, L. (1972). *Lagerungen in der Ebene, auf der Kugel und im Raum*. Springer, 2nd edition.
Fejes Tóth, L., Gritzmann, P., and Wills, J. (1989). Finite sphere packing and sphere covering. *Discrete & Computational Geometry*, 4:19–40.
Fliege, J. (1997). *Effiziente Dimensionsreduktion in Multilokationsproblemen*. PhD thesis, Universität Dortmund, Fachbereich Mathematik, Universität Dortmund, 44221 Dortmund. Shaker Verlag, Aachen, Germany. (An english translation is available from the author on request.).
Fliege, J. and Maier, U. (1999). The distribution of points on the sphere and corresponding cubature formulae. *IMA Journal of Numerical Analysis*, 19:317–334.
Fowler, R., Paterson, M., and Tanimoto, S. (1981). Optimal packing and covering in the plane are NP-complete. *Information Processing Letters*, 12:133–137.
Francis, R., McGinnis, L., and White, J. (1992). *Facility layout and location: an analytical approach*. Prentice Hall, 2nd edition.
Frederickson, G. and Johnson, D. (1983). Finding k-th paths and p-centres by generating and searching good data structures. *Journal of Algorithms*, 4:61–80.
Frenk, J., Gromicho, J., and Zhang, S. (1996). General models in min-max continuous location — theory and solution. *Journal of Optimisation Theory and Applications*, 89:39–63.
García-López, J., Ramos, P., and Snoeyink, J. (1998). Fitting a set of points by a circle. *Discrete and Computational Geometry*, 820:389–402.
Gass, S., Witzgall, C., and Harary, H. (1998). Fitting circles and spheres to coordinate measuring machine data. In Giannessi,F. et al., editor, *New trends in Mathematical Programming*, pages 65–91. Kluwer Academic Publishers.
Gómez, F., Ramaswami, S., and Toussaint, G. (1997). On removing non-degeneracy assumptions in computational geometry. In Bongiovanni, G., Bovet, D., and Di Battista, G., editors, *Algorithms and Complexity (Proceedings CIAC'97)*, volume 1203 of *Lecture Notes in Computer Science*. Springer.
Graham, R., Lubachevsky, B., Nurmela, K., and Östergård, P. (1998). Dense packings of congruent circles in a circle. *Discrete Mathematics*, 181:139–154.
Guibas, L. and Herschberger, J. (1989). Optimal shortest path queries in a simple polygon. *Journal of Computer and System Sciences*, 39:126–152.
Hales, T. (1998). The Kepler conjecture. Technical report, Department of Mathematics, University of Michigan.
http://www.math.lsa.umich.edu/ hales.
Hamacher, H. and Nickel S. (1995). Restricted planar location problems and applications. *Naval Research Logistics*, 42:967–992.
Hansen, P., Jaumard, B., and Lu, S. (1992). Global optimization of univariate Lipschitz functions II: new algorithms and computational comparison. *Mathematical Programming*, 55:273–292.

Hansen, P., Jaumard, B., Meyer, C., and Thisse, J.-F. (1998). New algorithms for product positioning. *European Journal of Operational Research*, 104:154–174.

Hansen, P., Jaumard, B., and Tuy, H. (1995). Global optimization in location. In (Drezner, 1995), pages 43–68.

Hansen, P., Peeters, D., and Thisse, J.-F. (1981). On the location of an obnoxious facility. *Sistemi Urbani*, 3:299–317.

Hearn, D. and Vijay, J. (1982). Efficient algorithms for the (weighted) minimum circle problem. *Operations Research*, 30:777–795.

Herschberger, J. and Suri, S. (1999). An optimal algorithm for euclidean shortest paths in the plane. *SIAM Journal on Computing*, 28:2215–2256.

Hiriart-Urruty, J.-B. and Lemaréchal, C. (1993). *Convex analysis and minimization algorithms*. Springer. 2 volumes.

Hsiang, W.-Y. (1993). On the sphere packing problem and the proof of Kepler's conjecture. *International Journal of Mathematics*, 4:739–831.

Hsu, W. and Nemhauser, G. (1979). Easy and hard bottleneck location problems. *Discrete Applied Mathematics*, 1:209–216.

Hurtado, F., Sacristán, V., and Toussaint, G. (2000). Some constrained minimax and maximin location problems. *Studies in Locational Analysis*, 15. forthcoming.

Idrissi, H., Lefebvre, O., and Michelot, C. (1989). Applications and numerical convergence of the partial inverse method. In *Optimisation – Fifth French-German Conference, Castel Novel 1988*, volume 1405 of *Lecture Notes in Mathematics*, pages 39–54. Springer.

Imai, H., Lee, D., and Yang, C. (1992). 1-segment center problem. *ORSA Journal on Computing*, 4:426–434.

Jacobsen, S. (1981). An algorithm for the minimax Weber problem. *European Journal of Operational Research*, 6:144–148.

Karkazis, J. (1988). The general unweighted problem of locating obnoxious facilities on the plane. *Belgian Journal of Operations Research Statistics and Computer Science*, 28:3–49.

Karkazis, J. and Karagiorgis, P. (1986). A method to locate the maximum circle(s) inscribed in a polygon. *Belgian Journal of Operations Research Statistics and Computer Science*, 26:3–36.

Klein, R., editor (1989). *Concrete and abstract Voronoi diagrams*, volume 400 of *Lecture Notes in Computer Science*. Springer.

Ko, M., Lee, R., and Chang, J. (1990). Rectilinear m-center problem. *Naval Research Logistics*, 37:419–427.

Konno, H., Thach, P., and Tuy, H., editors (1997). *Optimization on low rank nonconvex structures*. Kluwer Academic Publishers.

Love, R., Morris, J., and Wesolowsky, G. (1988). *Facilities location: models & methods*. North-Holland.

Maffioli, F. and Righini, G. (1994). An annealing approach to multi-facility location problems in euclidean space. *Location Science*, 2:205–222.

Masuyama, S., Ibaraki, T., and Hasegawa, T. (1981). The computational complexity of the m-centre problems on the plane. *Transactions of the IECE of Japan*, E64:57–64.

Megiddo, N. (1983a). Linear-time algorithms for linear programming in $\mathbb{R}^3$ and related problems. *SIAM Journal on Computing*, 12:759–776.

Megiddo, N. (1983b). The weighted euclidean 1-center problem. *Mathematics of Operations Research*, 8:498–504.

Megiddo, N. and Supowit, K. (1984). On the complexity of some common geometric location problems. *SIAM Journal on Computing*, 18:182–196.

Megiddo, N. and Tamir, A. (1982). On the complexity of locating linear facilities in the plane. *Operations Research Letters*, 1:194–197.

Megiddo, N., Tamir, A., Zemel, E., and Chandrasekaran, R. (1981). An $o(n \log^2 n)$ algorithm for the k-th longest path in a tree with applications to location problems. *SIAM Journal on Computing*, 10:328–337.

Mehrez, A., Sinuany-Stern, Z., and Stulman, A. (1986). An enhancement of the Drezner-Wesolowsky algorithm for single facility location with maximin of rectilinear distance. *Journal of the Operational Research Society*, 37:971–977.

Melachrinoudis, E. (1988). An efficient computational procedure for the rectilinear maximin location problem. *Transportation Science*, 22:217–223.

Melachrinoudis, E. and Cullinane, P. (1985). Locating an undesirable facility within a geographical region using the maximin criterion. *Journal of Regional Science*, 25:115–127.

Melachrinoudis, E. and Cullinane, P. (1986). Locating an undesirable facility with a minimax criterion. *European Journal of Operational Research*, 24:239–246.

Melachrinoudis, E. and MacGregor Smith, J. (1995). An $o(mn^2)$ algorithm for the maximin problem in E^2. *Operations Research Letters*, 18:25–30.

Mitchell, J. and Papadimitriou, C. (1991). The weighted region problem: Finding shortest paths through a weighted planar subdivision. *Journal of the ACM*, 38:18–73.

Mladenović, N., Labbé, M., and Hansen, P. (2000). Solving the p-center problem with tabu search and variable neighborhood search. Technical report, Service de Mathématiques de la Gestion, Université Libre de Bruxelles. http://smg.ulb.ac.be/Preprints/Labbe00_20.html.

Morris, J. (1973). A linear programming approach to the solution of constrained multifacility minimax location where distances are rectangular. *Operational Research Quarterly*, 24:419–435.

Morris, J. and Norback, J. (1983). Linear facility location —- solving extensions of the basic problem. *European Journal of Operational Research*, 12:90–94.

Nickel, S. (1998). Restricted center problems under polyhedral gauges. *European Journal of Operational Research*, 104:343–357.

Nurmela, K. and Östergård, P. (1999). More optimal packings of equal circles in a square. *Discrete & Computational Geometry*, 22:439–457.

Ohsawa, Y. (2000). Bicriteria euclidean location associated with maximin and minimax criteria. *Naval Research Logistics*, 47:581–592.

Ohsawa, Y. and Imai, A. (1997). Degree of locational freedom in a single facility euclidean minimax location model. *Location Science*, 5:29–45.

Okabe, A., Boots, B., and Sugihara, K. (1992). *Spatial tesselations, concepts and applications of Voronoi diagrams*. John Wiley & Sons, Chichester.

Okabe, A. and Suzuki, A. (1997). Locational optimization problems solved through Voronoi diagrams. *European Journal of Operational Research*, 98:445–456.

Orlowski, M. (1990). A new algorithm for the largest empty rectangle problem. *Algorithmica*, 5:65–73.

Patel, M. (1995). Spherical minimax location problem using the euclidean norm: formulation and optimisation. *Computational Optimization and Applications*, 4:79–90.

Pelegrín, B. (1991). Heuristic methods for the p-center problem. *RAIRO Recherche Opérationnelle/Operations Research*, 25:65–72.

Pelegrín, B. and Cánovas, L. (1998a). The minimum covering l_{pb}-hypersphere problem. *Computational Optimization and Applications*, 9:85–97.

Pelegrín, B. and Cánovas, L. (1998b). A new assignment rule to improve seed points algorithms for the continuous k-center problem. *European Journal of Operational Research*, 104:366–374.

Pelegrín, B., Michelot, C., and Plastria, F. (1985). On the uniqueness of optimal solutions in continuous location theory. *European Journal of Operational Research*, 20:327–331.

Plastria, F. (1987). Solving general continuous single facility location problems by cutting planes. *European Journal of Operational Research*, 29:98–110.

Plastria, F. (1990). The continuous p-center sum problem. *Studies in Locational Analysis*, 1:29–38.

Plastria, F. (1991). Optimal gridpositioning or single facility location on the torus. *Recherche Opérationnelle/ Operations Research*, 25:19–29.

Plastria, F. (1992). On destination optimality in asymmetric distance Fermat-Weber problems. *Annals of Operations Research*, 40:355–369.

Plastria, F. (1994). Fully geometric solutions to some planar minimax location problems. *Studies in Locational Analysis*, 7:171–183.

Plastria, F. (1995). Continuous location problems. In (Drezner, 1995), pages 225–262.

Plastria, F. (1997). Profit maximising single competitive facility location in the plane. *Studies in Locational Analysis*, 11:115–126.

Plastria, F. (2000). Static competitive facility location: an overview of optimisation approaches. *European Journal of Operational Research*. forthcoming.

Plastria, F. and Carrizosa, E. (1999). Undesirable facility location in the euclidean plane with minimal covering objectives. *European Journal of Operational Research*, 119:158–180.

Preparata, F. and Shamos, M. (1985). *Computational Geometry: An Introduction.* Springer-Verlag.

Rayco, B., Francis, R., and Lowe, T. (1997). Error-bound driven demand point aggregation for the rectilinear distance p-center model. *Location Science*, 4:213–235.

Rayco, B., Francis, R., and Tamir, A. (1999). A p-center grid-positioning aggregation procedure. *Computers and Operations Research*, 26:1113–1124.

Rivlin, T. (1979). Approximation by circles. *Computing*, 21:93–104.

Rodriguez-Ramos, W. and Francis, R. (1983). Single crane location optimization. *Journal of Construction Engineering and Management*, 109:387–397.

Rogers, C. A. (1964). *Packing and Covering.* Cambridge University Press.

Rolland, E., Schilling, D., and Current, J. (1996). An efficient tabu search procedure for the p-median problem. *European Journal of Operational Research*, 96:329–342.

Rosing, K. and ReVelle, C. (1997). Heuristic concentration: a two stage solution construction. *European Journal of Operational Research*, 97:75–86.

Rosing, K., ReVelle, C., and Schilling, D. (1999). A gamma heuristic for the p-median problem. *European Journal of Operational Research*, 117:522–532.

Rousseeuw, P. (1984). Least median of squares regression. *Journal of the American Statistical Association*, 79:871–880.

Roy, U. and Zhang, X. (1992). Establishment of a pair of concentric circles with the minimal radial separation for assessing rounding error. *Computer-Aided Design*, 24:161–168.

Rubinstein, R. and Samorodnitsky, G. (1986). Optimal coverage of convex regions. *Journal of Optimization Theory and Applications*, 51:321–343.

Ryzhkov, A. (1973). On certain covering problems. *Engineering Cybernetics*, II:543–548.

Sarkar, A. and Chaudhuri, P. (1996). Solution of equiweighted minimax problem on a hemisphere. *Computational Optimization and Applications*, 6:73–82.

Schmalensee, R. and Thisse, J.-F. (1988). Perceptual maps and the optimal location of new products: an integrative essay. *International Journal of Research in Marketing*, 5:225–249.

Schöbel, A. (1999). *Locating lines and hyperplanes*. Kluwer Academic Publishers.

Sharir M. and E.Welzl (1992). A combinatorial bound for linear programming and related problems. In *Proc. 9th Symp. Theoretical Aspects of Computer Science*, volume 577 of *Lecture Notes in Computer Science*, pages 569–579. Springer.

Späth, H. (1978). Explizite Lösung des dreidimensionales Minimax-Standortproblems in der City-Block-Distanz. *Zeitschrift für Operations Research*, 22:229–237.

Spiekerman, K. and Wegener, M. (1994). The shrinking continent. new time space maps of Europe. *Enviroment and Planning: Planning and Design*, 21:653–673.

Suzuki, A. and Drezner, Z. (1996). On the p-center location problem in an area. *Location Science*, 4:69–82.

Sylvester, J. (1857). A question in the geometry of situation. *Quarterly Journal of Mathematics*, 1:79.

Sylvester, J. (1860). On Poncelet's approximate linear valuation of Surd forms. *Philosophical Magazine*, 20 (Fourth series):203–222.

Toussaint, G. (1983). Computing largest empty circles with location constraints. *International Journal of Computer and Information Sciences*, 12:347–358.

Vijay, J. (1985). An algorithm for the p-center problem in the plane. *Transportation Science*, 19:235–245.

Watson-Gandy, C. (1982). Heuristic procedures for the m-partial cover problem on a plane. *European Journal of Operational Research*, 11:149–157.

Welzl, E. (1991). Smallest enclosing disks (balls and ellipsoids). In H.Maurer, editor, *New results and new trends in Computer Science*, volume 555 of *Lecture Notes in Computer Science*, pages 359–370. Springer.

Woeginger, G. (1998). A comment on a minmax location problem. *Operations Research Letters*, 23:41–44.

Xu, C. and Shiue, W. (1993). Parallel algorithms for least median of squares regression. *Computational Statistics & Data Analysis*, 16:349–362.

Xue, G. and Sun, S. (1995). The spherical one-center problem. In Du, D.-Z. and Pardalos, P., editors, *Minimax and applications*, pages 153–156. Kluwer Academic Publishers.

Zwick, D. (1994). A planar minimax algorithm for analysis of coordinate measurements. *Advances in Computational Mathematics*, 2:375–391.

3 Discrete Network Location Models

John Current[1], Mark Daskin[2], and David Schilling[1]

[1] Department of Management Sciences, Fisher College of Business, The Ohio State University, 2100 Neil Avenue, Columbus, Ohio 43210. e-mail: Current.1@osu.edu; Schilling@cob.ohio-state.edu
[2] Department of Industrial Engineering and Management Sciences, Northwestern University, 2145 Sheridan Road, Evanston, IL 60208. e-mail: m-daskin@northwestern.edu

3.1 Introduction

Undoubtedly, humans have been analyzing the effectiveness of locational decisions since they inhabited their first cave. We use the term "facility" here in its broadest sense. That is, it is meant to include entities such as air and maritime ports, factories, warehouses, retail outlets, schools, hospitals, day-care centers, bus stops, subway stations, electronic switching centers, computer concentrators and terminals, rain gages, emergency warning sirens, and satellites, to name but a few that have been analyzed in the research literature.

The ubiquity of locational decision-making has led to a strong interest in location analysis and modeling within the operations research and management science community. The long and voluminous history of location research results from several factors. First, location decisions are frequently made at all levels of human organization from individuals and households to firms, government agencies and even international agencies. Second, such decisions are often strategic in nature. That is, they involve large sums of capital resources and their economic effects are long term. In the private sector they have a major influence on the ability of a firm to compete in the market place. In the public sector they influence the efficiency by which jurisdictions provide public services and the ability of these jurisdictions to attract households and other economic activity. Third, they frequently impose economic externalities. Such externalities include pollution, congestion, and economic development, among others.

Fourth, location models are often extremely difficult to solve, at least optimally. Even the most basic models are computationally intractable for large problem instances. In fact, the computational complexity of location models is a major reason that the widespread interest in formulating and implementing such models did not occur until the advent of high-speed digital computers. Finally, location models are application specific. That is, their structural form (the objectives, constraints and variables) is determined by the particular location problem under study. Consequently, there does not exist a general location model that is appropriate for all potential or existing applications.

Facility Location: Applications and Theory.
Edited by Z. Drezner and H.W. Hamacher
© 2002 Springer-Verlag, ISBN 3-540-42172-6

3.1.1 Examples of Facility Location Decision Contexts

Much of the literature on facility location modeling has not been directed to specific applications (i.e., case studies). Rather, it has been directed to formulating new models and modifications to existing models which have many potential applications, and to developing efficient solution techniques for existing or newly formulated models.

There are several causes for this bias away from reporting specific applications in the literature. First, applications frequently employ existing models and solution techniques. Consequently, they are not viewed as scientific advances by the research community, but rather as applications of existing technology. Second, specific applications are frequently analyzed by consultants and planners; two professions which are rarely motivated to publish in research journals. Third, private sector advances in location modeling are often viewed as proprietary because they give the firm a competitive advantage; consequently, they are not shared with the larger community. In spite of this bias, there are still many articles which relate directly to a specific application or application area.

Table 3.1 lists some of the applications that have appeared in the literature. It is not intended to be exhaustive, but rather to demonstrate their diversity. A single citation is given for each application. This is done because of space limitations and is not meant to imply that the citation given is either the only or the most important article addressing the topic. In fact, for several of these applications, such as emergency medical services (EMS) siting, there exists a rather extensive literature.

Undoubtedly, the multitude of applications is a major reason for the multidisciplinary interest in location modeling. In addition to the more traditional applications listed in Table 3.1 there have also been less obvious ones. Some of these are listed in Table 3.2. The reader is referred to Eiselt (1992) for a review of facility location applications.

3.2 Basic Facility Location Models

In this section we present eight basic facility location models: set covering, maximal covering, p-center, p-dispersion, p-median, fixed charge, hub, and maxisum. In all of these models, the underlying network is given, as are the locations of the demands to be served by the facilities and the locations of existing facilities (if pertinent.) The general problem is to locate new facilities to optimize some objective. Distance or some measure more or less functionally related to distance (e.g., travel time or cost, demand satisfaction) is fundamental to such problems. Consequently, we have classified them according to their consideration of distance. The first four are based on maximum distance and the second four are based on total (or average) distance.

Table 3.1. Applications of Facility Location Models

Application	Citation
Airline hubs	O'Kelly, 1987
Airports	Saatcioglu, 1982
Auto Emission testing stations	Swersey and Thakur, 1995
Blood bank	Price and Turcotte, 1986
Brewery depots	Gelders, et al., 1987
Bus stops	Gleason, 1975
Bus garages	Maze et al., 1981
Coal handling facilities	Osleeb and Ratick, 1983
Computer concentrators	Pirkul, 1987
Computer service centers	Ghosh and Craig, 1986
Day-Care Centers	Holmes, et al., 1972
Electric power generating plants	Cohon, et al., 1980
Emergency medical services	ReVelle, et al., 1977
Emergency equipment for oil spills	Belardo et al., 1984
Essential air services	Flynn and Ratick, 1988
Fast-food restaurants	Min, 1987
Fire stations	Schilling et al., 1980
Forest harvesting sites	Hodgson, et al., 1987
Franchise outlets	Pirkul, et al., 1987
Hazardous waste disposal sites	ReVelle et al., 1991
Grain subterminals	Hilger, et al., 1977
Public swimming pools	Goodchild and Booth, 1980
Railroad sidings	Higgins, et al., 1997
Rain gauges	Hogan, 1990
Regional health facilities	Abernathy and Hershey, 1972
Rural health workers	Bennett, et al., 1982
Satellite homing stations	Helme and Magnanti, 1989
Satellite orbits	Drezner, 1988
Schools	Tewari and Sidheswar, 1987
Social Service Centers	Patel, 1979
Solar power system design	Birge and Malyshko, 1985
Solid waste collection	Marks and Liebman, 1971
Telecommunication switching centers	Hakimi, 1965
Truck terminals	Love, et al., 1985
Vehicle Inspection Stations	Hodgson, et al., 1996
Warehouses	Kuehn and Hamburger, 1963

Table 3.2. Additional Applications of Facility Location Models

Application	Citation
Apparel sizing	Tryfos, 1986
Archaeological settlement analysis	Bell and Church, 1985
Chip manufacturing	Cho and Sarrafzadeh, 1994
Data base management	Pirkul, 1986
Flexible Manufacturing System tool selection	Daskin, et al., 1990
Ingot size selection	Vasko, et al., 1988
Location of bank accounts	Cornuejols, et al., 1977
Medical diagnosis	Reggia, et al., 1983
Metallurgical grade assignment	Vasko, et al., 1989
Placement of dampers	Kincaid and Berger, 1993
Political party platform	Ginsberg, et al., 1987
Product positioning in feature space	Gavish et al., 1983
Product procurement and standardization	Watson, 1996
Production lot sizing	Van Oudheusden and Singh, 1988
Vehicle routing	Bramel and Simchi-Levi, 1995

3.2.1 Maximum Distance Models

In some locations problems, a maximum distance exists *a priori*. For example, in many school districts elementary school students within a mile of their school must walk to school. Public transportation must be provided for those not within this maximum distance. In the private sector, some businesses guarantee service within a pre-determined time (e.g., 20 minute pizza delivery). In the former case, a school district might want to locate schools to minimize the number of students who must be bussed at public expense. In the latter example, a pizza chain might want to locate its outlets to maximize the number of potential customers within 20 minutes of one of the outlets.

In the facility location literature, *a priori* maximum distances such as these are known as "covering" distances. Demand within the covering distance of its closest facility is considered "covered." An underlying assumption of this measure of maximum distance is that demand is fully satisfied if the nearest facility is within the coverage distance and is not satisfied if the closest facility is beyond that distance. That is, being closer to a facility than the maximum distance does not improve satisfaction.

Set covering location model The first location covering location problem was the set covering problem (Toregas et al., 1971). Here the objective is to locate the minimum number of facilities required to "cover" all of the demand nodes. To formulate this problem we define the following inputs and sets

I = the set of demand nodes indexed by i
J = The set of candidate facility locations, indexed by j
d_{ij} = distance between demand node i and candidate site j
D_c = distance coverage
$N_i = \{j \mid d_{ij} \leq D_c\}$
  = the set of all candidate locations that can cover demand point i

and the following decision variable

$$x_j = \begin{cases} 1 \text{ if we locate at site } j \\ 0 \text{ if not} \end{cases}$$

With this notation, the set covering location problem (SCLP) can be formulated as follows:

$$\text{Minimize} \quad \sum_{j \in J} x_j \qquad (3.1)$$

subject to:

$$\sum_{j \in N_i} x_j \geq 1 \ \forall i \in I \qquad (3.2)$$

$$x_j \in \{0,1\} \ \forall j \in J \qquad (3.3)$$

The objective function (3.1) minimizes the number of facilities located. Constraint set (3.2) ensures that each demand node is covered by at least one facility. Constraint set (3.3) enforces the yes or no nature of the siting decision. The objective function can be generalized by including site-specified costs as coefficients of the decision variables. In this case, the objective would be to minimize the total fixed cost of the siting configuration rather than the number of facilities sited. Both versions of the set covering problem are NP-hard (Garey and Johnson, 1979). However, the linear programming relaxation of the set covering location problem as formulated above often results in an all-integer solution. Typically, only a few branches in a branch and bound algorithm are needed to obtain an optimal all-integer solution when the LP relaxation is fractional.

A variety of row and column reduction rules have been developed to reduce the size of the problem considerably (see Daskin (1995) for a discussion of such rules). For example, variable x_k can be eliminated from the formulation if $M_k \subset M_j$, where $M_j = \{i | d_{ij} \leq D_c\}$ and $M_k = \{i | d_{ik} \leq D_c\}$. This column reduction is possible because a facility at j would cover all of the demand nodes that a facility at k would cover and possibly additional ones as well; therefore, location j "dominates" location k. Individual constraints, say h, of constraint set (3.2) can be eliminated if there is some covering set, say N_i, such that $N_i \subset N_h$. This row reduction is possible because the constraint in (3.2) for demand node h is redundant. That is, if the coverage constraint for demand node i is satisfied, then the constraint for demand node h is also satisfied.

Formulation (3.1)-(3.3) assumes that the candidate facility sites are located at the nodes of the network. A lower cost facility siting scheme might be possible if the facilities could be located along the arcs of the network as well. This is illustrated by Figure 3.1. If the coverage distance is 10 and facilities can only be located at the nodes, then two facilities are needed: one at node A and one at either node B or C. If we could locate along the arcs as well as at the nodes, then a single facility located ten units to the right of node A would cover all three demand nodes. Church and Meadows (1979) present a method to modify the original network to permit siting along the arcs but still solve the problem using formulation (3.1)-(3.3). This method augments the original network with a finite (albeit potentially large) number of nodes located along the arcs of the network. The inclusion of these additional nodes in set J will result in a solution as good as one permitting locations anywhere along the arcs.

Fig. 3.1. Example network

Maximal covering location problem An underlying assumption of the set covering location problem is that all of the demand nodes must be covered. In essence, there is no budget constraint. However, in many facility planning situations, a budget does exist. For example, many school districts would like to have an elementary school within walking distance of all of its elementary age students. However, satisfying such a requirement may require more schools than the district is prepared to build. The maximal covering location problem (MCLP, Church and ReVelle, 1974) was formulated to address planning situations which have an upper limit on the number of facilities to be sited. The objective of the MCLP is to locate a predetermined number of facilities, p, in such a way as to maximize the demand that is covered. Thus, the MCLP assumes that there may not be enough facilities to cover all of the demand nodes. If not all nodes can be covered, the model seeks the siting scheme that covers the most *demand*.

To formulate the maximal covering problem, we augment the definitions used in the SCLP with:

h_i = demand at node i
p = the number of facilities to locate
$z_i = \begin{cases} 1 \text{ if demand node } i \text{ is covered} \\ 0 \text{ if not} \end{cases}$

The maximal covering problem may now be formulated as follows:

$$\text{Maximize} \quad \sum_{i \in I} h_i z_i \tag{3.4}$$

subject to:

$$\sum_{j \in N_i} x_j - z_i \geq 0 \ \forall i \in I \tag{3.5}$$

$$\sum_{j \in J} x_j = p \tag{3.6}$$

$$x_j \in \{0,1\} \quad \forall j \in J \tag{3.7}$$

$$z_i \in \{0,1\} \quad \forall i \in I \tag{3.8}$$

The objective function (3.4) maximizes the total demand covered. Constraint set (3.5) ensures that demand at node i is not counted as covered unless we locate at one of the candidate sites that covers node i. Constraint (3.6) limits the number of facilities to be sited. Constraint sets (3.7) and (3.8) reflect the binary nature of the facility siting decisions and demand node coverage, respectively. Interestingly, constraint sets (3.5) and (3.7), allow us to replace constraint set (3.8) with $z_i \leq 1$, $\forall i \in I$, without loss of generality.

As with the SCLP, if the facilities can be sited anywhere along the arcs of the network, the network can be modified as proposed by Church and Meadows (1979) and the problem solved with (3.4)-(3.8). By systematically varying p from 1 to k, where k is the minimum number of facilities required to cover the entire demand, one can use (3.4)-(3.8) to determine the marginal benefits associated with additional facilities.

The maximal covering problem is also NP-hard (Megiddo, Zemel and Hakimi, 1983), but it can generally be solved effectively using heuristics of the sort outlined later. Particularly useful is Lagrangean relaxation embedded within a branch and bound algorithm (Daskin, 1995; Daskin and Owen, 1998; Galvão and ReVelle, 1996). As mentioned earlier, some planning scenarios exist where there is a desired coverage distance and some maximum distance beyond which service is unacceptable. For problems like this, Church and ReVelle (1974) formulated the maximal covering location problem with mandatory closeness constraints.

Given the previous definitions and D_m = the maximum distance that a demand node may be from an opened facility and $M_i = \{j | d_{ij} \leq D_m\}$, the MCLP with mandatory closeness constraints may be formulated by adding the following constraint set to formulation (3.4)–(3.8).

$$\sum_{j \in M_i} x_j \geq 1 \ \forall i \in I \tag{3.9}$$

The SCLP and MCLP assume that the covering distance, D_c, is a fixed, predetermined standard. This is certainly true in many location planning situations. However, in other situations D_c may be a goal, or target, rather than a fixed standard. For example, in siting facilities such as public libraries and recreational facilities, public agencies may desire to minimize the maximum distance that a citizen is from such a facility for equity reasons (Marsh and Schilling, 1994). Other facilities, such as schools or fire stations, may have a desired distance (e.g., less than 1 mile or 3 minutes travel time) and another distance (e.g., 5 miles or 10 minutes travel time) beyond which service is unacceptable. The following sections address planning situations of this nature.

p-center problem The p-center problem (Hakimi, 1964,1965) addresses the problem of minimizing the maximum distance that demand is from its closet facility given that we are siting a pre-determined number of facilities. There are several possible variations of the basic model. The "vertex" p-center problem restricts the set of candidate facility sites to the nodes of the network while the "absolute" p-center problem permits the facilities to be anywhere along the arcs. Both versions can be either *weighted* or *unweighted*. In the unweighted problem, all demand nodes are treated equally. In the weighted model, the distances between demand nodes and facilities are multiplied by a weight associated with the demand node. For example, this weight might represent a node's importance or, more commonly, the level of its demand.

Given our previous definitions and the following decision variables

W = the maximum distance between a demand node and the facility to which it is assigned

$y_{ij} = \begin{cases} 1 \text{ if demand node } i \text{ is assigned to a facility at node } j \\ 0 \text{ if not} \end{cases}$

the p-center problem can be formulated as follows:

$$\text{Maximize} \quad W \tag{3.10}$$

subject to:

$$\sum_{j \in J} x_j = p \tag{3.11}$$

$$\sum_{j \in J} y_{ij} = 1 \quad \forall i \in I \tag{3.12}$$

$$y_{ij} - x_j \leq 0 \quad \forall i \in I, j \in J \tag{3.13}$$

$$W - \sum_{j \in J} h_i d_{ij} y_{ij} \geq 0 \ \forall i \in I \tag{3.14}$$

$$x_j \in \{0,1\} \quad \forall j \in J \tag{3.15}$$

$$y_{ij} \in \{0,1\} \quad \forall i \in I, j \in J \tag{3.16}$$

The objective function (3.10) minimizes the maximum demand-weighted distance between each demand node and its closest open facility. Constraint (3.11) stipulates that p acilities are to be located. Constraint set (3.12) requires that each demand node be assigned to exactly one facility. Constraint set (3.13) restricts demand node assignments only to open facilities. Constraint (3.14) defines the lower bound on the maximum demand-weighted distance, which is being minimized. Constraint set (3.15) established the siting decision variable as binary. Constraint set (3.16) requires the demand at a node to be assigned to one facility only. Constraint set (3.16) can be replaced by $y_{ij} \geq 0\ \forall i \in I; j \in J$ because constraint set (3.13) guarantees that $y_{ij} \leq 1$. If some y_{ij} are fractional, we can simply assign node i to its closest open facility.

For fixed values of p, the vertex p-center problem can be solved in $O(N^p)$ time since we can enumerate each possible set of candidate locations in this amount of time. Clearly, even for moderate values of N and p, such enumeration is not realistic and more sophisticated approaches are required. For variable values of p, the problem is NP-hard (Garey and Johnson, 1979.)

If integer-valued distances can be assumed, the unweighted vertex or absolute p-center problem is most often solved using a binary search over a range of coverage distances (Handler and Mirchandani, 1979; Handler, 1990). For each coverage distance, a set covering problem is solved. When the solution to the SCP equals p, the minimum associated coverage distance is the solution to the p-center problem. Daskin (2000) has recently shown how the maximal covering model can be used effectively in place of the set covering model as a sub-problem in solving the unweighted vertex p-center problem.

3.2.2 The p-Dispersion Problem

For all of the models discussed above the concern is with the distance between demand and new facilities. Also, an unspoken assumption is that being close to a facility is desirable. The p-dispersion problem (PDP) differs from those problems in two ways (Kuby, 1987). First, it is concerned only with the distance between new facilities. Second, the objective is to maximize the minimum distance between any pair of facilities. Potential applications of the PDP include the siting of military installations where separation makes them more difficult to attack or locating franchise outlets where separation reduces cannibalization among stores.

To formulate this model we require an additional input (M) and a decision variable (D):

M = a large constant (e.g., $\max_{i \in I, j \in J} \{d_{ij}\}$)
D = the minimum separation distance between any pair of facilities

With this notation, the p-dispersion model may be formulated as follows:

$$\text{Maximize} \quad D \tag{3.17}$$

subject to:

$$\sum_{j \in J} x_j = p \tag{3.18}$$

$$D + (M - d_{ij})x_i + (M - d_{ij})x_j \leq 2M - d_{ij}$$
$$\forall i, j \in J, i < j \tag{3.19}$$

$$x_j \in \{0, 1\} \quad \forall j \in J \tag{3.20}$$

The objective function (3.17) maximizes the distance between the two closest facilities. Constraint (3.18) requires that p facilities are located. Constraint (3.20) is a standard integrality constraint. Constraint (3.19) defines the minimum separation between any pair of open facilities. Note that if either x_i or x_j is zero, the constraint will not be binding. However, if both are equal to 1, then the constraint is equivalent to $D \leq d_{ij}$. Therefore, maximizing D has the effect of forcing the smallest inter-facility distance to be as large as possible.

3.2.3 Total or Average Distance Models

Many facility location planning situations in the public and private sections are concerned with the total travel distance between facilities and demand nodes. An example in the private sector might be the location of production facilities that receive their inputs from established sources by truckload deliveries. In the public sector, one might want to locate a network of service providers such as licensing bureaus in such a way as to minimize the total distance that customers must traverse to reach their closest facility. This approach may be viewed as an "efficiency" objective as opposed to the "equity" objective of minimizing the maximum distance, which was mentioned earlier.

p-median problem One classic model in this area is the p-median model (Hakimi, 1964, 1965) which finds the locations of p facilities to minimize the demand-weighted total distance between demand nodes and the facilities to which they are assigned. This model may be formulated as follows:

$$\text{Minimize} \sum_{i \in I} \sum_{j \in J} h_i d_{ij} y_{ij} \qquad (3.21)$$

subject to:

$$\sum_{j \in J} x_j = p \qquad (3.22)$$

$$\sum_{j \in J} y_{ij} = 1 \quad \forall i \in I \qquad (3.23)$$

$$y_{ij} - x_j \leq 0 \quad \forall i \in I, j \in J \qquad (3.24)$$

$$x_j \in \{0, 1\} \quad \forall j \in J \qquad (3.25)$$

$$y_{ij} \in \{0, 1\} \quad \forall i \in I, j \in J \qquad (3.26)$$

The objective function (3.21) minimizes the demand-weighted total distance traveled. Constraint set (3.22) through (3.24) are identical to (3.11) through (3.13) of the p-center problem. Constraints sets (3.25) and (3.26) are identical to (3.15) and (3.16). Constraint set (3.26) can be eliminated following the same arguments as were used for constraint set (3.16). Toregas and ReVelle (1972) show that this formulation also minimizes the average travel distance between the sited facilities and the demand.

This formulation (3.21–3.26) assumes that the potential facility sites are nodes on the network. Hakimi (1964) proved that relaxing the problem to allow facility locations on the arcs of the network would not reduce total travel cost. Consequently, this formulation will yield an optimal solution, even if the facilities could be located anywhere on an arc. Like the p-center problem, the p-median problem can be solved in polynomial time for fixed values of p, but is NP-hard for variable values of p (Garey and Johnson, 1979).

Fixed Charge Location Problem The p-median problem makes three important assumptions that may not be appropriate for certain siting scenarios. First, it assumes that each potential site has the same fixed costs for locating a facility at it. Secondly, it assumes that the facilities being sited do not have capacities on the demand that they can serve. In the parlance of the literature, it is an "uncapacitated" problem. Finally, it also assumes that one knows, *a priori*, how many facilities should be opened (i.e., p)

The fixed charge location problem (FCLP) relaxes all three of these assumptions. The objective of the FCLP is to minimize total facility and transportation costs. In so doing, it determines the optimal number and locations of facilities, as well as the assignments of demand to a facility. Given the fact that the facilities have capacities, demand may not be assigned to its closest facility, as was the case in the previous models presented in this chapter.

Given the previous definitions and the following inputs

f_j = fixed cost of locating a facility at candidate site j
C_j = capacity of a facility at candidate site j
α = cost per unit demand per unit distance

the capacitated fixed charge location problem can be formulated as follows (Balinski, 1965):

$$\text{Minimize} \sum_{j \in J} f_j x_j + \alpha \sum_{i \in I} \sum_{j \in J} h_i d_{ij} y_{ij} \qquad (3.27)$$

subject to:

$$\sum_{j \in J} y_{ij} = 1 \qquad \forall i \in I \qquad (3.28)$$

$$y_{ij} - x_j \leq 0 \qquad \forall i \in I, j \in J \qquad (3.29)$$

$$\sum_{j \in J} h_i y_{ij} - C_j x_j \leq 0 \qquad \forall i \in I \qquad (3.30)$$

$$x_j \in \{0,1\} \qquad \forall j \in J \qquad (3.31)$$

$$y_{ij} \in \{0,1\} \qquad \forall i \in I, j \in J \qquad (3.32)$$

The objective function (3.27) minimizes the sum of the fixed facility location costs and the total travel costs for demand to be served. The second set of terms in (3.27) is often referred to as demand-weighted distance. Constraint (3.30) prohibits the total demand assigned to a facility from exceeding the capacity of the facility, C_j. Constraint sets (3.28), (3.29), (3.31) and (3.32) function as similar constraint sets in the previous problems. Relaxing constraint set (3.32) allows demand at a node to be assigned (partially) to multiple facilities. We also note that constraint (3.29) is not needed in this integer programming formulation since constraint set (3.30) will also force demands to be assigned only to open facilities. However, including constraint set (3.29) in the formulation significantly strengthens the linear programming relaxation of the model.

There are several other features of the FCLP that may not be initially apparent. The inclusion of (3.32) as a binary constraint requires that all demand points be singly sourced. That is, all demand at a particular location is assigned to one facility. Note also that, due to the facility capacities, demand may be served by a facility which is not its closest one. If constraint set (3.30) is removed, the model becomes the uncapacitated fixed charge location problem (UFCLP). In this case, each demand can be completely served by its closest facility and (3.32) can be replaced by non-negativity constraints on the assignment variables, y_{ij}.

Hub location problems Many logistics systems such as less-than-truckload carrier networks, airline networks, and inter-modal carriers, employ hub and spoke systems. These systems are designed to utilize larger capacity or faster vehicles or modes over the long-haul portion of an origin to destination delivery. Consequently, these systems reduce average per mile transportation cost or total delivery time. Numerous models (e.g. O'Kelly, 1986a, 1986b; and Campbell, 1990, 1994) have been formulated to locate the hubs and delivery routes of hub and spoke systems. Most of these models attempt to minimize total cost (as a function of distance.) We refer the reader to chapter 12 of this book for a detailed discussion of hub location problems. The basic p-hub location model can be formulated using the following notation inputs

h_{ij} = number of units of flow between nodes i and j
c_{ij} = unit cost of transportation between nodes i and j
α = discount factor for transport between hubs

and the following decision variables

$$x_j = \begin{cases} 1 \text{ if a hub is located at node } j \\ 0 \text{ if not} \end{cases}$$

$$y_{ij} = \begin{cases} 1 \text{ if demands at node } i \text{ are assigned to a hub located at node } j \\ 0 \text{ if not} \end{cases}$$

as follows (O'Kelly, 1987):

$$\text{Minimize} \sum_{i \in N} \sum_{j \in N} h_{ij} \left(\sum_{k \in N} c_{ik} y_{ik} + \sum_{m \in N} c_{jm} y_{jm} + \alpha \sum_{k \in N} \sum_{k \in N} c_{km} y_{ik} y_{jm} \right) \quad (3.33)$$

subject to:

$$\sum_{j \in N} x_j = p \quad (3.34)$$

$$\sum_{j \in J} y_{ij} = 1 \quad \forall i \in I \quad (3.35)$$

$$y_{ij} - x_j \leq 0 \quad \forall i \in I, j \in J \quad (3.36)$$

$$x_j \in \{0,1\} \quad \forall j \in J \quad (3.37)$$

$$y_{ij} \in \{0,1\} \quad \forall i \in I, j \in J \quad (3.38)$$

The objective function (3.33) minimizes the sum of the cost of moving items between a non-hub node and the hub to which the node is assigned, the cost of moving from the final hub to the destination of the flow, and the inter-hub movement cost which is discounted by a factor of α. The model assumes that the hub portion of the network is a complete graph and therefore flows between any pair of nodes i and j will pass through at most two different hub nodes. Constraints (3.34) through (3.38) are identical to constraints (3.22) through (3.27) for the p-median model above. In particular, constraints (3.35) stipulate that each node should be assigned to exactly one hub. In practical contexts, it may be valuable to relax this constraint and to allow flows from particularly large nodes to be served directly by two or more hub nodes.

Despite the similarity between the constraints of the two models, it is worth noting a number of important differences. First, the demands in the

p-hub location model are node-to-node flows and not simply demands at a particular node. Second, and perhaps most importantly, the objective function is quadratic in the assignment variables. Third, it may not be optimal to assign a node to the nearest hub since the objective function is measured in terms of node-to-node flows and not simply in terms of the cost of accessing the hub system.

Given the difficulties in solving even moderate sized hub location problems optimally, Ernst and Krishnamoorthy (1996) propose heuristic (or approximate) efficient algorithms to attack such problems. Kuby and Gray (1993) analyzed an express air carrier's network with a hub location model. For a more complete review of hub location models, the reader is referred to O'Kelley and Miller (1994) and chapter 12 of this book.

The maxisum location problem The average distance models discussed above assume that locating facilities as close as possible to demands is desirable. For many facilities this is the case. However, for undesirable facilities (e.g., prisons, power plants, and solid waste repositories) at least one objective involves locating facilities far from demand nodes.

The maxisum location problem seeks the locations of p facilities such that the total demand-weighted distance between demand nodes and the facilities to which they are assigned is maximized. This model may be formulated as follows:

$$\text{Maximize} \quad \sum_{i \in I} \sum_{j \in J} h_i d_{ij} y_{ij} \tag{3.39}$$

subject to:

$$\sum_{j \in J} x_j = p \tag{3.40}$$

$$\sum_{j \in J} y_{ij} = 1 \quad \forall i \in I \tag{3.41}$$

$$y_{ij} - x_j \leq 0 \quad \forall i \in I, j \in J \tag{3.42}$$

$$\sum_{k=1}^{m} y_{i[k]_i} - x_{[m]_i} \geq 0 \; \forall i \in I, m = 1, \ldots, N-1 \tag{3.43}$$

$$x_j \in \{0,1\} \quad \forall j \in J \tag{3.44}$$

$$y_{ij} \in \{0,1\} \quad \forall i \in I, j \in J \tag{3.45}$$

This formulation is identical to that of the p-median problem with two notable exceptions. First, the objective (3.39) is to maximize the demand-weighted total distance and not to minimize it. The unfortunate impact of this objective is that it forces demands to be assigned to the most remote facility. Thus, the formulation has been extended with constraint (3.43), which ensures that demands are assigned to the nearest facility. In this constraint, $[k]_i$ is the index of the kth farthest candidate location from demand node i.

Constraint (3.43) then states that if the mth closest facility to demand node i is opened then demand node i must be assigned to that facility or to a closer facility.

3.3 Location-Routing Models

The basic models discussed above assume that the demand is served directly from a facility. This is valid for many siting scenarios such as those involving direct shipments from factories to distribution centers or emergency medical services. However, many facilities such as solid waste collection substations and distribution centers provide collection and/or distribution functions in which demand is served by multiple drop off and/or pickup routes. In such cases, the overall effectiveness of the facility siting scheme depends not only upon the distances from the individual demands but also upon the efficiency of the vehicle routes needed to serve multiple demands. Such problems are referred to as location-routing problems. It is well established that modeling distribution cost as the cost of a simple round trip from a facility to a customer may significantly misrepresent the actual costs and may, as a consequence, result in the selection of sub-optimal facility sites when multi-stop tours are used. (Webb, 1968, Eilon, Watson-Gandy and Christofides, 1971, and Perl and Daskin, 1985) As Perl and Daskin pointed out, location-routing problems involve three inter-related, fundamental decisions: where to locate the facilities, how to *allocate* customers to facilities, and how to *route* the vehicles to serve customers.

Many variations of location-routing problems exist. For example, Laporte, Nobert, and Taillefer (1988) consider three variants of location-routing problems, including (1) capacity constrained vehicle routing problems, (2) cost constrained vehicle routing problems, and (3) cost constrained location-routing problems. The authors examine multi-depot, asymmetrical problems and develop an optimal solution procedure that enables them to solve problems with up to 80 nodes. Berger (1997) formulates a location routing problem for perishable commodities as a variant of a fixed charge facility location problem. Her model is applicable in cases in which the routes are constrained to be short (since the commodity is perishable) and the vehicle does not have to return to the original depot within a time window. Current and Schilling (1994) formulate the median and maximal covering tour problems. These problems determine which nodes should have facilities to serve demand as well as the route (a cycle) connecting the nodes with facilities. The interested reader is referred to Laporte (1988) for a discussion of application, formulations and solution approaches to a wide variety of location-routing problems.

3.4 Facility Location-Network Design Models

It is obvious that the underlying network is important in facility location and location-routing problems. The underlying network was assumed to be given in the models presented in the previous two sections. However, in many location problems, one must determine which arcs should be included in the network as well as where facilities should be located. Examples of such problems include the design of subway or rail systems, electricity distribution systems, and computer networks. The facilities in such systems are stations, transformers, or concentrators and the network arcs are rail lines, power lines, or fiber optic cable. One objective in these problems is to minimize total cost, which includes facility and arc costs.

Current (1988) and Current and Pirkul (1991) develop a model and solution procedure for problems where the desired network is a path and the facilities are "entry locations" (e.g., station) for demand to enter the path (e.g., rail line). The objective is to minimize facility and network costs as well as the cost of arcs needed for demand to reach a facility on the path. The models in Current and Schilling (1989) can be used in a similar fashion when the network design requires a cycle.

Many hub location problems also fall into this category. For example, in studying airline hub and spoke networks, we must simultaneously determine the location of the hubs, the assignment of non-hub airports to the hubs (i.e., the links to be used in the spoke part of the network), and the connectivity of the hubs. Similar problems arise in telecommunication, power transmission, and computer networks. This is especially true when they address problems where the spoke nodes are not necessarily assigned to their nearest hub or when the hub network is not a complete graph. Melkote (1996) proposed a number of novel formulations of the network design/facility location problem and outlined heuristic and optimal solution procedures. In particular, he extended a number of coverage-based models to include network design decisions. Such formulations are particularly appropriate for decisions in developing regions.

3.5 Multiobjective Models

Facility location decisions are often strategic in nature as they frequently involve large capital outlays and long-term planning horizons. In general, they are the least flexible component of a firm's supply chain or a government's provisions of services. Factories, distribution centers, libraries and sewage treatment plants have expected life times of 20 to 50 years. They impact not only the providers of the facility but also their users and neighbors. They impact the human resources, finance, accounting, marketing, production, and logistics functions of a firm. Consequently, facility location decisions often involve many stake holders and must consider multiple, often conflicting, objectives.

There are essentially two approaches to multiobjective problems: generating techniques and preference-based techniques (Cohon, 1978). Preference-based techniques employ some method to "rank" the objectives and then find the solution that optimizes this ranking. Ranking may be done by a variety of means from a simple ordinal weighting of the objectives to a complex utility function. In general, generating techniques identify the pareto optimal siting configurations from which the decision makers select the one that they prefer.

The importance of distance and the ways it can be addressed in facility location objectives is demonstrated by the development of models that include multiple distance-related objectives. For example, Halpern (1976) included a maximum distance (center) and an average distance objective (median) for locating a single facility. Schilling, et al. (1980) included several different maximum covering objectives for fire equipment location. Church, et al. (1991) included a maximum distance (covering) objective and an average distance (median) objective.

Other researchers have included new objectives in facility location models. Many of these have appeared in models to locate obnoxious or undesirable facilities such as waste disposal sites, prisons, and power plants (e.g., Ratick and White, 1988; Erkut and Newman, 1989, 1992; and Erkut and Verter, 1995). These models often consider objectives related to risk and equity as well as efficiency.

Multiple objectives have also been considered in location-routing and location-network design models. For example, List et al. (1991) and Current and Ratick (1995) formulated hazardous material facility location-routing models which include objectives related to cost, risk, and the equity of risk imposed resulting from the facilities and the the routing of hazardous materials to them. Current et al (1985) formulated location-network design problem in which one objective minimizes the cost of a path network and the second mazimizes the demand covered by facilities at the nodes of the path. Current, et al. (1987) replaced the second objective with one to minimize access cost for demand to reach a node on the path.

The models mentioned so far in this section are designed primarily to generate pareto optimal solutions to multiobjective facility location problems. That is, they are generating techniques. If preference weights on the objectives are known *a priori*, these models can be used to find the solution which optimizes a convex combination of the objectives based on these weights.

Another preference-based method employs a lexicographical ordering of the objectives. This approach optimizes a primary objective and then, from among the alternate optima for the primary objective, optimizes a secondary (or tertiary) objective. Daskin and Stern (1981) proposed a hierarchical objective approach to locating emergency medical services. Their primary objective is that of the set covering model, to which they append a secondary objective of maximizing a measure of backup coverage. Plane and Hendrick (1977) proposed a hierarchical objective location problem for locating fire stations in Denver. Their primary objective is also to minimize the number of required facilities, while their secondary objective is to maximize the number of existing facilities in the solution. Benedict (1983) outlines a number of other hierarchical objective covering models.

For reviews of multiobjective facility location problems, the reader is referred to Current, et al. (1990), and Erkut and Verter (1995).

3.6 Dynamic Location Models

Since the pioneering work of Manne (1961, 1967), researchers have been interested in dynamic location problems. As Ballou (1968) states: "the effect of the future time dimension cannot be neglected in location analysis." (p. 271) The basic models presented in section 3.2.3 have ignored time; that is, they are static. Dynamic models incorporate time. Current et al. (1998) define two categories of dynamic models: "implicitly" dynamic and "explicitly" dynamic. Implicitly dynamic models are "static" in the sense that all of the facilities are to be opened at one time and remain open over the planning horizon. They are dynamic because they recognize that problem parameters (e.g., demand) may vary over time and attempt to account for these changes in the facility location scheme generated. Examples of implicitly dynamic models include Mirchandani and Odoni (1979), Weaver and Church (1983), Drezner and Wesolowsky (1991), and Drezner (1995), which consider problems where demand and travel times change over time.

Explicitly dynamic models are those designed for problems where the facilities will be opened (and possibly closed) over time. Typically, explicitly dynamic models extend the basic, static models with the addition of temporal subscripts to the facility location and assignment variables and constraints linking these variables over time. Early examples of such problems include Roodman and Schwarz (1975), Wesolowsky and Truscott (1976), Schilling (1980), Van Roy, Erlenkotter (1982), and Campbell (1990). The decision to open and close facilities over time is related to changes in the problem parameters over time. Examples of parameters that might change include demand, travel time/cost, facility availability, fixed and variable costs, profit, and the number of facilities to be opened. Multiobjective approaches to dynamic problems also have been developed (e.g., Schilling, 1980; Gunawardane 1982, and Min 1988).

The interested reader is referred to Current et al. (1998), and Owen and Daskin (1998a) for reviews of dynamic location problems.

3.7 Stochastic Location Models

The basic facility location models presented in section 2 assume that the parameters of the problem are known with certainty. Many of the dynamic models discussed in the previous section assume that the changes over time are known with certainty. However, there is considerable uncertainty in most facility location problems. This is particularly true given the long life spans of most facilities. Demand, travel time, facility costs, and even distance may change. These changes are often random. Uncertain parameters which have been addressed in the literature include demand (e.g., Frank, 1966; Manne, 1961; Carbonne, 1974; Berman, 1985), travel time (e.g., Mirchandani and Odoni, 1979; Mirchandani, 1980; Berman and Odoni, 1982; Weaver and Church, 1983; Berman and LeBlanc, 1984; and Louveaux, 1986) the availability of the facility for service (e.g., Daskin, 1982, 1983; ReVelle and Hogan, 1989a, b; Marianov and ReVelle, 1992, 1996), and the number of facilities to be sited (Current, et al., 1998).

There have been four basic approaches to stochastic facility location problems. The first, approximates the uncertainty via a deterministic surrogate. For example, Bean, et al. (1992) formulated an equivalent deterministic problem by "replacing stochastic demand by its deterministic trend and discounting all costs by a new interest rate that is smaller than the original, in approximate proportion to the uncertainty in the demand." (p. S210) They cite seven articles since Manne (1967) that have used a similar approach.

The second approach develops chance constrained models (Chapman and White, 1974). For example, Daskin (1982, 1983) formulated a probabilistic extension of the maximal covering problem in which facilities are assumed to be busy with probability ρ. If busy, they cannot serve demand. The objective of these models is to maximize the number of demands that are covered by an available (i.e., not busy) facility. ReVelle and Hogan (1989a) formulated a similar model in which they maximized the number of demands that are covered at least b times, where b is the number of coverages needed to ensure that a demand is covered by an available facility with probability β. ReVelle and Hogan (1989b) formulated a set covering version of the problem which minimizes the number of facilities required to ensure that all demands are covered with probability β. Other articles using this approach include Marianov and ReVelle (1992) which incorporated multiple vehicle types housed at the facilities and Marianov and ReVelle (1996) which incorporated a M/G/s-loss queuing model to compute the minimum number of facilities needed to ensure coverage of a node within some minimum probability α. Daskin, Hogan and ReVelle (1988) summarize these and other related models.

The third approach explicitly accounts for the queuing interactions that occur in a spatially distributed queuing system with facilities at multiple locations in a network. Larson (1974) formulated a hypercube queuing model with 2^N states to account for all possible combinations of N facilities being available or unavailable. The resulting model has either 2^N linear equations or N non-linear equations in an approximate model (Larson, 1975), thus making it very difficult to embed in an optimization algorithm. Batta, Dolan and Krishnamurthy (1989), however, used the model to show that the implicit assumption of server independence in Daskin's expected covering model is often violated. Berman, Larson and Chiu (1985) used an M/G/1 queuing model to explore the location of a single facility on a network as a function of the demand intensity when demands could wait for service. At very low and very high demands, they showed that the facility would be located at the 1-median location; for intermediate demand intensities, alternate locations were shown to be optimal. When no queuing was permitted, they showed that the optimal location was always the 1-median.

The fourth approach utilizes scenario planning (van der Heijden, 1994; Vanston et al., 1977). Scenarios represent possible values for parameters that may vary over the planning horizon. One of the first applications of scenario planning to facility location problems was Sheppard (1974) which minimized the expected cost over all scenarios. Ghosh and McLafferty (1982) used scenarios to locate retail stores. Schilling (1982) extended the maximal covering location problem to incorporate scenarios by maximizing the number of covered demands over all possible scenarios in an EMS siting situation. In this model some facilities must be common to all scenarios, while others can be located in a scenario-specific manner. Daskin, Hopp and Medina (1992) demonstrated that this approach can lead to the selection of the worst possible sites under certain conditions. Serra and Marianov (1997) used scenarios to incorporate varying travel times and demand conditions over the course of a day. One of their objectives was to find locations that minimize the maximum average travel time over seven defined scenarios as well as locations that minimize the maximum regret over the scenarios. Carson and Batta (1990) used scenarios to describe demand conditions at different times of the day in an ambulance location problem. Jornsten and Bjorndal (1994) formulated an uncapacitated dynamic fixed charge facility location model using scenario planning to minimize the expected cost across all scenarios and time periods.

In recent years there have been several facility location articles which minimize regret or expected opportunity loss in scenario planning. Generally, regret in these articles is defined as the difference between the optimal solution (once the future is known) and the siting configuration selected when the future is not known. For example, Serra et al. (1996) incorporated a minimax regret approach in which demands vary over the scenarios. Serra and Marianov (1997) included a minimax regret objective for average travel times. Current, et al. (1998) considered the problem where demand, travel cost, and the number of facilities sited may vary over the different scenarios. Averbakh (1997) and Averbakh and Berman (1997a, 1997b, 1997c) have focused on the development of polynomial time algorithms for specially structured instances of these problems.

Daskin, Hesse and ReVelle (1987) extend the minimax regret approach by allowing the analyst or decision-maker to specify a reliability level, α. The model then endogenously picks a subset of the scenarios whose total probability of occurrence exceeds α and whose maximum regret is as small as possible. In other words, the model minimizes the maximum regret over an endogenously determined subset of the scenarios whose total probability is at least α. Clearly, if $\alpha = 1$, the model reduces to the standard minimax regret model. Owen (1998) formulates a variety of related models that do not necessitate the specification of scenario probabilities. These models can be solved effectively using evolutionary algorithms (Owen and Daskin, 1998b).

3.8 Solution Approaches for Location Models

As we have seen from the previous sections, discrete location models are generally constructed as mixed-integer linear programs. However, formulating an appropriate model is only one step in analyzing a location problem. Another (and often larger) challenge is identifying the optimal solution. Typically, the first approach to finding the optimal solution to such problems is to apply one of the well-known algorithms such as branch and bound or cutting planes. While these methods work on at least some instances of most location models, they are typically only useful on small problems. Realistically scaled location models can easily have thousands even hundreds of thousands of constraints and variables. Attempting a solution with these standard optimization methods will quite often consume unacceptable computational resources in terms of both computer memory and time and with no guarantee of success. The reason is that even the most basic location models are classified as NP-Hard (Garey and Johnson, 1979).

As a result, the location analyst must devise other methods to identify optimal solutions and, failing that, at least find very good solutions. A method of the latter type is known as a heuristic, which is an algorithm, that can find good solutions to a decision problem, but will not guarantee finding the optimal solution. In the remainder of this section, we will explore several of the most common solution approaches used by location analysts. Throughout, we will use the p-median model to describe the solution process. However, the methods discussed are generally applicable to a wide variety of location models.

Greedy heuristics When faced with selecting a subset of things (in the case of the p-median, facilities to open) that will optimize some objective, there are numerous tactics or "rules-of-thumb" that quickly suggest themselves. The most common is a sequential approach that begins by evaluating each site individually and selecting the one facility site that yields the greatest impact on the objective. That facility site is then fixed open. The location of the next facility is then identified by enumerating all remaining possible locations and choosing the site that provides the greatest improvement in the objective. Each subsequent facility is located in an identical manner. The method stops when the required number of facilities, p, have been sited.

For obvious reasons, this approach is known as a greedy heuristic. More specifically it is called *Greedy-Add* since there is a reverse approach known as *Greedy-Drop*. Greedy-Drop starts with facilities located at all potential sites, and then removes (drops) the facility that has the least impact on the objective function. We continue to drop facilities until p facilities remain.

Improvement heuristics While both the Greedy-Add and the Greedy-Drop heuristics are effective at identifying a feasible solution with modest computational effort, neither can be relied upon to consistently produce good solutions. Therefore, several different approaches have been developed that begin with a good (or at least feasible) solution and seek to improve upon it. Not surprisingly, these are known as improvement or search heuristics.

One of the earliest improvement heuristics is the neighborhood search algorithm (Maranzana, 1964). In this method, we begin with any feasible solution or specifically a set of p facility sites. Demand nodes are then assigned to their nearest facility. The set of nodes assigned to a facility constitutes a "neighborhood" around that facility. Within each neighborhood, the 1-median problem can be solved optimally by simply evaluating each potential site in the neighborhood and selecting the best. The facilities are then relocated to the optimal 1-median locations within each neighborhood. Then, if any facility sites are relocated, new neighborhoods can be defined and the algorithm is repeated. This cycle continues until there are no further changes in the facility sites or neighborhoods.

The most widely known improvement method was introduced by Teitz and Bart (1968). The basic idea is to move a facility from the location it occupies in the current solution to an unused site. Each unused location is tried in turn and when a move produces a better objective function value, then that relocation is accepted and we have a new (improved) solution. When an improved solution is obtained, the search process is repeated on the new solution. The procedure stops when no better solution can be found via this method. This heuristic is known as an "interchange", "exchange" or "substitution" approach, because it can be thought of as exchanging an open site with one of the unused sites. Although commonly used as an algorithm for the p-median problem, this approach has been found useful in innumerable facility location models.

While seemingly straightforward in concept, the exchange heuristic has a number of alternative approaches that can be used in implementing it. One, of course, is the process described above, where every time an exchange is found that yields a better solution, the search process is restarted and applied to improve this new solution. Alternatively, we could select the best solution after considering all possible moves for a given facility site, or even choose the best after all possible exchanges for all sites are examined. There are many other variations possible, and these often influence the computational speed of the heuristic. The most efficient implementation of the exchange algorithm was presented by Whitaker (1983). His "Fast Interchange" method is described in detail in Hansen and Mladenovic (1997).

One issue in using improvement heuristics is to decide how the initial solution is generated. An obvious choice is to use the result of another heuristic, such as one of the greedy heuristics mentioned earlier. However, since the interchange heuristic is relatively fast, many analysts have applied it to a series of randomly generated solutions, selecting the best solution among all of the local optima found as the one to be implemented.

The variety of local optima identified in repeated applications of an interchange heuristic often have an interesting characteristic the solutions are quite similar in the set of sites chosen for open facilities. Rosing and ReVelle (1997) used this characteristic in developing their Heuristic Concentration approach. In this methodology, the selected sites from a number of locally optimal solutions are used as a reduced set of potential facility sites. This reduced set produces a smaller model that can then be to be solved more efficiently using conventional optimization technology. In most search heuristics, each iteration of the search focuses on a "neighborhood" of the solution it is trying to improve (the current set of open facilities). Rosing and ReVelle, however, treat a neighborhood not as the set of nodes assigned to a facility, but rather as the set of solutions that can be examined around the current solution. For the interchange heuristic, the neighborhood is defined as that set of solutions that can be reached by a single exchange.

In a strategy designed to escape local optima, Hansen and Mladenovic (1997) present a "variable neighborhood" search algorithm for solving the p-median problem. In their approach, they extend the notion of neighborhood with a distance measure. That is, the neighborhood at a distance of k from a current solution is the set of solutions that can be obtained by moving k facilities. The algorithm performs an intensive local search (similar to the interchange algorithm outlined above) on the current solution until it settles in a local optima. It then diversifies the search by randomly selecting a solution from a neighborhood at a distance of k from the current best solution. The process continues, incrementing k, until some exogenously specified maximum value of k is attained. The algorithm compares very well with conventional heuristics as well as the enhancements provided by tabu search.

The problem with many search heuristics is that, instead of yielding the sought-after optimal solution, they become 'stuck' in local optima. More recently, researchers have sought to apply heuristics in a more thoughtful or "intelligent" manner. The strategy is to use what is known as a metaheuristic to guide the application of a core search heuristic. The intent is to help them break out of local optima and explore other regions of the solution space.

One of the earliest metaheuristics is Tabu search (Glover, 1989, 1990; Glover and Laguna, 1997; Hansen, 1986). The basic procedure employs "tabu" restrictions, which inhibit certain moves (exchanges). The tabu restrictions are generally implemented with a short-term memory function to make them time-dependent. That is, after a certain number of iterations, the moves are no longer tabu. There are also "aspiration criteria" which allow very good solutions to overcome any tabu status.

Designing tabu search heuristics involves defining what types of moves to restrict and the nature of the aspiration criteria and short-term memory to utilize. In addition to these features, most tabu search designs include other strategies such as a long-term memory function to diversify the search into other areas of the solution space. Examples in the literature of tabu search applied to location problems include: Hub location (Klincewicz, 1992; Skorin-Kapov and Skorin-Kapov, 1994), the location-routing problem (Tazun and Burke, 1999), multicommodity location (Gendron and Potvin, 1999), and the p-median (Rolland, et al, 1997).

Lagrangean relaxation When using any heuristic, we are trading off savings in solution time against the quality of the solution. While the heuristics discussed above often find good solutions to a variety of location problems and do so relatively quickly, it is difficult to evaluate the tradeoff since we have no way of knowing how far from optimality those solutions are. Without having the optimal value of the objective function available for comparison, we can sometimes approximate the difference between a heuristic's solution and the optimal solution by finding bounds. Worst-case bounds are one type that has been the focus of considerable research. Such bounds are the greatest distance from optimality the heuristic solution will be when solving the theoretically worst of all possible problem instances. Unfortunately, while they are theoretically precise, these bounds are often quite far from optimality (thus providing little insight). Fortunately, the average performance of heuristics is often far better than these worst-case bounds would indicate.

One of the primary attractions of the technique known as Lagrangean relaxation is that it provides both upper and lower bounds on the value of the objective function (Fisher, 1981). That is, we know the optimal objective function value lies between the value of the best feasible solution found and a value that it can be no better than. The difference between the bounds is known as the "gap."

Lagrangean relaxation replaces the original problem with an associated Lagrangean problem whose optimal solution will provide a bound on the objective function of the original problem. This is done by eliminating (i.e. relaxing) one or more of the constraints of the original model and adding these constraints, multiplied by an associated Lagrange multiplier, to the objective function. The idea is to relax constraints that will result in a relaxed problem that, given values of the multipliers, is much easier to solve optimally. The role of these multipliers is to drive the Lagrangean problem toward a solution that satisfies the relaxed constraints. Unfortunately, determining the values of the optimal Lagrangean multipliers is generally very difficult. In essence, the Lagrangean relaxation approach replaces the problem of identifying the optimal values of all of the decision variables with one of finding optimal or good values for the Lagrange multipliers. Most Lagrangean-based heuristics use a search heuristic to identify the optimal multipliers. The most common routine will be discussed below.

A major benefit of Lagrangean-based heuristics is that they generate bounds (i.e. lower bounds on minimization problems and upper bounds on maximization problems) on the value of the optimal solution of the original problem. For discussion purposes, we will limit ourselves to minimization problems like the four basic models presented in section 3.2. For any set of values for the Lagrangean multipliers, the solution to the Lagrangean model is less than or equal to the solution to the original model. Therefore, the Lagrangean solution is a lower bound on the solution to the original problem. However, all lower bounds are not created equal; the higher the bound, the smaller the gap in which the true optimal solution can be found.

The solution to the Lagrangean problem for any given values of the Lagrange multipliers will generally violate one or more of the relaxed constraints. Many Lagrangean based algorithms incorporate additional heuristics to convert these infeasible solutions to feasible ones. In this way, they can produce good solutions to the original model. The best feasible solution among those found by the procedure at any point, represents the upper bound on the value of the true optimal solution. The difference between the upper and lower bounds is referred to as the "gap." If the gap reaches zero (or some minimum value based on the integer properties of the model) then we have found the optimal solution. Otherwise, when the gap gets sufficiently small (e.g. less than 1%), the analyst may stop the procedure and be satisfied that the current best solution is within 1% of optimality.

The primary challenge in applying Lagrangean relaxation is in selecting which constraints to relax. The goal is to end up with a relaxed problem that can be solved very easily, and result in good lower bounds. Since the relaxed model may have to be solved hundreds or thousands of times in the search for the best multiplier values, the ease of solution is critical to the success of the approach. Ideally, the relaxed problem ought to be solvable by inspection or by a simple sorting the objective function coefficients.

An excellent tutorial on the general application of Lagrangean relaxation can be found in Fisher (1985). An exposition of its use in location models is in the text by Daskin (1995). To make our discussion of Lagrangean relaxation more concrete, consider its application to the p-median problem, which follows.

Recall the p-median formulation given above. Suppose we relaxed the constraints (3.22). When these constraints are multiplied by Lagrange multipliers, we obtain the following model:

$$\max_{\lambda} \min_{x,y} L = \sum_{i \in I} \sum_{j \in J} h_i d_{ij} y_{ij} + \sum_{i \in I} \lambda_i (1 - \sum_{j \in J} y_{ij})$$
$$= \sum_{i \in I} \sum_{j \in J} (h_i d_{ij} - \lambda_i) y_{ij} + \sum_{i \in I} \lambda_i \qquad (3.46)$$

subject to:

$$\sum_{j \in J} x_j = p \qquad (3.47)$$

$$y_{ij} - x_j \leq 0 \qquad \forall i \in I, j \in J \ (3.48)$$

$$x_j \in \{0, 1\} \qquad \forall j \in J \qquad (3.49)$$

$$y_{ij} \in \{0, 1\} \qquad \forall i \in I, j \in J \ (3.50)$$

Note that the objective function L in (3.46) is minimized with respect to the original (location and assignment) variables (x_j and y_{ij}, respectively) and is maximized with respect to the Lagrange multipliers (λ_i). The largest value of L over all iterations of the procedure represents a lower bound on the objective function for the original p-median model.

As we discussed earlier, the overall approach is to, iteratively, 1) set values of the multipliers, 2) solve the Lagrangean model and 3) adjust the multipliers. To do this effectively, we must construct a procedure which efficiently solves for x_j and y_{ij} when we are given values for the multipliers. In (3.46), when the λ_i are known, the last term is a constant and can be ignored. The objective is thus seen to be a direct function of the y_{ij}, but, as in many location problems, if we have the values of the facility siting variables (x_j in this case), we can derive the values of the remaining variables. That is, when $x_j = 0$, the y_{ij} must $= 0$, and there is no impact on the objective. When $x_j = 1$, then the y_{ij} can be either $= 0$ or 1. To minimize the first term of the objective function we would like to set any $y_{ij} = 1$, if the associated $(h_i d_{ij} - \lambda_i)$ is < 0. The remaining y_{ij} can be 0. Thus, to see the overall influence on the objective function of setting a particular $x_j = 1$, we calculate V_j, which is defined as $\sum_i \min\{0, h_i, d_{ij} - \lambda_i\}$. If we rank order the V_j values from smallest to largest, we can identify the first p values and set the corresponding x_j variables $= 1$. Then $y_{ij} = 1$, if $x_j = 1$ and $(h_i d_{ij} - \lambda_i)$ is < 0. This procedure yields the minimum objective function value for a given set of λ_i.

The Lagrangean solution found above may not be feasible for the original p-median model since the constraint we relaxed may be violated (demand nodes i may not be allocated to only one facility.) We can, however, develop a feasible solution by simply assigning the demand points to their nearest open facility. The resulting value of the p-median objective function represents an upper bound on the optimal solution. The best of the feasible solutions found over all iterations would also have the best (lowest) upper bound.

The final task is to modify the multipliers based on the solutions just obtained. (See Bazaara and Goode, 1979, for a survey of various methods.)

A common approach is subgradient optimization. Briefly, after any iteration t, the Lagrangean multipliers are updated as follows:

$$\lambda_i^{t+1} = \max\left\{0, \lambda_i^t - T^t\left(\sum_j y_{ij}^t - 1\right)\right\} \quad \forall i$$

where

$$T^t = \Delta\left[\frac{\overline{Z} - Z_L^t}{\sum_i\left(\sum_j y_{ij}^t - 1\right)^2}\right]$$

t = an index of iterations = 1,2,3, ...
$\overline{Z}$ = best (smallest) feasible solution value
Z_L^t = Lagrangean value from current iteration (t)

Typically, Δ is initially set to 2. Then, if there has been no improvement in the lower bound over some pre-specified number of iterations (e.g. 200), then Δ is replaced with $\frac{\Delta}{2}$. All λ_i are sometimes then reset to the values used to obtain the current best lower bound. Since the procedure is not guaranteed to terminate at optimality, it is usually stopped after reaching a certain number of iterations (e.g. 1000) or when Δ becomes sufficiently small. If the gap is not sufficiently small at the end of the Lagrangean procedure, the entire process can be embedded in a standard branch and bound algorithm.

For a numerical example of this procedure, the interested reader is directed to Daskin (1995).

3.9 Conclusions

Discrete network location problems have attracted the attention of both researchers and practitioners for several decades due to (1) the practical value of such models in both private and public-sector decision-making contexts, (2) the ever-present need and desire to incorporate increasingly realistic constraints and objectives into the models, (3) the challenges associated with solving the models and (4) the ability of the basic formulations to represent important decision-making issues in contexts far removed from a facility location environment. These four factors continue to this day and are likely to be present for years to come. We anticipate continued research, development and application of the models we have outlined.

Several areas of potential research and development warrant particular attention. Since location decisions are inherently strategic and long-term in nature, they often entail balancing conflicting objectives held by multiple constituents. This suggests four important developmental needs. First, multi-objective modeling will become increasingly important. Since most of the single objective problems are NP-hard, their multi-objective extensions are also NP-hard. Thus, it will be important to develop efficient and effective heuristic algorithms for identifying non-inferior solutions for multi-objective location problems. Methods for evaluating the quality of the solutions attained by such algorithms will need to be developed as well.

Second, long-term location decisions impact and constrain the shorter-term tactical and operational decisions made after facilities are in place. Such decisions include production scheduling and planning, inventory management and vehicle routing. Thus, the development of models that integrate facility location decisions with approximations of such tactical and operational decisions will be another important area for future work. One promising approach to such integrated problems is stochastic programming (Birge and Louveaux, 1999).

Third, as the famous philosopher (and baseball player) Yogi Berra once observed, "It's tough to make predictions, especially about the future." Thus, it is impossible to predict the future conditions under which facilities will operate. Therefore, it is important that models and solution algorithms be developed that account for future uncertainty explicitly and that identify solutions that are robust with respect to this uncertainty (Kouvelis and Yu, 1996).

Finally, public and private sector facilities interact with other parts of the infrastructure (e.g., highways, airports, rail lines, ports, production facilities and equipment). Embedding facility location modeling approaches in more general network design algorithms will also be an important and challenging area for future work. For example, in locating production facilities, managers need to consider not only the location of the plants, but the equipment that will be housed in the plant (ReVelle and Laporte, 1996).

References

Abernathy, W.J. and J.C. Hershey (1972) "A Spatial-Allocation Model for Regional Health Services Planning," *Operations Research*, 20, 629-642.

Averbakh, I. (1997) "On the complexity of a class of robust location problems," Working Paper, Western Washington University.

Averbakh, I. and Berman, O. (1997a) "Algorithms for the robust 1-center problem. Working Paper," Western Washington University.

Averbakh, I. and Berman, O. (1997b) "Minimax regret p-center location on a network with demand uncertainty," *Location Science*, 5, 247-254.

Averbakh, I. and Berman, O. (1997c) "Minimax regret robust median location on a network under uncertainty," Working Paper. Western Washington University.

Balinski, M.L. (1965) "Integer Programming: Methods, Uses, Computation," *Management Science*, 12, 253-313.

Ballou, R. H. (1968) "Dynamic warehouse location analysis," *Journal of Marketing Research*, 5, 271-276.

Batta R., Dolan J. M. and Krishnamurthy, N. N. (1989) "The maximal expected covering location problem: Revisited," *Transportation Science*, 23, 277-287.

Bazaara, M.S. and Goode, J.J. (1979) "A Survey of Various Tactics for Generating Lagrangean Multipliers in the context of Lagrangean Duality," *European Journal of Operational Research*, 3, 322-338.

Bean, J. C., Higle, J. L. and Smith, R. L. (1992) "Capacity expansion under stochastic demands," *Operations Research*, 40, 210-216.

Belardo, S., Harrald, J., Wallace, W. A. and Ward, J. (1984) "A partial covering approach to siting response resources for major maritime oil spills," *Management Science*, 30, 1184-1196.

Bell, T. and R. Church (1985) "Location-Allocation Modeling in Archaeological Settlement Pattern Research: Some Preliminary Applications," *World Archaeology*, 16, 354-371.

Benedict, J. M. (1983) "Three hierarchical objective models which incorporate the concept of excess coverage to locate EMS vehicles or hospitals," M.S. thesis, Department of Civil Engineering, Northwestern University, Evanston, IL 60208.

Bennett, V.L., D.J. Eaton and R.L. Church (1982) "Selecting Sites for Rural Health Workers," *Social Science and Medicine*, 16, 63-72.

Berger, R. T. (1997) "Location-Routing Models for Distribution System Design," Ph.D. dissertation, Department of Industrial Engineering and Management Sciences, Northwestern University, Evanston, IL 60208.

Berman O., Larson, R. C. and Chiu, S. S. (1985) "Optimal server location on a network operating as an M/G/1 queue," *Operations Research*, 33, 746-771.

Berman, O. (1985) "Locating a facility on a congested network with random lengths," *Networks*, 15, 275-294.

Berman, O. and LeBlanc, B. (1984) "Location-relocation of mobile facilities on a stochastic network," *Transportation Science*, 18, 315-330.

Berman, O. and Odoni, A. R. (1982) "Locating mobile servers on a network with Markovian properties," *Networks*, 12, 73-86.

Birge, J. R. and F. Louveaux (1999) *Introduction to Stochastic Programming*, Springer Series in Operations Research, New York.

Birge, J.R. and V. Malyshko (1985) "Methods for a Network Design Problem in Solar Power Systems," *Computers and Operations Research*, 12, 125-138.

Bramel, J. and Simchi-Levi, D. (1995) "A location based heuristic for general routing problems," *Operations Research*, 43, 649-660.

Campbell, J. F. (1990) "Locating transportation terminals to serve an expanding demand," *Transportation Research*, 24B, 173-192.

Campbell, J. F. (1994) "Integer programming formulations of discrete hub location problems," *European Journal of Operational Research*, 72, 387-405.

Carbone, R. (1974) "Public facilities location under stochastic demand," *INFOR*, 12, 261-270.

Carson, Y. M. and Batta, R. (1990) "Locating an ambulance on the Amherst campus of the State University of New York at Buffalo," *Interfaces*, 20:5, 43-49.

Chapman, S. C. and White, J. A. (1974) "Probabilistic formulations of emergency service facilities location problems," Paper presented at the 1974 ORSA/TIMS Conference, San Juan, Puerto Rico.

Cho, J. and Sarrafzadeh, M. (1994) "The Pin Redistribution Problem in Multi-Chip Modules," *Mathematical Programming*, 63, 297-330.

Church, R. L. and Meadows, M. (1979) "Location modeling utilizing maximum service distance criteria," *Geographical Analysis*, 11, 358-373.

Church, R. L. and ReVelle, C.S. (1974) "The maximal covering location problem," *Papers of the Regional Science Association*, 32, 101-118.

Church, R.L., Current, J.R. and Storbeck, J.E. (1991), "A Bicriterion Maximal Covering Location Problem Which Considers the Satisfaction of Uncovered Demand," *Decision Sciences*, 22, 38-52.

Cohon, J. L. (1978) *Multiobjective Programming and Planning*, Academic Press, New York, NY.

Cohon, J.L, C.S. ReVelle, J.R. Current, T. Eagles, R. Eberhart and R.L. Church (1980) "Application of a Multiobjective Facility Location Model to Power Plant Siting in a Six-State Region of the U.S.," *Computers and Operations Research*, 7, 107-124.

Cornuejols, G., M. Fisher and G. Nemhauser (1977) "Location of Bank Accounts to Optimize Float: An Analytic Study of Exact and Approximate Algorithms," *Management Science*, 23, 789-810.

Current, J., Min, H. and Schilling, D. (1990) "Multiobjective analysis of facility location decisions," *European Journal of Operational Research*, 49, 295-307.

Current, J., Ratick, S. and ReVelle, C. (1998) "Dynamic facility location when the total number of facilities is uncertain: A decision analysis approach," Forthcoming in the *European Journal of Operational Research*.

Current, J., ReVelle, C. and Cohon, J. (1985) "The maximum-covering/shortest-path problem: A multiobjective network design and routing problem," *European Journal of Operational Research*, 21, 189-199.

Current, J.R. (1988) "The Design of a Hierarchical Transportation Network with Transshipment Facilities," *Transportation Science*, 22, 270-277.

Current, J.R. and D.A. Schilling (1989) "The Covering Salesman Problem," *Transportation Science*, 23, 208-212.

Current, J.R. and D.A. Schilling (1994) "The Median Tour and Maximal Covering Tour Problems: Formulation and Heuristics," *European Journal of Operations Research*, 73, 114-126.

Current, J.R., and H. Pirkul (1991) "The Hierarchical Network Design Problem with Transshipment Facilities," *European Journal of Operational Research*, 51, 338-347.

Current, J.R., and S. Ratick (1995) "A Model to Assess Risk, Equity, and Efficiency in Facility Location and Transportation of Hazardous Materials," *Location Science*, 3, 187-202.

Current, J.R., C.S. Revelle and J.L. Cohon (1987) "The Median Shortest Path Problem: A Multiobjective Approach to Analyze Cost Vs. Accessibility in the Design of Transportation Networks," *Transportation Science*, 21, 188-197.

Daskin, M. S. (1982) "Application of an expected covering model to emergency medical service system design," *Decision Sciences*, 13, 416-439.

Daskin, M. S. (1983) "A maximum expected covering location model: Formulation, properties and heuristic solution," *Transportation Science*, 17, 48-70.

Daskin, M. S. (1995) *Network and Discrete Location: Models, Algorithms and Applications*, John Wiley, New York.

Daskin, M. S. (2000) "A new approach to solving the vertex p-center problem to optimality: Algorithm and computational results," *Communications of the Japanese OR Society*, 9, 428-436.

Daskin, M. S. and Owen, S. H. (1998) "Two new location covering problems: The partial covering p-center problem and the partial set covering problem," *Geographical Analysis*, 31, 217-235.

Daskin, M. S. and Stern, E. (1981) "A hierarchical objective set covering model for emergency medical service vehicle deployment," *Transportation Science*, 15, 137-152.

Daskin, M. S., Hesse, S. M. and ReVelle, C. S. (1997) "α-reliable p-minimax regret: A new model for strategic facility location modeling," *Location Science*, 5, 227-246.

Daskin, M. S., Hogan, K. and ReVelle, C. (1988) "Integration of multiple, excess, backup, and expected covering models," *Environment and Planning B*, 15, 15-35.

Daskin, M. S., Jones, P. C. and Lowe, T. J. (1990) "Rationalizing tool selection in a flexible manufacturing system for sheet metal products," *Operations Research*, 38, 1104-1115.

Daskin, M. S., W. J. Hopp and B. Medina. (1992) "Forecast Horizons and Dynamic Facility Location," *Annals of Operations Research*, 40, 125-151.

Drezner, Z. (1988) "Location Strategies for Satellites' Orbits," *Naval Research Logistics*, 35, 503-512.

Drezner, Z. (1995) "Dynamic facility location: The progressive p-median problem," *Location Science*, 3, 1-7.

Drezner, Z. and Wesolowsky, G. O. (1991) "Facility location when demand is time dependent," *Naval Research Logistics*, 38, 763-777.

Eilon, S., Watson-Gandy, C. D. T. and Christofides, N. (1971) *Distribution Management: Mathematical Modelling and Practical Analysis*. Griffin, London.

Eiselt, H.A., (1992), "Location Modeling in Practice," *American Journal of Mathematical and Management Sciences*, 12, 3-18.

Erkut, E. and Neuman, S. (1989) "Analytical models for locating undesirable facilities," *European Journal of Operational Research*, 40, 275-291.

Erkut, E. and Neuman, S. (1992) "A multiobjective model for locating undesirable facilities," *Annals of Operational Research*, 40, 209-227.

Erkut, E. and Verter, V. (1995) "Hazardous materials logistics," *Facility Location: A Survey of Applications and Methods*, ed. Z. Drezner, pp. 467-506, Springer-Verlag, New York.

Ernst, A. T. and Krishnamoorthy, M. (1996) "Efficient algorithms for the uncapacitated single allocation p-hub median problem," *Location Science*, 4, 139-154.

Fisher, M. L. (1981) "The Lagrangian relaxation method for solving integer programming problems," *Management Science*, 27, 1-18.

Fisher, M. L. (1985) "An applications oriented guide to Lagrangian relaxation," *Interfaces*, 15:2, 2- 21.

Flynn, J. and Ratick, S. (1988) "A multiobjective hierarchical covering model for the essential air services program," *Transportation Science*, 22, 139-147.

Frank, H., (1966) "Optimum Locations on a Graph with Probabilistic Demands," *Operations Research*, 14, 409-421.

Galvão, R. D. and ReVelle, C. (1996) "A Lagrangean heuristic for the maximal covering location problem," *European Journal of Operational Research*, 88, 114-123.

Garey, M. R. and Johnson, D. S. (1979) *Computers and Intractability: A Guide to the Theory of NP-Completeness*, W. H. Freeman and Co., New York.

Gavish, B., Horsky, D., and Srikanth, K. (1983) "An Approach to the Optimal Positioning of a New Product," *Management Science*, 29, 1277-1297.

Gelders, L.F., L.M. Pintelon and L.N. VanWassenhove (1987) "A Location-Allocation Problem in a Large Belgian Brewery," *European Journal of Operation Research*, 28, 196-206.

Gendron, B. and J.-Y. Potvin (1999) "Tabu search with exact neighborhood evaluation for multicommodity location with balancing requirements," *INFOR*, 37, 255-269.

Ghosh, A. and C.S. Craig (1986) "An Approach to Determining Optimal Locations for New Services," *Journal of Marketing Research*, 23, 354-362.

Ghosh, A. and McLafferty, S. L. (1982) "Locating stores in uncertain environments: A scenario planning approach," *Journal of Retailing*, 58:4, 5-22.

Ginsberg, V., P. Pestieau and J.F. Thisse (1987) "A Spatial Model of Party Competition with Electoral and Ideological Objectives," In *Spatial Analysis and Location-Allocation Models*, edited by A. Ghosh and G. Rushton, 101-117, New York: Van Nostrand Reinhold Company.

Gleason J. (1975) "A Set Covering Approach to Bus Stop Location," *Omega*, 3, 605-608.

Glover, F. (1989) "Tabu Search Part I," *ORSA Journal on Computing*, 1, 190-206.

Glover, F. (1990) "Tabu Search Part II," *ORSA Journal on Computing*, 2, 4-32.

Glover, F. and M. Laguna (1997) *Tabu Search*, Kluwer Academic Publishers, Boston, MA

Goodchild, M.F. and P.J. Booth (1980) "Location and Allocation of Recreational Facilities: Public Swimming Pools in London, Ontario," *Ontario Geography*, 15, 35-51.

Gunawardane, G. (1982) "Dynamic versions of set covering type public facility location problems," *European Journal of Operational Research*, 10, 190-195.

Hakimi, S. (1964) "Optimum location of switching centers and the absolute centers and medians of a graph," *Operations Research*, 12, 450-459.

Hakimi, S. (1965) "Optimum location of switching centers in a communications network and some related graph theoretic problems," *Operations Research*, 13, 462-475.

Halpern, J. (1976) "The Location of Center-Median Convex Combination on an Undirected Tree," *Journal of Regional Science*, 16, 237-245.

Handler, G. Y. (1990) "p-center problems," *Discrete Location Theory*, eds. P. B. Mirchandani and R. L. Francis, pp. 305-347, John Wiley, New York.

Handler, G. Y. and Mirchandani, P. B. (1979) *Location on Networks*, M.I.T. Press, Cambridge, MA.

Hansen, P. (1986) "The steepest ascent mildest descent heuristic for combinatorial programming," Paper presented at the Congress On Numerical Methods In Combinatorial Programming, Capri, Italy.

Hansen, P. and Mladenovic, N. (1997) "Variable neighborhood search for the p-median," *Location Science*, 5, 207-226.

Helme, M.P. and T.L. Magnanti (1989) "Designing Satellite Communication Networks by Zero-One Quadratic Programming," *Networks* 19, 427-450.

Higgins, A., Kozan, E., and Ferreira, L. (1997) "Modeling the number and location of sidings on a single line railway," *Computers and Operations Research*, 24, 209-220.

Hilger, D.A., B.A. McCarl and J.W. Uhrig (1977) "Facilities Location: The Case of Grain Subterminals," *American Journal of Agricultural Economics*, 6, 674-684.

Hodgson, J.M., Rosing, K.E., and Zhang, J. (1996) "Locatiing Vehicle Inspection Stations to Protect a Transportation Network," *Geographical Analysis*, 28, 299-314.

Hodgson, J.M., R.T. Wong and J. Honsaker (1987) "The p-Centroid Problem on an Inclined Plane," *Operations Research*, 35, 221-233.

Hogan, K. (1990), "Reducing Errors in Rainfall Estimates Through Rain Gauge Location," *Geographical Analysis*, 22, 33-49.

Holmes, J., F.B. Williams and L.A. Brown (1972) "Facility Location Under a Maximum Travel Restriction: An Example Using Day Care Facilities," *Geographical Analysis*, 4, 258-266.

Jornsten, K. and Bjorndal, M. (1994) "Dynamic location under uncertainty," *Studies in Regional and Urban Planning*, 3, 163-184.

Kincaid, R. and Berger, R. (1993) "The Damper Placement Problem on Space Truss Structures," *Location Science*, 1, 219-234

Klincewicz, J.G. (1992) "Avoiding Local Optima in the p-Hub Location Problem Using Tabu Search and Grasp," *Annals of Operations Research*, 40, 121-132

Kouvelis, P. and Yu, G. (1996) *Robust Discrete Optimization and Its Applications*, Kluwer Academic Publishers, Boston, MA.

Kuby, M. (1987) "The p-dispersion and maximum dispersion problems," *Geographical Analysis*, 19, 315-329.

Kuby, M. J. and Gray, R. G. (1993) "The hub network design problem with stop feeders: The case of Federal Express," *Transportation Research*, 27A, 1-12.

Kuehn, A.A. and M.J. Hamburger (1963) "A Heuristic Program for Locating Warehouses," *Management Science*, 9, 643-666.

Laporte, G. (1988) "Location-routing problems," *Vehicle Routing: Methods and Studies*, eds. B. L. Golden and A. A. Assad, pp. 163-197, Elsevier Science Publishers, North Holland, Amsterdam.

Laporte, G., Nobert, Y. and Taillefer, S. (1988) "Solving a family of multi-depot vehicle routing and location-routing problems," *Transportation Science*, 22, 161-172.

Larson, R. C. (1974) "A hypercube queuing model for facility location and redistricting in urban emergency services," *Computers and Operations Research*, 1, 67-95.

Larson, R. C. (1975) "Approximating the performance of urban emergency service systems," *Operations Research*, 23, 845-868.

List, G. F., Mirchandani, P. B., Turnquist, M. A. and Zografos, K. G. (1991) "Modeling and analysis for hazardous materials transportation: Risk analysis, routing/scheduling and facility location," *Transportation Science*, 25, 100-114.

Louveaux, F. V. (1986) "Discrete stochastic location models," *Annals of Operations Research*, 6, 23-34.

Love, R.F., W.G. Truscott, and J.H. Walker (1985) "Terminal Location Problem: A Case Study Supporting the Status Quo," *Journal of the Operation Research Society*, 36, 131-136.

Manne, A. S. (1961) "Capacity expansion and probabilistic growth," *Econometrica* 29, 632-649.

Manne, A. S. (1967) *Investments for Capacity Expansion: Size, Location and Time Phasing*, MIT Press, Cambridge, MA.

Maranzana, F. E. (1964) "On the location of supply points to minimize transport costs," *Operational Research Quarterly*, 15, 261-270.

Marianov, V. and ReVelle, C. S. (1992) "A probabilistic fire-protection siting model with joint vehicle reliability requirements," *Papers in Regional Science: The Journal of the RSAI*, 71, 217- 241.

Marianov, V. and ReVelle, C. S. (1996) "The queueing maximal availability location problem: A model for siting of emergency vehicles," *European Journal of Operational Research*, 93, 12- 120.

Marks, D.H. and J.L. Liebman (1971) "Location Models: Solid Waste Collection Example," *Urban Planning and Development Division of ASCE Journal*, 97, 15-30.

Marsh, M. and Schilling, D.A. (1994) "Equity Measurement In Facility Location Analysis: Review and Framework," *European Journal of Operations Research*, 74, 1-17.

Maze, T. H., Khasnabis, S., Kapur, K. and Poola, M. S. (1981) "Proposed approach to determine the optimal number, size and location of bus garage additions," *Transportation Research*, Record 781, Washington, D.C.

Megiddo, N., Zemel, E. and Hakimi, S. L. (1983) "The maximal coverage location problem," *SIAM Journal of Algebraic and Discrete Methods*, 4, 253-261.

Melkote, S. (1996) *Integrated Models of Facility Location and Network Design*, Ph.D. dissertation, Department of Industrial Engineering and Management Sciences, Northwestern University, Evanston, IL 60208.

Min, H. (1987) "A Multiobjective Retail Service Location Model for Fast Food Restaurants," *OMEGA, The International Journal of Management Science*, 15, 429-441.

Min, H. (1988) "The dynamic expansion and relocation of capacitated public facilities: A multi- objective approach," *Computers and Operations Research*, 15, 243-252.

Mirchandani, P. and Odoni, A. (1979) "Location of medians on stochastic networks," *Transportation Science*, 13, 85-97.

Mirchandani, P. B. (1980) "Locational decisions on stochastic networks," *Geographical Analysis*, 12, 172-183.

O'Kelly, M. E. (1986a) "Activity levels at hub facilities in interacting networks," *Geographical Analysis*, 18, p343-356.

O'Kelly, M. E. (1986b) "The location of interacting hub facilities," *Transportation Science*, 20, 92- 106.

O'Kelly, M. E. (1987) "A quadratic integer program for the location of interacting hub facilities," *European Journal of Operational Research*, 32, 393-404.

O'Kelly, M. E. and Miller, H. M. (1994) "The hub network design problems: A review and synthesis," *The Journal of Transport Geography*, 2, 31-40.

Osleeb, J. and S. Ratick (1983) "A Mixed-Integer and Multiple Objective Programming Model to Analyze Coal Handling in New England," *European Journal of Operational Research*, 12, 302-313.

Owen, S. H. (1998) *Scenario Planning Approaches to Facility Location: Models and Solution Methods*, Ph.D. dissertation, Department of Industrial Engineering and Management Sciences, Northwestern University, Evanston, IL 60208.

Owen, S. H. and Daskin, M. S. (1998a) "Strategic facility location: A review," *European Journal of Operational Research*, 111, 423-447.

Owen, S. H. and Daskin, M. S. (1998b) "Strategic facility location via evolutionary programming," Working paper, Department of Industrial Engineering and Management Sciences, Northwestern University, Evanston, IL, 60208.

Patel, N. (1979) "Locating Rural Social Service Centers in India," *Management Science*, 25, 22-30.

Perl, J. and Daskin, M. S. (1985) "A warehouse location-routing model," *Transportation Research*, 19B, 381-396.

Pirkul, H. (1986) "An Integer Programming Model for the Allocation of Databases in a Distributed Computer System," *European Journal of Operational Research*, 26, 401-11.

Pirkul, H. (1987) "Efficient Algorithms for the Capacitated Concentrator Location Problem," *Computers and Operations Research*, 14, 197-208.

Pirkul, H., S. Narasimhan and P. De (1987) "Firm Expansion Through Franchising: A Model and Solution Procedures," *Decision Sciences*, 18, 631-645.

Plane, D. R. and Hendrick, T. E. (1977) "Mathematical programming and the location of fire companies for the Denver fire department," *Operations Research*, 25, 563-578.

Price, W.L. and M. Turcotte (1986) "Locating a Blood Bank," *Interfaces*, 16, 17-26.

Ratick, S. J. and White, A. L. (1988) "A risk-sharing model for locating noxious facilities," *Environment and Planning B*, 15, 165-179.

Reggia, J., D. Nau and P. Wang (1983) "Diagnostic Expert Systems Based on a Set Covering Model," *International Journal of Man-Machine Studies*, 19, 437-460.

ReVelle, C. S. and G. Laporte. (1996) "The Plant Location Problem: New Models and Research Prospects," *Operations Research*, 44, 864-874.

ReVelle, C. S. and Hogan, K. (1989a) "The maximum availability location problem," *Transportation Science*, 23, 192-200.

ReVelle, C. S. and Hogan, K. (1989b) "The maximum reliability location problem and α-reliable p-center problem: Derivatives of the probabilistic location set covering problem," *Annals of Operations Research*, 18, 155-174.

ReVelle, C., Cohon, J. and Shobrys, D. (1991) "Simultaneous siting and routing in the disposal of hazardous wastes," *Transportation Science*, 25, 138-145.

ReVelle, C., D. Bigman, D. Schilling, J. Cohon and R. Church (1977) "Facility Location: A Review of Context-Free and EMS Models," *Health Services Research*, 129-146.

Rolland, E., Schilling, D.A., Current, J.R. (1997) "An Efficient Tabu Search Heuristic for the p- Median Problem," *European Journal of Operational Research*, 96, 329-342.

Roodman, G. M. and Schwarz, L. B. (1975) "Optimal and heuristic facility phase-out strategies," *AIIE Transactions*, 7, 177-184.

Rosing, K.E. and ReVelle, C.S. (1997) "Heuristic Concentration: Two Stage Solution Construction," *European Journal of Operational Research*, 97, 75-86.

Saatcioglu, O. (1982) "Mathematical Programming Models for Airport Site Selection," *Transportation Research B*, 16B, 435-447.

Schilling, D. A. (1980) "Dynamic location modeling for public sector facilities: A multicriteria approach," *Decision Sciences*, 11, 714-724.

Schilling, D. A. (1982) "Strategic facility planning: The analysis of options," *Decision Sciences*, 13, 1-14.

Schilling, D. C. ReVelle, J. Cohon and D. Elzinga (1980) "Some Models for Fire Protection Locational Decisions," *European Journal of Operational Research*, 5, 1-7.

Serra, D. and Marianov, V. (1997) "The p-median problem in a changing network: The case of Barcelona," Submitted for publication to *Location Science*.

Serra, D., Rattick, S. and ReVelle, C. (1996) "The maximum capture problem with uncertainty," *Environment and Planning B*, 23, 49-59.

Sheppard, E. S. (1974) "A conceptual framework for dynamic location-allocation analysis," *Environment and Planning A*, 6, 547-564.

Skorin-Kapov, D. and Skorin-Kapov, J. (1994) "On tabu search for the location of interacting hub facilities," *European Journal of Operational Research*, 73, 502-509.

Swersey, A. and Thakur, L. (1995) "An Integer Programming Model for Locating Vehicle Emissions Testing Stations," *Management Science*, 41, 496-512.

Tazun, D. and L. I. Burke (1999) "A two-phase tabu search approach to the location-routing problem," *European Journal of Operational Research*, 116, 87-99.

Teitz, M. B. and Bart, P. (1968) "Heuristic methods for estimating generalized vertex median of a weighted graph," *Operations Research*, 16, 955-961.

Tewari, V.K. and J. Sidheswar, (1987) "High School Location Decision Making in Rural India and Location-Allocation Models," in: A. Ghosh and G., Rushton (eds.), *Spatial Analysis and Location-Allocation Models*, Van Nostrand Reinhold Co., New York 137-162.

Toregas C. and C. ReVelle (1972) "Optimal Location Under Time or Distance Constraints," *Papers of the Regional Science Association*, 28, 133-143.

Toregas, C., Swain, R., ReVelle, C. and Bergman, L. (1971) "The location of emergency service facilities," *Operations Research*, 19, 1363-1373.

Tryfos, P. (1986) "An Integer Programming Approach to the Apparel Sizing Problem," *Journal of the Operational Research Society*, 37, 1001-1006.

van der Heijden, K. (1994) "Probabilistic planning and scenario planning," *Subjective Probability*, eds. G. Wright and P. Ayton, pp. 549-572, Wiley, New York.

Van Oudheusden, D.L. and K.N. Singh (1988) "Dynamic Lot Sizing and Facility Location," *Engineering Costs and Production Economics*, 14, 267-273.

Van Roy, T. J. and Erlenkotter, D. (1982) "A dual-based procedure for dynamic facility location," *Management Science*, 28, 1091-1105.

Vanston, J. H., Frisbie, W. P., Lopreato, S. C. and Poston, D. L. (1977) "Alternate scenario planning," *Technological Forecasting and Social Change*, 10, 159-180.

Vasko, F., F. Wolf and K. Scott (1988) "Optimal Selection of Ingot Sizes Via Set Covering," *Operations Research*, 35, 346-353.

Vasko, F., F. Wolf and K. Scott (1989) "A Set Covering Approach to Metallurgical Grade Assignment," *European Journal of Operational Research*, 38, 27-34.

Watson, M. (1996) "A Standardization Analysis Process Applied to Steel Coils in the Automotive Industry," Ph.D. dissertation, Department of Industrial Engineering and Management Sciences, Northwestern University, Evanston, IL 60208.

Weaver, J. and Church, R. (1983) "Computational procedures for location problems on stochastic networks," *Transportation Science*, 17, 168-180.

Webb, M. H. J. (1968) "Cost functions in the location of depot for multiple-delivery journeys," *Operational Research Society*, 19, 311-320.

Wesolowsky, G. O. and Truscott W. G. (1975) "The multiperiod location-allocation problem with relocation of facilities," *Management Science*, 22, 57-65.

4 Location Problems in the Public Sector*

Vladimir Marianov[1] and Daniel Serra[2]

[1] Department of Electrical Engineering, Pontificia Universidad Catòlica de Chile, Santiago, Chille. e-mail: marianov@ing.puc.cl
[2] Department of Economics and Business, Universitat Pompeu Fabra, Trias Fargas 25-27, 08005 Barcelona, Spain. e-mail: daniel.serra@econ.upf.es

4.1 Introduction

The past four decades have witnessed an explosive growth in the field of network-based facility location modeling. As Krarup and Pruzan (1983) point out, this is not at all surprising since location policy is one of the most profitable areas of applied systems analysis and ample theoretical and applied challenges are offered. Location-allocation models seek the location of facilities and/or services (e.g., schools, hospitals, and warehouses) so as to optimize one or several objectives generally related to the efficiency of the system or to the allocation of resources. There are several ways of classifying network-based location models and problems. A good taxonomy of this type of problems can be found in Daskin (1995). The dichotomy between public versus private sector problems is a common way of classification.

This chapter concerns the location of facilities or services in discrete space or networks, that are related to the public sector, such as emergency services (ambulances, fire stations, and police units), school systems and postal facilities. This does not mean that these type of services necessarily belong strictly to the public sector, e.g., a medical emergency service may be owned by a private firm but regulated by a public health agency. So the question is, What is the main difference between the location of public facilities and private facilities? The answer lies in the nature of the objective or objectives that decision makers are considering. Public and private sector applications are different, because of the optimization criteria used in both cases. Profit maximization and capture of larger market shares from competitors are the main criteria in private applications, while social cost minimization, universality of service, efficiency and equity are the goals in the public sector. Since these objectives are difficult to measure, they are frequently surrogated by minimization of the locational and operational costs needed for full coverage by the service, or the search for maximal coverage given an amount of available resources. Note that although it is not usual, it is perfectly allowed for a public service planner to use some of the tools that are typical to private investors. For example, a public health service could compete with private

---

* This research has been partially funded by the Fundaciòn BBVA and MEC grant BEC2000-1027.

providers, and so reduce the subsidy needed from the state for maintaining the service (Marianov and Taborga, 2000).

An additional problem is that in public sector location models there is no one overriding objective, and a variety of responses may be given to a simple question on the "best" locational configuration of some service. For example, when locating ambulances we may be interested in siting them so as to minimize the weighted average response time of the system, or to cover the population at risk within a given time or distance. The first approach corresponds to what is known in location literature as a p-median problem, and the second one is a covering problem (Location Set Covering Problems, LSCP, or Maximal Covering Location Problems, MCLP). Most public facility located models use one of these approaches (or a combination of both) to set the foundations of the formulation at hand. In fact, both p-median and covering problems can be considered benchmarks in the development of location models, and as such, we will classify our examples as belonging to one of these two broad categories.

This chapter is structured as follows: In the next section we will focus on public facility location models that use some type of coverage criterion, with special emphasis in emergency services. The third section will examine models based on the p-median problem and some of the problems faced by planners when implementing this formulation in real world situations. Finally, the last section will examine new trends in public sector facility location problems.

4.2 Covering Models in the Public Sector

4.2.1 The Notion of Coverage

In this subsection we will refer to public sector applications of covering models. Being proximity (distance or travel time[1]) one of the fundamental aspects of location analysis, many models (as the p-median, analyzed in the next subsection) seek to minimize the distance or travel time between a customer and the facility at which she/he receives a service. As opposed to those models, covering models are based on the concept of acceptable proximity. In covering models, a maximum value is preset for either distance or travel time. If a service is provided by a facility located within this maximum, then the service is considered adequate[2] or acceptable; the service is equally good if provided by facilities at different distances, as long as both distances are smaller than this maximum value. Then, a customer is considered *covered* by the service,

---

[1] Most of the location models are related to geographical location. In this case, proximity refers to a distance or time metric. However, proximity can be defined also in other spaces; for example, two persons can have proximity in terms of similar opinions.

[2] Note that the distance requirement is one of the necessary conditions for an adequate service; it may not be sufficient to guarantee a good overall service.

or just covered, if she/he has a facility sited within the preset distance or time. An example of this is the case in which it is desired that the population in a rural area have access to a health care center within, say, 2 miles. It is said that a customer in this area is *covered* if she/he has a health center within 2 miles of her/his home. Another example appears when dealing with fire fighting services. The Insurance Services Office (ISO), is an organization which rates cities according to their fire protection capability (ISO, 1974). They establish distance standards for fire-fighting response. If the distance standards are not fulfilled in a city, the rating decreases, indicating that the risk of property loss is higher in those cities. Thus, it is reasonable to design fire fighting systems in such a way to assure attention of all calls within the time standard or, equivalently, to have an available server within a standard distance of each and every customer.

Covering models can be classified according to several criteria. One of such criteria is the type of objective, which allows us to distinguish two types of formulations. In the first place, those seeking to minimize the number of facilities needed for full coverage of the population (Set Covering Models) and secondly, those that maximize covered population, given a limited number of facilities or servers (Maximum Covering Models). Covering models can also be classified in formulations for systems with fixed servers and systems with mobile servers. Examples of the former are schools, hospitals, and other systems in which customers travel to the facility to receive service. Examples of systems with mobile servers are emergency services, in which servers are initially located at depots, and whenever a call is received, they travel to the location of the call and back to the depot. In turn, any of these can be classified as capacitated or uncapacitated, depending on the capacity limits of the facilities or servers to be sited. These capacity limits can be for example the number of children that a primary school can accept in a particular year, or the number of customers that can be attended by an ambulance system within a reasonable waiting time.

The notion of coverage can be extended in several ways. For example, a single policeman can not control alone some police emergencies. Coverage, then, must be defined as the response to the emergency by, say, p policemen. If fewer than p policemen attend the call, the emergency is not counted as covered. Similarly, the usual fire emergency puts people *and* property at risk. Then, engine fire companies and ladder fire companies are both needed at the scene of the fire, in order to protect property and people. Furthermore, different numbers of companies are needed in different cases. ISO defines standard response to a fire in medium size cities as response by three engine companies and two ladder companies. In this case, coverage is defined as attendance by three engine companies and two ladder companies, within their respective (response) time standard. Finally, coverage could mean *availability* of a service within certain time limits, as opposed to just *location* of a server within these time limits. For example, in an emergency service, a customer could be considered as covered if all her/his calls find an idle server with probability, say, 98%. Or, a customer might be considered as covered if, whenever she/he arrives at the health care center, there is a waiting line shorter than 5 people, with probability 95%. Note that, in this case, availability is somehow related to the capacity of the servers. Covering models have been used profusely in both private and public sectors (Schilling et al, 1993).

4.2.2 Basic Covering Models

There are two basic covering models. The first one is the Location Set Covering Model (LSCP), cast as a linear programming formulation by Toregas et al (1971), and Toregas and ReVelle (1973). This model seeks to locate a minimum number of servers needed to obtain mandatory coverage of all demands. In other words, each and every demand point has at least one server located within some distance or time standard S. The first application of this model was in the area of emergency services (ReVelle et al. 1976). In this context, the model positions the minimum possible number of emergency vehicles in such a way that the entire population has at least one of these vehicles initially located within the time or distance standard. Note that coverage is not affected by the fact that the servers (vehicles) may be busy at times. The formulation of the model is as follows:

$$\text{Minimize} \left\{ Z = \sum_{j \in J} x_j \right\} \quad (4.1)$$

Subject to:

$$\sum_{j \in N_i} x_j \geq 1 \quad \forall i \in I \quad (4.2)$$

$$x_j \in \{0, 1\} \quad \forall j \in J \quad (4.3)$$

where

J = set of eligible facility sites (indexed by j) ;
I = set of demand nodes (indexed by i) ;
$x_j = \begin{cases} 1 \text{ if a facility is located at node } j \\ 0 \text{ Otherwise} \end{cases}$
$N_i = \{j | d_{ji} \leq S\}$; with d_{ji} = shortest distance from potential facility location j to demand node i, and S = distance standard for coverage.

Note that N_i is the set of all those sites that are candidates for potential location of facilities, that are within distance S of the demand node i. If a facility is located in any of them, demand node i becomes covered. The objective (4.1) minimizes the number of facilities required. Constraints (4.2) state that the demand at each node i must be covered by at least one server located within the time or distance standard S.

The solution to this model can be easily found solving its linear programming relaxation, with occasional branch and bound applications. Before solving, its size can be reduced by successive row and column reductions, as proposed by Toregas and ReVelle (1973).

Church and ReVelle (1974) and White and Case (1974) formulated the second basic covering model, the Maximal Covering Location Problem (MCLP). Although public services should be available to everybody, as modeled by the LSCP, the MCLP recognizes that mandatory coverage of all people in all occasions and no matter how far they live, could require excessive resources. Thus, MCLP does not force coverage of all demand but, instead, seeks the location of a fixed number of facilities, most probably insufficient to cover all demand within the standards, in such a way that population or demand covered by the service is maximized. The fixed number of facilities is a proxy for a limited budget. Its integer programming formulation is the following:

$$\text{Maximize} \left\{ Z = \sum_{i \in I} a_i y_i \right\} \quad (4.4)$$

Subject to:

$$y_i \leq \sum_{j \in N_i} x_j \quad \forall i \in I \quad (4.5)$$

$$\sum_{j \in J} x_j = p \quad (4.6)$$

$$x_j, y_i \in \{0, 1\} \quad \forall j \in J, i \in I$$

where additional notation is

$p =$ the number of facilities to be deployed;
$a_i =$ the population at demand node i.

and all other variables and parameters are the same as defined for LSCP. The objective (4.4) maximizes the weighted sum of covered demand nodes. Constraints (4.5) state that the demand at node i is covered whenever at least one facility is located within the time or distance standard S. Constraint (4.6) gives the total number of facilities that can be sited. Church and ReVelle (1974) used relaxed linear programming, supplemented by occasional use of branch and bound, to provide solutions to this problem. Other solving procedures include Greedy or Myopic heuristics (Daskin, 1995, Schilling et al, 1993), Lagrangean Relaxation (Daskin, 1995, Galvão and ReVelle, 1996) and Heuristic Concentration (Rosing, 1997, Rosing and ReVelle, 1997).

Applications of these models in the public sector range from emergency services to location of archeological sites (Bell and Church, 1987). The set covering model has been used to allocate bus stops (Gleason, 1975). The maximal covering location model, and different variants of it, has been used for the location of health clinics (Eaton et al, 1981), hierarchical health services (Moore and ReVelle, 1982), and many other applications. The MCLP has also been used in several non-locational problems, as for example, the determination of test points in the human eye, to diagnose vision loss in glaucoma suspects (Kolesar, 1980), or for the allocation of marketing resources in journals (Dwyer and Evans, 1981).

4.2.3 Models for Mobile Emergency Services

In the case of most emergency systems, a fundamental issue is the amount of time a customer waits for service. This is the case of any public emergency services, either medical, fire fighting or police related. In the case of medical emergencies, there is a correlation between life loss risk and response time. Thus, it seems to be a good approach to assure medical attention of all calls within a time standard or, equivalently, have an available server within a standard distance of each and every customer. The same happens in the case of fire fighting services. Since it can be expected that loss of property increase with time, each type of company has to respond within its standard time. Police emergencies are not the exception. Again, the best model for these services is a covering model.

Many issues have to be considered in order to determine the performance of an emergency service. Response time is one of them. From the point of view of the geographical design of such a system, an important issue is the location of the depots, that is, the initial location of the emergency vehicles (servers). Another one is the number of servers. A third issue is the *availability* of servers, as opposed to just their initial location within time standard. Availability, in this case, is defined as the actual percentage of time the server is idle, as opposed to being on repair, or attending other calls. Finally, the dispatching policy has also an influence on the efficiency of the system.

Several approaches have been presented to attack the design of systems with mobile servers under congestion (or availability less than 100%). These can be classified in descriptive and prescriptive. Descriptive methods originate from a seminal paper by Larson (1974), in which a descriptive model (Hypercube) is presented for analysis of emergency systems. The hypercube model builds on previous developments by Carter, Chaiken and Ignall (1972) and Larson and Stevenson (1972) for two servers, and describes a spatially distributed queuing system with distinguishable servers. The model can be used, either in its complete version or in approximate versions (Larson, 1975), for testing the responsiveness of an emergency system and all its parameters. Many iterative methods have derived from the hypercube, as Jarvis' (1985) and Burwell, Jarvis and McKnew (1993) . Other descriptive models have been used for location of one mobile server, and can also be used in heuristics that locate multiple servers. Among them, the models and methods by Berman, Larson and Chiu (1985), Batta (1988), Batta, Larson and Odoni (1988), Batta (1989), Berman, Larson and Parkan (1987), Berman and Larson (1985) and Berman and Mandowsky (1986). The interested reader can refer to these papers or a review in Marianov and ReVelle (1995). However, we do not focus on these models, but rather on models derived from the basic coverage formulations.

Prescriptive models are based on optimization, and derive from the basic models outlined above. Good reviews of optimization models presented before 1990 are included in the articles by Daskin, et al (1988), and ReVelle (1989). We will include some of them, which are representative of different classes of formulations.

An interesting generalization of the maximal covering model, because it considers the simultaneous location of several types of facilities, is the FLEET (Facility Location and Equipment Emplacement Technique) by Schilling et al. (1979). This model was used for locating fire-fighting services in the city of Baltimore. The goal of this formulation is to locate simultaneously two different types of fire-fighting servers (pump or engine brigades and ladder or truck brigades), as well as the depots housing them. The objective of the FLEET model was coverage of the maximum number of people by both an engine company sited within an engine company distance standard *and* a truck company sited within the truck company distance standard. Other objectives included in a multi-objective formulation of the FLEET model were the maximum coverage of fire frequency, maximum coverage of property value and maximum coverage of population at risk.

The maximum population coverage version can be stated mathematically as

$$\text{Maximize} \quad \left\{ Z = \sum_{i \in I} a_i y_i \right\} \tag{4.7}$$

Subject to:

$$y_i \leq \sum_{j \in N_i} x_j^E \quad \forall i \in I \tag{4.8}$$

$$y_i \leq \sum_{j \in N_i} x_j^T \quad \forall i \in I \tag{4.9}$$

$$x_j^E \leq x_j^S \quad \forall j \in J \tag{4.10}$$

$$x_j^T \leq x_j^S \quad \forall j \in J \tag{4.11}$$

$$\sum_{j \in J} x_j^E + \sum_{j \in J} x_j^T = p^{E+T} \tag{4.12}$$

$$\sum_{j \in J} x_j^S = p^S \tag{4.13}$$

$$x_j^E, x_j^T, x_j^S, y_i \in \{0, 1\} \quad \forall j \in J, i \in I$$

where

$$x_j^E = \begin{cases} 1 \text{ if an engine company positioned in a fire house at site } j \\ 0 \text{ Otherwise} \end{cases}$$

$$x_j^T = \begin{cases} 1 \text{ if a truck company positioned in a fire house at site } j \\ 0 \text{ Otherwise} \end{cases}$$

$$x_j^S = \begin{cases} 1 \text{ if a fire station or depot is established at site } j \\ 0 \text{ Otherwise} \end{cases}$$

$N_i^E = \{j|t_{ji} \leq E\}$; set of potential engine sites j which can cover node i by virtue of being within the engine distance standard E

$N_i^T = \{j|t_{ji} \leq T\}$; set of potential truck sites j which can cover node i by virtue of being within the truck distance standard T

p^{E+T} = number of fire companies, and
p^S = number of fire stations or depots.

The first two constraints define coverage as achievable only if *both* one or more engine companies are sited within the engine distance standard *and* one or more truck companies are sited within the truck distance standard. The third and fourth constraints allow housing of companies only at nodes where a depot has been sited. The fifth constraint limits the total number of companies, and the sixth constraint limits the number of stations. Schilling et al (1979) solved the linear relaxation of the problem. If two or more servers of each type are needed, because the attendance of only one of each is not enough (as in police or fire emergencies), the model can be modified, as in Marianov and ReVelle, (1991) and (1992). For example, if three engine brigades are needed at the site of the emergency, the second constraint of the preceding model can be changed to:

$$y_i + w_i^E + u_i^E \leq \sum_{j \in N_i^E} \sum_{k=1}^{C_i} \quad \forall i \in I, \tag{4.14}$$

and the ordering constraints:

$$y_i \leq w_i^E \quad \forall i \in I \tag{4.15}$$
$$w_i^E \leq u_i^E \quad \forall i \in I \tag{4.16}$$

added. Here, $w_i^E, u_i^E = 1$ if demand node i is covered by a second engine, first engine, respectively, and 0 otherwise. Constraint (4.14) does not allow variables w_i^E, u_i^E and y_i to be one unless there are three engines located within standard time of node i. Constraints (4.15) and (4.16) force coverage by one engine before coverage of two engines and coverage by two before coverage by three.

Note that, as opposed to the models in which one server was enough for coverage, when multiple coverage is sought, co-location of servers at the same depots might become convenient. Models which limited the number of stations but allowed more than one server at the same site, up to a certain capacity, were suggested by Bianchi and Church, (1988), for one type of vehicle. Also, Marianov and ReVelle, (1991), propose a tighter set of constraints for siting up to C_j engines plus trucks in a depot:

$$x^E_{(C_j-k+1)j} + x^T_{kj} \leq x^S_j \quad \forall j \in J, k \leq C_j \qquad (4.17)$$

$$x^l_{(k+1)j} \leq x^l_{kj} \quad \forall j \in J, k \leq C_j - 1, l = E \text{ or } T \qquad (4.18)$$

where

$$x^E_{kj} = \begin{cases} 1 \text{ if a } k^{th} \text{ engine company is located at site } j \\ 0 \text{ Otherwise} \end{cases}$$

$$x^T_{kj} = \begin{cases} 1 \text{ if a } k^{th} \text{ truck company is located at site } j \\ 0 \text{ Otherwise} \end{cases}$$

These constraints, together, have three effects: first, they allow the siting of servers only at depots. Second, they limit to C_j the number of servers at each depot (no more than $\{C_j - k\}$ engines plus k trucks) at a site. Finally, they state that a $(k+1)^{th}$ server must be located at a site after the kth server.

In most cases, a server can not attend more than one call at a time. This means that, when a call arrives, a server can be busy attending other calls or on repair. This leads to congestion, which is the dynamic equivalent to a limited capacity. When there is the possibility of congestion, different approaches can be used. When congestion is not expected to be severe, there is an approach that does not need any analysis of the probabilistic characteristics of the system. This approach consists in seeking redundancy in the servers able to attend calls originating from a demand node. In other words, to allocate more than one server (say, two) to cover each demand within the standard time. The same construct used for attendance of more than one server to an emergency can be used when redundant coverage is needed, as did Hogan and ReVelle (1986) in their BACOP 2 model, which trades off first coverage versus backup coverage. Its formulation is

$$\text{Maximize} \left\{ Z_1 = \sum_{i \in I} a_i y_i \right\} \quad (4.19)$$

$$\left\{ Z_2 = \sum_{i \in I} a_i r_i \right\} \quad (4.20)$$

Subject to:

$$r_i + y_i \leq \sum_{j \in N_i} x_j \quad \forall i \in I \quad (4.21)$$

$$r_i \leq y_i \quad \forall i \in I \quad (4.22)$$

$$\sum_{j \in J} x_j = p \quad (4.23)$$

$$x_j, r_i, y_i \in \{0,1\} \quad \forall j \in J, i \in I$$

where r_i is one if a second coverer is sited within standard time of node i. The objectives maximize first and second coverage, respectively. The first constraint says that coverage by a first and second server is not possible unless at least two servers are initially located in the neighborhood. The second constraint reflects the fact that backup coverage can not be fulfilled without first coverage. The next constraint limits the number of servers to be deployed. The authors report that marginal reductions in first coverage improve strongly backup coverage. They solved the linear relaxation of the model, with occasional branch and bound.

Berlin (1972), Daskin and Stern (1981), Benedict (1983) and Eaton et al (1986) used a different approach, based on the LSCP for mandatory first coverage. They solve the LSCP first, and, using a number of servers that is at least the needed for full first coverage, maximized redundant coverage maintaining mandatory first coverage for all demands.

When congestion is expected to be more severe, a frankly probabilistic approach provides safer and more efficient system designs. In this case, a probabilistic modeling of the system is required, as well as the use of this probabilistic models in the developments of objectives or constraints of the optimization model.

Two approaches have been used for probabilistic models: The first consists in maximizing expected coverage of each demand node. The second, in either constraining the probability of at least one server being available (to each demand node) to be greater than or equal to a specified level α, or to count demand nodes as covered if this probability is at least α.

The maximization of expected coverage was proposed by Daskin (1983), who utilized the notion of a server *busy fraction*, or probability of being busy, to formulate the Maximum Expected Covering Location Problem, (MEX-CLP). Daskin assumed a single system-wide busy fraction (probability of a server being busy or fraction of the time during which it is busy) q, as well as independence between the probabilities of different servers being busy, which leads to a binomial distribution of the probability of k servers being busy. The MEXCLP maximized the expected value of population coverage within the time standard, given that p facilities are to be located on the network. Daskin computed the increase in the expected coverage of a demand, when a kth server is added to its neighborhood, which turns out to be just $(1-q)q^{k-1}$. Then, the expected coverage for all possible number of servers k at each neighborhood, and for all demand nodes weighted by their demand, is maximized:

$$\text{Maximize} \left\{ Z = \sum_{i \in I} \sum_{k=1}^{n_i} a_i (1-q) q^{k-1} y_{ik} \right\} \quad (4.24)$$

Subject to:

$$\sum_{k=1}^{n_i} y_{ik} \leq \sum_{j \in N_i} x_j \quad \forall i \in I \quad (4.25)$$

$$\sum_{j \in J} x_j = p \quad (4.26)$$

$$y_{ik} \in \{0,1\} \quad \forall i, k,$$

$$x_j \text{ integers} \quad \forall j \in J$$

where

$$y_{ik} = \begin{cases} 1 \text{ if node } i \text{ has at least } k \text{ servers in its neighborhood} \\ 0 \text{ Otherwise} \end{cases}$$

x_j = the number of servers at site j, and
n_i = the maximum number of servers in N_i.

The first constraint says that the number of servers covering demand i is bounded above by the number of servers sited in the neighborhood. The second constraint limits the number of servers to be deployed. Declining weights $(1-q)q^{k-1}$ on the variables y_{ik} make unnecessary any ordering constraints for these variables, and help to the integrality of these variables in the solution, if the linear relaxation of the model is solved. Daskin proposed a heuristic method of solution of the MEXCLP, which gives solutions for the system for different ranges of values of q. More details on this important model, can be found in Daskin (1995).

Later, Bianchi and Church (1988), modified MEXCLP to consider location of vehicles and depots as well, referring to their model as the Multiple cover, One-unit Facility Location, Equipment Emplacement Technique (MOFLEET). Recognizing the need for relaxing the assumption of independence between probabilities of servers being busy, Batta et al (1989) proposed a modified MEXCLP, in which the factors $(1-q)q^{k-1}$ are corrected by an approximation to a queuing system, based on the approximated hypercube of Larson (1975). In that model, called AMEXCLP, the busy fraction of servers is still assumed to be the same over the whole system. Another model, by Goldberg and Paz (1991), maximizes the expected number of calls reached within a set time threshold. The model is nonlinear, based on Jarvis' (1975) mean service time computation. The service time depends on call location, and independence is assumed between probabilities of servers being busy. The authors present a heuristic method of solution.

Instead of maximizing expected value of coverage, Chapman and White (1974) formulated a probabilistic version of the LSCP in which the probability that at least one server being available to each demand node was constrained to be greater than or equal to a reliability level α. To compute such a probability, they make use of estimates derived from simulations of the busy fraction q. Again, each server's busy fraction is assumed to be independent of the probability of other servers being busy. Chapman and White's model could not be solved to convergence because busy fractions of individual servers were difficult to estimate. Later, ReVelle and Hogan (1988, 1989a) formulated a new form of Probabilistic LSCP (PLSCP), basically a LSCP with an added constraint on the availability of servers to each demand node, which utilized region-specific (local) estimates of the busy fraction, and binomial distribution. Unfortunately, it is not possible to determine the busy fraction of the servers before knowing the final locations of all of them. To go around this problem, ReVelle and Hogan computed a local estimate of the busy fraction in the neighborhood of demand node i, as the demanded service time in the region, divided by the available service time in the region. Detailed description of the model are given in Chapter 10 in this book.

In the fire protection arena, ReVelle and Marianov (1991), formulated a comparable model, the Probabilistic Facility Location, Equipment Emplacement Technique, PROFLEET. This model considered the deployment of several types of vehicles, simultaneously covering each emergency, as well as the siting of depots or stations. The model considered independence between availabilities of engines and trucks, so Marianov and ReVelle (1992) presented a second version, in which availabilities of engines and trucks were no longer independent. Later, Marianov and ReVelle (1994) formulated a new version of MALP (the QMALP), in which the independence assumption is relaxed through a treatment of each neighborhood as a queuing system, keeping a neighborhood-specific busy fraction.

Meanwhile, Ball and Lin (1993), formulated a new version of the PLSCP, in which a desired level of reliability is mandatory for each demand, condition that is achieved by an upper bound of the "uncoverage probability" of each demand. In their model, Ball and Lin consider the worst case of busy fraction, which occurs when each server is attending all calls from its neighborhood, as if it was alone in the system. In their model, independence is assumed between probabilities of servers being busy.

4.2.4 Models for Fixed Services

Although the LSCP and MCLP models have been used more frequently for locating mobile servers, there are some exceptions. For example, Goodchild and Lee (1989), locate the minimum number of observation points for monitoring an entire geographical region. Meyer and Brill (1988), locate the least number of monitoring wells for detecting contamination in ground water. Bell and Church (1987), locate archeological settlements. Other fixed server application is the location of bus stops that minimize the walking distance for customers (Gleason, 1975). A review of these and other applications can be found in Schilling et al (1993). An interesting feature of the LSCP, is that it can be used for solving the p-center problem, which consists on finding the locations of p facilities in such a way as to minimize the maximum distance between a customer and its allocated facility (Daskin, 1995). This problem is adequate for its use in applications in the public sector, because it tends to generate certain equity in the access to facilities by their users.

Besides the applications, it is interesting to give some attention to the variations that have been made to the basic covering models. Set covering models can be formulated for covering arcs, as well as nodes. Also, they can be rewritten for situations in which the demand changes or is uncertain over time. Coverage models can be merged with other models, as Current and Schilling (1989) did with routing, in their covering salesman problem. Coverage, in its usual sense of proximity, can also be reversed, for example when locating obnoxious facilities, which should be located as far as possible from population. Capacity of the facilities is an important issue, and several authors have presented capacitated versions of covering models. Among them, Pirkul and Schilling (1991) propose a Lagrangian Relaxation method for the solution of a maximal covering problem with a capacity constraint.

In some cases, fixed facilities can also suffer from congestion. This is the case of health care services, including hospitals, and, in general, public services of any nature that have fixed offices serving users or customers. The latest developments in capacitated covering models for fixed servers are due to Marianov and Serra (1998 and 2001). In these papers, they develop several probabilistic maximal covering location-allocation models with constrained waiting time for queue length in order to consider service congestion. The first paper addresses the issue of the location of the least number of single-server centers such that all the population is served within a standard distance, and nobody stands in line for a time longer than a given time-limit, or with more than a predetermined number of other clients. They then formulate several maximal coverage models, with one or more servers per service center. In the second paper they address the issue of locating hierarchical facilities in the presence of congestion. Two hierarchical models are presented, where lower level servers attend requests first, and then, some of the served customers are referred to higher level servers. In the first model, the objective minimizes the number of servers and finds their locations so that they will cover a given region with a distance or time standard. The second model is cast as a Maximal Covering Location formulation. In both models they develop a capacity-like constraint to control for congestion in the second level of the hierarchy.

4.3 p-Median Models in Public Facility Location

The p-Median Problem belongs to a class of formulations called minisum location models. This class of problems was first formulated in its discrete form by Kuehn and Hamburguer (1963), Hakimi (1964), Manne (1964) and Balinski (1965). The problem can be stated as:

> Find the location of a fixed number of p facilities so as to minimize the weighted average distance of the system.

The first explicit formulation of the p-Median Problem is attributed to Hakimi (1964). Hakimi not only stated the formulation of the problem but he also proved that in a connected network were the triangle inequality is observed, optimal locations can always be found at the nodes. So it is only necessary to consider as potential locations the nodes of a given network under certain geometric conditions. The model formulated by Hakimi was not applied to a public sector location problem, since was used in the field of telecommunications, more precisely in the location of switching centers on a graph. Four years later, ReVelle and Swain (1970) gave the formulation as an integer linear program and studied its integer properties when solving it with linear programming and branch and bound. As mentioned before, even though the first known proposed application of Hakimi was not for public services and facilities, the p-median Problem has been since then extensively used as a basis to build problems related to public sector facility location-allocation modeling. A very similar model, the Uncapacitated or Simple Plant Location Problem has been used in private sector location settings[3].

The integer programming formulation of the p-Median Problem is as follows:

$$\text{Minimize} \left\{ Z = \sum_{i=1}^{m} \sum_{j=1}^{n} a_i d_{ij} x_{ik} \right\} \quad (4.27)$$

Subject to:

$$\sum_{j=1}^{n} x_{ij} = 1 \qquad i = 1, \ldots, m \quad (4.28)$$

$$x_{ij} \leq x_{jj} \qquad i = 1, \ldots, m;\ j = 1, \ldots, n \quad (4.29)$$

$$\sum_{j=1}^{n} x_{jj} = p \quad (4.30)$$

$$x_{ij} \in \{0, 1\} \qquad i = 1, \ldots, m;\ j = 1, \ldots, n$$

---

[3] An excellent presentation of classical Minisum Location Problems can be found in Krarup and Pruzan (1983)

where

i = Index of demand points
m = Total number of demand points in the space of interest
j = Index of potential facility sites
n = Total number of potential facility locations
a_i = Weight associated to each demand point.
d_{ij} = Distance between demand area i and potential facility at j.
$x_{ij} = \begin{cases} 1 \text{ if demand area } i \text{ is assigned to a facility at } j \\ 0 \text{ Otherwise} \end{cases}$

In this formulation it is assumed that all demand points are also potential facility sites ($m = n$). The first set of constraints forces each demand point to be assigned to only one facility. The second set of constraints allows demand point i to assign to a point j only if there is an open facility in this location. Finally, the last constraint sets the number of facilities to be located. The second set of constraints is known as the "Balinski" constraints, since he was the first to write them in this form in 1965, when studying the Simple Plant Location Problem. An alternative condensed version of the problem can be formulating by substituting the "Balinski" constraints with the following set:

$$\sum_{j=1}^{n} x_{ij} \leq m x_{jj} \qquad i = 1, \ldots, m \qquad (4.31)$$

This constraint states that no demand node can assign to point j, unless there is a facility open there. While this set of constraints substantially reduces the size of the problem, when solving it using linear programming without any integer requirements will nearly produce all x_{jj} fractional. On the other hand, the "Balinski" set of constraints makes the problem at hand quite large as the number of constraints required together with the number of variables are very large even in relatively small problems. Nevertheless, when solving the p-Median Problem in its extended form using linear programming relaxation, most solutions are integer. ReVelle and Swain (1970) observed that when branch-and-bound was required to resolve fractional variables produced by linear programming, the extent of branching and bounding needed was very small, always less than 6 nodes of a branch-and-bound tree. Therefore, the expanded form of the constraint makes integer solutions far more likely. Morris (1978), solved 600 randomly generated problems of the very similar Simple Plant Location Problem with the extended form of the constraint and found that only 4% did require the use of branch-and-bound to obtain integer solutions. Rosing et al. (1979c) proposed several ways to reduce both the number of variables and constraints in order to make the p-Median Problem more tractable. An extended discussion of "integer friendly" location formulations can be found in ReVelle (1993).

The p-Median Problem, due to its mathematical structure, is NP-hard[4], and therefore cannot be solved in polynomial time. Our experience shows that complete enumeration can be used in a network with up to 50 nodes and 5 facilities in reasonable computer time. Even though the size of the problems that can be solved by using Linear Programming and Branch and Bound (LP+BB), as proposed by ReVelle and Swain (1970), has been rapidly increasing with the advances in hardware and software technologies and algorithmic sophistication, there is still a strong need for exact and heuristic methods for large and realistic p-Median Problems. Therefore, since its early formulation in 1964, the p-Median Problem has been a fertile ground for innovative approaches and algorithms to obtain solutions. Garfinkel et al. (1974) and Swain (1974) used the Dantzig-Wolfe decomposition to obtain solutions. Another approach, lagrangian relaxation, was used by Gnarl et al. (1977) and Cornuejols et al. (1977) and extended with the use of the linear programming dual by Galvão (1980). See Galvão (1993) for an excellent review on Lagrangian Relaxation applied to uncapacitated facility location problems.

Another class of heuristics, and most widely used in applications to large problems are the ones based in interchange methods. Maranzana (1964) presented the first known local search procedure that was extremely fast, but with a weak search strategy. His heuristic begins by finding a feasible solution, that is, locating p facilities and then dividing the space into p subsets, each one associated with a specific location. Successive relocation within the subset, followed by redivision of the points into clusters, produced stable solutions.

The most widely used heuristic for the problem is the Teitz and Bart (1968). The procedure starts with an initial solution to obtain the initial facility set (for example the Maranzana procedure can be used to obtain the p initial locations) and the p-median objective is computed. The second phase of the heuristic seeks the improvement of the initial solution by exchanging members of the facility set for members of the non-facility set. Each exchange is evaluated by computing the new objective value. Trades are only allowed if the objective improves. The heuristic terminates when, after a full cycle of exchanges, no improvement in the objective is found. Rosing et al. (1979a, 1979b) and Cornuejols et al.. (1977), among others, extensively analyzed the performance of the Teitz and Bart heuristic in relatively small networks and obtained excellent results. Nevertheless, later on, Rosing (1997) showed that the Teitz and Bart heuristic did not behave in large networks as efficient as in small ones.

Other heuristics with similar search strategy have been proposed by Goodchild and Noronha (1983), with similar results. Whitaker (1983) modified the Teitz and Bart heuristic and developed a greedy stepwise exchange heuristic for which he claims good results. Nevertheless, the heuristic developed cannot be used with different random starts, since they always end in the

---

[4] For a discussion of NP hardness, see Krarup and Pruzan (1990).

same solution. That is, either heuristic will produce a single local optimum when applied to any given problem setting (Rosing et al. 1999). Densham and Rushton (1992a, 1992b) developed a very efficient version in computer time of the Teitz and Bart heuristic for very large p-Median Problems, namely, the global/regional interchange algorithm (GRIA). Despite its speed, GRIA was not as good as the Teitz and Bart in finding optimal solutions, since some exchanges are missed (Horn 1996). Rolland et al. (1997) designed a tabu search heuristic to solve the p-median problem that improved both speed and results efficiency over the existing heuristics. In essence, Tabu search is an interchange heuristic that tries to escape from a local optimum, and then continues on towards the global optimum by employing a memory of where it has been already. This memory makes specific, already investigated, interchanges illegal in the hope that a possible short-term degradation of the objective function will lead to an uninvestigated region of the solution space and hence to further improvement of the objective function (Rosing et al. 1998). Full details of this metaheuristic can be found in Glover (1986, 1989, 1990; Glover and Laguna, 1993) and details of its implementation in the p-median context can be found in Rolland et al. (1996). Another very similar Tabu Search approach for Uncapacitated Facility Location Problems (and therefore p-median problems) has been proposed by Al-Sultan and Al-Fawzan (1999).

Another recent heuristic, baptized as Heuristic Concentration, has been developed by Rosing and ReVelle (1997). Basically, this heuristic has two phases. In the first phase, several random trials of an interchange heuristic such as Teitz and Bart are executed. This allows in the second phase the development of a construction set as the union of the sets of facilities (each consisting of p nodes) found in each of the random trials. Then, the best set of facilities is obtained from the concentrated set, by means of LP+BB. In other words, in the second phase the p-median is solved to optimality using the nodes in the concentrated set as the potential facility locations. Rosing et al. (1998) compared the Heuristic Concentration with the Tabu Search developed by Rolland et al. (1997) and concluded that the first one was superior in finding optimal solutions. On the other hand, it was not clear its efficiency in terms of computer time. Rosing et al. (1999) modified the second phase of the Heuristic Concentration by using a 2-opt algorithm. Genetic algorithms have also been proposed to solve the problem. A excellent review and a new proposed genetic algorithm can be found in Chapter 6 in this book.

Since its formulation in the late sixties - early seventies, the p-median problem has been modified to be adapted to specific location problems or to allow a better "real world" implementation in the public sector. Services such as public libraries, schools, pharmacies, primary health care centers have benefited from this model. Nevertheless, in most cases, when implementing the location of such services, it has been necessary to modify the p-median in relation both to its parameters and its basic formulation.

One of the first rigidities of the p-median problem is that it presents a complete inelastic demand with respect to distance. People travel to the closest facility regardless of the distance or time traveled. As early as 1972, Holmes et al. presented a formulation that considered that people would not travel beyond a given a distance or time threshold. In essence the p-Median objective was replaced by the following one:

$$\text{Maximize} \left\{ \sum_{i=1}^{m} \sum_{j=1}^{n} a_i (S - d_{ij}) x_{ij} \right\} \quad (4.32)$$

where S is the threshold distance beyond no one will travel. It is also necessary to re-write constraint number 2 with a "$\leq$" sign, since not everyone will be assigned to a facility. The model was applied to locate public day care facilities in Columbus, Ohio. In this work, Holmes et al. also introduced the Capacitated p-median problem. In this model, facilities have a limited capacity and therefore the following constraint needs to be added:

$$\sum_{i=1}^{m} a_i x_{ij} \leq C \quad j = 1, \ldots, n \quad (4.33)$$

where C is the maximum capacity level. Computational experience shows that that by adding this constraint the number of fractional variables increases considerably when using linear programming and branch and bound.

Another problem when implementing the p-median problem is related to the distance parameter. The model supposes that distances (or travel times) do not change with time. But, what happens when we want to locate, for example, fire stations in a city? Travel times change during the day and therefore an optimal location during traffic peak hours may be very deficient in valley hours. On the other hand, the demand may also change during the day. CBD areas may be crowded during daytime while residential areas are empty, and vice-versa during night time. Serra and Marianov (1998) introduced the concept of regret and minmax objectives when locating fire stations in Barcelona (Spain) taking into account what they called "changing networks". Basically, uncertainty was treated using the classic scenario approach, in which different patterns of demand or travel times are realized in different scenarios. Frist, over a range of possible demand scenarios, facilities are deployed to site in such a way to minimize the maximum average travel time in a given scenario (minmax approach). Second, over that same range of scenarios, facilities are positioned in such a way as to minimize the maximum regret. Regret is defined as the difference between (1) the optimal average travel time that would be obtained had the decision maker planned its sites

4 Location in the Public Sector   137

for the scenario that actually occurred; and (2) the value of average travel time that was actually obtained (regret approach).

The minmax p-median problem formulation is as follows:

Minimize $\{ M \}$ (4.34)

Subject to:

$$\sum_{i=1}^{m}\sum_{j=1}^{n}\frac{a_{ij}d_{ij}^{k}}{W_{k}}x_{ij}^{k} \leq M \quad k=1,\ldots,K \tag{4.35}$$

$$\sum_{j=1}^{n} x_{ij}^{k} = 1 \quad i=1,\ldots,m;\ k=1,\ldots,K \tag{4.36}$$

$$x_{ij}^{k} \leq w_{j} \quad \begin{array}{l} i=1,\ldots,m;\ j=1,\ldots,n; \\ k=1,\ldots,K \end{array} \tag{4.37}$$

$$\sum_{j=1}^{n} w_{j} = p \tag{4.38}$$

$$x_{ij}, w_{j} \in \{0,1\} \quad i=1,\ldots,m;\ j=1,\ldots,n$$

where additional notation is defined as follows:

K = number of scenarios
k = index of scenarios
a_{ik} = the population at node i in scenario k
W_{k} = the total population in scenario k
d_{ijk} = the travel time between i and j in scenario k.
$x_{ij}^{k} = \begin{cases} 1 \text{ if node } i \text{ is assigned to node } j \text{ in scenario } k \\ 0 \text{ Otherwise} \end{cases}$
$w_{j} = \begin{cases} 1 \text{ if there is a facility at node } j \\ 0 \text{ Otherwise} \end{cases}$

The first constraint is directly related to the objective. Since the maximum average travel time across scenarios is to be maximized, we want to find a set of locations that will give the smallest maximum average travel time possible when evaluated for all scenarios. The left side of each constraint (one for each scenario) represents the demand weighted average travel time that will be achieved in the corresponding scenario. The right-hand side, M, is the same in each constraint. The objective of the model is to minimize M. That is, the model will try to find a set of locations that minimizes the largest total travel time achieved in each scenario. The rest of the constraints are very similar to the constraint set of the p-median problem.

If the regret objective is used, constraint set (4.35) is replaced by the following:

$$\sum_{i=1}^{m}\sum_{j=1}^{n}\frac{a_{ij}d_{ij}^{k}}{W_{k}}x_{ij}^{k} - Z_{k} \leq M \quad k=1,\ldots,K \tag{4.39}$$

where Z_k is the optimal objective, a known value, found when p facilities are located optimally in each scenario. Its value is found by applying the original p-median formulation to each scenario individually. The unknown variable M represents the largest regret evaluated over all scenarios.

The authors developed a one-opt exchange heuristic to find solutions to both formulations. The regret model was used to locate fire stations in the city of Barcelona.

Another problem related to the p-median is data aggregation (see Chapter 7 for a complete discussion of the aggregation issue). When selecting locations for facilities, the p-median takes (as most location models do) into account the demand for the service provided by the facility. When implementing the discrete problem in a real world setting, it is necessary to identify the demand areas that will be modeled as "nodes" or "points". Therefore, some spatial aggregation of the demand is performed. This is especially true when locating facilities in urban areas. In general, census tracks are aggregated to form demand areas. In the location of fire stations above mentioned, Barcelona has around 1800 census tracks. An aggregation was performed to reduce the problem to 200 demand areas. This aggregation leads to three types of source errors (Hillsman and Rhoda, 1978). Source A errors are a direct result of the loss of locational information. When performing the aggregation, the distance or travel time is modified since it is considered only from the centroid of the area to the potential facility site. Therefore, an over or under estimation of this parameter may occur. Source B errors are a special case of source A errors. If a facility is positioned in a given aggregated demand area, the corresponding weight in the objective function is set to zero since the corresponding distance is set to 0. But in reality it should not be 0, since at the desaggregated level some areas will have to travel to the closest facility. Source B errors always yield a measured weighted travel distance less than the true weighted travel distance. Finally, type C source errors appear when part of an aggregated area is not assigned to its closest facility. Several methods have been proposed to reduce or eliminate these errors (Hillsman and Rhoda 1978; Goodchild 1979; Bach 1981; Current and Schilling 1987; Bowerman et al. 1997 among others). For an excellent overview and new methods of reduction see Erkut and Bozkaya (1999). In this reference, demand point aggregation is examined in detail for the planar p-median problem.

When planning public facilities it may be necessary not only to obtain a good location, but to achieve also a balanced demand assignment level. Sometimes, in order to be efficient, facilities need to have a minimum demand threshold level. An area of application where the concept of threshold is relevant involves to the provision of services that are considered merit goods, but that are serviced by the private sector. This is specially relevant for merit services that have been publicly owned or controlled in several countries and are being transferred to the private sector, such as postal services, gas stations, fire departments and pharmacies. While the planner seeks to maintain good service quality by keeping a balanced spatial distribution of services, these need to have a minimum service threshold level that will allow them to survive. In the p-median formulation, this is achieved by adding to the original problem the following constraint set:

$$\sum_{i=1}^{m} a_i x_{ij} \geq C \qquad j = 1, \ldots, n \qquad (4.40)$$

Carreras and Serra (1999) used this formulation to examine the impact of the spatial deregulation of pharmacies in a region of Spain. They showed that by de-regulating the sector, the number of pharmacies would increase by at least 20%.

Sometimes, the location of new facilities is conditioned by the existence of districts in the region of interest. Demand areas within a given district can be assigned only to a facility within the same district. This problem would arise in the allocation of schools in a county, in voting-machine siting among voting districts, or large-scale facility-siting studies for regions encompassing many counties or states. The problem has two aspects. First, to decide how many facilities are assigned to each district. Secondly, where to locate these facilities. ReVelle and Elzinga (1989) developed an algorithm that solved optimally this problem.

The p-median model assumes that facilities are alike or of a single type. Nevertheless, it is widely accepted that many facility systems and institutions are hierarchical in nature, providing several levels of service. More specifically, a hierarchical system is one in which services are organized in a series of levels that are somehow related to one another in the complexity of function/service.

The organizational structure of hierarchical systems may vary considerably. There may be institutional ties between levels, whereby lower levels are administratively subordinate to higher ones (e.g., health care delivery systems, banking systems). On the other hand, there are several hierarchical systems that have no such inter-level linkages, different levels being distinguished solely by the range of goods and/or services they provide (e.g., educational systems, production-distribution systems, waste collection systems) (Hodgson 1986).

The p-median model locates p facilities such that the average distance from the users to their closest facility is minimized. In a hierarchical setting it has been generally used to locate a given number of facilities for each level, one at a time. Several hierarchical models based on the p-median have been formulated.

Calvo and Marks (1973) constructed a multiobjective integer linear model to locate multi- level health care facilities: the model minimized distance (travel time), user costs, and maximized demand or utilization, and utility. It was based on assumptions that (1) users go to the closest appropriate level;

(2) there is no referral to higher levels; and (3) all facilities offer lower level services.

Tien et al. (1983) argued that the approach taken by Calvo and Marks resulted in deficient organization across hierarchies. In order to resolve this deficiency, they presented models derived from Calvo and Mark's formulation: nested and non-nested models. They also introduce a new feature whereby a demand cannot assign to a place more than once even if additional service levels may available at that point. Both models, unlike Calvo and Marks', can be solved by standard integer programming solution procedures. Mirchandani (1987) extended the hierarchical p-median formulation of Tien et al. model, allowing various allocation schemes by redefining the cost parameter in the objective.

Harvey et al. (1974) used a p-median formulation to determine the number and optimal locations of intermediate level facilities in a central place hierarchy. The p-median model was used in one-level problem, but consideration was given on the interaction among lower and higher levels.

Narula et al. (1975) developed a nested hierarchical health care facility location model that located on a network a fixed number of facilities. At each level, the objective was to minimize patients' total travel. They considered referrals between levels, based on the proportion of patients treated at each level. Narula and Ogbu (1979) gave some heuristic procedures for the solution of the problem. Later, Narula and Ogbu (1985) solved a two- level mixed-integer p-median problem using the same objective and referral pattern.

Berlin et al. (1976) studied two hospital and ambulance location problems. The first one focused on patient needs by minimizing (1) average ambulance response time from ambulance bases to demand areas and (2) average distance to hospitals from demand areas. The second model added a new objective to take into account the efficiency of the system: minimization of (3) distance from ambulance bases to hospitals. It was named the "dual- facility" location problem: the locations of both hospitals and ambulance depots were basic to determine response times. It is interesting to note that although two levels are defined (stations or depots where ambulances sit, and hospitals), the formulation can be decomposed into independent hospital and ambulance location problems and solved optimally. It is not a clear hierarchical model since relations among levels differ from the traditional regionalized models.

Fisher and Rushton (1979) and Rushton (1984) used the average and maximum distance from any demand area to its closest health care center to study and compare actual and optimal hierarchical location patterns in India. The Teitz and Bart heuristic was used in three ways to determine hierarchies: constructing top-down hierarchical procedure (same as Banergi and Fisher 1974); constructing a bottom-up hierarchical procedure (opposite of top- down); and constructing a hierarchical procedure where the first step was to locate a middle-level of the hierarchy optimally, and then proceed as the bottom heuristic for upper levels, and use the top-down heuristic for lower levels.

Tien and El-Tell (1984) defined a two-level hierarchical LP model consisting of village and regional clinics. It is a top-down formulation in the sense that the flow patterns start at the hospitals. That is, health professionals go from hospitals to village centers. Both village and regional clinics are located using a criterion of minimizing the weighted distance of assigning villages to clinics and village clinics to regional clinics. The model was applied to 31 villages in Jordan.

Hodgson (1984) demonstrated that the use of top-down or bottom-up techniques to locate hierarchical systems generally leads to suboptimal locational patterns. By a top-down (bottom-up) technique is meant the location first of the highest (lowest) level of the hierarchy and then successive location of facilities in the following level. Hodgson used both the p-median model and a formulation based on Reilly's gravitational law (Reilly, 1929) to compare both techniques with the simultaneous location of all hierarchies.

As mentioned before, sometimes it is necessary to obtain not only good locations but also an efficient district for each facility. Serra and ReVelle (1993) introduced the concept of coherence in hierarchical models. A coherence in a hierarchical system is defined as follows: all areas assigned to a particular facility at one hierarchical level should belong to one and the same district in the next level of the hierarchy. The authors developed the pq- median model. This formulation locates two types of facilities by combining two p-median formulations. Each level has the objective of minimizing the average distance or travel time from the demand areas to the nearest facility whilst ensuring coherence. Hence, a trade off between access to each hierarchical level is expected. The model was used to design the location and districting hierarchical primary health care services in Barcelona (Serra 1996).

The implementation of location-allocation problems in third world areas may present different problems that the ones implemented in developed countries. Perhaps the most notorious problem involves data limitations. An excellent review of application of p-median and covering problems in the real word can be found in Oppong (1996). The work by Oppong also examines the location of hierarchical Primary Health Care Centers in Suhum District, Ghana. This region is affected by strong climate differences during the year. There is a strong seasonal variation in road surface conditions. Therefore, this problem is similar to some extent to the one of locating fire stations in a city: there are different networks according to the season of the year. He developed a decision support tool to improve the solutions given by the p-median formulations.

4.4 Conclusions

The development of models and methodological frameworks to design or reconfigure emergency and non-emergency systems has taken place over the span of a quarter century. It has developed alongside and in concert with the evolution of the modern computer as it transited from a room-filling behemoth to a desk-top associate. And like the computer on which the models must rely, the models and methods are not done evolving. The shape of the next models can be predicted by simply observing how the current generation still falls short of perfectly describing reality. We will focus on seven areas.

First, we should begin to see a new generation of models that deal with the issue of co- location of servers from different emergency systems. ReVelle and Snyder (forthcoming) introduce this line of research in the FAST (fire and ambulance siting technique) model that examines the link between ambulance and fire company siting. In the United States, ambulance deployment has traditionally taken place either at hospitals or at fire stations or both, but rarely have ambulances been positioned at free standing ambulance stations. ReVelle and Snyder, in a deterministic covering model, examine the consequences of allowing the ambulances to be sited free of constraints on the location of other services. These models should eventually develop all the probabilistic sophistication and nuances of the models discussed above.

Second, we can expect that the estimate of server or region-specific busy fraction will be refined. Although we have moved from deterministic to redundant to probabilistic models, and although within this last category we have moved from a system-wide busy fraction to a region-specific busy fraction using queuing concepts, we still have not precisely matched the busy fractions estimated by simulation. Unless the challenge exceeds the imaginative powers of investigators, we will soon see server-specific busy fractions or more refined region-specific busy fractions.

Third, we should see focus developing on workload issues, a topic that has largely been ignored till now and one that greatly concerns emergency and non emergency system planners. The issues of busy fraction, workload and threshold levels are tightly connected so progress on the former should bring achievement on the latter as well.

Fourth, we should see a gradual melding of the two lines of evolution, queuing and location. It is hard to predict how this will take place, but certainly the use of heuristics offer the descriptive queuing models an opportunity to compete with the location models in the arena of design. And the introduction of queuing concepts by Marianov and ReVelle, as well as Batta, into location models, suggest movement from the other side as well.

Fifth, most of the models we have examined consider that customers always patronize the closest facility. That is, distance (or travel time) is the only parameter considered by customers. But there may be other decision parameters related to service quality such as service speed, cleanness or efficiency that may influence customers' decisions. New models have been developed to examine this issue (Serra et al. 1999, Colome and Serra 1999).

Sixth, the evolution of public sector deregulation is gradually introducing competition between providers. For example, the deregulation of a health care system may introduce some level of competition among providers at the primary (and secondary) level to attract patients. Therefore, the dichotomy between public and private location modeling is being diffused and Location Capture Models can be adapted to accommodate public sector issues.

Last but not least, this chapter has not addressed the development of public sector location-allocation models that consider the siting of undesirable facilities. There is a considerable volume of literature on this topic. A good starting point to interested readers can be found in Murray et al. (1998).

References

Al-Sultan K. and M. Al-Fawzan (1999) "A tabu search approach to the uncapacitated facility location problem," *Annals of Operations Research*, 86, 91-103.

Bach, L. (1981) "The problem of aggregation and distance for analyses of accessibility and access opportunity in location-allocation models," *Environment and Planning A*, 13, 955-978.

Balinski, M.L. (1965) "Integer Programming: Methods, Uses and Computation," *Management Science*, 12, 253-313.

Ball M, and F. Lin (1993) "A Reliability Model Applied to Emergency Service Vehicle Location," *Operations Research*, 41, 18-36.

Banergi, S. and H. Fisher (1974) "Hierarchical Location Analysis for Integrated Area Planning in Rural India," *Papers of the Regional Science Association*, 33, 177-194.

Batta R. (1988) "Single Server Queueing-Location Models with Rejection," *Transportation Science*, 22, 209-216.

Batta R. (1989) "A Queueing-Location Model with Expected Service Time dependent Queueing Disciplines," *European Journal of Operational Research*, 39, 192-205.

Batta R., Dolan J. and N. Krishnamurthy (1989) "The Maximal Expected Covering Location Problem: Revisited," *Transportation Science*, 23, 277-287.

Batta R., Larson R. and A. Odoni (1988) "A Single-Server Priority Queueing-Location Model," *Networks*, 8, 87-103.

Bell T. and R. Church (1987) "Location-Allocation Modeling in Archeology", in A Ghosh and G Rushton (eds): *Spatial Analysis and Location-Allocation Models*, Van Nostrand Reinhold.

Benedict J. (1983) "Three hierarchical objective models which incorporate the concept of excess coverage for locating EMS vehicles or hospitals," Ms.C. Thesis, Northwestern University.

Berlin G. (1972) "Facility Location and Vehicle Allocation for Provision of an Emergency Service," PhD Dissertation, The Johns Hopkins University, Baltimore, MD.

Berlin, G., ReVelle, C. and J. Elzinga (1976) "Determining Ambulance-Hospital Locations for On-Scene and Hospital Services," *Environment and Planning A*, 8, 553-561.

Berman O. and R. Larson (1985) "Optimal 2-Facility Network Districting in the Presence of Queuing," *Transportation Science*, 19, 261-277.

Berman O., Larson R. and S. Chiu (1985) "Optimal Server Location on a Network Operating as a M/G/1 Queue," *Operations Research*, 12, 746-771.

Berman O., Larson R. and C. Parkan (1987) "The Stochastic Queue p-Median Location Problem," *Transportation Science*, 21, 207-216.

Berman O. and R. Mandowsky (1986) "Location-Allocation on Congested Networks," *European Journal of Operational Research*, 26, 238-250.

Bianchi C. and R. Church (1988) "A hybrid FLEET model for emergency medical service system design," *Social Sciences in Medicine*, 26, 163-171.

Bowerman R., Calamai, P. and G. Hall (1997) "The demand partitioning method for reducing aggregation errors in p-median problems," *Computers and Operations Research*, 26, 1097-1111.

Burwell T., Jarvis J. and M. McKnew (1993) "Modeling Co-located Servers and Dispatch Ties in the Hypercube Model," *Computers and Operations Research*, 20, 113-119.

Calvo, A. and H. Marks (1973) "Location of Health Care Facilities: An Analytical Approach," *Socio-economic Planning Sciences*, 7, 407-422.

Carreras M. and D. Serra (1999) "On Optimal Location con Threshold Requirements," *Socio-economic Planning Sciences*, 33, 91-103

Carter G. M., Chaiken J. M. and E. Ignall (1982) "Response areas for two emergency units," *Operations Research*, 20, 571-594.

Chapman S. and J. White (1974) "Probabilistic formulations of emergency service facilities location problems," ORSA/TIMS Conference, San Juan, Puerto Rico.

Church R. and C. ReVelle (1974) "The Maximal Covering Location Problem," *Papers of the Regional Science Association*, 32, 101-118.

Chung C. (1986) "Recent Applications of the Maximal Covering Location Planning (MCLP) Model," *Journal of the Operational Research Society*, 37, 735-746.

Cornuejols, G., Fisher, M. and G. Nemhauser (1977) "Location of bank accounts to optimize float: An analytic study of exact and approximate algorithms," *Management Science*, 23, 789-810.

Current, J. and D. Schilling (1987) "Elimination of source A and B errors in p-median location problems," *Geographical Analysis*, 19, 95-110.

Current J, and D. Schilling (1989) "The covering salesman problem", *Transportation science*, 23, 208-213.

Daskin M, (1983) "A maximum expected covering location model: formulation, properties and heuristic solution," *Transportation Science*, 17, 48-70.

Daskin M. (1995) *Network and Discrete Location: Models, Algorithms and Applications*, Wiley Interscience Series in Discrete Mathematics and Optimization, Toronto.

Daskin M., Hogan K. and C. ReVelle (1988) "Integration of multiple, excess, backup, and expected covering models," *Environment and Planning B: Planning and Design*, 15, 15- 35.

Daskin M. and E. Stern (1981), "A hierarchical objective set covering model for emergency medical service vehicle deployment," *Transportation Science*, 15, 137-152.

Densham P. and G. Rushton (1992a) "Strategies for solving large location-allocation problems by heuristic methods," *Environment and Planning A*, 24, 289-304.

Densham P. and G. Rushton (1992b) "A more efficient heuristic for solving large location- allocation problems," *Papers in Regional Science*, 71, 307-329.

Dwyer, E. and E. Evans (1981) "A Branch and Bound Algorithm for the List Selection Problem in Direct Mail Advertising," *Management Science*, 29, 658-667.

Eaton D., Church R., Bennet V. and B. Namon (1981) "On deployment of health resources in rural Columbia," *TIMS Studies in the Management Sciences*, 17, 331-359.

Eaton D, Hector M., Sanchez V, Lantigua R and J. Morgan (1986) "Determining ambulance deployment in Santo Domingo, Dominican Republic," *Journal of the Operational Research Society*, 37, 113.

Erkut E. and B. Bozkaya (1999) "Analysis of aggregation errors for the p-median problem," *Computers & Operations Research*, 26, 1075-1096.

Fischer, H. and G. Rushton (1979) "Spatial Efficiency of Service Locations and the Regional Development Process," *Papers of the Regional Science Association*, 42, 83-97.

Galvão, R. (1980) "A dual-bounded algorithm for the p-median problem," *Operations Research*, 28, 1112-1121.

Galvão, R. (1993) "The use of lagragean relaxation in the solution of uncapacitated facility location problems," *Location Science*, 1, 57-70.

Galvão, R and C. ReVelle (1996) "A Lagrangean heuristic for the maximal covering location problem," *European Journal of Operational Research*, 88, 114-123.

Garfinkel, R. S., Neebe, A. W., and Rao, M. R. (1974) "An algorithm for the m-median plant location problem," *Transportation Science*, 8, 217-236.

Gleason, J. (1975) "A set covering approach to bus stop location," *Omega*, 3, 605-608.

Glover, F. (1986) "Future paths for integer programming and links to artificial intelligence," *Computers and Operations Research*, 5, 533-549.

Glover, F. (1989) "Tabu Search, part I," *ORSA Journal of Computing*, 1, 190-206.

Glover. F. (1990) "Tabu Search, part II," *ORSA Journal of Computing*, 2, 4-32.

Glover F. And M. Laguna (1993) "Tabu Search", in C. Reeves (ed.) *Modern Heuristic Techniques for Combinatorial Problems*, Blackwell Publishing

Goldberg J., and L. Paz (1991) "Locating Emergency Vehicle Bases When Service Time Depends on Call Location," *Transportation Science*, 25, 264-280.

Goodchild M. and J. Lee (1989) "Coverage Problems and Visibility Regions on Topographic Surfaces," *Annals of Operations Research*, 18, 175-186.

Goodchild, M. (1979) "The aggregation problem in location-allocation," *Geographical Analysis*, 11, 240-255.

Goodchild M. and V. Noronha (1983) "Location-Allocation for Small Computers," Monograph 8, Department of Geography, The University of Iowa, Iowa City, IA.

Gunawardane, G. (1982) "Dynamic Versions of Set Covering Type Public Facility Location Problems," *European Journal of Operational Research*, 10, 190-195.

Hakimi, S. (1964) "Optimum locations of switching centres and the absolute centres and medians of a graph," *Operations Research*, 12, 450-459.

Harvey, M., Hung M., and J. Brown (1974) "The Application of a p-median algorithm to the Identification of Nodal Hierarchies and Growth Centers," *Economic Geography*, 50, 187-202.

Hillsman, E. and R. Rhoda (1978) "Errors in measuring distance from populations to service centers," *Annals of the Regional Science Association*, 12, 74-88.

Hogan K. and C. ReVelle (1986) "Concepts and applications of backup coverage", *Management Science*, 32, 1434-1444.

Hodgson, J. (1984) "Alternative approaches to hierarchical location-allocation systems," *Geographical Analysis*, 16, 275-275.

Hodgson, J. (1986) "A Hierarchical location-allocation model with allocation based on facility size," *Annals of Operations Research*, 6, 273-289.

Holmes, J., Williams, F. and L. Brown (1972) "Facility location under maximum travel restriction: An example using day care facilities," *Geographical Analysis*, 4, 258-266.

Horn, M. (1996) "Analysis and computational schemes for p-median heuristics," *Environment and Planning A*, 28, 1699-1708.

Insurance Services Office (1974) Grading Schedule for Municipal Fire Protection, Insurance Services Office, New York.

Krarup J. and P.M. Pruzan (1983) "The Simple Plant Location Problem: Survey and Synthesis," *European Journal of Operational Research*, 12, 36-81.

Jarvis J. (1975) "Optimization in stochastic systems with distinguishable servers," Technical Report No. 19-75, Operations Research Centre, M. I. T. (June), 1975.

Jarvis J. (1985) "Approximating the equilibrium behavior of multi-server loss systems," *Management Science*, 31, 235-239.

Kolesar P. (1980) "Testing for vision loss in glaucoma suspects," *Management Science*, 26, 439-450.

Kuehn, A. and M. Hamburger (1960) "A heuristic program for locating warehouses," *Management Science*, 9, 643-666.

Larson, R. (1974) "A Hypercube Queuing Model for Facility Location and Redistricting in Urban Emergency Services," *Computers and Operations Research*, 1, 67-95.

Larson, R. (1975) "Approximating the Performance of Urban Emergency Service Systems," *Operations Research*, 23, 845-868.

Larson R. and A. Odoni (1981) *Urban Operations Research*, Prentice-Hall, Englewood Cliffs, NJ.

Larson R. and K. Stevenson (1972) "On insensitivities in urban redistricting and facility location," *Operations Research*, 20, 613-618.

Manne, A. (1964) "Plant location under economies of scale, Decentralization and computation," *Management Science*, 11, 213-235.

Maranzana, F. (1964) "On the location of supply points to minimize transport costs," *Operations Research Quarterly*, 15, 261-270.

Marianov V. and C. ReVelle (1991) "The standard response fire protection siting problem," *INFOR*, 29, 116-119.

Marianov V. and C. ReVelle (1992) "The capacitated standard response problem: deterministic and probabilistic models," *Annals of Operations Research*, 40, 303-322.

Marianov V. and C. ReVelle (1992) "A probabilistic fire protection siting model with joint vehicle reliability requirements," *Papers in Regional Science*, 71, 217-241.

Marianov V. and C. ReVelle (1994) "The Queuing Probabilistic Location Set Covering Problem and Some Extensions," *Socio-Economic Planning Sciences*, 28, 167-178

Marianov V. and C. ReVelle (1995) "Siting of Emergency Services," In *Facility Location: A Survey of Applications and Methods*, Zvi Drezner (ed), 570 pp., Springer Verlag, New York NY.

Marianov V. and D. Serra (1998) "Probabilistic Maximal Covering Location-Allocation for Congested Systems," *Journal of Regional Science*, 38, 401-424.

Marianov V. and D. Serra D. (2001) "Hierachical location-allocation models for congested systems," *European Journal of Operational Research*, forthcoming.

Marianov V. and P. Taborga (2000) "A Model for the Location of Public Services Providing Both Competitive and Subsidized Services," *Journal of the Operational Research Society*, (forthcoming)

Meyer P. and D. Brill (1988) "A Method for Locating Wells in a Groundwater Monitoring Network under Conditions of uncertainty," *Water Resources Research*, 24, 1277-1282.

Mirchandani, P. (1987) "Generalized hierarchical facility location," *Transportation Science*, 21, 123-125.

Moore G. and C. ReVelle (1982) "The hierarchical service location problem," *Management Science*, 28, 775.

Morris, J. (1978) "On the extent to which certain fixed-charge depot location problems can be solved by LP," *Journal of the Operational Research Society*, 29, 71-76.

Murray, A., Church, R., Gerrard, R. and W. Tsui (1998) "Impact models for siting undesirable facilities," *Papers in Regional Science*, 77, 19-36.

Narula, S. and U. Ogbu, U. I. (1979) "An hierarchical location-allocation problem," *OMEGA*, 7, 137-143.

Narula, S. C. and U. Ogbu (1985) "Lagrangean relaxation and decomposition in an uncapacitated 2-hierarchal location-allocation problem," *Computers and Operations Research*, 12, 169-180.

Narula, S., Ogbu U. and H. Samuelson (1975) "Location of Health Facilities in Developing Countries," presented at the ORSA/TIMS meeting, Chicago, working paper no.221, School of Management, State University of New York, Buffalo.

Narula, S. C., Ogbu, U., and H. Samuelsson (1977) "An algorithm for the p-median problem," *Operations Research*, 25, 709-713.

Oppong, J. (1996) "Accommodating the rainy-season in third world location-allocation applications," *Socio-Economic Planning Sciences*, 30, 121-137.

Pirkul H.. and D. Schilling (1991) "The Capacitated Maximal Covering Location Problem with capacities on total workload," *Management Science*, 37, 233-248.

Reilly, M. (1929) "Methods for the Study of Retail Relationships," Austin, Texas: Bureau of Business Research, Monograph No. 4, Texas University.

ReVelle, C. (1989) "Review, extension and prediction in emergency service siting models," *European Journal of Operational Research*, 40, 58-69.

ReVelle, C. (1993) "Facility siting and integer-friendly programming," *European Journal of Operational Research*, 65, 147-158.

ReVelle, C. and D. Elzinga (1989) "An algorithm for facility location in a districted region," *Environment and Planning B*, 16, 41-50.

ReVelle C. and K Hogan (1988) "A reliability-constrained siting model with local estimates of busy fractions," *Environment and Planning B: Planning and Design*, 15, 143-152.

ReVelle C. and K. Hogan (1989a) "The maximum reliability location problem and α- reliable p-center problem: Derivatives of the probabilistic location set covering problem," *Annals of Operations Research*, 18, 155-174.

ReVelle C. and K. Hogan (1989b) "The maximum availability location problem," *Transportation Science*, 23, 192-200.

ReVelle C. and V. Marianov (1991) "A Probabilistic FLEET Model with Individual Vehicle Reliability Requirements," *European Journal of Operational Research*, 53, 93-105.

ReVelle C. and S. Snyder (2001) "Integrating Emergency Services: The fire and Ambulance Siting Techniques," *Socio-Economic Planning Sciences*, forthcoming.

ReVelle, C. S. and R. Swain (1970) "Central facilities location," *Geographical Analysis*, 2, 30-42.

ReVelle C., Toregas C., and L. Falkson (1976) "Applications of the Location Set Covering Problem," *Geographical Analysis*, 8, 65-76.

Rolland E., Schilling D., and J. Current (1997) "An efficient tabu search procedure for the p-median problem," *European Journal of Operational Reserach*, 96, 329-342

Rosing. K. (1997) "An empirical investigation of the effectiveness of a vertex substitution heuristic," *Environment and Planning B*, 24, 59-67.

Rosing K., Hillsman E and H. Rosing-Vogelaar (1979a) "A note comparing optimal and heuristic solutions to the p-median problem," *Geographical Analysis* 11, 86-89.

Rosing K., Hillsman E and H. Rosing-Vogelaar (1979b) "The robustness of two common heuristics for the p-median problem," *Environment and Planning A*, 11, 373-380

Rosing K., Hillsman E and H. Rosing-Vogelaar (1979c) "The p-median and its linear programming relaxation: an approach to large problems," *Journal of Operational Research Society*, 30, 815-823.

Rosing, K. and C. ReVelle (1997) "Heuristic concentration: Two stage solution construction," *European Journal of Operational Research*, 97, 75-86.

K. Rosing, C., ReVelle, E. Rolland, D., Schilling D. and J. Current (1998) "Heuristic concentration and Tabu search: a head to head comparison," *European Journal Of Operational Research*, 104, 93-99

Rosing, K., ReVelle C. and D. Schilling (1999) "A gamma heuristic for the p-median problem," *European Journal of Operational Research*, 117, 522-532.

Rushton, G. (1984) "Use of Location-Allocation Models for Improving the Geographical Accessibility of Rural Services in Developing Countries," *International Regional Science Review*, 9, 217-240.

Schilling D., Elzinga D., Cohon J., Church R. and C. ReVelle (1979) "The TEAM/FLEET models for simultaneous facility and equipment siting," *Transportation Science*, 13, 163-175.

Schilling D., Jayaraman V. and R. Barkhi (1993) "A Review of Covering Problems in Facility Location," *Location Science*, 1, 25-55.

Serra, D. (1993) "La organizaciòn de los servicios de salud: una aproximaciòn cuantitativa a la división territorial sanitaria de Barcelona," *Hacienda Publica Espanola*, 1, serie monografias, 81-100

Serra D., Eiselt H., Laporte G. and C. ReVelle (1999) "Market Capture Models under Various Customer Choice Rules," *Environment and Planning B*, 26, 141-150

Serra D. and V. Marianov (1999) "The p-median Problem in a Changing Network: The Case of Barcelona," *Location Science*, 4, 383-394

Serra D. and C. ReVelle (1993) "The pq-median problem. Location and districting of hierarchical facilities," *Location Science*, 1, 299-312.

Swain, R. (1971) "A Decomposition Algorithm for a Class of Facility Location Problems", Ph. D. Dissertation, Cornell University, Ithaca, N. York.

Swain, R. (1974) "A parametric decomposition approach for the solution of uncapacitated location problems," *Management Science*, 21, 189-198.

Teitz, M. and P. Bart (1968) "Heuristic methods for estimating the generalized vertex median of a weighted graph," *Operations Research*, 16, 955-961.

Tien, J. M. and K. El-Tell (1984) "A Quasi-Hierarchical Location-Allocation Model for Primary Health Care Planning," *IEEE Transactions in Systems Management and Cybernetics*, SMC-14, 373-380.

Tien, J. M., El-Tell, K. and G. Simons (1983) "Improved Formulations to the Hierarchical Health Facility Location-Allocation Problem," *IEEE Transactions in Systems Management and Cybernetics*, SMC-13, 1128-1132.

Toregas C. and C. ReVelle (1973) "Binary Logic Solutions to a Class of Location Problems," *Geographical Analysis*, 5, 145-155.

Toregas C., Swain R., ReVelle C. and L. Bergmann (1971) "The location of emergency service facilities," *Operations Research*, 19, 1363-1373.

Weaver, J. and R. Church (1991) "The nested hierarchical median facility location model," *INFOR*, 29, 100-115.

Whitaker, R. (1983) "A fast algorithm for the greedy interchange for large-scale clustering and median location problems," *INFOR*, 21, 95-108.

White J. and K. Case (1974) "On covering problems and the central facility location problem," *Geographical Analysis*, 6, 281-293.

5 Consumers in Competitive Location Models

Tammy Drezner[1] and H. A. Eiselt[2]

[1] College of Business and Economics, California State University, Fullerton, CA 92834. e-mail: tdrezner@fullerton.edu
[2] Faculty of Administration, University of New Brunswick, Fredericton, NB, Canada. e-mail: haeiselt@unb.ca

5.1 Introduction

Location models have been discussed by researchers from a wide variety of disciplines, among them mathematicians, geographers, marketing and retail specialists. Virtually all location models have a common basic structure: a metric space in which customers (representing demand) are positioned, and in which facilities are to be located by a decision maker. Most authors classify models based on the space of the model (e.g., Euclidean space or a network), the number of facilities to be located, the objectives the facility planners follow, and the type of behavior customers are assumed to exhibit. This contribution takes the last route and focuses on consumer behavior in competitive facility location modeling.

In general, whenever more than a single facility is given in some space (regardless whether it already exists in space or it has to be located in the process), any location model needs a rule for assigning demand to the facilities. Here, we can distinguish between *allocation models* and *choice models*. In allocation models, the facility planner assigns customers to facilities, while in choice models, customers are free to choose the facility. The latter is the case when modeling competitive facility locations, particularly in the context of retail location. Here, planners, must understand how consumers behave and how they make spatial choices. The importance of understanding the choice models of consumers was pointed out by Golledge and Simpson (1987).

One of the problems in competitive facility location models is to estimate the market share captured by each facility. This market share depends on three factors:

1. Customer characteristics,
2. Facility attributes, and
3. The spatial separation of customers and facilities.

Typical customer characteristics include the consumers' disposable income as an expression of their buying power, the customers' age, and their level of education. All characteristics will determine what potential customers consume, enjoy, and, more generally, their life-style. Facility attributes include

Facility Location: Applications and Theory.
Edited by Z. Drezner and H.W. Hamacher
© 2002 Springer-Verlag, ISBN 3-540-42172-6

the variety offered by a facility, such as the number of books in a library, the number and types of services provided by a medical facility, or the goods offered in a store or a shopping mall. Finally, the spatial separation, i.e., the distance between a customer and a facility, will determine the level of support the facility will enjoy from the customer. In general, a customer will consider a facility less attractive the greater the distance between them. This is typically referred to as a transportation cost, a term that is to be understood in the wide sense, including actual transport costs, the time of transport, aggravation due to long waiting times, and similar factors. Recent reviews of competitive location models are Drezner (1995), and Plastria (2001).

In order to demonstrate the differences in location models based on different customer behavior, consider the location of gas stations, given a typical suburban bedroom community and a major town, both connected by a road. First assume a case in which customers always make a special trip from their home to the station. In such a case, the planner of a new station will locate his facility on either side of the road as close as possible to the suburb in order to capture its demand. Suppose now, somewhat more reasonably, that consumers will fill up their cars on the way to or from work. If they do so on the way to town in the morning, they will prefer to have a gas station on the right-hand side anywhere along the road, as left turns are more time-consuming and require more attention than right turns. More importantly, a station on the left side of the road requires the customers to cross one lane after they turn out of the station and rejoin the traffic flow. The opposite will apply if customers can be expected to fill up their cars on their way back home. As customers are usually more hurried on their way to work in the morning than on their way back in the afternoon, they will tend to fill up in the afternoon, so that gas stations will often be located on the right side of the road as seen from town towards the suburb.

Figures 5.1 and 5.2 depict the decision-making processes of customers and facility planners, respectively. In both Figures, the arcs represent the sequence of decisions as well as external factors that influence facility planning.

In the remainder of this chapter we will focus on consumers and (retail) stores. This is, of course, not to imply that stores or shopping malls provide the only application of customers who choose facilities. The modeling concerns three questions:

1. what to buy (the commodity),
2. where to buy it (the facility), and
3. how much to buy (the quantities).

In order to formalize our discussion, it is useful to first define a *rational consumer* as an individual who has full information, constructs a rational and logical utility function for each of a set of alternative facilities, optimizes it by comparing his expected satisfaction at each facility, and then behaves accordingly. It will become abundantly clear that such a utility function is

Fig. 5.1. The Customer Choice Process

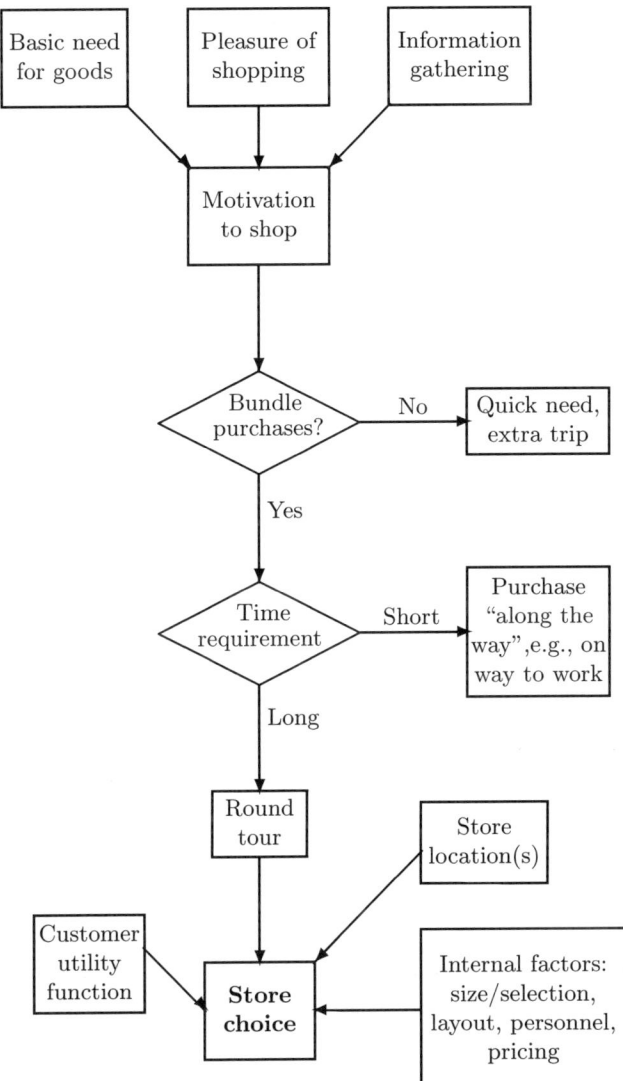

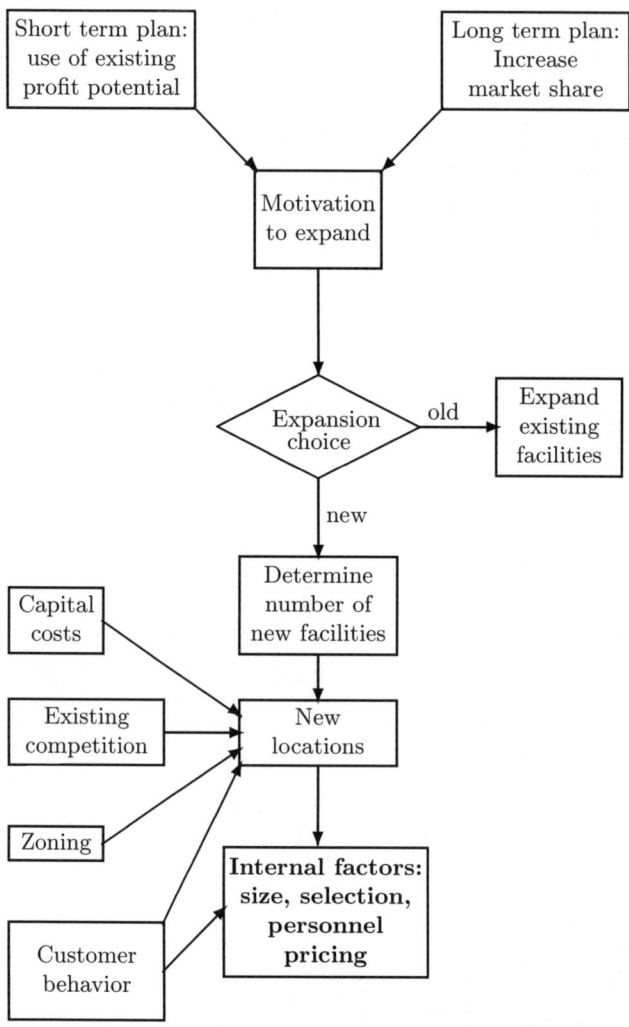

Fig. 5.2. Facility Decisions

by no means restricted to monetary components. In our basic approach, a typical rational consumer would set up a utility function including all the available information about himself, the facilities in question, the degree of spatial separation between customer and facilities, as well as the products, and then make decisions on the basis of these utilities. We suppose that all customers have complete (and accurate) information about themselves, the facilities, and the spatial separation between them and the facilities. Note that this information includes prices and distances to all facilities. To simplify further, assume that consumers demand only a single product (answering the "what" question), and have fixed demand, such as may be reasonable in the case of essential goods (taking care of the "how much" question). This leaves the question "where" consumers purchase the goods. To formalize, define a utility function for each combination of a customer C_i at a distance of d_{ij} from facility F_j as $u_{ij} = f(C_i, d_{ij}, F_i)$. Eiselt and Laporte (1998) have shown that if all products and facilities are truly homogeneous, a rational customer will satisfy his entire demand at the facility with the highest utility, resulting in "winner-takes-all" behavior that we will encounter repeatedly below. Note that this is essentially a deterministic framework; random factors relating to how much to buy and where to buy may be added.

By definition, all spatial models must include the facility – customer distances, and for the basic approach, we delete all other factors, leaving us with $u_{ij} = f(-d_{ij})$, where the "-" sign in front of the distances indicates that the utility function is non-increasing in the distance. This function simply states that customers will satisfy their entire demand at the closest source, a behavior known as *proximity models* first suggested by Hotelling (1929). In n-dimensional real space, the facilities can be mapped as points P_j, and with each point P_j we can associate a Voronoi set $V(P_j)$ which includes all points that are closer to P_j than to any $P_k, k \neq j$ according to some predefined metric. The union of all Voronoi sets forms a tessellation of space known as a *Voronoi diagram*. For an in-depth discussion of Voronoi diagrams, readers are referred to Boots et al. (1992).

5.2 Incorporating Facilities Features

So far, all facilities in the model were assumed to be homogeneous and differed only by their location. This assumption is usually not satisfied even for highly standardized facilities such as post offices or government-run tourist information centers: not all post-office outlets offer parcel service, and only some of the bigger information centers include features such as stores selling local books and crafts. It is, however, not necessarily true that a bigger and cleaner facility with friendlier staff necessarily has a stronger pull: for example, customers who patronize a facility solely for purchases of a standardized good such as a newspaper, parcel, or mail order delivery, will consider all facilities that offer such service as basically equivalent to each other and make their choice based on their convenience. On the other hand, a grocery store, library, or hospital with a greater selection of goods and services will, in general, be considered more attractive and thus offer a stronger "pull" to customers than a similar facility with fewer goods and services. The main purpose of this Section is the modeling of non-homogeneous facilities.

Non-homogeneous facilities generally differ from each other in a variety of attributes. As an example, consider attributes of shopping malls, which are among the most researched facilities in the literature. In this context, Downs (1970) has defined nine individual attributes that represent the main components of shopping centers. Similarly, Timmermans (1981) has undertaken comprehensive studies of consumer behavior in Eindhoven, the Netherlands. He investigated 12 shopping centers in addition to the city center in Eindhoven. In his study, he focused on the variables that customers use to discriminate between shopping centers. Table 5.1 shows the features of shopping malls that consumers consider the most important. For simplicity we only include factors that are mentioned more than once; other factors mentioned include the structure and design of the mall, the opening hours of the mall, the cleanliness of the mall and its stores, and its ambience.

Table 5.1. The main factors of mall patronage

Reference	Service Quality	Prices	Parking	Product quality	Product variety	Distance from home	Convenience	Store variety
Downs(1970)	x	x				x		
Burnett (1973)			x	x	x	x		
Hudson (1974)		x		x		x		
Pacione (1975)		x		x	x	x		
Spencer (1978)		x			x	x	x	
Patricios (1979)	x				x	x	x	
Schuler (1979)		x	x	x		x		
Bloomstein et al.(1980)	x	x			x	x	x	
Hudson (1981)		x		x	x	x		x
Timmermans et al. (1982)			x			x		x
Finn & Louviere (1990)	x	x		x	x	x		
Drezner & Furihata (2000)								x
Yavas (2000)		x		x	x			x

A wide variety of researchers has reported research with respect to shopping center features that are deemed important by customers. Among them are Samli et al. (1983), Lord (1986), Finn and Louviere (1990), Roy (1994), Bloch et al. (1994), Labich (1995), Gips (1996), Haynes and Talpade (1996), Klara (1997), Kasrel (1998). Additional research in consumer behavior focuses on consumer characteristics and personality (Swinyard (1998), Shim and Eastlick (1998), the motivation for mall visit (Bloch et al., 1994), and the role of promotion in mall visits (Roy, 1994). Other research in the area focuses on grocery shopping (Bell et al., 1996, 1998; Ho (1998), Langston et al., 1997).

In a recent investigation, Yavas (2000) found that among mall shoppers who visit a mall four or more times a month, the following attributes are considered the most important ones (by rank order): cleanliness, variety of stores, price competitiveness, merchandise quality in stores, product selection in stores. Drezner and Furihata (2000) found that, for both the U.S. and Japan, the three most important components of mall attractiveness (in order) are variety of stores, favorite brand names, and mall appearance. Surprisingly, prices and distance did not prove to be important. Chen (2001) investigated consumer decision making based on habitual domains and competence sets.

The main question is now how to incorporate the multitude of relevant attributes in a model that includes consumer behavior. One way or another, it is necessary to aggregate the utilities of individual attributes in some composite expression. There are two main approaches to this task, an additive and a multiplicative model. The basic idea in additive models is that the overall utility of a facility can be expressed as the weighted sum of its rated attributes, whereas multiplicative models assume that the weighted product of rated attributes multiplied by a descending function of the distance is an appropriate measure of the attractiveness of the facility. Without further investigating the problem, we would like to point out the difficulty of quantifying some of the qualitative individual attributes, e.g., the friendliness of staff, quality of service, atmosphere, and others. The main difference between additive and multiplicative models is that in additive models, an attribute that is either completely lacking in a facility or exists only in a rudimentary fashion, can be compensated for by other factors that exist in abundance. This is not the case in multiplicative models; here, a nonexistent factor will cause the entire expression to assume a utility of zero, see, e.g., Meyer et al. (1980). There are certain situations, in which multiplicative models are appropriate. Eiselt and Eiselt (1998) report about such a situation in the hospitality industry, where the attribute "cleanliness" is of such overriding importance in the semi- official grading of accommodations, that it being deficient (and thus having a small score), the attractiveness of the entire facility and its utility will be greatly affected. Based on his findings in shopping centers in the Netherlands, Timmermans (1982) found that consumer behavior neither firmly supports a multiplicative nor an additive choice rule. From a theoretical point of view, additive models have long been popular with decision analysts, while multiplicative models are frequently used by economists and marketers. Typical examples of the former are multi-attribute utility functions and other decision analysis techniques such as the analytic hierarchy process, or any of the ELECTRE and PROMETHEE methods. On the other hand, economists often use production functions of the Cobb- Douglas type, which combine effects of capital and labor in a multiplicative fashion.

Formally, we can define a general "attractiveness index" w_j that is associated with the jth facility and that includes all relevant features of that facility. The utility in an additive model can then be written as $u_{ij} = -d_{ij}^k + w_j$, while the utility in a multiplicative model can be expressed as $u_{ij} = w_j/d_{ij}^k$ where $k \geq 1$ is a factor that is further alluded to below.

Additive models go back to Launhardt (1885) and Fetter (1924), who coined the term "economic law of market areas". The original idea was to delineate market areas based on customers who minimize the full price of a homogeneous good, i.e., the sum of the store price plus transport cost. In our context, this is an additive model with a weight w_j denoting the price at facility F_j and d_{ij} being the transportation costs from customer C_i to facility F_j, resulting in the full price $d_{ij} + w_j$ at C_i. In the simple Hotelling case of equal weights and Euclidean distances, since the w_j's are the same at both facilities, the line of equal full prices charged by two facilities F_j and F_k (and therefore the set of points of equal facility attractiveness) is the perpendicular bisector of the plane between the two facilities. In the case of unequal weights, this line is a hyperbola. Researchers investigating competitive locations using additive models are Ben-Akiva et al. (1989), Capozza and van Order (1989), or Economides (1986) in the deterministic context, while Braid (1988), DePalma et al. (1985, 1987), and Tabuchi (1988) have examined pertinent probabilistic models.

Drezner (1994) proposed that the competing retail outlets are not equally attractive, and that consumers construct a hypothetical utility function for each. This utility function is an additive composite measure, where different stores are rated on various consumer-specific attributes and each attribute is multiplied by its relative importance (weight) to form one composite index of attractiveness for each store. Each store, then, has its own utility measure. A trade-off between distance and attractiveness expresses the fact that consumers are willing to travel greater distances to more attractive stores up to a "breakeven" distance, at which point the two facilities are equally attractive. Consumers then choose to patronize the store with the highest utility value.

Multiplicative models were first mentioned by Reilly (1929, 1931) and his famed law of retail gravitation. It states that the degree to which a customer is attracted to a facility is proportional to the facility's weight and inversely proportional to the square of the distance between facility and customer. Formally, we can write a utility $u_{ij} = w_j/d_{ij}^2$ for each facility – customer combination. Gravity models are based on their physical equivalents, such as gravity, optics, and acoustics, as, e.g., gravitational force, the intensity of light and sound also diminish with the square of the distance. Reilly also performed empirical tests by varying the power of the distance. These powers measure the degree to which the distance has an effect on the utility. They are called *distance decay* and measure the rate at which a customer's utility for a facility declines, or, in other words, the likelihood of patronage diminishes.

In general, large powers indicate that the attractiveness of a facility diminishes rapidly with distance, whereas facilities with small powers retain large attractiveness even at significant distances. Reilly determined that for most goods the power ranges from 1.5 to 2.75 with an average of 2. Some goods, however, exhibited a distance decay as low as 0, while others showed values as high as 12.5. Huff (1964) found values of 2.723 for furniture and 3.191 for clothes.

Deterministic gravity models with linear distance decay have been described by Eiselt (1991) and Eiselt and Laporte (1991). Regardless of the decay function, these approaches result in market areas that are not necessarily connected. In other words, it is possible for a customer, located at some given point, to be most attracted to a facility F_j but, as the customer approaches the facility on a "reasonable" (e.g., the shortest) path, he is suddenly attracted more to a different facility F_k; a phenomenon discussed by Drezner et al. (1996). The reason for this counterintuitive behavior is based on the fact that the multiplicative model has a nonlinear decay with respect to distance, so that the utilities of two customer - facility pairs may intersect, which is not the case in the additive model.

One of the criticisms leveled against Reilly's model is based on what it is designed to do, i.e., outline market areas. Many practitioners claim that there exists no breaking point beyond which customers switch their alliance to a different facility. This feature is also referred to as the "winner-takes-all" property which essentially states that all customers at any given site patronize the same facility or, equivalently, a single customer satisfies his entire demand at a single facility. While practical observations clearly indicate that many trading areas overlap, it is not trivial to discard Reilly's model from a theoretical point of view. If customers have full knowledge about distances and facility attributes and their valuation of these attributes, i.e., their utility function does not change over time, why would a customer patronize a facility other than the one with the highest utility? One possible explanation could use the argument that customer demand changes over time, i.e., the utility of a facility diminishes each time a customer patronizes it. Rather than including the decay of attractiveness over time in a dynamic model, one could do away with the winner-take-all assumption and consider instead a model in which customers patronize the given facilities in relation to their utilities.

This view resulted in the class of "proportional models" advocated by Huff (1964). Similar to Reilly, Huff considers the absolute utility of a customer C_i for a facility F_j as $u_{ij} = w_j/d_{ij}^2$, where he uses travel time (rather than distance) for the parameter d_{ij}. The relative utility is then calculated by dividing the absolute utility by the sum of utilities the customer has to all facilities in the model. Formally, we can write

$$u_{ij} = \frac{\frac{w_j}{d_{ij}^k}}{\sum_t \frac{w_t}{d_{it}^k}}$$

where k denotes the parameter that measures distance decay and the weights w_j symbolize the floor space of a facility (as a proxy expression for selection of goods). Customers are then assumed to patronize facilities in proportion to these utilities. The expected values of the sales of a facility can then be calculated. In this model, a customer will satisfy some portion of his demand at each of the existing facilities, so that there are no exclusive market areas in which all demand is satisfied at one facility. Nakanishi and Cooper (1974) suggested the multiplicative competitive interaction (MCI) model as an extension to Huff's (1964) shopping center floor area as a surrogate expression for facility attractiveness. The MCI model uses a weighted composite multiplicative index of attributes.

Drezner (1998) investigated the optimal number of new retail facilities to be opened in a market area. Is it a preferred startegy to establish one "big" facility or several "smaller" facilities? She concludes that the best market entry strategy depends on the nature of the relationship between investment and attractiveness. New franchises entering a mature market with established competition have an increasing marginal return investment-attractiveness curve and should invest the total budget in one new facility rather than open a number of smaller facilities simultaneously. Established chains which have a decreasing marginal return on investment should consider equally dividing the available budget among several new facilities.

Drezner and Drezner (1996) proposed a probabilistic "random utility model" to the standard competitive location model. The assumption in their model is that customers patronize various competing facilities according to some probability distribution, rather than confining their buying to one store or one mall. In the resulting model, consumers spread their buying power between various retail outlets and, in doing so, the "all or nothing" concern is resolved.

5.3 The Lack of Rationality

All models presented in the previous two sections make assumptions concerning customer perception and behavior. In particular, they assume that customers have some kind of utility function based on which a logical decision-making process takes place. Tversky and Kahneman (1974) were among the first to question the assumption of *rational man*. They claim that the majority of individuals have rather limited information processing capabilities, and because of these limitations are unlikely to make decisions that optimize some function. This echoed the earlier thoughts of Simon (1957), whose concept of *satisficing* assumes that decision makers are content once they have reached a certain level of utility. This is, however, not a contradiction to human optimizing capabilities: the multi-level optimization structure of goal programming reflects satisficing behavior. March (1978) and Simon (1979) coined the phrase of *bounded rationality* which discards the assumption that customers have perfect information and, for that matter, neither have the information they need nor know what information they need.

In the context of grocery shopping, research suggests that consumers do not generally know the prices in each store for each product on their shopping list. Rather, they have knowledge of the price distribution as pointed out by Dickson and Sawyer (1990), Assuncao and Meyer (1993), Ho et al. (1996), and Lal and Rao (1997). Research by marketers such as Winer (1986), Lattin and Bucklin (1989), Kalwaniet et al. (1990), and Kalyanaram and Winer (1995) support the notion that consumers develop knowledge for product prices and the pricing environment in different stores over time, and that this knowledge influences choice behavior. Typically, consumers have some sense of *relative* mean price levels in different stores. Bell and Tang (1996) argue that such knowledge is acquired through exposure to television and newspaper advertising, and shopping experience, a repetitive and frequent activity for supermarket shopping. Alba et al. (1994) showed that consumers retain strong impressions about relative price levels across supermarkets, and that these impressions are consistent with actual price levels at the stores.

There are, of course, reasons other than price that influence store choice. One of them is habitual store visit behavior, which generates two kinds of store loyalty:

1. store loyalty, i.e., habitual preference for a store analogous to brand loyalty, and
2. category-specific store loyalty which is the association of specific brand-name products with specific stores, which can vary from trip to trip.

Habitual behavior is obviously relevant to store choice. Habitual visits to one store facilitate the development of familiarity with the store's characteristics, such as service, parking, location of products, etc., and thus implicitly reduce the cost of subsequent visits (mostly in terms of shopping time) to that same store; see, e.g., Ho et al. (1997).

An additional dimension of store choice is captured by the concept of "basket size thresholds". This threshold is defined as the breakeven quantity above which one store is preferable to the other, and below which the reverse is true. Simply put, consumers will patronize nearby, and possibly more expensive, stores for small quantities, while they will accept longer drives if the quantity they are interested in buying, justifies it. Ho et al. (1997) demonstrated that the threshold has a linear expected disutility over the total shopping cost. The authors also found a pattern of patronage of supermarkets that runs counter to conventional retailing wisdom that location explains most of the variance in store choice. In particular, the average number of trips received by a particular store, from a given segment, is not necessarily correlated to distance. Rather, store choice is better explained by consumer response to shopping costs, in which locational differences are only part of the total cost of shopping. Clearly, distinct customer segments are associated with different response profiles and shopping patterns which, in turn, are closely linked to demographics. For instance, larger, younger families with lower income per person typically prefer "every-day-low-price" stores. Drezner and Furihata (1999) also found distance not to be correlated with mall patronage in both Japan and the U.S.

Another attempt to explain the idiosyncrasies of human decision making was made by Lindblom (1959) who observed that many people appeared to prefer incremental improvements with small increases in utility over dramatic changes, even if the latter were to result in greatly improved utilities. He referred to the behavior that leads to this phenomenon as *muddling through*. In the field of optimization, it would be similar to a local search technique that attempts to achieve local improvements rather than global ones.

For our further discussion it is convenient to set up a model for the human decision making process. Again, this will be an idealized version designed to pinpoint areas in which simplifying assumptions were made. The general idea is that the first step in decision-making is the collection of information. This information enters the central part of the decision-making system, where the information is processed. One outcome of the processing is the formation of a *cognitive map*. An individual's cognitive map is an incomplete, augmented, schematized, and somewhat idiosyncratic representation of objective reality. In the context of location choice this is a representation of the physical structure of the city including its shopping opportunities and facilities. The pre-processed information together with the individual's "preference structure", which consists of the individual's hierarchies of objectives and acceptable tradeoffs, leads to a choice decision. This would again be a model of rational man, if it were not for the limitations of the processing unit, the fuzziness of the preference structure, and the contamination of the optimizer by random effects. The model proposed by Louviere and Henley (1977) is similar.

It is important to realize that the individual components in the above process are by no means independent. For instance, as an individual observes reality, his preferences may change. A typical example is advertising: a shopper, strolling through a shopping mall, may be interested in having a seafood meal, but changes his mind upon seeing an advertising of a nearby Chinese restaurant in the mall's food court.

It is readily apparent that the model presented above is but a very crude approximation to real human decision making. However, it includes three of the most important areas in which location models make simplifying assumptions about human behavior: the processor, the preference structure, and the optimizer. For instance, the proximity-based models assume that the processor recognizes only distance to facilities, and the preference structure consists of a single utility component called distance and the optimizer is a simple minimization finder.

Most research that deals with decision makers' preferences and the optimizer is found in the literature on decision analysis. For a good survey of a variety of modeling approaches, readers are referred to Olsen (1996). In the context of customer behavior, most research that deals with cognitive maps focuses on customer – facility distances. This is the aspect we focus on below.

Spatial separation can be expressed in many ways. Among them are actual distance, travel time between points, or the cost to get from one point to another. All of these terms are measurable, quantifiable entities, which is not to say that they are deterministic: while travel distances usually are, travel time and cost may vary with the time of day the trip is undertaken and the ensuing congestion. Thompson (1963) argued that rather than the objective distance (or, similarly, the expected travel time or cost), customers usually make decisions on the basis of *perceived distances*. He then used these perceived distances in conventional gravity models in an attempt to help explain store choice.

Perceived or cognitive distances form a part of a person's cognitive map, an individual's mental representation of the physical environment. Cadwallader (1975) tested the hypothesis that human decision making is more closely linked to cognitive rather than objective distance. In his work on shopping centers he found that a greater proportion of sampled shoppers believed that they patronized the nearest supermarket than actually did.

An interesting issue is that of directional bias. It refers to the fact that many individuals would judge the length of a path from a point A to a point B differently from the length of the same path from B to A. As an example, Lee (1962) found that distances outward from the center of the British city of Dundee were overestimated more than inward distances. Similarly, Golledge et al. (1969) found that the distances toward downtown in U.S. cities were overestimated, while those toward the periphery were underestimated. Not all of these differences were biases, though: experiences with past trips regarding congestion and one-way streets contributed to the differences. The work by Adams (1969) and Zannaras (1976) follows along similar lines. Given that many people live in suburban areas that surround a larger city, people from these suburbs tend to move around in, and be familiar with, that part of their metropolitan area that is on the same side of the city as their home location, resulting in a wedge-shaped bias. It has been noted in many empirical studies that reasons for such biases include a resistance to travel through or around the center of large cities, the tendency for individuals to select activity locations nearest them when they are duplicated elsewhere, and the radial nature of some transportation lines. This bias does not, however, usually hold for individuals living near the center of cities.

Golledge et al. (1969) used arguments such as these in their study to help explain the decline of patronage of central business districts, whose declining image encouraged patrons to turn their trade towards suburban shopping centers. Their argument is that if a consumer's cognitive map is incomplete, distorted, or fragmented, or includes substantial distance, directional, or temporal biases, the individual may behave "irrationally" such as choosing routes or time paths that are not even close to minimal path solutions, or they may patronize stores or centers that are much further away than more obvious, closer ones. Along similar lines, Jackle et al. (1976) have noted linkages between the frequency of patronage of a facility and the proximity of the facility to a customer's place of residence. It is hardly surprising that activities with the greatest frequency of participation are generally located close to the home, while an inverse relationship is found between the distance traveled to an activity and the frequency of participation in that activity. This does, however, pose a chicken-or-egg question: does the proximity of a facility to a customer's place of residence imply a higher frequency of patronage, or does a higher propensity for patronage result in a facility locating close to customers' homes?

Another interesting bias of cognitive distances has been linked by Lee (1970) to sex, to the nature and frequency of intervening barriers by Lowrey (1970), and to the nature of the endpoints between which cognitive distances are estimated by Sadalla and Staplin (1980). There is a general consensus that cognitive distances may be asymmetric and that, in some cases, the distances may be interpreted in a functional, proximity, or similarity context rather than in a geometrical one.

It is apparent that the modeling of a customer's processing unit and preference structure poses formidable difficulties. One approach uses pairwise comparisons, a principle suggested already by Fechner (1860). Starting with customer-generated pair-wise comparisons of preferences, Briggs (1969) uses multidimensional scaling techniques to translate this customer input into a utility scale that measures attractiveness. His results for predicting consumer spatial behavior compare favorably with the traditional gravity model formulations.

5.4 More Complex Customer Behavior

This section deals with a variety of consumers' behavioral patterns that do not fall into the mold of the somewhat simplistic models presented in the preceding sections. These extensions are of four different types:

1. Non-homogeneity of consumers
2. Non-homogeneity of goods and services,
3. Uncertainty of various types, and
4. Dynamic considerations.

First consider dropping the assumption of homogeneity of all customers at one location. So far, we have assumed that all customers at one location, say at some point P_i are as attracted to a facility at point P_j as customers at P_k, as long as the distances or travel costs $d_{ij} = d_{jk}$, i.e., the only parameter that distinguishes customers is their distance to a facility. At equidistance, a facility is equally attractive to customers at both demand points. This assumption is, however, not necessarily justified. For non-homogeneous customer populations, a facility equidistant from two different demand points will not have the same appeal. Consider, for example, a facility such as a gourmet restaurant. The restaurant will be very attractive to residents of an up scale neighborhood, while it will have little appeal in a blue-collar neighborhood. The exact opposite will be the case for a facility such as a bowling alley. One possibility to model non-homogeneities of demand is to introduce weights that do not only indicate the basic attractiveness of a facility by attribute as compared to other facilities, but does so with respect to an individual demand point (homogeneity within neighborhoods). Formally, we can introduce weights w_{ij} indicating the relative attractiveness of the facility at point P_j to consumers at point P_i.

Other non-homogeneities of customer demand occur when some customers face constraints regarding the time frame during which they can patronize a facility, e.g., due to working hours. Sheppard (1980) suggests to model such constraints by differentiating those subsets of the population that may be characterized as having more or less free choice from those that have severely constrained choices. Often the free choice subset involves higher socio-economic mobile subgroups. In contrast, the constrained choice is more typical of lower income, less mobile, or culturally, ethnically constrained groupings. Burnett and Hanson (1982) and Desbarates (1983) have focused on the constrained choice problem and suggest that many of the mismatches between spatial cognition of choice situations and objective descriptions of the same situation are the result of existing constraint situations.

Consider now dropping the assumption of homogeneity of products. Non-homogeneous products can easily be redefined as multiple products. If, in case of multiple products, customers make individual trips to satisfy each of their needs separately, then the problem is decomposable and can be solved as a number of separate and unrelated single-product trips. If this is, however, not the case, we have to deal with multi-purpose trips. The advantage of such trips can easily be demonstrated. Suppose that a customer wishes to acquire three goods or services. There are two ways to do so: either obtain them from single-product stores F_1, F_2, and F_3 that are one mile West, South, or East of the customer, respectively, or purchase all of them at a multi-product facility F^* such as a shopping mall two miles North of the customer C. This situation is depicted in Figure 5.3. A two-way trip to the mall is 4 miles long; two-way independent trips to each of the three stores are 6 miles long, and a single trip to the three stores (sequentially from one store directly to another store) from home and back home is 4.8 miles long.

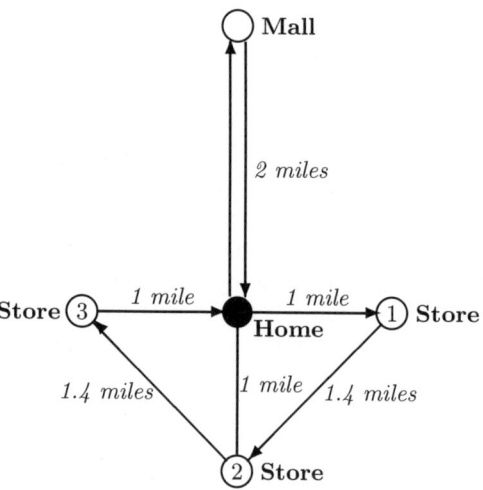

Fig. 5.3. Alternative Trips

It is apparent that while a multi-purpose shopping trip to the North will be longer than a trip to any of the single-product facilities, the overall length of the multi-purpose shopping trip is shorter than any of the other options. The locational effect of multi-purpose shopping is apparent: due to the reduction in the number of shopping trips, customers are prepared to drive to clusters of stores (i.e., bigger malls) even if they are located farther away from the customers. Facility locations outside city limits are also fostered by lower real estate costs faced by mall developers. As a result, multi-purpose shopping trips have led to the development of higher order regional shopping centers providing a range of goods and services in one location at sometimes significant distances from population centers. In addition, they have also led to interesting new phenomena. In particular, there is a tendency for shoppers to bypass the nearest commercial center and to shop at more distant ones, even if the nearer center provided some of the required goods and/or services, a behavioral pattern pointed out and modeled by Drezner et al. (1996).

Given that customers engage in multipurpose shopping, one important question is how suppliers of goods and services can take this behavior into account and capitalize on it. Consider again the retail market in shopping malls. If customers do engage in multipurpose shopping, then they will opt for convenience and try to do all their shopping in a single location. Consequently, a successful shopping mall will attempt to supply combinations of goods and services that frequently appear together in shopping baskets of customers. For instance, if groceries, 1-hour photo services and banking services frequently are on the lists of many shoppers, a mall would benefit greatly by offering those services under one roof, as customers would be quite unwilling to make an extra stop for one or two of these services elsewhere. Goods and services that are frequently demanded together by individual customers could be the basis of a plan for the mix of stores in a mall. Such a concept is reminiscent of Nelson's (1957) famous compatibility tables. In a study of household travel conducted in Uppsala, Sweden, Hansen (1971) showed the relations between multi-purpose stops on a trip, and the goods and services acquired on that trip. The author looked at four different types of goods and services to illustrate behavior for distinctly different purposes, viz., convenience (e.g., food store) shopping, shopping for non-convenience goods such as clothes, conducting personal business such as at a bank or post office, and socializing or visiting. The study determined that only 27% of food shoppers confined their shopping to one store, while 60% of post office users went to a single location, i.e., the purpose of the travel affects the number of destinations used. In a follow-up study, Hansen (1980) showed that there are groups, or bundles, of functions that are frequently used in combination on the same multi-purpose trip. For instance, on trips from home bundles of functions include other peoples' homes (for visits), doctor, dentist, shoe repair, barber, and post office. From the work place, multi-purpose trips were made to restaurants, car repair shops, banks, and grocery stores, etc.

However, the assertions that shoppers who patronize malls always engage in multi-purpose and multi- store shopping does not appear to be justified. Finn and Rigby (1992) have demonstrated that 53.2% of shoppers who patronize regional malls make only a single purchase, a finding corroborated by Eiselt et al. (1992). On the other hand, Finn and Rigby determine that shoppers behave differently when they visit mega malls, such as the West Edmonton Mall with its approximately 4 million square feet of floor space. Here, almost of the shoppers make two or more (planned and unplanned) purchases.

Another generalization of customer behavior concerns the presence of uncertainty. Consumers almost never have complete and correct knowledge of the products, their prices, the availability of the goods and services at the facilities, and the distances to the facilities. Issues relating to the uncertainty about distances was discussed in Section 5.3 and here we discuss some of the basic issues pertaining to uncertainty about prices. In the context of retailing, Potter (1976, 1977, 1978, 1979) demonstrated that consumers do go through some spatial search process before making decision concerning purchases. He also provided evidence that higher income groups had a tendency to travel greater distances to a greater variety of retail outlets than did lower income groups. A recent account of results in this area can be found in Harwitz et al. (1998, 2000).

A simple example for some of the mathematics involved in spatial search processes can be stated as follows. Suppose that a customer is interested in a product that is offered by a variety of facilities in a homogeneous quality, e.g., a brand name good. While the customer does not know the exact prices at individual stores, he is aware of the distribution of the prices. For simplicity, assume that different stores charge prices of $p = 2, 3, 5$, and 6 with probabilities of .3, .2, .4, and .1, respectively. Assume that the first store offers the product for a price of $p^* = 4$. The probability of realizing a saving with further search equals $P(p < 4) = .5$, and the probability that a further search does not result in savings equals $P(p \geq 4) = .5$. In particular, we now have possible savings of 0 (for prices $p \geq 4$), 1 (for $p = 3$), and 2 (for $p = 2$) with probabilities of .5, .2, and .3 with an expected savings of $.2 + .6 = .8$. A decision rule would then prescribe to purchase the product for $p = 4$ at the present store if the search cost to the next store is at least .8, and to continue the search if the travel cost is strictly less than .8. Reasonably realistic situations are, of course, considerably more complicated. They can involve uncertainties of travel costs, the original probability distributions, and many others.

The final extension of consumer behavior surveyed here deals with dynamic aspects. In particular, we want to address the aspect of learning over time as it applies to customers when they choose the facilities they will patronize in the future. From an empirical point of view, Rogers (1970) determined that consumer behavior, even with respect to frequent purchases such as groceries, took 6 to 8 weeks to form. Prior to that, considerable search activity and experimentation was undertaken and the funneling of behavior took place. Funneling involved successive elimination of unfavorable alternatives as a result of evaluation of a particular choice, such that over time the average number of places visited decreased and then became relatively stable. This confirmed the theoretical suppositions made by Golledge (1967), who had developed the consumer behavior, or market decision process, in the form of a dynamic Markov chain, which eliminated a number of alternatives after feedback had shown that expected reward thresholds were not reached.

The 1970s and 1980s have seen an increasing emphasis on stochastic choice models which can account for dynamic aspects of behavior. Pertinent references are Martin et al. (1978), Tardiff (1980), Crouchley, et al. (1982), Bowers (1978a, 1978b), and Arbib and Cornelis (1981).

The probability of customers patronizing a specific facility depends on the date of the last purchase, the brand last bought, size of family, level of income, exposure to advertising, and similar factors. In general, we can distinguish between two different types of models: *brand choice* models predict which specific brand a customer is going to purchase, while *purchase incidence* models predict how many purchases will occur over given time intervals. The latter include exponential market penetration models and logistic

In essence, it is a two-dimensional stochastic model that describes and predicts major regularities in consumer purchasing patterns. It does so for a single object (e.g., a single brand) or for a single event (e.g., patronizing a single store or shopping center). The model has shown to be capable of providing information on market penetration of stores over time, the proportion and significance of repeat buying and the total patronage pattern, information particularly relevant for location models.

5.5 Conclusions

The location of retail facilities in the presence of competition is a highly complex task. One of the many features that influence the location decision concerns the behavior of consumers. Consumer behavior, in turn, is the result of a multitude of influencing forces. When consumers select a retail facility to patronize, they consider distance, attractiveness, cost, as well as other factors. In modeling consumer behavior and spatial choice, it is neither correct to assume that consumers are rational, nor that they have complete information about their choice set. Furthermore, consumer behavior varies considerably across market segments and retail categories. As an example, the choice of a supermarket is most likely motivated by factors different from those affecting mall choice. Attempting to model customer behavior is, therefore, a very complex problem.

For modeling purposes, researchers commonly assume that customers select the facility closest to them. This assumption simplifies the resulting models and makes them more tractable, but at the cost of providing but a crude approximation of actual consumer behavior. This paper has surveyed some more sophisticated approaches that incorporate some components of real customer behavior. It is apparent that in order to remain computationally tractable, even the most sophisticated models will have to make serious limiting assumptions such as rational customer behavior and complete information. While they have their own limitations, the inclusion of parameters such as a store's pricing policies and its perceived attractiveness results in computationally tractable models that are considerably closer to reality than simple proximity-based models.

Acknowledgments This research was in part supported by a grant from the Natural Sciences and Engineering Research Council of Canada under grant number OGP00009160. This support is gratefully acknowledged.

References

Adams, J.S. (1969) "Directional Bias in Intra-Urban Migration," *Economic Geography*, 45:302-323.

Alba, J.W., S. M. Broniarczyk, T.A. Shimp and J.E. Urbany (1994) "The influence of Prior Beliefs, Frequency Cues, and Magnitude Cues on Consumers' Perceptions of Comparative Price Data," *Journal of Consumer Research*, 21, 219-235.

Arbib, M.A., and A. Cornelis (1981) "The Role of Systems Theory in Social Sciences: An Interview," *Journal of Social and Biological Structures*, 4, 375-386.

Assuncao, J., and R.J. Meyer (1993) "The Rational Effect of Price Promotions on Sales and Consumption," *Management Science*, 39, 517-535.

Bell, D.R., T.-H. Ho and C.S. Tang (1998) "Determining Where to Shop: Fixed and Variable Costs of Shopping," *Journal of Marketing Research*, 35, 352-370.

Bell, D.R. and J.M. Lattin (1998) "Grocery Shopping Behavior and Consumer Response to Retailer Price Format: Why 'Large Basket' Shoppers Prefer EDLP," *Marketing Science*, 17, 66-89.

Bellenger, D.N., D.H. Robertson and B.A. Greenberg (1977) "Shopping Center Patronage Motives," *Journal of Retailing*, No.2, 29-38.

Bettman, J. (1997) *An Information Processing Theory of Consumer Choice*, Addison-Wesley, Reading, Ma.

Bloch, P.H., N.M. Ridgway and S.A. Dawson (1994) "The Shopping Mall as Consumer Habitat," *Journal of Retailing*, 70, 23-42.

Bloomstein, H., P. Nijkamp, and W. van Veenendaal (1980) "Shopping Perceptions and Preferences: A Multidimensional Attractiveness Analysis of Consumer and Entrepreneurial Attitudes," *Economic Geography*, 56, 155-174.

Briggs, R. (1969) "Scaling of Preferences For Spatial Location: An Example Using Shopping Centers," Unpublished MA Thesis, Department of Geography, Columbus, Ohio: Ohio State University.

Bucklin, R.E., and J.M. Lattin (1992) "A Model of Product Category Competition Among Grocery Retailers," *Journal of Retailing*, 68, 271-293.

Burnett, K.P. (1973) "The Dimensions of Alternatives in Spatial Choice Processes," *Geographical Analysis*, 5, 181-204.

Burnett, K.P., and S. Hanson (1982) "The Analysis of Travel as an Example of Complex Human Behavior in Spatially-Constrained Situations: Definition and Measurement Issues," *Transportation Research A*, 16, 87-102.

Cadwallader, M. (1975) "A Behavioral Model of Consumer Spatial Decision-Making," *Economic Geography*, 51, 339-349.

Cadwallader, M. (1977) "Frame Dependence in Cognitive Maps: An Analysis Using Directional Statistics," *Geographical Analysis*, 9, 284-291.

Cadwallader, M. (1981) "Towards a Cognitive Gravity Model: The Case of Consumer Spatial Behavior," *Regional Studies*, 15, 275-284.

Chatfield C., A.S.C. Eherenberg, and G.J. Goodhardt (1966) "Progress on a Simplified Model of Stationary Purchasing behavior," *Journal of the Royal Statistical Society (Series A)*, 129, 317-367.

Chen T.Y. (2001) "Using Competence Sets to Analyze the Consumer Decision Problem," *European Journal of Operational Research*, 128, 98-118.

Christaller, W. (1933) *Die zentralen Orte in Süddeutschland (trans. Baskin, C.W., 1963, as Central Places in Southern Germany)*, Prentice-Hall, Englewood Cliffs, N.J..

Clark, W.A.V. (1969) "Information Flows and Intra-Urban Migration: An Empirical Analysis," *Annals of the Association of American Geographers*, 1, 38-42.

Clark, W.A.V., and G. Rushton (1970) "Models of Intra-Urban Consumer Behavior and Their Applications For Central Place Theory," *Economic Geography*, 46, 486-497.

Costanzo, C.M., W.C. Halperin, N. Gale, and G.D.R. Richardson (1982) "An Alternative Method for Assessing Goodness-of-fit for Logit Models," *Environment and Planning, A*, 14, 963-971.

Crouchley, R., A. Pickles, and R. Davies (1982) "Dynamic Models of Shopping Behavior: Testing the Linear Learning Model and Some Alternatives," *Geografiska Annaler B*, 63, 27-33.

Demko, D., and R. Briggs (1970) "An Initial Conceptualization and Operationalization of Spatial Choice Behavior: A Migration Example Using Multidimensional Unfolding," *Proceedings, Canadian Association Of Geographers*, 1, 79-86.

Desbarates, J. (1983) "Spatial Choice and Constraints on Behavior," *Annals of the Association of American Geographers*, 73, 340-357.

Dickson, P. and A. Sawyer (1990) "The Price Knowledge and Search of Supermarket Shoppers," *Journal of Marketing*, 54, 42-53.

Downs, R.M. (1970a) "The Cognitive Structure of an Urban Shopping Center," *Environment and Behavior*, 2, 13-39.

Downs, R.M., (1970b) "Geographic Space Perception: Past Approaches and Future Prospects," In C. Board, R.J. Chorley, P. Haggett, and D.R. Stoddart (eds.): *Progress in Geography*, Vol. 2, 65-108, Edward Arnold, London.

Drezner, T. (1994) "Locating a Single New Facility Among Existing, Unequally Attractive Facilities," *Journal of Regional Science*, 34, 237-252.

Drezner, T. (1995) "Competitive Facility Location in the Plane," in Z. Drezner (Ed.) *Facility Location: A Survey of Applications and Methods*, Springer, NY, 285-300.

Drezner T. (1998) "Location of Multiple Retail Facilities with a Limited Budget," *Journal of Retailing and Consumer Services*, 5, 173-184.

Drezner, T. and Z. Drezner (1996) "Competitive Facilities: Market Share and Location with Random Utility," *Journal of Regional Science*, 36, 1-15.

Drezner, T., Z. Drezner, and G.O. Wesolowsky (1998) "On the Logit Approach to Competitive Facility Location," *Journal of Regional Science*, 38, 313-327.

Drezner, T., Z. Drezner and H.A. Eiselt (1996) "Consistent and Inconsistent Rules in Competitive Facility Choice," *The Journal of the Operational Research Society*, 47, 1494-1503.

Drezner T. and Z. Drezner (2000) "On the Attractiveness of Shopping Malls," *Proceedings of the 29th Western DSI Conference*, Kapalua, Maui, Hawaii, 766-768.

Drezner T. and Furihata T. (2000) "Components of attractiveness of shopping malls – U.S. and Japan," Working paper, College of Business and Economics, California State University-Fullerton.

Ehrenberg, A.S.C. (1972) *Repeat Buying: Theory and Application*, North Holland Publishing, Amsterdam.

Eiselt, H.A., and M. Eiselt (1998) "Grading Procedures in the Tourism Industry: The Case of New Brunswick", *Canadian Journal of Administrative Sciences*, 15, 65-75.

Eiselt, H.A., M. Gendreau, and G. Laporte (1992) "Tracking Down Elusive Customers The Fredericton Mall Survey", *OR Insight*, 5, 18-22.

Eiselt, H.A., and G. Laporte (1998) "Demand Allocation Functions", *Location Science*, 6, 175-187.

Fechner, G.T. (1860) *Elemente der Psychophysik*, Breitkopf & Hörtel, Leipzig, Germany.

Finn, A. and J. Louviere (1990) "Shopping-Center Patronage models: Fashioning a Consideration set Segmentation Solution," *Journal of Business Research*, 21, 259-275.

Finn, A., and J. Rigby (1992) "West Edmonton Mall: Consumer Combined-Purpose Trips and the Birth of the Mega-Multi-Mall?", *Canadian Journal of Administrative Sciences*, 9, 134-145.

Gaile, G.L., and J.E. Burt (1980) "Directional Statistics," *Concepts and Techniques in Modern Geography*, No. 25, University of East Anglia.

Gips, M. (1996) "Shopping for Security," *Security Management*, January, 12.

Golledge, R.G. (1967) "Conceptualizing the Market Decision Process", *Journal of Regional Science*, 7, 239-258.

Golledge, R.G., R. Briggs, and D. Demko (1969) "The Configurations of Distances in Intra-Urban Space," *Proceedings of the Association of American Geographers*, 1, 60-65.

Golledge, R.G., and L.A. Brown (1967) "Search, Learning and the Market Decision Process," *Geografiska Annaler*, 49B, 116-124.

Golledge, R.G., and G. Rushton (1972) "Multidimensional Scaling: Review and Geographical Applications," *AAG Commission on College Geography*, Technical Paper, 10.

Golledge, R.G., G. Rushton, and W.A.V. Clark (1966) "Some Spatial Characteristics of Iowa's Farm Population and their Implications for the Grouping of Central Place Functions," *Economic Geography*, 43, 261-272.

Golledge, R.G. and R.J. Stimson (1987) *Analytical Behavioral Geography*, Croom Helm, New York.

Halperin, W.C. (1985) "The Analysis of Panel Data for Discrete Choices," in P. Nijkamp, H. Leitner, and N. Wrigley (eds.): *Measuring the Unmeasurable*, Martinus Nijhoff Publishers, The Hague, 561-586.

Hanson, S. (1978) "Measuring the Cognitive Levels of Urban Residents," *Geografiska Annaler*, B59, 67- 81.

Hanson, S. (1980) "Spatial Diversification and Multipurpose Travel: Implications for Choice Theory," *Geographical Analysis*, 12, 245-257.

Hanson, S. (1984) "Environmental Cognition and Travel Behavior," In D.T. Herbert and R.J. Johnston (eds.): *Geography and the Urban Environment: Progress in Research and Application*, Vol. 6, John Wiley, London, 95-126.

Hanson, S., and D.F. Marble (1971) "A Preliminary Typology of Urban Travel Linkages," *East Lakes Geographer*, 7, 49-59.

Harwitz, M., B. Lentnek, P. Rogerson, and T.E. Smith (1998) "Optimal Search on Spatial Paths with Recall, Part I: Theoretical Foundations", *Papers in Regional Science*, 77, 301-327.

Harwitz, M., B. Lentnek, P. Rogerson, and T.E. Smith (2000) "Optimal Search on Spatial Paths with Recall,Part II: Computational Procedures and Examples", *Papers in Regional Science*, 79, 293-305.

Haynes, J.B., and S. Talpade, 1996, "Does Entertainment Draw Shoppers? The Effect of Entertainment Centers on Shopping Behavior in Malls," *Journal of Shopping Center Research*, No. 2, 29-48.

Ho T.-H., C.S. Tang and D.R., Bell (1998) "Rational Shopping Behavior and the Option Value of Variable Pricing", *Management Science*, 44, S145-160.

Hoch, S.J., X. Dreze and M.E. Purk (1994) "EDLP, Hi-Lo, and Margin Arithmetic," *Journal of Marketing*, 58, 16-27.

Hoch, S.J., B.-D. Kim, A.M. Montgomery, and P.E. Rossi (1995) "Determinants of Store-Level Price Elasticity," *Journal of Marketing Research*, 32, 17-29.

Hotelling, H. 1929 "Stability in Competition", *Economic Journal*, 39, 41-57.

Huff, D.L. (1962) "Determination of Intra-Urban Retail Trade Areas, Real Estate Research Program," Ph.D. dissertation, University of California, Los Angeles, CA.

Huff, D.L. (1964) "Defining and Estimating a Trade Area," *Journal of Marketing* 28, 34-38.

Huff, D.L. (1966) "A Programmed Solution for Approximating an Optimum Retail Location," *Land Economics*, 42, 293-303.

Jackle, J.A., S. Brunn, and C.C. Roseman (1976) *Human Spatial Behavior*, Duxbury Press, North Scituate, Ma.

Jain A.K. and V. Mahajan (1979) "Evaluating the competitive environment in retailing using multiplicative competitive interactive models," in Sheth J. (Ed.), *Research in Marketing*, JAI Press, Greenwich, Conn. 217-235.

Kahn, B., and D.C. Schmittlein (1989) "Shopping Trip Behavior: An Empirical Investigation," *Marketing Letters*, 1, 55-69.

Kahn, B., and D.C. Schmittlein (1992), "The Relationship Between Purchases Made on Promotion and Shopping Trip Behavior," *Journal of Retailing*, 68, 294-315.

Kalwani, M.U., C.H. Yim, H.J. Rinne, and Y. Sugita (1990), "A Price Expectations Model of Customer Brand Choice," *Journal of Marketing Research*, 27, 251-262.

Kalyanaram, G., and R.S. Winer (1995) "Empirical Generalizations from Reference Price Research," *Marketing Science*, 14, G161-179.

Kasrel, D. (1998) "Feeding Frenzy at the Malls," *Philadelphia Business Journal*, July 31, 3-4.

Klara, R. (1997) "Courting the Future: Once the Ugly Ducklings of the Industry, Mall Eateries are Going Through Some Amazing Transformations," *Restaurant Business*, January 15, 25-27.

Labich K., (1995), "What It Will Take to Keep People Hanging Out at the Mall," *Fortune*, May 29, 102- 106.

Lal, R., and C. Matures (1994) "Retail Pricing and Advertising Strategies," *Journal of Business*, 67, 345-370.

Lal, R., and R. Rao (1997) "Supermarket Competition: The Case of Every Day Low Pricing," *Marketing Science*, 16, 60-80.

Langston P., G.P. Clarke, D.B. Clarke (1997) "Retail Saturation, Retail Location, and Retail Competition: an Analysis of British Grocery Retailing," *Environment and Planning A*, 29, 77-104.

Lattin, J.M., and R.E. Bucklin (1989) "Reference Effects of Price and Promotion on Brand Choice Behavior," *Journal of Marketing Research*, 26, 299-310.

Lee, T.R. (1962) "Brennan's Law of Shopping Behavior," *Psychological Reports*, 11, 662.

Lee, T.R. (1970) "Perceived Distance as a Function of Direction in the City," *Environment and Behavior*, 2, 40-51.

Lloyd, R., and D. Jennings (1978) "Shopping Behavior and Income: Comparisons in an Urban Environment," *Economic Geography*, 54, 157-167.

Looman, J.D. (1969) "Consumer Spatial Behavior: A Conceptual Model and an Empirical Case Study of Supermarket Patronage," unpublished M.A. Thesis, Department of Geography, The Ohio State University, Columbus, Ohio.

Lord, J.D. (1986) "Cross Shopping Flows Among Atlanta's Regional Shopping Centers," *International Journal of Retailing*, 1, 33-54.

Louviere, J. (1974) "Predicting the Response to Retail Stimulus Objects from an Abstract Evaluation of their Attributes: The Case of Trout Streams," *Journal of Applied Psychology*, 59, 572-577.

Louviere, J., and H.D. (1977) "Information Integration Theory Applied to Student Apartment Selection Decisions," *Geographical Analysis*, 9, 130-141.

Louviere, J., and R. Meyer (1976) "A Model for Residential Impression Formation," *Geographical Analysis*, 8, 479-486.

Louviere, J., and R. Meyer (1981) "A Composite Attitude-Behavior Model of Traveler Decision- Making," *Transportation Research*, B15, 411-420.

Lowrey, R.A. (1970) "Distance Concepts of Urban Residents," *Environment and Behavior*, 2, 52-73.

MacKay, D.B., and R. Olshavsky (1975) "Cognitive Maps of Retail Location: An Investigation of Some Basic Issues," *Journal of Consumer Research*, 2, 197-205.

Martin, R.L., N.J. Thrift, and R.J. Bennett (1978) "Towards the Dynamic Analysis of Spatial Systems," Pion, London.

Meyer, R.J., I.P. Levin, J. Louviere, and D. Henley (1980) "Issues in Modeling Travel Behavior in Simulated Choice Environments: A Review," Department of Geography, University of Iowa, Iowa City.

Morwitz, V. and L. Block (1996), "An Analysis of Planned Grocery Shopping: Influences on Consumer Goal Setting and Goal Fulfillment," Stern School, NYU Working Paper.

Mulhern, F.J., and R.P. Leone (1990) "Retail Promotional Advertising: Do the Number of Deal Items and Size of Deal Discounts Affect Store Performance?" *Journal of Business Research*, 21, 179-194.

Nakanishi, M., and L.G. Cooper (1974) "Parameter Estimate for Multiplicative Interactive Choice Model: Least Squares Approach", *Journal of Marketing Research*, 11, 303-311.

Nevin, J.R., and M.J. Houston (1980) "Image as a Component of Attraction to Intraurban Shopping Areas," *Journal of Retailing*,1, 77-93.

Okabe A., B. Boots, and K. Sugihara (1992) *Spatial Tessellations: Concepts and Applications of Voronoi Diagrams*, John Wiley & Sons, Chichester.

Olsen, D.L. (1996) *Decision Aids for Selection Problems*, Springer Verlag, New York, NY.

Pacione, M. (1976) "A Measure of the Attraction Factor: A Possible Alternative," *Area*, 6, 279-282.

Pipkin, J. (1981) "The Concept of Choice and Cognitive Explanations of Spatial Behavior," *Economic Geography*, 57, 315-331.

Pipkin, J.S. (1981) "Cognitive Behavioral Geography and Repetitive Travel," in K.R. Cox and R.G. Golledge (eds.): *Behavioral Problems in Geography Revisited*, Methuen, New York, 145-181.

Plastria, F. (2001) "Static Competitive Facility Location: An Overview of Optimisation Approaches," *European Journal of Operational Research*, 129, 461-470.

Potter, R.B. (1976) "Directional Bias within the Usage and Perceptual Fields of Urban Consumers," *Psychological Reports*, 38, 988-990.

Potter, R.B. (1977) "Spatial Patterns of Consumer Behavior and Perception in Relation to the Social Class Variable," *Area*, 9, 153-156.

Potter, R.B. (1978) "Aggregate Consumer Behavior and Perception in Relation to Urban Retailing Structure: A Preliminary Investigation", *Tidschrift voor Economiche en Sociale Geografie*, 69, 345- 352.

Potter, R.B. (1979) "Perception of Urban Retailing Facilities: An Analysis of Consumer Information Fields," *Geografiska Annaler*, 61B, 19-29.

Rickard, L. (1995), "Shopping lists Tell Just Half the Real Story," *Advertising Age*, January 9, 20.

Rogers, D.S. (1970) "The Role of Search and Learning in Consumer Space Behavior: The Case of Urban In-Migrants," Unpublished M.A. Thesis, Department of Geography, University of Wisconsin, Madison, Wisconsin.

Roy, A. (1994) "Correlates of Mall Visit Frequency," *Journal of Retailing*, 70, 139-161.

Rushton, G. (1965) "The Spatial Pattern of Grocery Purchases in Iowa, Ph.D. Dissertation, Department of Geography, University of Iowa, Iowa City, Iowa.

Sadalla, E.K., and L.J. Staplin (1980a) "The Perception of Traversed Distance: Intersections," *Environment and Behavior*, 12, 167-182.

Samli, A.C., G. Riecken, and U. Yavas (1983) "Intermarket Shopping Behavior and the Small Community: Problems and Prospects of a Widespread Phenomenon," *Journal of the Academy of Marketing Science*, 2, 1-14.

Schuler, H.J. (1979) "A Disaggregate Store Choice Model of Spatial Decision Making," *The Professional Geographer*, 31, 146-16.

Sheppard, E.S. (1980) "The Ideology of Spatial Choice," *Papers of the Regional Science Association*, 45, 197-213.

Simester, D.I. (1995) "Signaling Price Image Using Advertised Prices," *Marketing Science*, 14, 166-188.

Shim, S., and M.A. Eastlick (1998) "The Hierarchical Influence of Personal Values on Mall Shopping Attitude and Behavior," *Journal of Retailing*, No. 1, 139-160.

Simon, H.A. (1957) *Models of Man* Wiley, New York.

Singson, R. (1975) "Multidimensional Scaling Analysis of Store Image and Shopping Behavior," *Journal of Retailing*, 38-52.

Spencer, A.H. (1978) "Deriving Measures of Attractiveness for Shopping Centers," *Regional Studies*, 12, 713-726.

Spencer, A.H. (1980) "Cognition and Shopping Choice: A Multidimensional Scaling Approach," *Environment and Planning A*, 12, 1235-1251.

Stanley, T., and M. Sewall (1976) "Image Inputs to Probabilistic Model: Predicting Retail Potential," *Journal of Marketing*, 40, 48-53.

Swinyard, W.R. (1998) "Shopping Mall Customer Values: the Rational Mall Shopper and the List of Values," *Journal of Retailing and Consumer Services*, 5, 131-197.

Tardiff, T. (1980) "Definition of Alternatives and Representation of Dynamic Behavior in Spatial Choice Models," *Transportation Research Record*, 723, 25-30.

Thompson, D.L. (1963) "New Concept: Subjective Distance," *Journal of Retailing*, 39, 1-6.

Timmermans, H. (1981) "Spatial Choice Behavior in Different Environmental Settings: An Application of the Revealed Preference Approach," *Geografiska Annaler B*, 63, 57-67.

Timmermans, H. (1982) "Consumer Choice of Shopping Center: An Information Integration Approach," *Regional Studies*, 16, 171-182.

Timmermans, H. (1983) "Non-Compensatory Decision Rules and Consumer Spatial Choice Behavior: A Test of Predictive Ability," *Professional Geographer*, 35, 449-455.

Timmermans, H., R.V. van der Heijden, and H. Westerveld, *1982a* "Cognition of Urban Retailing Structures: A Dutch Case Study," *Tijdschrift voor Economiche en Sociale Geografie*, 73, 1-12.

Timmermans, H., R.V. van der Heijden, and H. Westerveld (1982b) "The Identification of Factors Influencing Destination Choice: An Application if the Repertory Grid Methodology," *Transportation*, 11, 189-203.

Tversky, A., and D. Kahneman (1974) "Judgment Under Uncertainty: Heuristics and Biases," *Science* 185, 1124-1131.

Wilson, A.G. (1962) "Livability of the City: Attitudes and Urban Development," in F.S. Chapin and S.F. Weiss (eds.): *Urban Growth Dynamics*, John Wiley, New York, 359-399.

Wilson, A.G. (1972) "Some Recent Developments in Microeconomic Approaches to Modeling Household Behavior, With Special Reference to Spatiotemporal Organization," In A. Wilson (ed): *Papers in Urban and Regional Analysis*, Pion, London.

Winer, R.S. (1986) "A Reference Price Model of Brand Choice for Frequently Purchased Consumer Products', *Journal of Consumer Research*, 13, 250-56.

Wrigley N. (1985) "Categorical Data Methods and Discrete Choice Modelling in Spatial Analysis: Some Directions for the 1980's," in P. Nijkamp, H. Leitner, and N. Wrigley (eds.): *Measuring the Unmeasurable*, Martinus Nijhoff, Dordrecht, 115-137.

Wrigley N., and R. Dunn (1984a) "Stochastic Panel-Data Models of Urban Shopping Behavior 3: The Interaction of Store Choice and Brand Choice," *Environment and Planning A* 16, 1221-1236.

Yavas U. (2000) "Visit Frequency to Malls: A Comparison of Four Shopper Groups," *Proceedings of the 29th Western DSI Conference*, Kapalua, Maui, Hawaii, 298-300.

Zannaras, G. (1976) "The Relation Between Cognitive Structure and Urban Form," In G.T. Moore and R.G. Golledge (eds.): *Environmental Knowing*, Dowden, Hutchinson and Ross, Stroudsburg, Pa., 336-350.

6 An Efficient Genetic Algorithm for the *p*-Median Problem

Burçin Bozkaya[1], Jianjun Zhang[2], and Erhan Erkut[3]

[1] ESRI Transportation/Logistics Services 380 New York St. Redlands, CA 92373. e-mail: bbozkaya@esri.com
[2] Telus Geomatics 16A, 10020 100 St. Edmonton, AB, Canada T5J 0N5. e-mail: j.zhang@sk.sympatico.ca
[3] School of Business University of Alberta Edmonton, AB, Canada T6G 2R6. e-mail: Erhan.Erkut@Ualberta.ca

6.1 Introduction

One of the most popular models in the facility location literature is the *p*-median model. This model locates *p* facilities in some space (such as Euclidean plane or a road network) that will serve *n* demand points. The objective of the model is to minimize the total (weighted) distance between the demand points and the facilities. The *p*-median problem is known to be NP-Hard and this has resulted in the development of heuristic solution techniques in an effort to solve large-scale problems to near-optimality with a reasonable computational effort. Perhaps the most popular *p*-median heuristic was developed by Teitz and Bart (1968). This is a node-exchange procedure that attempts to improve the objective function value at each iteration. It is simple to implement and it produces relatively good solutions to smaller problems when applied with multiple starting solutions.

There are more sophisticated heuristics for the *p*-median problem. For example, Narula et al. (1977) proposed a mathematical programming-based heuristic, which uses Lagrangian relaxation. Densham and Rushton (1992) developed an efficient two-phase search heuristic. More recently, Rolland et al. (1996) designed a Tabu search algorithm and Murray and Church (1996) developed a simulated annealing algorithm for this problem. Rosing and ReVelle (1997) showed that results from multiple runs of an exchange heuristic can be combined to construct a solution that is better than any of the best local optima previously found.

It seems that researchers have focused recently on the design of modern heuristics to solve the *p*-median problem. Modern heuristics can generate better results than the simple node–exchange procedure, and each one of the three most recent references above offers an empirical comparison demonstrating the superiority of the proposed approach to the node exchange heuristic. However, the empirical effectiveness of one modern heuristic, namely the genetic algorithm, has not been demonstrated adequately in the *p*-median literature. Genetic Algorithms (GAs) are heuristic techniques, which are usually referred to as "modern" heuristics, even though they were first developed over 30 years ago. GA is a generic method that can be applied to any problem if the feasible solutions of the problem can be represented as strings that correspond to genetic encoding of the solutions. The technique is designed to imitate the selective breeding of organisms (such as animals, plants) in the context of problem solving. This technique has been adapted to many problems in the operations research literature, such as the traveling salesman problem and scheduling problems. However there are very few applications to location-allocation problems. In this paper, we review the previous studies in this area and discuss how one can design an effective GA for this problem. We outline a GA approach to the p-median problem that involves three crossover operators for comparison and the concept of "invasion". We also report the results of an experimental study on random problems using our GAs.

We intend to demonstrate that one can design a GA which generates good solutions to the p-median problem, comparable in quality to those of the node exchange heuristics. We suspect that the non-promising results reported in the only journal article that describes a GA application to the p-median problem (Hosage and Goodchild, 1986) might have discouraged research in this area. The GAs outlined in this paper far outperform the one suggested by Hosage and Goodchild (1986), and further exploration of GAs for the p-median problem may be justified.

The organization of this chapter is as follows. In Section 6.2 we formally define and discuss the p-median problem, in Section 6.3 we provide an overview of the GA concept and discuss its important features. In Section 6.4 we summarize and critique past attempts in designing a GA heuristic for the p-median problem. In Section 6.5 we outline our GA approach and in Section 6.6 we report the results of an empirical comparison of the proposed approach with the node exchange algorithm. Finally, Section 6.7 contains our conclusions.

6.2 The p-Median Problem

The p-median model identifies locations for p new facilities in some space (such as the Euclidean plane or a road network) to serve n demand points so that the total weighted distance (or cost) between the facilities and the demand points they serve is minimized. It is a multi-facility extension of the Weber problem (Weber, 1909) which is widely accepted as the first formalized facility location problem in the literature. The first treatments of the p-median problem are due to Cooper (1963) and Hakimi (1965). ReVelle and Swain (1970) provided an integer programming formulation for the discrete

p-median problem, which is given below.

$$\min \left\{ Z = \sum_{i=1}^{n} \sum_{j=1}^{n} w_i d_{ij} x_{ij} \right\}$$

s.t.:

$$\sum_{j=1}^{n} x_{ij} = 1 \qquad \forall i$$

$$x_{ij} \leq x_{jj} \qquad \forall i,j \text{ pairs}$$

$$\sum_{j=1}^{n} x_{jj} = p$$

$$x_{ij} \in \{0,1\} \qquad \forall i,j \text{ pairs}$$

where

$x_{ij} = 1$ if point i is assigned to facility located at point j, 0 otherwise
$x_{jj} = 1$ if a facility is located at point j, 0 otherwise
$w_i =$ demand at point i
$d_{ij} =$ travel distance between points i and j
$p =$ number of facilities to be located

The solution to the above mathematical programming model identifies the locations of the facilities and the allocations of the demand points to the facilities. In the optimal solution, every demand point is served from the closest facility – demand is not split. The model assumes that there are no capacity constraints at the facilities. There is a separate trip for every demand point-facility pair; trips to several demand points cannot be combined. While these assumptions seem somewhat limiting, the p-median model is, in fact, a very useful planning tool for many single-level distribution (or collection) systems. It can be used for locating production or storage facilities to service a large region, or for locating public facilities (such as libraries) in a city. Applications of the p-median model are reported in the practice literature. For example, Fitzsimmons and Austin (1983) locate audit offices for a comptroller in the continental USA, and Erkut et al. (2000) locate service teams for a utility company serving a large portion of a Canadian province using this model.

The p-median model is perhaps the most popular facility location model among researchers and practitioners. Furthermore, Hillsman (1984) showed that a large number of facility location problems can be reformulated as p-median problems through data processing. Hence, it is important to develop algorithms to solve the p-median problem effectively. Since this problem is NP-Hard (Kariv and Hakimi, 1979), most of the algorithmic research on this problem has been devoted to developing heuristic solution procedures. Perhaps the best-known heuristic for the p-median problem is the interchange algorithm first proposed by Teitz and Bart (1968) – T&B for short. Given a solution to the problem, T&B explores solutions that differ from the current solution by only one facility location. If at least one improved solution is detected, the algorithm takes the best of all such improved solutions, makes it the current solution, and continues searching. Since the method considers interchanging only one facility at a time, it may not converge to an optimal solution. Nevertheless, multiple applications of the algorithm increase the likelihood of finding an optimal (or near-optimal) solution.

We choose to compare the performance of our GAs with the performance of T&B, since T&B is very simple to understand and implement, and it produces good solutions with limited computational effort. Furthermore, it belongs to a popular family of heuristics (neighborhood search/exchange) which has proven to be effective for many different combinatorial optimization problems. We recognize, however, that it is possible to generate better solutions for some instances of the p-median problem using other heuristics (such as Lagrangian relaxation or Tabu search). For an extensive treatment of the p-median problem and solution methods, we refer the reader to Mirchandani (1990).

6.3 Genetic Algorithms

GA is a problem-solving approach that utilizes the concepts of natural adaptation and selective breeding of organisms. The general idea was introduced by Holland and his associates in the 1960s, but it was not until recently that GA became more popular among the operations researchers. GAs are particularly useful when dealing with highly complex objective functions. The approach uses analogies between the biological process and problem solving. For example, a chromosome, which is the encoded representation of the genetic material of an organism, corresponds to a solution in the feasible solution set of a problem. In other words, each feasible solution of a problem is encoded as a string of characters from some alphabet and this string carries the information defining that particular solution. The encoding is usually a binary string, although another alphabet may be used. Each chromosome (i.e. solution) has a fitness value that corresponds to the objective function value of the associated solution. It is fair to say that the encoding of the solutions is perhaps the most crucial step in designing a GA. If a poor encoding is selected, the resulting GA is unlikely to perform better than random search regardless of the design of the rest of the algorithm.

The basic element of a GA, the chromosome, is the medium for carrying out the operations needed to perform a random search for a "good" solution over the solution space. These operations are designed to imitate the natural breeding process. Initially, there is a pool (or population) of solutions (or chromosomes) that are randomly generated. Random generation of these solutions is necessary for an unbiased representation of solutions from different parts of the solution space. Next, two solutions (or parents) from this pool are selected for "mating" in order to produce two new solutions (or offspring).

The selection of parents may be random or based on the fitness values with the expectation that good genetic material carried by the two parents will be transferred to the offspring.

The production of the offspring from the parents is the core part of a GA. It specifies explicitly how genetic material will be transferred from parents to the offspring. The chromosome of an offspring is generated by copying parts of the two parent chromosomes. In its most basic form, this is done by identifying a random crossover point in the parent chromosomes and exchanging the sections of the two parent chromosomes after the crossover point, which results in the generation of two offspring from two parents. This particular way of transferring genes from parents to the offspring is an example of what is called a crossover operator in the GA context and will be referred to as the basic operator in this paper. Clearly, there are many different ways to implement this step using different crossover operators and the performance of the algorithm depends on the selection of the operator. An offspring may be subject to mutation with a given probability. The objective of this step is to introduce genetic variety in the population and hence force the algorithm out of local optima. Following this step, the parents are replaced by the offspring in order to keep the population size constant. Alternatively, one can generate as many offspring as the population size, and then replace the entire population with the offspring. These two replacement strategies (i.e. the partial and complete replacement of the pool) are called overlapping and non-overlapping replacement respectively, and the set of new solutions that replace the parents is referred to as the new generation.

The above process can be repeated as many times as needed until a certain stopping criterion is satisfied. This criterion could be a fixed number of generations or could be related to the quality of the solutions discovered. For example, one may choose to trace the best solution found by the algorithm and terminate when this solution has not been improved upon for a given number of iterations. One immediate observation regarding GAs is the existence of a number of parameters involved in the process: the size of the initial pool, the method for selecting parents for mating, the crossover operator, the number of offspring to produce, the replacement technique, and the mutation rate. One disadvantage of GAs is the lack of existence of a single combination of these parameters that will work best for all problems, or even for different instances of the same problem. Common settings for some of the parameters, such as the mutation rate and the replacement technique, seem to work for different problems. However, other parameters may be problem-dependent. For example, a crossover operator that performs well for a particular problem may be ineffective for another problem, especially when the structure of the problems are significantly different. For this reason, one may have to perform an extensive computational study in order to fine-tune the parameters of a GA.

GA is a generic solution technique; it can be applied to any problem after developing an encoding scheme for the solutions of the problem and deciding the settings of the parameters mentioned above. However, although it is possible to design a GA for any optimization problem, GAs are not equally effective for all problems. Problems with many constraints may imply difficulties in identifying an effective encoding scheme. Also, constraints may slow down the algorithm significantly because of the additional effort needed to ensure that the constraints are not violated. In fact, GAs perform best for problems with a small number of (or no) constraints and a complicated objective function (Reeves, 1993). This makes the p-median problem a good candidate for solution via a GA, especially if the constraints of the problem can be handled through an effective encoding scheme.

Why genetic algorithms work is explained by the schema theory and the concept of intrinsic (or implicit) parallelism. These concepts basically refer to the ability of the algorithm to preserve the common sections of the chromosomes that have above-average fitness values. This occurs when, in the course of the algorithm, some components of the chromosome strings converge to certain values, which together form a particular schema. The algorithm consistently disregards schemata that correspond to inferior solutions and evaluates more and more of the schemata that correspond to solutions with high fitness values. Mutation enters into the picture occasionally (but with a very small probability) in order to divert the algorithm from possible local optima and re-direct it to untested schemata.

To summarize, GA is a random search technique designed to mimic selective breeding (evolution) in problem solving. It employs components like chromosome, crossover, mutation and utilizes the survival-of-the-fittest principle. For more detail on GAs, we refer the interested reader to Goldberg (1989) and Reeves (1993).

6.4 Review of the Relevant Literature

As described in Section 6.3, GA is a generic algorithm and, therefore, may be (in principle) applied to any problem. We find its application to the p-median problem of interest for one major reason: the technique has been tested for different types of combinatorial optimization problems and seems to perform well in some instances (Reeves, 1993). However, little effort has been directed towards designing GAs for location-allocation problems. We know of only one journal article that describes a GA for the p-median problem (Hosage and Goodchild, 1986) and the results reported in this paper are not encouraging. The only other references we can find in this area are in conference proceedings (Dibble and Densham, 1993, and Moreno-Perez et al., 1994). In this section, we review these studies.

Hosage and Goodchild (1986) – H&G for short – are the first to apply GA to the p-median problem. In their study, each solution to the problem is encoded as a string of n binary digits (genes). Each digit indicates whether or not a facility is opened at the corresponding candidate location. If the number of 1's in this string is not equal to p, the corresponding solution is infeasible and a penalty (proportional to the violation of the constraint) is imposed on the solution. The fitness value Z (i.e. the objective function value) is modified to incorporate this penalty effect. The initial population is generated randomly and the parents are selected based on the fitness values with fitter parents being associated with higher propagation probabilities. Both the crossover point and the surviving offspring are chosen randomly. The offspring selected is further subject to inversion and mutation. The best chromosome of each generation is carried into the next generation. The rest are replaced by $P - 1$ offspring, P being the population size. This process is repeated for a pre-specified number of generations and the best solution found is recorded.

The H&G algorithm works on a binary string of length n. This is an inefficient encoding scheme. Not only does it waste memory, but also it requires unnecessary operations for the crossover operator and for the computation of the fitness values. The authors test this algorithm on $_{20}C_3$ problems (i.e., $n = 20$, $p = 3$) with $P = 25$ and four different numbers of generations (120, 150, 180 and 210). They test 100 randomly generated problems and find optimal solutions 69, 85, 84, 89 times for the four generation levels. At first, it may seem that the GA has been relatively successful on these test problems finding an optimal solution between 70% and 90% of the time. However, upon closer look, the results seem not very impressive. Note that these $_{20}C_3$ problems have only 1140 distinct (feasible) solutions. Yet the algorithm generates and evaluates $25 + 24P$ solutions in each run (i.e. 2095 solutions – not necessarily distinct – for 120 generations and 5065 solutions for 210 generations). Although the problem has only 1140 solutions, this algorithm searches over a space of 2^{20} solutions defined by all combinations of the 20 binary digits. Clearly, the great majority (about 99.9%) of these solutions is infeasible for the p-median problem and the algorithm wastes computational effort by considering these infeasible solutions. Our computational experiment indicates that this algorithm is very ineffective, and even a random search over the feasible region performs considerably better than this GA.

In a conference proceeding, Dibble and Densham (1993) describe their application of GA to a multi-criteria facility location problem where they solve a $_{150}C_9$ problem with $P = 1000$ and 150 generations. They compare their results with those of T&B and report that their algorithm matches the T&B solution most of the time and always finds a solution within 0.5% of the T&B solution. However, their GA takes much longer to find the solution than T&B. To argue in favor of the GA, they suggest that GA produces a range of alternatives for multi-criteria decision problems while T&B can only produce one solution for each objective. This argument is flawed, since an exchange algorithm (such as T&B) produces a series of solutions, which can be useful in multi-criteria analysis.

The two studies summarized above imply that GAs are not competitive with the simple T&B algorithm for the p-median problem. However, we feel that the potential of GAs has not been fully exploited in these studies, and we describe a more effective GA in the next section.

Moreno-Perez et al. (1994) report on another application of the GA to the p-median problem. The most distinctive feature of their application is the existence of multiple population groups ("colonies") and the exchange of individuals between these groups ("migration") in a "parallelized" setting. This feature prevents premature homogenization in the population groups, and improves the performance of the algorithm. Unfortunately, it is difficult to judge the relative effectiveness of this algorithm against others since the authors do not provide an empirical comparison. We believe that there is room for improvement in this implementation. For example, their crossover operator allows for duplication of indices in a solution. As well, the forcing of the crossover operator to replace a facility location with another that is spatially close seems to restrict the search over the feasible region. While this may be an effective way to generate good solutions quickly (which is not necessarily desirable for a GA), it also makes it more difficult to search the entire solution space. Finally, these authors use a large population size (20 groups of 100 individuals each) which we believe is excessive for the size of the problems they solve.

6.5 The Proposed Genetic Algorithm

The basic structure of a GA was given in Section 6.3. In this section, we specify the details of the GA we propose for the p-median problem. We first define the notation and provide a generic outline of the algorithm. Then describe each aspect of the GA.

6.5.1 Notation and Outline:

The notation that will be used in the rest of the paper is as follows:

P = Size of the initial population
G = Number of generations
O = Number of overlapping solutions from one generation to the next
M = Mutation rate
C_i = Crossover operator i
Z_j = Fitness or objective function value of solution j
IF = Frequency of invasions
IR = % of population replaced by invasion.

The general outline of the proposed GA is as follows:

1. Initialize P, G, O, M.
2. Randomly generate an initial population of size P.
3. Repeat G times.
   (a) For each member of the population, compute its probability of selection as a parent.
   (b) Repeat $(P - O)$ times.
      i. Select two parents P_1 and P_2 based on the probabilities computed in Step 3a.
      ii. Use operator C_i to produce two offspring from P_1 and P_2. Randomly select one of them and keep it as the surviving offspring.
      iii. With probability M, apply mutation to a random digit of the surviving offspring.
   (c) Replace $(P-O)$ parents in the population with the $(P-O)$ offspring produced in Step 3b, while keeping the best solution found so far.
   (d) Every $\frac{1}{IF}$ generation, replace $IR\%$ of the population with new solutions (keeping the best solution of the generation)
4. Record the best solution found so far and compare it with the benchmark solution.

6.5.2 Encoding:

As discussed in the previous section, the binary string representation of p-median solutions is not effective. We use the encoding scheme proposed by Dibble and Densham (1993) which is based on strings (chromosomes) of length p and where the digits (genes) within a string correspond to the indices of open facilities. We avoid the duplication of a facility index in the string during the course of the algorithm. Hence, the encoding scheme enforces the requirement that p distinct facilities be selected.

6.5.3 Population Size:

We use the following formula to determine the value of P

$$P = \left\lceil k \times \frac{n}{p} \right\rceil. \tag{6.1}$$

where $k > 1$. This formula ties the population size to the problem parameters with the goal of including as many distinct indices in the initial gene pool as possible. This is necessary since the genes of the offspring are determined by mixing the genes of the parents, and if one of the genes of the optimal chromosome is missing from the original gene pool, the algorithm may never find the optimal chromosome. A population consisting of P members will have a total of $P \times p$ genes. Making this number a multiple of n will result in the appearance of each candidate location index, on average, k times as one of the $P \times p$ genes in the initial gene pool. While a missing index can still creep in later in the execution of the algorithm via mutation or invasion, one can use a large value of k to reduce the probability of missing an index in the initial gene pool. This probability is given by the formula

$$\Pr(\text{an index missing in the initial gene pool}) = \left(\frac{n-1}{n}\right)^{P \times p} \qquad (6.2)$$

and for the case $P = \frac{kn}{p}$, it reduces to $\left[\frac{n-1}{n}\right]^{kn}$. Since mutation and invasion occur with low frequencies, the chances for a missing index to appear later are low, and it is important to keep the above probability sufficiently small. Note that, for given k and n, we can evaluate (6.2) and compare it with a threshold value to decide on the value of P. More specifically, if the threshold probability is P_0, the associated population size is given by the formula

$$P = \left\lceil \frac{1}{p} \times \frac{\ln P_0}{\ln \frac{n-1}{n}} \right\rceil \qquad (6.3)$$

There are two ways to employ the above analysis to determine a population size: set different threshold probability values and compute P using (6.3); or to adopt (6.1) and compute P for different values of k. In our experiments, we have chosen to use the latter and used k values that result in sufficiently small probabilities of not selecting an index.

6.5.4 Parent Selection:

After initializing the parameters and generating the initial population, the algorithm first computes, for each member of the population in a particular generation, the probability of selection for mating. The literature suggests different methods for this step and the two most popular ones attempt to mate fitter members of the population, with the expectation that the resulting offspring will also be fit. The first one of these two methods computes, for solution j, the probability of selection for mating as proportional to $\frac{1}{Z_j}$ (for a minimization problem). The second method ranks the members of the population in decreasing order of Z_j values and computes the probability associated with the solution ranked in the ith position as

$$\frac{2i}{P(P+1)}.$$

Hence, the first method assigns probabilities proportional to the inverse of the objective function values, whereas the second method assigns probabilities based on the relative position of a solution in the ordinal ranking of all solutions. Note that, the second method may assign considerably different probabilities to solutions with approximately equal fitness values. This will create an artificial discrimination between solutions of similar quality for problems with a lot of near-optimal solutions (i.e. a fairly flat objective function around the optimum). Our early experiments indicated that the first method is slightly superior to the second for the p-median problem, and we chose the first method for our algorithm.

6.5.5 Crossovers:

The production of offspring is implemented by first selecting a pair of parents, then applying a crossover operator to produce two offspring and discarding one at random. This process is repeated $(P - O)$ times to produce $(P - O)$ offspring, where O is the number of solutions that are kept intact in the pool. We now give the details of the modified basic operator (for reference) and the three genetic operators used in the generation of offspring from parents.

Modified Basic Operator (MBO): The basic operator that exchanges the post-crossover-point segments of the two parent chromosomes needs some modifications due to the encoding scheme we used. A simple exchange of the two segments may cause some of the indices to be duplicated. For example, the basic crossover of the chromosomes [5,12,7,15,3] and [12,6,4,7,1] after the 3rd gene results in the offspring [5,12,7,7,1] which duplicates location index 7. To overcome this problem, we have modified the basic operator. After copying the first segment from Parent 1, a particular gene from the second segment of Parent 2 is copied only if it does not cause duplication of genes. In case of gene duplication, the operator searches the entire chromosome of Parent 2 from left to right to find a location index that is different from all of the previously copied locations indices. This ensures that we always copy a distinct gene from Parent 2. For the example given above MBO would produce the offspring [5,12,7,6,1].

String-of-Change Operator (SOC): This operator is the same as the basic operator except for a slight modification in the selection of the crossover point. With the basic operator, a random crossover point may result in offspring *identical* to the parents. For example, if the two parents are [10,9,12,24, 7,3] and [10,9,7,8,12,3], the two offspring will be identical to their parents if the crossover point is before the 3rd gene or after the 5th gene. There are two problems with the basic operator in this regard. First, it may waste crossovers and not always introduce a new genetic combination to the population. Second, when the algorithm replaces members of a generation with offspring, it may reduce the number of distinct solutions in the population by one (in case the chromosome being replaced is not a parent of the offspring), resulting in a decrease in the diversity of the population. This is undesirable; the less diversity in the population, the less likely a crossover will be able to produce a completely new offspring.

To avoid these problems, we adopted the string-of-change operator suggested independently by Booker (1987) and Fairley (1991). Before randomly selecting the crossover point, the two parents are first subjected to an "exclusive - or" (XOR) operator (the expression $aXORb$ is defined to be equal to 1 if $a \neq b$, and equal to 0 otherwise). In the example given above, the XOR operator is applied to [10,9,12,24,7,3] and [10,9,7,8,12,3] to get [0,0,1,1,1,0]. To avoid generation of identical offspring, only the genes between the first and the last 1 in the resulting string should be selected as the crossover point. The rest is the same as in the modified basic operator.

Template Operator (TEMP): This operator first creates a random template (a string of 0's and 1's). The 0's in this template correspond to the genes to be copied from Parent 1 and the 1's correspond to those to be copied from Parent 2. The operator traces the template from left to right and copies the genes accordingly as long as no previously copied index is duplicated. If TEMP encounters a gene with a duplicating index, it selects a random gene from Parent 2, making sure the above rule is not violated and then copies the index. This last step is repeated until such a gene is successfully identified.

To give a numerical example for TEMP, suppose the template [0,1,0,0,1,1] is applied to parents [10,9,12,24,7,3] and [10,9,7,8,12,3]. The first four genes are copied successfully from the parents (i.e. the location indices 10, 9, 12 and 24), however the 5th gene from Parent 2 violates the rule. At this point, TEMP randomly selects location index 3 from Parent 2 as the 5th index to be copied. For the 6th gene, the rule is again violated (since location index 3 has just been copied) and TEMP randomly selects location index 8 to be copied for the last position. The resulting offspring is [10,9,12,24,3,8].

During the course of the algorithm, templates can be generated in three levels of frequency.

1. A single random template for all the crossovers (in all generations).
2. A random template for each new generation. All crossovers in that generation use that template.
3. A random template for each new crossover.

In our preliminary experiments, we tested these three levels for a medium-sized problem and observed that the most frequent generation of templates (i.e. Level 3) gives the best performance. This might be due to the fact that more templates imply a more irregular fashion of copying genes from parents, meaning a more effective search for untested schemata. Hence, we chose to use the TEMP operator with Level 3 frequency in reporting our results.

Note that TEMP is a generalized version of the basic operator (except the effort for avoiding duplication of location indices) in the sense that the basic operator can be viewed as a special template composed of 0's up until the crossover point and 1's for the rest.

Backward-Crossover Operator (BACK): This operator first identifies a random crossover point and copies all the genes of Parent 1 to the offspring up until the crossover point. The remaining genes are copied from Parent 2 but starting from the last gene of Parent 2 and moving backwards. If the location index of the current gene from Parent 2 has previously been copied (a violation of the rule), the operator skips that gene and continues moving backwards until all the genes of the offspring are complete.

6.5.6 Mutation:

The role of mutation is to enrich the gene pool of the population through diversification. Any offspring may be subject to mutation. An offspring is mutated by closing a randomly selected facility and opening a facility at another random location. The mutation rate M is an important algorithm parameter. We assumed a default constant rate of 1% for M, implying 1% of all offspring is subjected to mutation. In our experimental study, we test other values for the mutation rate together with different invasion rates. While increasing mutation rate up to a certain level is expected to produce good results, high mutation rates are undesirable, as they are likely to disrupt the genetic process and reduce the effects of crossovers.

6.5.7 Replacement:

Once offspring are generated (and possibly mutated), they are used to replace some of the existing population. The overlapping level O indicates the number of common solutions in the population of two successive generations. For example, $O = 0$, implies a complete replacement of P parents with P offspring, whereas $O = P - 1$ means only 1 solution is replaced in each generation. We experimented with different overlapping levels and found that the best performance is achieved by using $O = 1$ where the single overlapping solution is the fittest solution of the previous generation.

6.5.8 Invasion:

The last component of the algorithm involves the concept of "invasion". Invasion could be viewed as a more intensive form of mutation where invaders (randomly generated solutions) invade the population every $\frac{1}{IF}$ generations and replace $IR\%$ of the population. (For a natural analogy, consider the invasion of an island by outsiders who replace a portion of the natives and habitate the island.) In general, invasion increases genetic variety. Specifically, it addresses a shortcoming of the encoding scheme. With the proposed scheme, notice that the crossovers create a new combination of the existing location indices, but they cannot introduce a new location index to the gene pool. Hence, if the initial gene pool excludes an index, the only way that index can enter the gene pool is through mutations and invasions.

The two parameters IF and IR together define the total number of solutions replaced by the invaders. For example, if invaders appear every 2nd generation and replace 10% of the population at every arrival ($IF = 0.5$, $IR = 10\%$), they will replace approximately $IF \times IR = 5\%$ of all solutions throughout the execution of the algorithm. To study the effects of invasion empirically (and in the presence of low/high mutation), we experimented with different invasion rates (including no invasion) combined with different mutation rates. We found that relatively high (low) invasion rates result in better performance with relatively low (high) mutation rates. We report our experimental results in the next section.

6.6 Computational Study

In Section 6.6.1 we report the results of our computational study for determining good values for the parameters of our algorithm. In Section 6.6.3 we compare the performance of our algorithm with that of H&G as well as T&B and random search on randomly generated test problems.

6.6.1 Experiments for Identifying Good Parameter Settings

The genetic algorithm we have outlined in the previous section has six parameters (in addition to the crossover operator chosen): P, population size; G, number of generations; O, overlapping level; M, mutation rate; IF and IR, invasion parameters. For the purposes of this section, we study the first three of these parameters (P, G, O) individually and the remaining three (M, IF, IR) in two different groups. The parameters in the first group determine together the number of solutions evaluated in an execution of the algorithm, whereas the parameters in the second group describe how solutions in a particular population are modified genetically. The following two subsections detail the experiments we perform on these two groups.

Experiment 1: Experiments on Population, Number of Generations and Overlapping Level

Experiment 1.1: Changing one parameter at a time Our first approach for identifying reasonable settings for P, G and O, is to vary *one parameter at a time* and analyze the best solutions found. We expect that increasing G or P while keeping the other two unchanged will result in better average performance. For O, lower overlapping levels imply more solutions generated, and are likely to produce better solutions.

While this approach is useful for testing the sensitivity of each parameter, it is clear that one cannot simply increase P and G indefinitely given the limitations of computational resources. In this subsection, our objective is to observe the sensitivity of the performance of the algorithm to the changes in

P and G (as well as identifying a good choice for O). We turn to the tradeoff analysis between P and G in Experiment 1.2.

Table 6.1. Combinations tested for P, G and O

Parameter	CASES								
	1	2	3	4	5	6	7	8	9
P	50	75	100	50	50	50	50	50	50
G	100	100	100	200	300	100	100	100	100
O	1A	1A	1A	1A	1A	0	1B	10	20

$O = 1A$: keep the best solution in the pool
$O = 1B$: keep a random solution instead of the
best solution for comparison purposes

Table 6.1 summarizes the 9 combinations of the values we test for each of the three parameters (we use $M = 1\%$ and $IF = 0$).

As shown in Table 6.1, we start with the base case of $P = 50$, $G = 100$ and $O = 1A$ and test the values $P = 50, 75, 100$, $G = 100, 200, 300$ and $O = 1A, 0, 1B, 10, 20$ by changing one parameter at a time. We use 100 randomly generated $_{50}C_5$ problems as our testbed. Note that, given this problem size, we are using $k = 5, 7.5, 10$ (from equation (6.1)) associated with $P = 50, 75, 100$, respectively. Using (6.2), we compute the corresponding probabilities of missing a location index in the initial gene pool as 0.0064, 0.0005, and 0.00004 which we consider to be sufficiently small. We test the nine cases in Table 6.1 for the three crossover operators described in the previous section. Table 6.2 reports the performance of the algorithm with each operator, relative to the operator's base case. The per cent figures in the table represent the average percentage deviation from the best solution found in the base case.

Table 6.2. Results for the combinations of the parameter values tested

Case	SOC	TEMP	BACK
2	-0.97%	-0.82%	-0.65%
3	-1.28%	-1.36%	-1.27%
4	-0.46%	-0.74%	-0.83%
5	-1.06%	-1.23%	-1.29%
6	6.82%	6.21%	6.49%
7	6.88%	6.45%	6.82%
8	0.84%	0.69%	0.65%
9	0.82%	0.75%	1.12%

Our first observation in Table 6.2 is that the performances of the three operators improve as P is increased (cases 2 and 3) and G is increased (cases 4 and 5). However, the marginal improvements are not substantial. For example, tripling G gives us only a maximum of 1.3% improvement. In Table 6.2, we also observe that the best setting for O is $1A$ (keep the best solution in the population) for all three operators. The differences for different values of O are rather significant and in all subsequent runs we keep the best solution in the population.

The three operators show practically the same performance in terms of the quality of the best solution found for the base case. Table 6.3 provides further evidence for this conclusion, where each operator is tested by combining the (individual) best of the values tested above, namely $P = 100$, $G = 300$ and $O = 1A$. The row operators are taken as the reference point for calculating the per cent differences.

Table 6.3. Comparison of the three operators with the combination of best parameter settings

	SOC	TEMP	BACK
SOC	–	0.03%	0.03%
TEMP	–	–	0.00%
BACK	–	–	–

Experiment 1.2: Keeping the total number of solutions evaluated constant As mentioned above, we cannot simultaneously increase P and G indefinitely. We may have a limited amount of computing time which is sufficient to evaluate a given number of solutions. The total number of solutions evaluated in an execution of the algorithm is given by

$$S = P + G(P - O). \tag{6.4}$$

Suppose this is an approximation of the computational resource to be allocated between P, G, and O. There is a spectrum of P and G values (for fixed O) and we expect the two extreme settings of P and G to perform poorly. In fact, if P is very large (and G is very small), the algorithm reduces to random search with P solutions. On the other hand, if G is very large (and P is very small), the algorithm relies on mutation and invasion (which also involve randomness) to generate new solutions. To explore the spectrum of P and G, we test different values while keeping S roughly constant according to Eq. (6.4). We limit our experiments to 30 random $_{50}C_5$ problems and to three levels of S: low, medium and high, with $S = 4950$, 14875 and 29800 respectively. We use $M = 1\%$ and $IF = 0$.

Fig. 6.1. Performance of the operators as a function of P at a low level of S

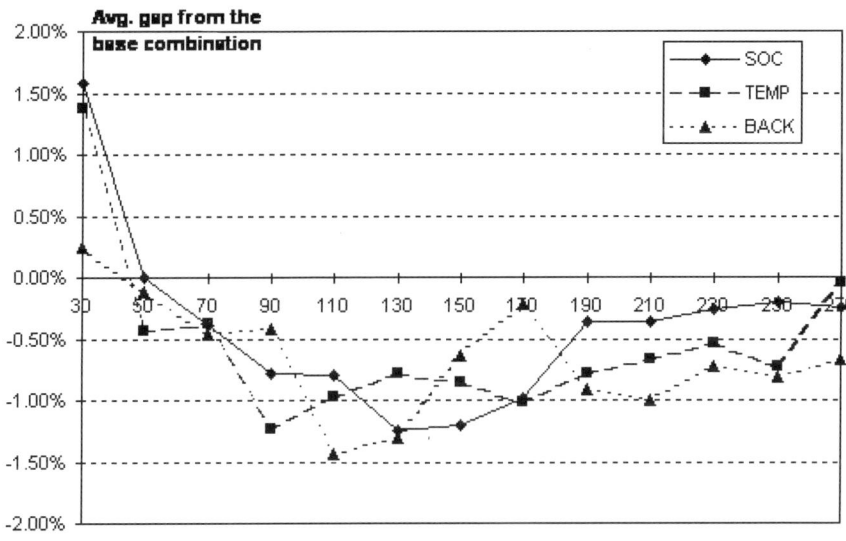

With $S = 4950$, the results suggest the existence of an intermediate "ideal" combination of P and G. In Figure 6.1, we plot for each operator the average deviation from SOC's base case ($P = 50, G = 100, O = 1A$) as a function of P. The most effective value of P seems to be in the (90-130) range. Furthermore, the three operators behave more or less similarly. Also we note that the best improvement is only around 1.50%, even when P is allowed to vary widely. This suggests that unless one chooses P extremely small or large, the performance difference between different values of P will be insignificant with all three operators.

Figure 6.1 supports our initial expectation that there should be a balanced combination of P and G in terms of sharing the total number of solutions to be evaluated. When S is at its medium or high level, however, we are unable to observe this behavior. For medium S, we tested P values ranging from 55 to 245 with increments of 10 and the performances of all three operators show a slightly increasing trend. In these tests, SOC also appears to outperform the other two operators by approximately 1.5%. For large S, the behavior is completely unpredictable. For all three operators, average deviation from the base case fluctuate over the range of $P = 80$ to $P = 270$ (with increments of 10). At this level of S, (for example $P = 100$, $G = 300$) we suspect that the algorithm evaluates so many solutions that almost always a good solution is discovered regardless of the value of P. However, when the algorithm evaluates fewer solutions (i.e. low S) the allocation of solutions between P and G can become critical. If computational resource is an issue for the problem on hand, then it may make sense to perform some experiments to determine an appropriate value for P.

We believe we have sufficient evidence that the three crossover operators perform very similarly. Although we carried out the rest of the tests with all three operators, to save space we report only the results associated with SOC from this point on.

6.6.2 Mutation and Invasion

We test different mutation rates and different invasion rates to see how mutation and invasion behave for producing good solutions. We expect higher mutation (or invasion) rates to produce better results, but only up to a certain level. When these rates are very high, the process is no longer governed by the crossovers, and the good genetic material in the pool is likely to be lost to mutations or invasions.

We test 100 random $_{50}C_5$ problems with mutation rate (M) changing from 0.5% to 25% and invasion frequency (IF) changing from 0.05 to 1 (with $IR = 10\%$). We also include the case where $IF = 1$ and $IR = 15\%$ to have some experience with a higher population replacement rate $(IF \times IR)$. The other parameters are set at $P = 50$, $G = 100$ and $O = 1A$ as in the base case in Section 6.6.1. Table 6.4 summarizes per cent improvements (in terms of the objective function value of the best solution found) relative to the base case $(M = 1\%, IF = 0)$ whereas Figure 6.2 displays the same results graphically.

Table 6.4. Mutation vs. Invasion: performance relative to the base case

$IF \times IR$	Mutation rate (M)						
0.5%	-0.64%	-0.15%	-1.02%	-2.10%	-1.95%	-2.01%	-2.40%
1%	-0.68%	-1.00%	-1.41%	-2.02%	-2.17%	-2.25%	-2.24%
2.5%	-1.32%	-1.18%	-1.55%	-2.01%	-2.12%	-1.99%	-2.10%
5%	-1.81%	-2.16%	-2.18%	-2.09%	-2.03%	-1.97%	-1.77%
10%	-2.16%	-2.00%	-1.71%	-2.16%	-2.13%	-1.68%	-1.75%
15%	-1.64%	-1.40%	-1.68%	-1.68%	-1.50%	-1.62%	-1.23%

Table 6.4 and Figure 6.2 illustrate that low-low as well as high-high combinations of mutation and invasion do not perform well together. In fact, the best results are obtained with low invasion and high mutation rates, while intermediate values of invasion (even with low mutation) also produce relatively good results. Based on our limited experimental results, we conclude that one should combine either high mutation rates with low invasion rates, or low mutation rates with reasonably high invasion rates.

Fig. 6.2. Mutation vs. Invasion: performance relative to the base case

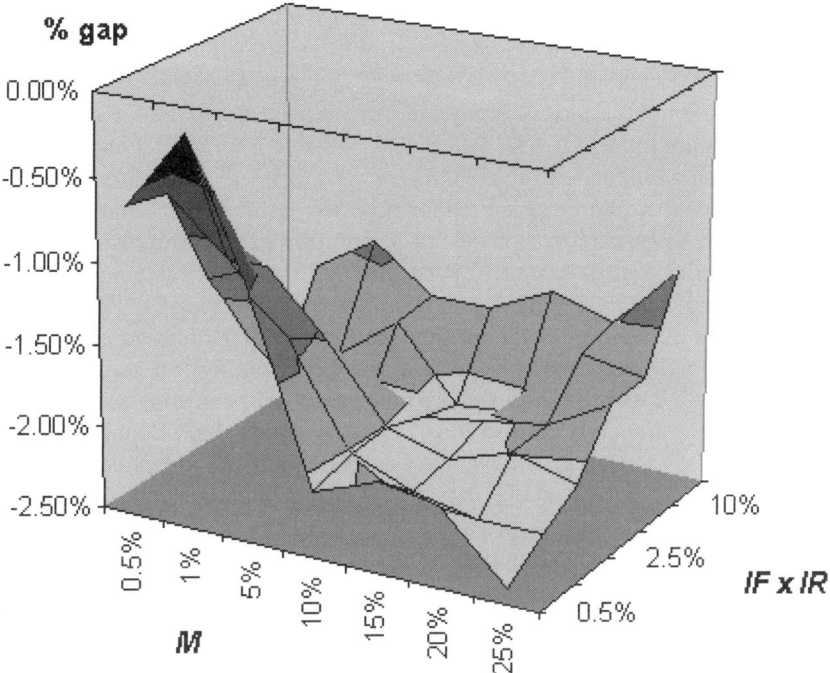

6.6.3 Comparison of the Genetic Algorithm with Other Algorithms

We first compare our implementation of GA with H&G and random search (RS) on $_{20}C_3$ problems. Then we test our GA on $_{50}C_5$ and $_{100}C_{10}$ problems and compare the results with T&B. For the larger problems, we first choose a small number of iterations in the interest of low run times, and then we increase the number of iterations to observe the improvement of solutions. We solve $_{20}C_3$ problems optimally by using complete enumeration and we use T&B solutions as a basis of comparison for $_{50}C_5$ and $_{100}C_{10}$ problems.

Comparing the results with H&G for small problems H&G report 89% of optimality for their 100 randomly generated $_{20}C_3$ problems with $P = 25$, $G = 210$ and $O = 1$. This choice of parameters implies the evaluation of 5,065 solutions. We use the same number of solutions for RS and SOC, and 20 starts for T&B. Table 6.5 lists the comparison of results from different algorithms with optimal solutions. We are unable to use H&G's problem set in our experiment, since their data are not available. However, our test problems are created using the same method employed by H&G, i.e. generating random demand points in a square on the Euclidean plane.

Table 6.5. Results for the three algorithms applied to 100 $_{20}C_3$ problems

	No. of times Optimum found	Average gap from optimality	Worst case gap from optimality
T&B	100	0.00%	0.00%
SOC	99	0.01%	0.92%
RS	99	0.01%	0.84%

As expected, T&B found optimal solutions for all problems with 20 random starts. RS and SOC found 99 optimal solutions by evaluating 5,065 feasible solutions (SOC with $P = 25$, $G = 210$, $O = 1$), the same number of solutions evaluated by H&G. RS is the fastest among these algorithms because there is virtually no processing overhead involved. RS performs so well in this case because the number of feasible (but not necessarily distinct) solutions evaluated is much more than the total number of solutions (5065 vs. 1140). However, RS is not likely to perform similarly well for larger problems due to size of the solution space. It is included here only to demonstrate that even a random search method can outperform H&G due to its inefficient encoding.

Comparing the results for larger problems We tested our GA implementations on larger problems: 100 randomly generated $_{50}C_5$ and $_{100}C_{10}$ problems. Each implementation is tested with population sizes of 50, 75 and 100 (or equivalently $k =$5, 7.5 and 10) which have associated probabilities of missing a location index as 0.0064, 0.0005 and 0.00004 for the $_{50}C_5$ problems, and 0.0066, 0.0005 and 0.00004 for the $_{100}C_{10}$ problems. Because it is time-consuming to find the optimal solutions for problems of this size, we use T&B as the basis of comparison for this test. We set S, the total number of solutions evaluated by GA, equal to the number of solutions evaluated by T&B in 20 random starts. For given P and S, we can find G using (6.4) and rounding to the nearest integer – i.e. using the formula $G = \text{ROUND}\left(\frac{S-P}{P-O}\right)$. (We considered fixing the amount of processing time for each algorithm and then compare their results. But since different algorithms have different programming efficiencies, fixing the processing time may not yield a fair comparison. Therefore, we use S as a proxy to control the processing time.) Tables 6.6 and 6.7 show the results for $_{50}C_5$ and $_{100}C_{10}$ problems respectively.

The upper part of these tables lists the parameters for each algorithm and the lower part contains a performance summary in comparison to T&B.

Table 6.6. Comparison of GA with T&B for $_{50}C_5$ problems

	GA(SOC)		
	$P = 50$	$P = 75$	$P = 100$
G	117	77	58
Solutions Evaluated(S)*	5783	5773	5842
Times GA < T&B	13	17	12
Times GA = T&B	15	16	8
Times GA > T&B	72	67	80
Avg. gap for GA < T&B cases	-1.31%	-0.92%	-1.20%
Avg. gap for GA > T&B cases	0.80%	1.22%	1.27%
Overall avg. gap	0.41%	0.66%	0.87%

* not necessarily distinct. For T&B, $S = 5807$, the average of 100 Problems

Table 6.7. Comparison of GA with T&B for $_{100}C_{10}$ problems

	GA(SOC)		
	$P = 50$	$P = 75$	$P = 100$
G	574	380	284
Solutions Evaluated(S)*	28176	28195	28216
Times GA < T&B	25	21	18
Times GA = T&B	1	1	0
Times GA > T&B	74	78	82
Avg. gap for GA < T&B cases	-1.04%	-1.16%	-0.90%
Avg. gap for GA > T&B cases	1.13%	1.37%	1.82%
Overall avg. gap	0.83%	1.33%	2.01%

* not necessarily distinct. For T&B, $S = 28199$, the average of 100 Problems

The first three rows in the second part list the number of times GA found better, the same, or worse solutions than T&B among 100 problems. The next two rows list the average difference in terms of objective function value between GA and T&B for the cases GA performed better or worse than T&B. In the case of $_{50}C_5$, GA found 42 solutions (out of 300 runs) which were better than T&B. The overall average gap between GA and T&B is 0.65%. In the case of $_{100}C_{10}$ problems, GA outperforms T&B in 64 of the 300 runs, and the overall average gap between GA and T&B is 1.39%. With no special programming tricks, GA solved each one of the $_{50}C_5$ problems in about three seconds and each one of the $_{100}C_{10}$ problems in under one minute on a Pentium 166 microcomputer with 32MB of RAM. On average, GA was about 60% slower than T&B. This is because GA has the extra burden of creating the population pool, and maintaining it by crossing parent chromosomes and replacing parents with offspring. T&B, on the other hand, spends little time on creating new solutions.

Comparing the algorithms with extended processing time In this set of tests, we let the algorithms run longer. The maximum value of S we use is 10,000 for $_{50}C_5$ problems and 50,000 for $_{100}C_{10}$ problems. These numbers are somewhat arbitrary but they are roughly twice the number of iterations carried out for producing the results in Tables 6.6 and 6.7. In addition to the final results from each algorithm, we also recorded the solutions at 50 checkpoints to observe how each algorithm behaves during the entire process. The results for SOC (with $P = 50$) at 10 evenly distributed points are listed in Tables 6.8 and 6.9 for $_{50}C_5$ and $_{100}C_{10}$ problems respectively. The last two columns in these two tables contain the final results for $P = 75$ and $P = 100$.

Table 6.8 shows that, as the number of solutions is increased, the performance of the GA improves in comparison to T&B. In fact, for a population size of 50, GA (barely) outperforms T&B for the first time in our experiment in terms of average overall gap when 10,000 solutions are generated.

Table 6.8. Comparison of GA(SOC) with T&B for 100 $_{50}C_5$ problems

Checkpoints	1,000	2,000	3,000	4,000	5,000	6,000	7,000	8,000	9,000	10,000	10,000	10,
Population size					$P = 50$						$P = 75$	$P =$
Times GA < T&B	3	4	6	7	11	11	14	17	19	21	22	
Times GA = T&B	1	2	4	9	19	22	31	39	42	48	33	2
Times GA > T&B	96	94	90	84	70	67	55	44	39	31	45	
Avg. gap for GA < T&B	-0.69%	-0.64%	-0.70%	-0.72%	-0.77%	-0.97%	-0.91%	-1.11%	-1.13%	-1.13%	-0.94%	-0.
Avg. gap for GA > T&B	4.48%	3.48%	2.51%	2.07%	1.62%	1.18%	0.97%	0.78%	0.65%	0.64%	0.58%	0.7
Overall avg. gap	4.28%	3.25%	2.22%	1.68%	1.05%	0.68%	0.40%	0.16%	0.04%	-0.04%	0.06%	0.2

Table 6.9. Comparison of GA(SOC) with T&B for 100 $_{100}C_{10}$ problems

Checkpoints	5,000	10,000	15,000	20,000	25,000	30,000	35,000	40,000	45,000	50,000	50,000	50,
Population size					$P = 50$						$P = 75$	$P =$
Times GA < T&B	0	3	4	10	11	17	22	29	34	37	31	2
Times GA = T&B	0	0	0	0	0	2	2	2	2	3	6	
Times GA > T&B	100	97	96	90	89	81	76	69	64	60	63	7
Avg. gap for GA < T&B	–	-0.57%	-0.98%	-0.95%	-1.17%	-1.14%	-1.13%	-1.06%	-1.07%	-1.13%	-1.11%	-1.0
Avg. gap for GA > T&B	6.62%	5.02%	3.84%	2.93%	2.17%	1.78%	1.37%	1.29%	1.11%	0.89%	0.12%	0.3
Overall avg. gap	6.62%	4.85%	3.64%	2.54%	1.80%	1.25%	0.80%	0.58%	0.35%	0.11%	0.74%	0.9

The relative performances of the two algorithms are displayed in Figures 6.3 and 6.4. Again, we only use the $P = 50$ case as an example. The X-axis of the chart is the number of solutions evaluated at each checkpoint. The Y-axis is the average objective function value for all $_{50}C_5$ or $_{100}C_{10}$ problems. The inset of the charts is a view re-scaled on the Y-axis. These figures show how each algorithm progresses in their search. T&B starts from a single random

Fig. 6.3. Average objective function values for all $_{50}C_5$ problems at each checkpoint

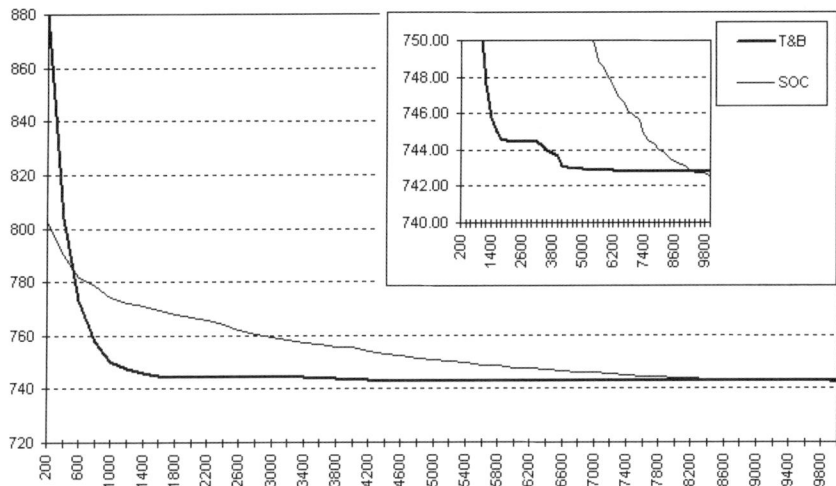

Fig. 6.4. Average objective function values for all $_{100}C_{10}$ problems at each checkpoint

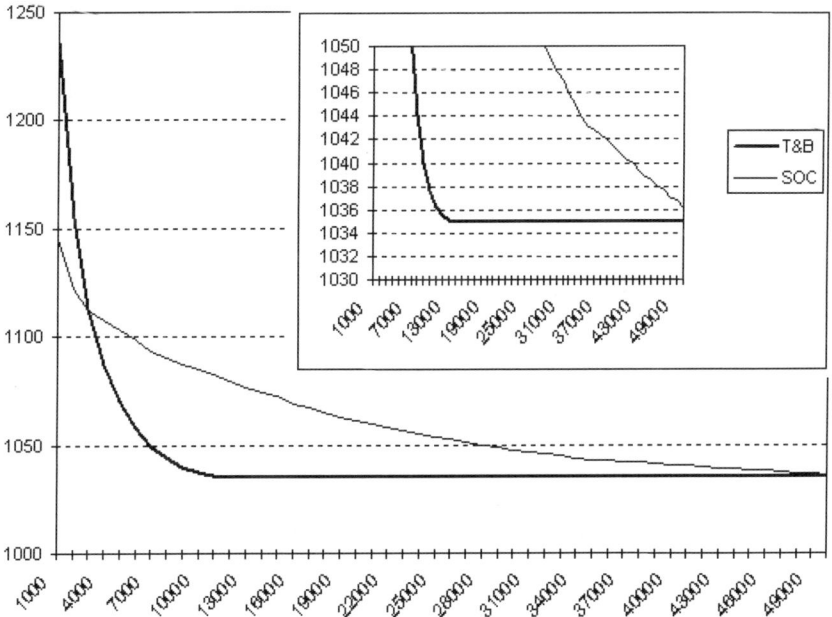

solution which is much worse than the starting point of GA because GA starts from the best of an initial population of P. T&B improves the solutions very quickly. After a short period of time it reaches a very good solution value level and stays at this level with only occasional minor improvements thereafter. On the other hand, GA improves the solutions slowly but steadily and approaches the T&B solutions. In fact, for $_{50}C_5$ problems SOC finds better solutions than T&B for many test problems. These results encourage us to further increase the number of iterations for $_{100}C_{10}$ problems to see whether the GAs can outperform T&B.

For our final test, we used $P = 50$ for each operator. We simply doubled S in the last test and used $S = 100,000$ for $_{100}C_{10}$ problems. The results for SOC are listed in Table 6.10 and Figure 6.5 shows the average solution value for all 100 problems for the second 50,000 solutions (since the results for the first 50,000 solutions are the same as in Figure 6.4). At the end of the run, all three crossover operators perform better than T&B in terms of overall average objective function value, and SOC finds better solutions more often than T&B does. Although a very large number of solutions are generated, these problems take an average of 198 seconds on a Pentium 166 to solve with GA.

Table 6.10. Results of the extended runs on 100 $_{100}C_{10}$ problems (with SOC)

Checkpoints	50,000	100,000
Times GA < T&B	37	49
Times GA = T&B	3	19
Times GA > T&B	60	32
Avg. gap for GA < T&B	-1.13%	-1.23%
Avg. gap for GA > T&B	0.89%	0.52%
Overall avg. gap	0.11%	-0.44%

6.7 Conclusions

In this paper, we applied the principles of genetic algorithms to solve the p-median problem. Our objective was to exploit the features of genetic algorithms and demonstrate that GAs can generate good solutions to location-allocation problems. Contrary to the unpromising previous results, we found that our implementation of GA performs well on test problems. We used a string of p integers to represent a feasible p-median solution, designed three genetic operators and compared our genetic algorithm with T&B. We tested the effect of individual parameters on the performance of GA and found that keeping the best solution in every generation is particularly important for the p-median problem. When the total number of solutions evaluated (or the amount of processing time) is small, the allocation of solutions to each generation (population size) becomes more important. However, GA is not very sensitive to the change of population size when the number of solutions is

Fig. 6.5. Average objective function values for all $_{100}C_{10}$ problems after checkpoint 50,000

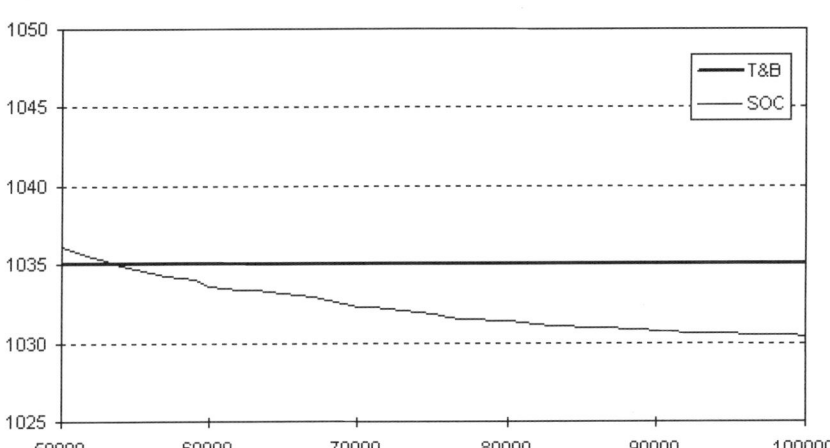

sufficiently large. We introduced and tested a new concept of invasion and observed that high (low) rates of invasion should be employed only with low (high) rates of mutation. We compared our GA with T&B algorithm using randomly generated problems and found that GA and T&B behave quite differently during the process of searching for an optimal solution. T&B finds good solutions very quickly but may be trapped in local optima because of its myopic nature. GA improves solutions slowly but steadily and is less likely to be trapped in local optima. When the algorithms are allowed to evaluate a small number of solutions, T&B produces better solutions than GA. When allowed to run for a sufficiently long time, GA catches up with and outperforms T&B. Our GA implementations take about 60% longer than T&B for each problem (for the same number of solutions generated). The longest GA run, which evaluated 100,000 solutions (or 2,040 generations for a population size of 50) for $_{100}C_{10}$ problems, took an average of 198 seconds per problem on a Pentium 166. Based on our computational tests, we believe that GA has the potential to be a useful heuristic for the p-median problem, and we hope to see more research on applications of genetic algorithms to solve the p-median and other facility location problems.

Acknowledgments: This research has been supported in part by the Natural Sciences and Engineering Research Council of Canada (OGP 25481, CPG 181834). This paper originated from a Ph.D. course on modern heuristics taught at the University of Alberta during academic year 1996-1997, and preliminary results were presented at the Fall 1997 INFORMS meeting in Dallas.

References

Booker, L.B. (1987), "Improving Search in Genetic Algorithms," in *Genetic Algorithms and Simulated Annealing* (Edited by L. Davis), pp. 61-73, Morgan Kauffmann, Los Altos, CA.

Cooper (1963), "Location-Allocation Problems," *Operations Research*, 11, 331-343.

Dibble, C. and P.J. Densham (1993), "Generating Interesting Alternatives in GIS and SDSS Using Genetic Algorithms," *GIS/LIS*.

Densham P.J. and G. Rushton (1992) "A More Efficient Heuristic for Solving Large p-Median Problems," *Papers in Regional Science*, 71, 307-329.

Erkut, E., T. Myroon and K. Strangway (2000), "TransAlta Redesigns Its Service Delivery Network," *Interfaces*, 30(2), 54-69.

Fairley, A. (1991), "Comparison of Methods of Choosing the Crossover Point in the Genetic Crossover Operation," Dept. of Computer Science, University of Liverpool.

Filho, V.J.M.F.F and R.D. Galvao (1996), "A Tabu Search Heuristic for the Concentrator Location Problem," *Studies in Locational Analysis: ISOLDE VII*, 9, 47-50.

Fitzsimmons, J.A. and A.L. Austin (1983), "A Warehouse Location Model Helps Texas Comptroller Select Out-of-State Audit Offices," *Interfaces*, 13(5), 40-46.

Goldberg, D.E. (1989), *Genetic Algorithms in Search, Optimization and Machine Learning*, Addison-Wesley, Menlo Park, CA.

Hakimi, S.L. (1965), "Optimum Distribution of Switching Centers in a Communication Network and Some Related Graph Theoretic Problems," *Operations Research*, 13, 462-475.

Hillsman, E.L. (1984), "The p-Median Structure as a Unified Linear Model for Location- Allocation Analysis," *Environment and Planning A*, 16, 305-318.

Hosage, C.M. and M.F. Goodchild (1986), "Discrete Space Location-Allocation Solutions from Genetic Algorithms," *Annals of Operations Research*, 6, 35-46.

Kariv, O. and S.L. Hakimi (1979), "An Algorithmic Approach to Network Location Problems. Part II: The p-Medians," *SIAM Journal on Applied Mathematics*, 37, 539-560.

Mirchandani, P. (1990), "The p-Median Problem and Generalizations", Ch. 2 in *Discrete Location Theory*, ed. P.B. Mirchandani and R.L. Francis, Wiley Interscience, New York.

Moreno-Perez, J.A., J.M. Moreno-Vega and N. Mladenovic (1994), "Tabu Search and Simulated Annealing in p-Median Problem," presented at the Canadian Operational Research Society Conference, Montreal.

Murray, A.T. and R.L. Church (1996) "Applying Simulated Annealing to Location-Planning Models," *Journal of Heuristics*, 2, 31-53.

Narula, S.C., U.I. Ogbu and H.M. Samuelsson (1997), "An algorithm for the p-median problem," *Operations Research*, 25, 709-712.

Reeves, C.R. (1993), "Genetic Algorithms," in *Modern Heuristic Techniques for Combinatorial Problems* (Edited by C.R. Reeves), pp. 151-196, John Wiley and Sons, New York.

ReVelle, C. and R. Swain (1970), "Central Facilities Location," *Geographical Analysis*, 2, 30-42.

Rolland, E., D.A. Schilling and J.R. Current (1996) "An Efficient Tabu Search Procedure for the p-Median Problem," *European Journal of Operational Research*, 96, 329-342.

Rosing, K.E. and C.S. ReVelle (1997) "Heuristic Concentration: Two Stage Solution Construction," *European Journal of Operational Research*, 97, 75-86.

Teitz, M.B. and P. Bart (1968), "Heuristic Methods for Estimating the Generalized Vertex Median of a Weighted Graph," *Operations Research*, 16, 955-961.

Weber, A. (1909), "Uber den Standort der Industrien," Tubingen, Germany. English translation (1957) *Theory of the Location of Industries* (C.J. Friedrich ed.), Chicago University Press, Chicago.

7 Demand Point Aggregation for Location Models

Richard L. Francis[1], Timothy J. Lowe[2], and Arie Tamir[3]

[1] Industrial and Systems Engineering Department, University of Florida, Gainesville, FL 32601. e-mail: francis@ise.ufl.edu
[2] Tippie College of Business, University of Iowa, Iowa City, IA 52242. e-mail: Timothy-lowe@uiowa.edu
[3] Department of Statistics and Operations Research, School of Mathematical Sciences, Tel Aviv University, Tel-Aviv, 69978, Israel. e-mail: atamir@post.tau.ac.il

7.1 Introduction

A location problem usually involves locating one or more facilities with respect to demand points, also called existing facilities. In urban modeling contexts, each private residence can be a demand point. Thus there can be millions of demand points to deal with. Demand point data may be readily available, available at some cost, or unavailable within the time and budget limitations imposed on solving the problem. Even if the data is readily available, it may be computationally impractical to make use of all of it. Thus, it is a very common practice in location modeling, and other related geographic modeling areas, to aggregate demand points and solve the problem using the reduced data set. For example, if a postal code area (PCA) has 1000 distinct residences, we might suppose all 1000 residences are at the centroid of the PCA. Centroids are commonly used, for example, with geographic information systems and CD-ROM phone books (Francis, Lowe, Rushton and Rayco 1999). Some U.S. Bureau of the Census data is organized by centroids. In addition to the centroid, other points which represent "central" locations can be used to represent an aggregated group. We make use of such an alternative later in this chapter.

The tax office location problem considered by Domich et al. (1991) is a good example of how aggregation methods can be employed to model and solve a large-scale location problem. The problem involved locating field offices (called Posts-of-Duty) for the Internal Revenue Service. It was essential that these offices be located close to centers of IRS activity in order to minimize IRS staff and customer travel costs. However, due to the fixed cost of establishing an office as well as office operating costs, the number of offices located had to be limited. In the problem, demand points were all residences in a given geographic area. In order to generate a model of reasonable size, they used centroids of PCAs to represent all residences in each PCA. The "weight" of each centroid was proportional to the anticipated "level of activity" in the PCA. This weight depended on the number of demand points

---

Facility Location: Applications and Theory.
Edited by Z. Drezner and H.W. Hamacher
© 2002 Springer-Verlag, ISBN 3-540-42172-6

in the area, as well as other factors. It was assumed each customer would visit the tax office to which the customer's centroid was closest. The objective was to locate the offices to minimize the sum of facility costs plus total travel cost. Although the model generated was the so-called uncapacitated fixed charge location-allocation model (Erlenkotter 1978), the problem was essentially solved by solving a sequence of N-median problems. Recall from Chapters 3 and 6 that the objective of the N-median problem is to locate N new facilities to minimize the sum of travel costs of each demand point to its closest new facility.

It is well-known that any spatial aggregation of dispersed demand introduces error into the value of the objective function and in the selection of optimal locations. Only a relatively small literature has addressed the question of estimating the magnitude of these errors, and devising methods for aggregating dispersed demand to reduce their effect.

Questions about errors due to aggregation of demand data exist irrespective of whether the analyst has done the aggregation from disaggregate data or has simply adopted aggregations prepared by others. The latter case is exemplified by the many facility location applications that use census data that, possibly by law, is only available for aggregated zones. The distinction is, however, important in that methods for estimating the error due to aggregation can capitalize on the increased information available in the first case, where the analyst has access to the disaggregated data. In the second case, estimates of error due to aggregation must be based on assumptions about the spatial pattern of the disaggregated data. In this case, the solver of the location problem must live with the uncertainty of never knowing the full extent of the error introduced by spatial aggregation of demand units.

7.2 The Aggregation Problem

In what follows, we will represent the locations of demand points by $P_1, \ldots, P_M$, and any location choice of one or more new facilities by the vector X. An original location model of interest will be denoted as $f(X : P)$. In this chapter we will deal with planar problems until Section 7.7, unless otherwise specified. The following contains a list of notation and abbreviations used in this section as well as later in this chapter.

7.2.1 Notation

$P_m = (a_m, b_m), m = 1, \ldots, M$ (demand points)
$X_n = (x_n, y_n), n = 1, \ldots, N$ (new facilities)
$d(U, V) =$ distance between any two points U and V.
$d_m(X_n) =$ distance between demand point m and new facility n
$X = (X_1, \ldots, X_N)$ an N-median or N-center
$D_m(X) =$ distance between demand point m and closest new facility in X

$P = (P_1, \ldots, P_M)$ a demand point vector
$f(X : P)$: the original location model
P'_m : the aggregate demand point replacing $P_m, m = 1, \ldots, M$
$P' = (P'_1, \ldots, P'_M)$ an aggregate demand point vector
$f(X : P') =$ the approximating location model

A useful way to think of an aggregation of demand points $P_1, \ldots, P_m$, is as a *replacement* of these points by fewer points, say $Z_1, \ldots, Z_q$ (called *aggregate demand points*). This replacement is defined by some rule, so that each P_m is replaced by one of the q aggregate demand point, say P'_m. (Each P'_m is some Z_j; the Z_j are distinct, but the P'_m are not. Thus it is helpful to have two notations for the aggregate demand points.) If P'_m replaces P_m, we customarily say that P_m is *aggregated into* P'_m. Often the rule used is to replace P_m by a closest aggregate demand point to P_m. Another possible rule derives from using some partition of the demand points into q disjoint and exhaustive sets; every demand point in the jth set is replaced by Z_j. For example, demand points might be partitioned according to the postal code area they are in; every demand point in each postal code area might be replaced by the centroid of the area (Domich et al. 1991). Thus, to obtain an aggregation we must decide on three things:

(D1): q, the number of aggregate demand points,
(D2): the locations of the aggregate demand points,
(D3): the replacement rule.

In principle, one could imagine an approach which determines (**D2**) and (**D3**) simultaneously instead of sequentially, but we do not know of one. The freedom of choice in the three decisions, **D1, D2, D3** allows numerous aggregation schemes. There are many ways in geography of partitioning demand points, including tiling, space filling curves, voter districting methods, (Williams 1995). Each of these partitioning methods can result in an aggregation method. A customary way of choosing each aggregate demand point location, given the partition, is to use the centroid of the demand points in each part of the partition. Medians can also be used, provided they are well-defined.

Subsequent to the choice of (**D1**)-(**D3**), one may want to construct a "weight" for each aggregate demand point. How one constructs the weight depends on the problem structure. Suppose, for example,

$$f(X : P) = w_1 d(X, P_1) + \ldots + w_4 d(X, P_4) \tag{7.1}$$

is a 1-median function, P_1 and P_2 are replaced by Z_1, and P_3 and P_4 are replaced by Z_2. An approximating 1-median function $f(X : P')$ then replaces $f(X : P)$. A sensible choice for $f(X : P')$ is

$$\begin{aligned} f(X : P') &= w_1 d(X, Z_1) + w_2 d(X, Z_1) + w_3 d(X, Z_2) + w_4 d(X, Z_2) \\ &= (w_1 + w_2) d(X, Z_1) + (w_3 + w_4) d(X, Z_2). \end{aligned} \tag{7.2}$$

Thus Z_1 and Z_2 have weights $w'_1 = w_1+w_2$ and $w'_2 = w_3+w_4$, respectively. Similarly, if

$$g(X:P) = \max\{w_1 d(X,P_1),\ldots,w_4 d(X,P_4)\} \qquad (7.3)$$

is a 1-center model, and we use the same replacement rule as for the 1-median example, then $g(X:P)$ would be replaced by the approximating 1-center model

$$g(X:P') = \max\{w'_1 d(X,Z_1), w'_2 d(X,Z_2)\}, \qquad (7.4)$$

where $w'_1 = \max\{w_1,w_2\}$ and $w'_2 = \max\{w_3,w_4\}$.

From the above two examples we see that it is the structure of the location model that dictates how the demand point weights should be treated to construct the approximating problem. Unfortunately, however, demand aggregations developed in practice tend to be used for multiple purposes.

When the replacement rule (decision **D3**) is to replace each demand point by a closest aggregate demand point, we note *that the three decisions, **D1**, **D2** and **D3**, are identical to ones made in solving a location problem*; to choose the number of new facilities, the locations of the new facilities, and the assignment of each demand point to a closest new facility. Thus it is not surprising that previous work in location theory can be helpful in doing aggregation. In particular, suppose we wish to aggregate demand points for the N-median model. *In principle*, we could solve a q-median model to do the aggregation. In practice, however, there may be difficulties, since the same reasons which require demand point aggregation for the N-median model will cause difficulties in solving the q-median model (this is what Francis and Lowe 1992 call the *paradox of aggregation*). With care, however, it is possible to devise an approach of low computational order to find a good solution to this q-median problem.

7.3 Aggregation Error

The result of aggregation is to replace the original location model, say $f(X:P)$, by an approximating location model, say $f(X:P')$. This replacement can cause an error for each choice of X. There are various ways of measuring the error. For any location model of interest, we define the error, at X, as

$$e(X) = f(X:P) - f(X:P') \qquad (7.5)$$

Hillsman and Rhoda (1978), in the context of the N-median problem, classify possible demand point aggregation errors into three error types. Instead of focussing on individual error types, we synthesize these errors into a "total error," and consider approaches to make this total error as small as possible. As an example, suppose

$$f(X:P) = \sum_{m=1}^{M} w_m D(X, P_m) \tag{7.6}$$

is the N-median model. For each demand point m we can consider the *demand point m error* (given X), defined by

$$e_m(X) = w_m D(X, P_m) - w_m D(X, P'_m) = w_m(D(X, P_m) - D(X, P'_m)). \tag{7.7}$$

This error is the product of the weight for demand point m and a distance error. Thus with $e(X)$ as the *total error*, we have

$$e(X) = \sum_{m=1}^{M} e_m(X) = f(X:P) - f(X:P'). \tag{7.8}$$

It is interesting to note that the demand point m errors can be positive or negative; many can be nonzero and yet it is possible to have $e(X) = 0$ due to self-canceling error with the N-median problem. Thus we consider it more important to have $e(X)$ nearly 0 than to analyze individual demand point errors in detail. Just as with a profit and loss statement, it is the "bottom line" which is most important. Another error measure, widely used in numerical analysis (see Yakowitz and Szidarovsky 1989), is denoted by $ae(X) = |e(X)|$, and is called the *absolute error*. (See Table 7.1 for a summary of the error terminology we consider.) Note if we have a "small" absolute error (one nearly zero in value) that $f(X:P)$ and $f(X:P')$ are nearly the same, whereas a small value of $e(X)$ does not mean that $f(X:P')$ is a good approximation of $f(X:P)$, since $e(X)$ can be negative. Assuming $f(X:P)$ is always positive, we can also define the *relative error*, $rel(X)$, where

$$rel(X) = 100 \frac{ae(X)}{f(X:P)}. \tag{7.9}$$

Yet another possible error is the *maximum absolute error*,

$$mae = \max\{ae(X) : X \in R\}, \tag{7.10}$$

where R is some region containing all of the demand points. In the field of numerical analysis function approximation, the mae criterion is quite commonly used, and there is a good theoretical justification for its use, due to Geoffrion (1977). It is well known that if the mae is small then various other types of average absolute errors which can be well-defined will be at least as small. Unfortunately, the mae is quite difficult to compute for most location models. However, one can use statistical sampling methods to approximate it. Also, it may be possible to find a reasonably tight upper bound on mae. We will address this upper bounding approach shortly.

We now find it useful to consider a numerical example to illustrate the above definitions and concepts. Consider Figure 7.1, which displays a one-dimensional location problem. In this problem, we contrast the behavior of both the 1-median problem and the 1-center problem. Customers are located along the line at locations 1, 5, 11, 12 , 21, and 25; the numbers above the line in the figure are the number of customers at these locations. The locations on the line marked "X" and "C" are the optimal locations for the median and (unweighted) center problems, respectively. (We assume for the center problem, all weights are equal to one.) That is, $X = 11$ is the optimal 1-median location for the 1-median problem posed on the original data. Similarly, $C = 13$ is the optimal 1-center for the 1-center problem posed on the original data.

Note that the objective value for the median problem (posed on the original data and evaluated at $X = 11$) is $f(X) = 10 + 8 \times 6 + 1 + 8 \times 10 + 14 = 153$. This value is obtained by noting that the single customer located at 1 is 10 units of distance from X, the eight customers located at 5 are 6 units of distance from X, ..., and the single customer located at 25 is 14 units of distance from X. Similarly, the objective value for the center problem (posed on the original data and evaluated at $C = 13$) is $g(C) = 12$ (units of distance for the customers located at either 1 or 25).

The point-sets $\{AM_1, AM_2, AM_3\}$, and $\{AC_1, AC_2, AC_3\}$ represent two different aggregations of the original problem data. Consider first the aggregation set $\{AM_1, AM_2, AM_3\} = \{5, 11, 21\}$. Supposing that each demand point is allocated to a closest aggregate demand point, the demands at points 1 and 5 are allocated to AM_1, while demands at points 11 and 12; and 21 and 25, are allocated to AM_2 and AM_3, respectively. If we then use these three aggregate points to represent the median problem, by our earlier discussion the three points would be assigned the weights 9, 3, and 9, respectively. The approximating median function value, evaluated at X on these three aggregate points, is $f'(X) = 9 \times 6 + 9 \times 10 = 144$. Thus the absolute error of the aggregate set $\{AM_1, AM_2, AM_3\}$ for the median problem evaluated at X is $ae(X) = |f(X : P) - f(X : P')| = |153 - 144| = 9$. The relative error would be $rel(X) = 100 \frac{9}{153} = 5.9\%$.

Suppose now that the point-set $\{AC_1, AC_2, AC_3\} = \{3, 11.5, 23\}$ is used as the aggregation set for the 1-median problem *instead of* $\{AM_1, AM_2, AM_3\}$. In this case, again assigning each original demand point to its closest aggregation point, and passing on the corresponding weight, we see that the weights for the points AC_1, AC_2, and AC_3 would be 9, 3, and 9, respectively. The corresponding 1-median objective function value, evaluated on these three aggregate demand points with the new facility at the point X would be 181.5. Thus the absolute error for this particular aggregation scheme (evaluated at X) would be $ae(X) = |153 - 181.5| = 28.5$. The relative error would be $rel(X) = 100 \frac{28.5}{153} = 18.6\%$.

This example illustrates our principle that *when choosing an aggregation scheme, the analyst must consider the objective function of the problem that is being solved.* Note that for the case with three aggregate points ($q = 3$), the aggregation points $\{AM_1, AM_2, AM_3\}$ are at the respective medians of the original demand points that they represent in the approximating problem. The aggregation set $\{AC_1, AC_2, AC_3\}$ does not possess this property. Thus it is not surprising that the aggregation error values for the one median problem are smaller for the first aggregation than for the second.

Table 7.1. Various Aggregation Error Measures and Notation

$e(X) = f(X : P) - f(X : P')$	Total error for an original location model, given any X
$ae(X) = \|e(X)\| = \|f(X : P) - f(X : P')\|$	Absolute error, given any X
$rel(X) = 100 \times \frac{ae(X)}{f(X:P)}$	Relative error, expressed in percent, given any X
$mae = \max\{ae(X) : X\}$	Maximum absolute error
$eb(Z)$	Error bound: a number depending on Z, that is an upper bound on the maximum absolute error
$e_m(X) = w_m D(X, P_m) - w_m D(X, P'_m)$	Demand point m error for N-median model, for any given X
$e(X) = e_1(X) + \ldots + e_M(X)$ $= f(X : P) - f(X : P')$	Total error for N-median model, given any X

Fig. 7.1. Illustration of Aggregation with Different Objectives

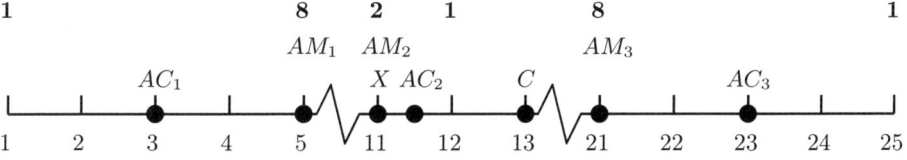

We now introduce another measure of aggregation error, one that we focus on in the remainder of this chapter. Recall that when P_m (the location of demand point m) is replaced by P'_m (the point P_m is aggregated into) in $f(X : P)$, one obtains the approximating function $f(X : P')$. Francis and Lowe (1992) have shown that when $d(.)$ is any distance measure, for the

N-median problem,

$$ae(X) \leq \sum_{m=1}^{M} w_m d(P_m, P'_m), \qquad (7.11)$$

for all choices of X. Thus if $Z = \{Z_1, \ldots, Z_q\}$ is a set of aggregation points, and P'_m is a closest point in Z to P_m, then

$$ae(X) \leq \sum_{m=1}^{M} w_m D(Z, P_m), \qquad (7.12)$$

for all choices of X. We denote the right hand side of (7.12) as the *error bound*, or the *error bound function*, say $eb(Z)$. Thus a means of reducing aggregation error is to locate the aggregation points in order to reduce the error bound value.

Francis, Lowe and Rayco (1996) consider the case where the distance measure $d(.)$ is the rectilinear distance, and the q aggregation points are arranged in a lattice structure with r rows and c columns, where $q = rc$. They give an algorithm to find the row and column spacings and to locate the q aggregation points in the lattice. The algorithm is motivated by (7.12) and the objective is to minimize a restriction of the error bound function for the rectilinear N-median problem. (The restriction is due to the requirement that the aggregate points are located in a lattice.) In Section 7.5, we outline this "row-column" algorithm, and in Section 7.6 we report computational experience with the algorithm. Before presenting those details, we first present some general guidelines for selecting and implementing an aggregation scheme.

7.4 Guidelines for Aggregation

It is clear that decisions made regarding aggregation can have an effect on all other aspects of the location process. Accordingly, it is of interest to list some evaluation criteria for an aggregation algorithm, as follows:

(**EC1**): aggregation error
(**EC2**): computational cost to
    a) get demand point data
    b) implement and run aggregation algorithm
    c) solve the approximating location model
(**EC3**): ease of explanation
(**EC4**): problem structure exploitation
(**EC5**): robustness (works for many different problems)
(**EC6**): GIS implementable

While these criteria seem mostly self-explanatory, their interactions deserve comment. Even if no explicit decision is made as to an acceptable level of aggregation error, *an implicit decision is being made*, since doing aggregation will definitely introduce error. Certainly the grosser the aggregation data the less the computational cost. Thus *there is a tradeoff between (EC1) and (EC2) which should be considered.* (EC1) and (EC3) may also involve a tradeoff, since achieving low error may require a more complicated aggregation algorithm. Likewise, there is a tradeoff between (EC4) and (EC5) since the more problem structure is exploited the less general is the aggregation algorithm. It is probably impossible to construct an aggregation algorithm that does well according to all these evaluation criteria. In any case, (EC1) and (EC2) appear to be the most important of the criteria. To judge from much of the literature, (EC2) is so much more important than (EC1) that there is no point in even trying to estimate errors. Figure 7.2 traces the process of identifying a location problem, getting data, constructing a model, making decisions (perhaps implicit) about acceptable aggregation error, and applying an aggregation algorithm. The outputs of the aggregation algorithm permit the construction of a location model which, irrespective of whether it is then solved by an exact or a heuristic algorithm, will still be an approximating algorithm since its solution will be based on the aggregate demand points. The solution should also ideally (see the dotted lines) provide information which can be used to modify the aggregation algorithm. What are the "implicit decisions" mentioned above, and what rules can guide them? We address these questions in the following section.

7.5 An Aggregation Algorithm

We consider now an aggregation algorithm for the planar rectilinear N-median problem, called MRC, due to Francis, Lowe and Rayco (1996) (hereafter denoted as FLR). The motivation for the algorithm is to seek an aggregation with a small error bound. Imagine we enclose all the demand points in a smallest rectangle, say R, having each edge parallel to an axis. Imagine any grid composed of dashed lines with r rows and c columns imposed on R (see Figure 7.3, with $r = 2$, $c = 3$); spacings between adjacent rows and columns need not be the same. Define the aggregate demand points to be the points where the lines of the grid intersect, and denote the set of these points by Z. We call Z an rc-median. Let $q = rc$. The algorithm of FLR finds an rc-median that minimizes the objective function value of the q-median problem with rectilinear distances over *all possible rc-medians*. If we construct a solid line from one edge of R to the other halfway between every pair of adjacent dashed lines (see Figure 7.4) then the solid lines partition R into $q = rc$ cells (6 cells in the example). Every demand point in each cell is aggregated into the aggregate demand point of the cell (illustrated with crosses in Figure 7.4, with dashed lines omitted).

Let $Z_{ij} = (u_j, v_i)$ be the coordinates of the aggregation point in the ith row and the jth column. Francis, Lowe and Rayco (1996) show that

Fig. 7.2. Visualization of Aggregation Approach for Location Problems

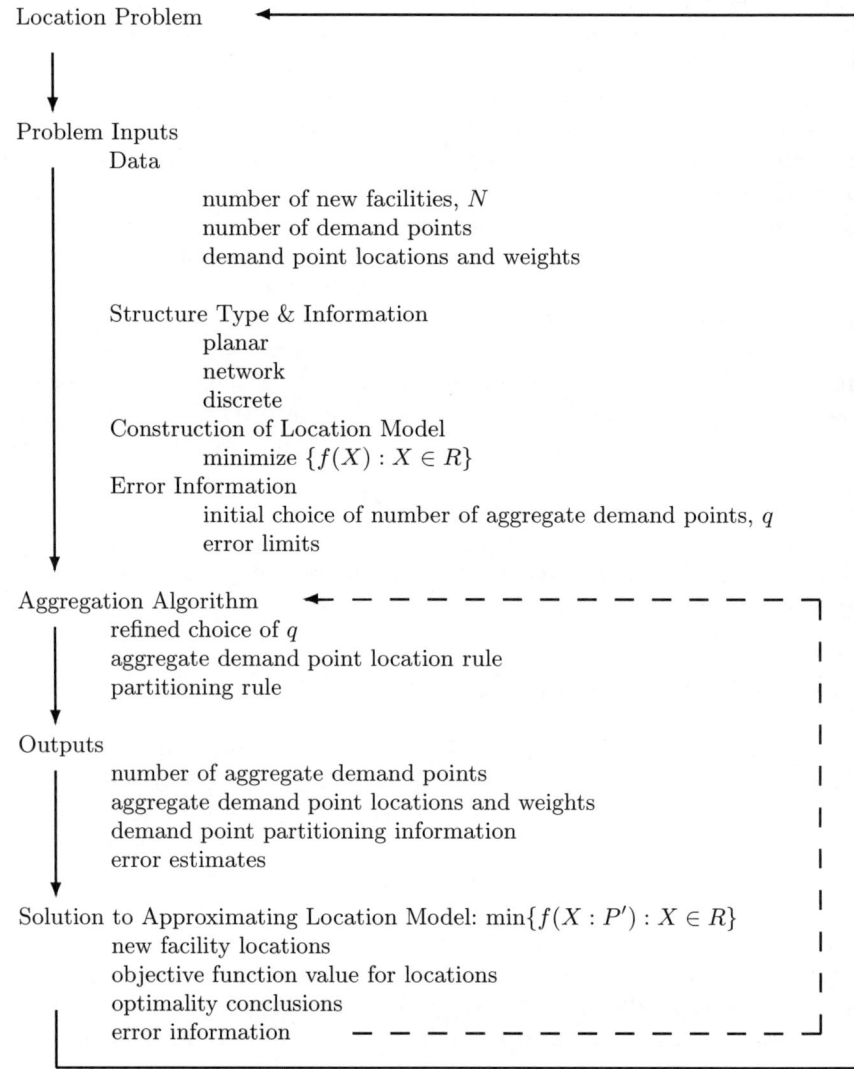

Fig. 7.3. Smallest rectangle R, and construction of dashed lines with coordinates of solutions to projected median problems.

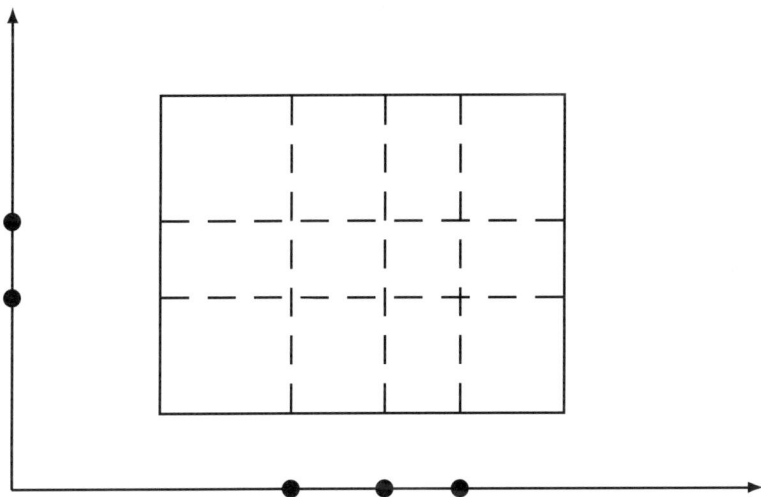

Fig. 7.4. Construction of solid lines half-way between dashed lines. Crosses show intersections of dashed lines, with one cross per cell.

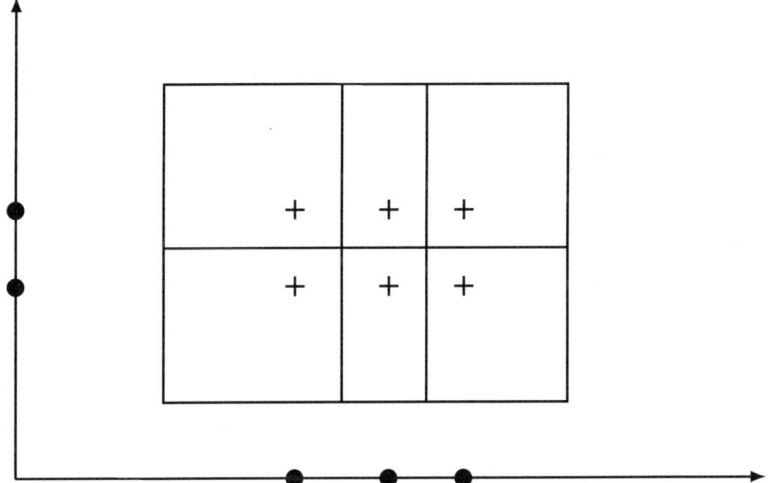

$$eb(Z) = \sum_{m=1}^{M} w_m \min\{|u_j - a_m| : \ j = 1, \ldots, c\} \qquad (7.13)$$
$$+ \sum_{m=1}^{M} w_m \min\{|v_i - b_m| : \ i = 1, \ldots, r\}.$$

This error bound for the rectilinear distance N-median problem, restricted to the case where the aggregation points are located in a grid of rows and columns, can be decomposed into two separate error bounds, each one posed on one of the coordinate axes.

One of the points we are stressing in this chapter is that the aggregation process should somehow reflect the original location problem. Note that the error bound function in (7.13) satisfies this property since the function is clearly a median function.

FLR determine the coordinates of the dashed lines by constructing c and r median problems which are "projected" onto the x axis and y axis respectively (see Francis, McGinnis and White 1992). These projected problems are solved to optimality using the algorithm of Hassin and Tamir (1991). The dots in the figures lying on the axes illustrate possible such solutions. This approach makes use of all the demand point data, and takes demand point density into consideration; regions of high demand point density typically have smaller cells than those of low density.

The algorithm MRC is of low computational order, $O(M(\log M + r + c))$, and has $O(M)$ storage requirements. It is therefore suitable for problems with many demand points, and has been run on problems with about 200,000 demand points. For most problems, it has been found that an extra "touchup" step is useful with MRC. For each cell, the aggregate demand point of the cell is recomputed as the 1-median for the demand points in that cell. This is an $O(M)$ computation and has no effect on the order of the algorithm.

Referring to the guidelines of the previous section, note that the objective of the algorithm is to reduce the error bound function, and thus to indirectly reduce $ae(X)$. Thus (EC1) is clearly a consideration. Because of the relatively low computational effort, (EC2) b) is a prime consideration for using MRC. MRC is relatively easy to explain (EC3), and certainly exploits problem structure (EC4). MRC is GIS implementable (EC6) since many GIS data bases provide (x, y) coordinates (EC2) a). Limited experiments with MRC on both real and simulated data indicate it may very well be robust (EC5), when the underlying problem is the rectilinear N-median problem. We report on some of these experiments in the next section. Regarding (EC2) c), experiments have shown (discussed in the next section) that the error bound (and thus several other error measures) can be reduced with a relatively small number of aggregation points. Since the computational effort of the best-known algorithm for solving the N-median problem is exponential in the number of "demand" points, this later observation is pertinent.

Fig. 7.5. Graphs of sample average absolute error, N-median model with row-column aggregation method, various N

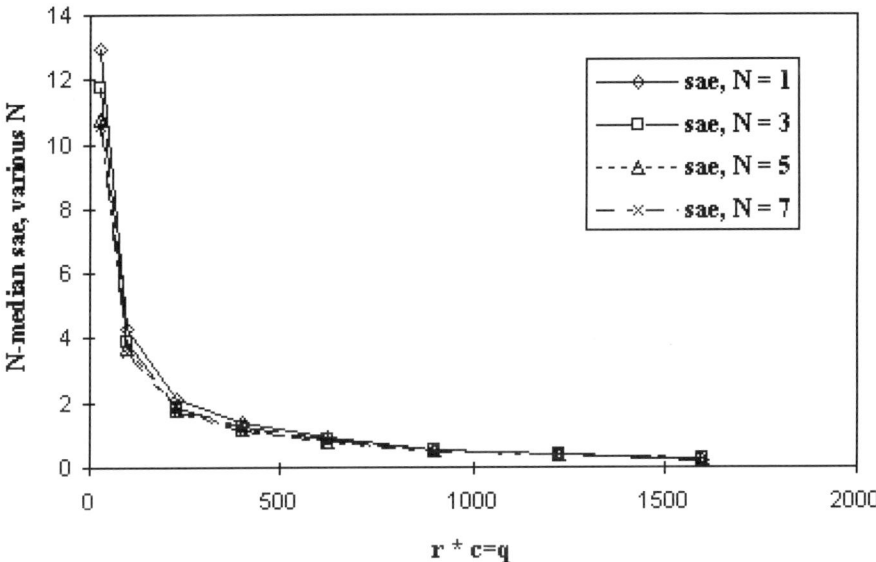

7.6 Computational Experience

Francis, Lowe and Rayco (1993,1996) carried out extensive computer experimentation with MRC for the rectilinear N-median model. (Figures 7.5, 7.6 and 7.7 are based on tables of numerical results from those experiments that appear in the 1993 research report.) They assumed, for part of their experimentation, that demand points were distributed according to a density function which declined in value with the distance from its center. MRC was used to generate a lattice of q aggregation points. They then used the Monte Carlo sampling method, repeatedly locating N facilities randomly in the given region to obtain sample values of absolute error. The sampling gave estimates of the average absolute error (sae) and average relative error (sare).

Figures 7.5 and 7.6 show graphs of, respectively, the sae and sare versus q for various N values. All the error graphs decrease as q increases, but at a decreasing rate, approaching zero for large q. Now consider what happens for fixed q, as N increases. For fixed q, sae decreases as N increases (Figure 7.5), whereas sare increases as N increases (Figure 7.6). For bigger N, the N-median function values (for both true and approximating models) are smaller, and subtracting smaller numbers from smaller numbers gives smaller numbers; this explains the behavior in Figure 7.5. However, the differences

Fig. 7.6. Graphs of sample average relative error, N-median model with row-column aggregation method, various N

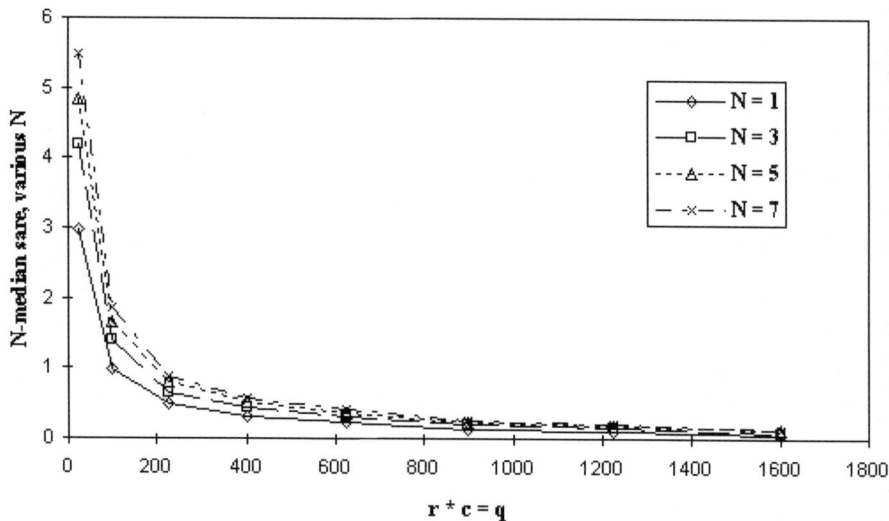

Fig. 7.7. Scatter plot of sample average relative error and $\frac{q}{N}$, N-median model with the row-column aggreagtion method.

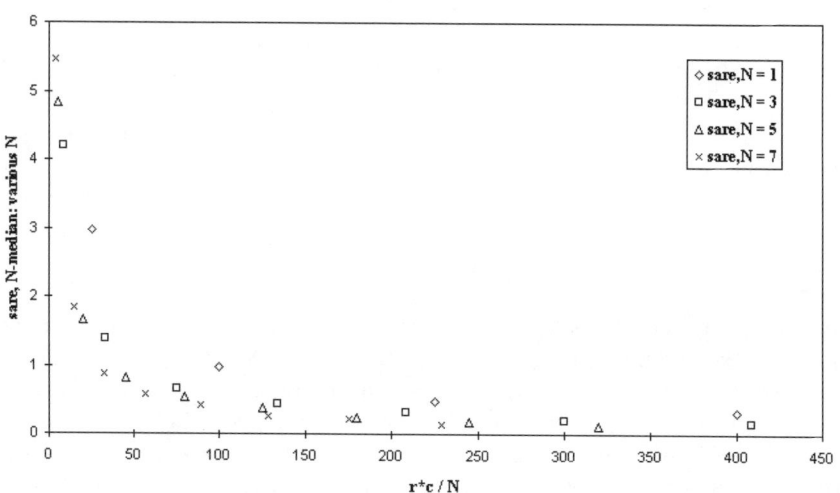

between function values decrease more slowly than the function values, which explains the behavior in Figure 7.6. Thus both N and q influence the error. The combined influence of N and q can be seen by considering the ratio $\frac{q}{N}$, the number of aggregate demand points per new facility. Figure 7.7, based on the same data as Figure 7.6, illustrates how the relative error depends on both N and q, by showing a scatter plot of sare and $\frac{q}{N}$. The larger the q value, and the smaller the N value, the better.

Francis, Lowe, Rayco and Tamir (2000) conducted another set of experiments on data distributed much the same as above. The objective of this latter set of experiments was to compare MRC with a competing aggregation approach, and also to further investigate the relationships between various error measures and the number of aggregation points, q. Experiments were run with 25,000 data points, and values of $N = 1, 3,$ and 5. Similar to the above, MRC was used to aggregate the data points for various values of q, and then sampling (randomly locating the N facilities) was implemented. Using the output of these runs, average values of sae and sare were computed. Finally, the Power Curve Fitting option in Excel was used to find parameters a and b where ERROR $= aq^b$, and ERROR is one of the error measures. Table 7.2 reports the results of some of these curve fits. The table contains the a, b, and R^2, for $N = 1, 3,$ and 5. Note that in each case, the exponent b is nearly -1, and that the R^2 values are quite close to 1. The implication here is that each of these four error measures is inversely proportional to q, the number of aggregate points. This is an important result since it implies that when q is small, increasing the number of aggregate points can significantly reduce error. However, as q gets large, the impact of adding additional aggregate points gets much smaller.

Table 7.2. Error Power Curve fits, aq^b, with central tendency demand point data for MRC, with $m = 25,000$ demand points, for $N = 1, 3$ and 5.

$N = 1$	sae	sare
a	249.73	0.5781
b	-0.9052	-0.9074
R^2	0.9992	0.9993

$N = 3$	sae	sare
a	214.89	0.7814
b	-0.8914	-0.8943
R^2	0.9983	0.9981

$N = 5$	sae	sare
a	202.8	0.9012
b	-0.8917	-0.891
R^2	0.9988	0.9989

7.7 Error Bound Generalizations

To this point we have seen an error bound for the N-median problem, as well as an algorithm for reducing this error bound for the rectilinear distance version of the problem. We now generalize our error bound approach to an entire class of location problems. Material in this section is all based on Francis, Lowe and Tamir (2000). Our location context now is more general. We are not restricted to the N-median problem nor are we restricted to the case where customers are located (and the new facilities are to be located) in the plane or on a transport network with shortest-path distances, etc. Since we will be studying location problems in a general way, we let $f(X : P)$ denote the location model to represent the general problem of determining the location of one or more facilities at points represented by X, to serve demands located at points represented by P.

Given two locations, say U and V, we denote by $d(U,V)$ the distance between the two points. We assume the distance has the following (metric space) properties: (1) $d(U,V) = d(V,U) \ \forall \ U,V$ (symmetry); (2) $d(X,Y) \le d(X,Z) + d(Z,Y) \ \forall \ X,Y,Z$ (triangle inequality); (3) $d(U,V) \ge 0 \ \forall \ U,V$, with $d(U,V) = 0$ meaning $U = V$ (positivity). As is usually the case, if the location problem is posed on an undirected transport network, then $d(.)$ is the shortest path distance. If the location problem is posed in the plane, then $d(.)$ is some relevant distance defined by a norm.

As before, by an error bound, we mean a number (hereafter denoted by eb) such that

$$|f(X : P) - f(X : P')| \le eb \quad \forall X. \tag{7.14}$$

That is, it is guaranteed that regardless of the locations X of the new facilities, the absolute difference of the objective function values of the original problem and the approximate problem will not exceed eb. As we will see, the exact form of eb depends upon the structure of the original problem, as well as the vector P' of aggregate locations. Thus, one can readily see from (7.14) that a good aggregation scheme can be derived by paying close attention to the error bound expression when doing the aggregation. In fact, in both Francis, et. al. (1996), and Rayco, et. al. (1997), aggregations are derived for the problems studied by first deriving the appropriate eb function, and then heuristically minimizing (by appropriately choosing P') the eb function.

We now demonstrate the usefulness of error bounds by showing how to use these error terms to bound other related important quantities. Specifically, we wish to bound the additional cost (error) due to implementing the optimal solution to the aggregate model and not the optimal solution to the original problem: the "cost" of approximation. These bounds can be used to control the error by adjusting the number of aggregate demand points. As above let $f(X : P)$ represent the location model with respect to the original customer vector P, and let $f(X : P')$ represent the model with respect to the aggregate vector P'. Let $X^*(X')$ be the vector of optimal facility locations for model $f(X : P)(f(X : P'))$. (Occasionally in what follows, we assume that we have the optimal solution to the aggregated location problem available to us.) Consider the following two differences: $|f(X^* : P) - f(X' : P')|$ and $|f(X^* : P) - f(X' : P)|$. The first is the difference between the optimal objective function value (OFV) using the optimal locations for the new facilities when serving the original demand point set, and the OFV using the optimal locations for the new facilities when serving the aggregate demand point set. The second is the difference between the OFV using the optimal locations for the new facilities when serving the original demand point set, and the OFV using the optimal new facility locations for the aggregated problem, but when these locations serve the original demand point set. Let eb denote the error bound defined in (7.14). (See Table 7.3 for appropriate expressions for a variety of location models.) Theorems 1 and 5 in Geoffrion (1977) can be used to show the following:

$$|f(X^* : P) - f(X' : P')| \leq eb, \qquad (7.15)$$

$$|f(X' : P) - f(X^* : P)| \leq 2eb. \qquad (7.16)$$

The difference $|f(X' : P) - f(X^* : P)|$ in (7.16) is the additional forgone savings due to implementing an approximate solution (based on a solution to the aggregate model), and not an optimal solution. Note that the bound $2eb$ in (7.16) can be computed for any aggregate vector P' before we solve the aggregate model to optimality and obtain X'.

Table 7.3 shows a partial list of the location models for which we have error bounds. Each model includes (nearest) distances between facilities explicitly in its objective function, and each has given demand points as well as new facility locations. Each distance can be multiplied by a nonnegative number, called a "weight." The operators involving the (weighted) distances are the maximum, summation, and minimum operators. The last column of the table shows the error bounds for the models. Since the error bounds hold for all X, we leave unstated the sets of feasible solutions for the models.

As a motivation for the error bounds, we state the following known result (Zemel 1985, Francis and Lowe, 1992) which is based on the triangle inequality. The lemma amounts to a sort of triangle inequality for nearest-distances, and is essential for deriving the aggregation error bounds.

Lemma 1: Suppose we are given any X. For $m = 1, \ldots, M$, we have the following.

$$D(X, P_m) \leq D(X, P'_m) + d(P'_m, P_m).$$

We now discuss briefly the models in the table, referring to models by their table numbers. The conditional N-median problem, # 1, (see Minieka, 1980, Berman and Simchi-Levi, 1990, and Drezner, 1995) is to locate N new facilities; the travel distance for each demand point can be either to a closest new facility location in X, or to a closest existing service facility in E, whichever of the two is closest. The nonnegative "weight" associated with each demand point is typically a measure of relative frequency of travel. Thus $f(X:P)$ is a total travel distance to closest facilities (new or existing). The conditional N-center model, # 6, (Minieka, 1980 and Drezner, 1989) has a similar interpretation but occurs in a worst-case context; thus it is the maximum nearest-distance that is of interest. We remark the error bounds for these two conditional models are the same as those for the conventional N-median and N-center models (Francis and Lowe, 1992) that is, the "unconditional" models. The multifacility minisum (MFMS) model, # 2, (see Francis, McGinnis and White, 1992) adds weighted distances between new facilities and demand points to weighted distances between pairs of new facilities. The analogous multifacility minimax (MFMM) model, # 7, (Francis, McGinnis and White, 1992) considers the maximum of such weighted distances.

The quadratic assignment problem, #3, abbreviated as QAP (see Burkhard, 1990 and Pardalos and Wolkowicz, 1994) is a form of the MFMS problem with a discrete possible choice of new facility locations. We denote the location of each new facility j by $a[j]$, and choose $a[j]$ from a finite set.

Goldman's two-stage processing model, #4, (Goldman, 1971) addresses the processing of mail. Mail originating at a demand point goes through processing at two facilities and then goes on to some other demand point. The model is set up so as to consider a "cheapest" choice of the two facilities for each demand point pair. For demand points h and m, the weights a_{hm} and c_{hm} are associated with the distances between demand points and new facilities, while the weight b_{hm} is associated with the distance between the new facilities.

The supporting median model, # 5, (Mirchandani and Odoni, 1979) considers travel between demand point locations (P_m) and airports (V_t). Travel can either be directly as measured by a metric d_2, or by means of a cheapest supporting median, as measured by a different metric d_1. The supporting medians could be public transit stops, such as subway stations. Note there is no need to aggregate the airport locations.

The round-trip model, # 9, (Elzinga and Hearn, 1972) considers responding to an emergency at some demand point P_m from a closest center, in a time $w_m D(X, P_m)$. From P_m there is then travel to some given location Q_m (perhaps a hospital) in a time of $\omega_m d(P_m, Q_m)$. The result is a "round-trip" time of $w_m D(X, P_m) + \omega_m d(P_m, Q_m)$. The weights w_m and ω_m reflect the possibility of different travel velocities depending upon the "leg" of the trip. The time to return from Q_m to the starting point of the round-trip is omitted from the model as being unimportant in comparison to the other times. Since the problem occurs in a worst-case context, it is the maximum time that is relevant.

The following scenario motivates the multistop model, # 8 (Handler and Mirchandani, 1979). A breakdown repair service is to be located at some point X. With probability α, a service unit is dispatched directly to some breakdown point P_m. With probability $1 - \alpha$ the unit is first dispatched to some point p_0 to collect existing equipment, and then to the point P_m. The maximum expected response time to a random call is of interest. Note the similarity in model structure to the round-trip problem.

The cent-dian model, # 10, (Halpern, 1978), also called the medi-center model (Handler, 1985), considers what amounts to a weighted average of an N-median and a N-center objective. The goal is to find a balance, using an adjustable "balancing" parameter β, of efficiency (least-cost) and equity (worst-case). The error bound reflects the model structure and combines N-median and N-center error bounds.

The 1-facility ℓ_p model, # 11, (Shier and Dearing, 1983) captures several different objective functions depending upon the value of p $(1 \leq p \leq \infty)$. When $p = 1$, the median problem results, whereas $p = \infty$ corresponds to the center problem. Note that the error bound inherits the structure of the model.

The obnoxious location model, # 12, (Erkut and Neuman, 1989) considers the minimum of the distances between each demand point and a closest obnoxious new facility. This minimum is to be maximized. This model is the only maximization model in the table. If we also include inside the min-operator the weighted minimum of the distances between the obnoxious new facilities, the error bound remains valid. Note the error bound does not reflect the "min" structure of the original model. All the models in the table but this one fit into the general class of models we discuss. Francis, Lowe and Tamir (2000) provide separate, but related, theory for this model.

It can be shown that if each weight in the table that multiplies a distance is replaced by a concave, nonnegative, and nondecreasing function of the distance, then the error bounds are still valid. Note that such a function can model economies of scale in distances.

Certain error bound properties are evident from the error bound column. The error bound expressions all include the weights involving the demand points, and omit the other weights. The error bounds all involve the distances between the P_m and the P'_m, and are nondecreasing in these distances. The error bounds are zero if and only if $P_m = P'_m$, for every positively weighted P_m. The error bounds all involve max-operators and/or sum-operators; they "inherit" these operators from the original model structure, with the one exception of the obnoxious location model. We consider all these properties quite reasonable. Indeed, it can be shown that each of the error bounds except the one for model # 9 defines a distance between P and P', given the reasonable assumption that if the error bound value is zero then every $d(P'_m, P_m) = 0$. Thus attempts to make the lower bound small can be viewed as attempts to make P' "close" to P.

The error bound expressions lead in a natural way to the definition of "second-order" location problems. Such second-order problems have previously been noted by Francis and Lowe (1992), when the original model is either an N-median or N-center model.

To see how such second-order problems are obtained, suppose Z is a collection of aggregate demand points, with $|Z|$ fixed. Suppose we choose as P'_m a closest point in Z to P_m, so that $d(P_m, P'_m) = D(Z, P_m)$. (Picking as P'_m a point in Z other than a closest one cannot give a smaller error bound value.) If we make the replacement of each $d(P_m, P'_m)$ by $D(Z, P_m)$ in the error bound expression we get an expression, say $eb(Z : P)$. We can consider the problem of minimizing $eb(Z : P)$ over Z, subject to $|Z|$ fixed (or an adjustable parameter), as the second-order location model.

For models 1-4 in the table, the second-order model is a $|Z|$-median model. For models 6-8, and 12 in the table, the second-order model is a $|Z|$-center model. The second-order models for 5 and 9 are new, as best we can tell. The second-order model for 10 is a cent-dian / medi-center model, while for 11 it is a sort of ℓ_p $|Z|$-median model. We believe that the predominance of resulting secondary models that are $|Z|$-center or $|Z|$-median models reinforces the central role these two models play in location theory.

The error bounds of Table 7.3 can be used to drive demand point aggregation heuristic algorithms. When the original model is the N-median or N-center model, Francis, Lowe and Rayco (1996) and Rayco, Francis and Lowe (1997) respectively have studied the use of low computational order "row-column" heuristics for minimizing the resulting $|Z|$-median or $|Z|$-center models to obtain a small error bound. Their point of departure is what Francis and Lowe (1992) call the "paradox of aggregation" discussed in section 7.2 of this chapter.

Essentially, the paradox states that the same reasons that require aggregation in the original model also limit what can be done computationally in minimizing the error bound model exactly. Certainly the paradox applies to all the models in the table. There are, however, several things that can be done. First, when the original distances are network shortest-path distances, the distances in the error bound model can be taken to be simpler planar distances, such as rectilinear distances. Second, $|Z|$ can be treated as an adjustable model parameter.

This second recourse, adjusting $|Z|$, is particularly important to note, because for many location problems the number of new facilities is relatively small, and can seldom be varied within a large range. In this sense the second-order model is fundamentally different from the first-order model. Indeed, this difference suggests that asymptotic analysis, as done by Zemel (1985), and by Rayco and Francis (1996) may be a fruitful approach to solving the secondary models.

Figure 7.8 illustrates fundamental features of all the location models in Table 7.3. All the models we consider in this section are constructed by beginning with a vector of new facility locations $X = (X_1, ..., X_N)$, and a vector of demand point locations, $P = (P_1, ..., P_M)$. Then various distances between

Table 7.3. Aggregation Error Bounds for a Family of Location Models

#	Model Name	Location Model $f(X:P)$	Error Bound
1.	conditional N-median	$\sum\{w_m \min\{D(E, P_m), D(X, P_m)\}\|m\}$	$\sum\{w_m d(P'_m, P_m)\|m\}$
2.	MFMS	$\sum\{w_{mj} d(X_j, P_m)\|m, j\}$ $+\sum\{v_{jk} d(X_j, X_k)\|j<k\}$	$\sum\{w_{mj} d(P'_m, P_m)\|m, j\}$
3.	QAP	$\sum\sum\{w_{mj} d(a[j], P_m)\|m, j\}$ $+\sum\{v_{jk} d(a[j], a[k])\|j<k\}$	$\sum\{w_{mj} d(P'_m, P_m)\|m, j\}$
4.	Goldman's 2-stage	$\sum\sum\{min\{a_{hm} d(P_h, X_j)$ $+b_{hm} d(X_j, X_k)$ $+c_{hm} d(X_k, P_m)\|j, k\}\|h, m\}$	$\sum\sum\{a_{hm} d(P_h, P'_h)$ $+c_{hm} d(P_m, P'_m)\|h, m\}$
5.	supporting median	$\sum\sum\{w_{mj} \min\{d_2(P_m, V_t),$ $\min\{d_1(P_m, X_j)$ $+d_1(X_j, V_t)\|j\}\|m, j\}$	$\sum\sum\{w_{mj} \max\{d_1(P_m, P'_m),$ $d_2(P_m, P'_m)\}\|m, j\}$
6.	conditional N-center	$\max\{w_m \min\{D(E, P_m), D(X, P_m)\}\|m\}$	$\max\{w_m d(P'_m, P_m)\|m\}$
7.	MFMM	$\max\{\max\{w_{mj} d(X_j, P_m)\|m, j\},$ $\max\{v_{jk} d(X_j, X_k)\|j<k\}\}$	$\max\{w_{mj} d(P'_m, P_m)\|m, j\}$
8.	multistop (α is a probability)	$\max\{\alpha d(X, P_m)+$ $(1-\alpha)[d(X, P_0)$ $+d(P_0, P_m)]\|m\}$	$\max\{d(P_m, P'_m)\|m\}$
9.	round-trip	$\max\{w_m D(X, P_m)$ $+\omega_m d(P_m, Q_m)\|m\}$	$\max\{w_m (P_m, P'_m)$ $+\omega_m \|d(P_m, Q_m)$ $-d(P'_m, Q_m)\|\ \|m\}$
10.	cent-dian/ medi-center (β is a positive constant)	$\sum\{w_m D(X, P_m)\|m\}$ $+\beta \max\{w_m D(X, P_m)\|m\}$	$\sum\{w_m d(P'_m, P_m)\|m\}+$ $\beta \max\{w_m d(P'_m, P_m)\|m\}$
11.	1-facility ℓ_p model	$[\sum\{d(x, P_m)^p\|m\}]^{1/p}$	$[\sum\{d(P'_m, P_m)^p\}]^{1/p}$
12.	obnoxious	$\min\{w_m D(X.P_m)\|m\}$	$\max\{w_m d(P'_m, P_m)\|m\}$

Fig. 7.8. Construction Process for SAND Location Models

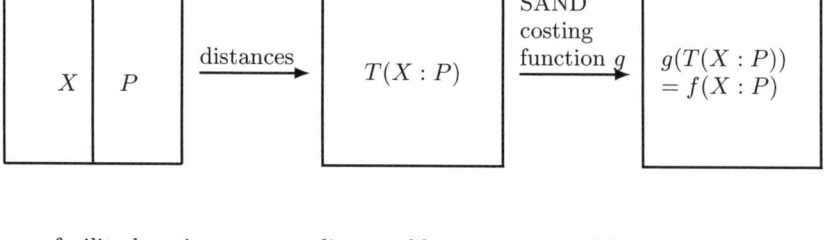

new facility locations   distance-like vector   SAND location model
demand point locations

elements in X and P are computed, resulting in a "distance-like" vector we denote by $T(X:P)$. Finally a "costing function" g is applied to $T(X:P)$ to give the location model $f(X:P) = g(T(X:P))$. For example, consider $T(X:P) = (D(X,p_1),...,D(X,p_M))$. For any nonnegative vector $U = (u_1,...,u_M)$, if the costing function is $g(U) = w_1u_1 + ... + w_Mu_M$, then $f(X:P) = g(T(X:P))$ is the N-median model. If $g(U) = \max\{w_1u_1,...,w_Mu_M\}$, then $f(X:P) = g(T(X:P))$ is the N-center model.

It is the choice of a costing function that gives the model much of its structure. Two properties of the costing function are essential to our approach: (1) subadditivity (SA) ($g(U+V) \leq g(U)+g(V)$), and (2) the nondecreasing (ND) property ($0 \leq U \leq V$ implies $g(U) \leq g(V)$). We refer to a costing function with both these properties as a SAND function. We call a location model constructed using the process the figure illustrates a SAND location model. The "distance-like" vector must also satisfy a type of triangle inequality. Namely, if P' denotes the vector of aggregate demand points, there is a nonnegative vector $S(P:P') = S(P':P)$ so that $T(X:P) \leq T(X:P') + S(P':P)$ and $T(X:P') \leq T(X:P) + S(P:P')$ for all X. The error bound for any such SAND model is then $g(S(P:P'))$. For example, for the N-median model, we have $g(U) = w_1u_1 + ... + w_Mu_M$, $T(X:P) = (D(X,p_1),...,D(X,p_M))$, and $S(P':P) = (d(P_1,P_1'),...,d(P_M,P_M'))$. Thus $g(S(P:P')) = w_1d(P_1,P_1') + ... + w_Md(P_M,P_M')$.

Properties of SAND functions provide basic "building blocks" for constructing the SAND location models we study. For example, because the SAND property is preserved under the addition and maximization operations, both the N-median and N-center models are SAND models. Because a certain type of concave function is a SAND function, replacing the weights of any model in Table 7.3 by concave, nonnegative and nondecreasing functions of the distances (thus capturing transportation economies of scale) results in a SAND location model. Knowing that certain compositions of SAND functions give a SAND function can be useful in combining two such location models to obtain another one.

A SAND costing function g^k permits capturing a family of location models in one model; the k-centrum model of Slater (1978), Andreatta and Mason (1985a, 1985b), and Tamir (2001). If $T(X:P) = (w_1 D(X, P_1), \ldots, w_M D(X, P_M))$, then $f(X:P) = g^k(T(X:P))$ is the sum of the k-largest entries in $T(X:P)$. Taking $k = 1$ gives the N-center model; taking $k = M$ gives the N-median model. The fact that the error bound for this model, $g^k(w_1 d(P_1, P_1'), \ldots, w_M d(P_M, P_M'))$, reflects the structure of the costing function is a strong indication that it is important to consider carefully the structure of the location model when doing demand point aggregation. Similarly, the ℓ_p-norm model of Shier and Dearing (1983), listed in Table I, captures a family of location models in one.

7.8 Summary

For location problems with hundreds of thousands of demand points, aggregation is often essential. This chapter has dealt with the topic of demand point aggregation for location models. We have pointed out that demand point aggregation causes error, and presented some possible error measurements. We have also presented (see Figure 7.2) the place of demand point aggregation in the entire location problem solving process, as a sort of feedback loop. We discussed in some detail one specific "row-column" demand point aggregation method, abbreviated as MRC. Graphs of error versus the number of aggregate demand points for aggregations obtained with MRC exhibit decreasing returns to scale - the law of diminishing returns is highly evident. We believe this decreasing returns phenomenon will occur with other aggregation approaches, and that it is very important for location modelers to be aware of it. Certainly using an aggregation with a small number of demand points can be a risky undertaking.

Theoretical difficulties in computing actual errors lead to the concept of an error bound, an upper bound on the maximum absolute error. This error bound can be used as a surrogate for the maximum absolute error, as is done in MRC for the N-median model. In fact, error bounds can be computed for many other location models, as Table 7.3 shows. All but the last of the models in the table are instances of a class of location models we call SAND models, which have subadditive and nondecreasing costing functions. These error bounds strongly suggest that the way to aggregate demand points in a location model should depend upon the structure of the model. In practice, centroids are commonly used as aggregate demand points, regardless of the model structure. Thus, there appear to be substantial opportunities to improve on existing aggregation approaches.

Acknowledgements This work is based mostly on research published in Francis, Lowe, Rushton and Rayco (1999) and Francis, Lowe and Tamir (2000), sponsored in part by the National Science Foundation, Grants DMI-9522882 and DMI-9908124.

References

Andreatta, G. and F. M. Mason. (1985) "k-eccentricity and Absolute k-centrum of a Probabilistic Tree," *European Journal of Operational Research*, 19, 114-117.

Berman, O. and D. Simchi-Levi (1990) "Conditional Location Problems on Networks," *Transportation Science*, 24, 77-78.

Burkard, R. (1990) "Locations with Spatial Interactions: the Quadratic Problem," Chapter 9 in *Discrete Location Theory*, editors P. B. Mirchandani and R. L. Francis, J. Wiley & Sons, N. Y., 387-438.

Domich, P. D., K. L. Hoffman, R. H. F. Jackson and M. A. McClain (1991) "Locating Tax Facilities: a Graphics-Based Microcomputer Optimization Model," *Management Science*, 37, 960-979.

Drezner, Z. (1989) "On the Conditional p-Center Problem," *Transportation Science*, 23, 51-53.

Drezner, Z. (1995) "On the Conditional p-Median Problem," *Computers and Operations Research*, 22, 525-530.

Elzinga, J. and D. W. Hearn (1972) "Geometrical Solutions for Some Minimax Location Problems," *Transportation Science*, 6, 379-394.

Erkut, E. and S. Neuman (1989) "Analytical Models for Locating Undesirable Facilities," *European Journal of Operational Research*, 40, 275-291.

Erlenkotter, D. (1978) "A Dual-based Procedure for Uncapacitated Facility Location," *Operations Research*, 26, 992-1009.

Francis, R. L. and T. J. Lowe (1992) "On Worst-Case Aggregation Analysis for Network Location Problems," *Annals of Operations Research*, 40, 229-246.

Francis, R. L., T. J. Lowe and M. B. Rayco (1993) "Row-Column Aggregation for Rectilinear Distance p-Median Problems." Research Report # 93-5, Department of Industrial and Systems Engineering, University of Florida, Gainesville, FL 32611

Francis, R. L., T. J. Lowe and M. B. Rayco (1996) "Row-Column Aggregation for Rectilinear Distance p-Median Problems," *Transportation Science*, 30, 160-174.

Francis, R. L., T. J. Lowe, M. B. Rayco and A. Tamir (2000) "Exploiting Self-Canceling Demand Point Aggregation Error for Some Planar Median Problems," working paper, Tippie College of Business, University of Iowa, Iowa City, IA.

Francis, R. L., T. J. Lowe, G. Rushton and M. B. Rayco (1999) "A Synthesis of Aggregation Methods for Multi-facility Location Problems: Strategies for Containing Error," *Geographical Analysis*, 31, 67-87.

Francis, R. L., T. J. Lowe and A. Tamir, (2000) "On Aggregation Error Bounds for a Class of Location Models," *Operations Research*, 48, 294-307.

Francis, R.L., L. F. McGinnis and J. A. White (1983) *Facility Location and Layout: An Analytical Approach*, 2nd Edition. Prentice Hall, Englewood Cliffs, New Jersey.

Geoffrion, A. (1977) "Objective Function Approximations in Mathematical Programming," *Mathematical Programming*, 13, 23-37.

Goldman, A. J. (1971) "Optimum Location for Centers in a Network," *Transportation Science*, 5, 212-221.

Halpern, J. (1978) "Finding Minimal Center-Median Convex Combination (Cent-Dian) of a Graph," *Management Science*, 24, 535-544.

Handler, G. Y. (1985) "Medi-Centers of a Tree," *Transportation Science*, 19, 246-260.

Handler, G. Y. and P. B. Mirchandani (1979) *Location on Networks: Theory and Algorithms*, The MIT Press, Cambridge, MA.

Hassin, R. and A. Tamir (1991) "Improved Complexity Bounds for Location Problems on the Real Line," *Operations Research Letters*, 10, 395-402.

Hillsman, E. and R. Rhoda (1978) "Errors in Measuring Distances from Populations to Service Centers," *Annals of Regional Science*, 12, 74-88.

Minieka, E. (1980) "Conditional Centers and Medians on a Graph," *Networks*, 10, 262-272.

Mirchandani, P. B. and A. R. Odoni (1979) "Locating New Passenger Facilities on a Transportation Network," *Transportation Research*, 13B, 113-122.

Pardalos, P. M and H. Wolkowicz, editors (1994) *Quadratic Assignment and Related Problems*, DIMACS Series in Discrete Mathematics and Theoretical Computer Science, 16, American Mathematical Society: Chapter 1, "The Quadratic Assignment Problem", pp. 1-42.

Rayco, M. B. and R. L. Francis (1996) "Asymptotically Optimal Aggregation for Some Unweighted p-Center Problems with Rectilinear Distances," *Studies in Locational Analysis*, 10, 25-36.

Rayco, M. B., R. L. Francis and T. J. Lowe (1996) "Error-bound Driven Demand Point Aggregation for the Rectilinear Distance p-center Model," *Location Science*, 4, 213-236.

Shier, D. R. and P. M. Dearing (1983) "Optimal Locations for a Class of Nonlinear, Single- Facility Location Problems on a Network," *Operations Research*, 31, 292-303.

Slater, P. J. (1978) "Centers to Centroids in a Graph," *Journal of Graph Theory*, 2, 209-222.

Tamir, A. (2001) "The K-centrum multi-facility Location Problem," *Discrete Applied Mathematics*, 109, 293-307.

Williams, J.C., Jr. (1995) "Political Redistricting: a Review," *Papers in Regional Science*, 74, 13-40.

Yakowitz, S. and F. Szidarovsky (1989) *An Introduction to Numerical Computations*, 2nd Edition. Macmillan Publishing Co. New York.

Zemel, E. (1985) "Probabilistic Analysis of Geometric Location Problems," *SIAM J. Alg. Disc. Meth.* 6, 189-200.

8 Location Software and Interface with GIS and Supply Chain Management

Thorsten Bender[1], Holger Hennes[2], Jörg Kalcsics[3], M. Teresa Melo[3], and Stefan Nickel[3]

[1] SAP AG, Walldorf, Germany. e-mail: thorsten.bender@sap.com
[2] University of Kaiserslautern, Kaiserslautern, Germany. e-mail: hennes@mathematik.uni-kl.de
[3] Fraunhofer Institut für Techno- und Wirtschaftsmathematik, Kaiserslautern, Germany. e-mail: kalcsics@itwm.uni-kl.de; melo@itwm.uni-kl.de; nickel@itwm.uni-kl.de

8.1 Introduction

A distinguishing peculiarity of location theory is that a very active research community exists (see Chap. 1–3, and the homepages of EWGLA (EWGLA, 2000) and SOLA (SOLA, 2000)) consisting of people with a very diverse background. Mathematicians, computer scientists, operation researchers, industrial engineers, regional planners, marketing researchers as well as practitioners from various branches work on location problems, but each group gives emphasis to different aspects. This implies that location problems are typically solved with methods from different disciplines. As a result, research articles do not tend to use a common language, and it is very hard to find out if a certain problem has already been addressed because it may have been solved under a completely different name. To cope with this problem, several classification schemes for location models have been proposed (see Handler and Mirchandani, 1979; Brandeau and Chiu, 1989; Eiselt et al., 1993; Carrizosa et al., 1995), and in this book also a common notation according to Hamacher and Nickel (1998) is used.

However, there is another even more severe problem (at least from a practical point of view): It is very difficult to establish a continuous knowledge transfer between the various groups dealing with location problems since, for example, mathematicians have a completely different way of describing their solution approaches than regional planners. As a consequence, improved and new solutions to important location problems are not used because the people who are in need of a practical solution do not understand the language in which the solution has been stated.

A possible way of approaching this problem is to develop software for facility location problems and make it available to the location community. This is of considerable assistance since the software focuses only on the required input and gives a specified output. The solution procedure itself does not need to be completely clear to a first–time user. However, especially for researchers, it is absolutely necessary that not only the executables are available, but also that the source code can be used and — if needed — modified. From the argumentation given so far it might seem that this is a one way road just giving the practitioners an easier access to new theoretical results. This is not quite true since also only exhaustive comparisons between different solution methods for a specific problem class allow to choose a good solution. Not too rarely a simply good idea from a practitioner may yield a very good algorithm and give the more theory oriented people some work to do. This occurs because researchers would like to explain why a good algorithm is "good" (and also the meaning of "good").

At this point, the reader should be convinced that developing software for location problems is particularly useful (otherwise, you should either stop reading now or reread this introduction until you are fully convinced). Now the question changes from why develop software for location problems to **how** to develop software for location problems. We have many different potential users with completely different requirements for such software:

Teachers would like to have software with a good user interface and an implementation of all the algorithms that are described in their course text books. The software package should have some graphical *Frontend* that visualizes results and shows illustrative examples. Finally, the software should be cheap.

Students would like to have software which both helps them to understand the theoretical material better and also supports them in doing their assignments. They would like to run the software on their home computers and that software should be free.

Researchers would like to have software with as many implemented algorithms as possible. They need the software to construct examples and counter–examples. Moreover, the software should provide a library which can be used for implementing new algorithms faster than those made from scratch. The software should be free and the source code should be available.

Practitioners would like to have software which can handle the exact problem class they need to solve. It should be able to solve very large problem instances in a reasonable amount of time. Finding an optimal solution is not so important. In most situations a good solution suffices. The software should have good interfaces for data import and export. The price is not so important.

Software companies would like to develop software which can be easily configured to handle several problem classes. It should be possible to use only the parts of the software that apply to the user's specific problems. The software should have a modern design and provide the possibility of linking third party software. Therefore, interfaces on different levels are essential. The price is not so important.

In this chapter we will describe the software components that we have developed to fulfill the needs of the different groups listed above and indicate how to obtain that software. Other software packages for solving facility location problems can be found on the EWGLA (EWGLA, 2000) and SOLA (SOLA, 2000) homepages. However, these packages either specialize on solving a specific problem or their functionality is already contained in the software described below.

The remainder of the chapter is organized as follows. First, LoLA (Library of Location Algorithms) is presented in Sect. 8.2. This software library was partially funded by the Deutsche Forschungsgemeinschaft (DFG) and has been the starting point of an organized implementation of location algorithms (Hamacher et al., 1999). In the next section, we describe how LoLA and geographical information tools can be linked in order to access the enormous amount of demographic data available. In Sect. 8.4, we describe how location software supports strategic supply chain management decisions. An actual implementation in the framework of the Advanced Planner and Optimizer business application developed by SAP AG, Walldorf, Germany is presented. The chapter ends with some conclusions and an outlook to future activities.

8.2 LoLA– Library of Location Algorithms

8.2.1 Motivation and Aim

Our main motivation for developing LoLA was, as the name already hints, to build a powerful collection of algorithms for location theory, encompassing planar, network and discrete problems, and to use these algorithms for solving various problems in a fast and easy way. LoLA can be downloaded free of charge from http://www.mathematik.uni-kl.de/~lola. Moreover, all source codes are available in the above address.

To detect if a given location problem was already treated in the literature and (in the case of LoLA) to apply the appropriate algorithm we need a uniform language to classify the problem. LoLA utilizes the classification scheme of Hamacher and Nickel (1998) for easy access to the implemented algorithms. The scheme consists of the following five positions

$$pos1/pos2/pos3/pos4/pos5 \ .$$

The meaning of each position along with some examples is given in Table 8.1. If no special assumptions are made in a position, this is indicated by a •. For example, a • in position 4 means that any distance function is considered.

LoLA is designed to address several different user groups. It should accompany lectures in the field of location theory and provide the possibility for students to apply the presented methods to specific problems and "see" location theory in action. Schools make up another application field. A vivid

Table 8.1. Classification scheme for facility location problems

position	meaning	usage (examples)	
1	number of new facilities		
2	type of problem	P	planar problem
		D	discrete problem
		G	problem on a general undirected graph
3	special assumptions and restrictions	$w_m = 1$	all weights are equal
4	type of distance function	γ	a general gauge
5	type of objective function	Σ	median problem
		max	center problem

alternative to the classical topics of high school mathematics should be provided in order to bring mathematics and especially location theory closer to pupils. Finally, LoLA enables researchers and software developers to compare their new results with those that already exist or to incorporate some of the available algorithms into their own applications.

LoLA provides a graphical user interface that allows its simple application in industrial projects as well as for demonstrations in high school and university teaching. In addition, a *Text–based* user interface is available to call algorithms of LoLA from other applications. To solve individual facility location problems, a programming interface allows the direct incorporation of specific algorithms of the program library into the implementation of extended routines (*Callable Library*).

To present the different alternatives available in LoLA for solving facility location problems, we will use the following two examples throughout this section.

Example 1. We consider a 1–facility planar minisum problem with the squared Euclidean distance and a convex polyhedron as a forbidden region inside. According to the classification scheme of Hamacher and Nickel (1998), this problem is denoted by $1/P/\mathcal{R} = convpoly/l_2^2/\Sigma$.

Example 2. The second example we will focus on, is a 4–facility undirected network minisum problem where the solutions are searched on the nodes. In the classification scheme of Hamacher and Nickel (1998) this problem is described by $4/G/ \bullet /d(V,V)/\Sigma$. We will solve the problem with the interchange heuristic of Teitz and Bart (1968).

8.2.2 System Design

The main component of LoLA is the open source C++ Routine Library which uses the software packages LEDA (Library of Efficient Data types and

Algorithms, Mehlhorn et al., 2000) and LP–Solve (Berkelaar, 1995) as shown in Fig. 8.1.

In order to be applicable for teaching, LoLA is available as a stand-alone program. For this purpose there exist two different ways of executing algorithms. One is with the help of the graphical user interface (GUI) of LoLA, called *Frontend*, which is directly attached to the library to load data files, select algorithms and view the results of the optimization. In order to guide the user to the appropriate solution of his/her problem, the *Frontend*, which is implemented in Tcl/Tk (Ousterhout, 2000), reflects the classification scheme of Hamacher and Nickel (1998) and provides a detailed help manual. The second possibility is to run algorithms *Text–based* from a console (e.g. from MS-DOS[1], a Unix[2] or Linux Shell). Therefore, specific algorithms can be called from other applications, e.g. a geographical information system (see Sect. 8.3), providing input for the chosen method and showing the results of the optimization. Data is transferred between the invoking application and LoLA using text files based on a descriptive language.

To develop new applications using algorithms available in LoLA, a *Programming Interface* was devised for researchers and software developers. The *Callable Library* enables the incorporation of LoLA algorithms into other applications at an implementational level. This design allows fast access to the available functions and the direct transfer of data to and from these functions thus avoiding the detour via ASCII files.

The above mentioned software packages LEDA, Tcl/Tk and LP–Solve are available for the platforms Windows[3] 95/98/NT as well as for Unix and Linux systems. Hence, LoLA is independent of these platforms.

8.2.3 The Components of LoLA

Graphical User Interface. The GUI of LoLA, called *Frontend*, is based on the 5–position classification scheme which was introduced by Hamacher and Nickel (1998) and briefly described in Sect. 8.2.1. If Tcl/Tk is available on your system, calling LoLA creates the window shown in Fig. 8.2.

Following the classification scheme, the menu of the *Frontend* contains the buttons listed below:

Number has the options 1–facility, N–facility and 1–line.

Type is used to choose among the problem types, P: planar, G: graph, D: discrete and T: tree.

Specials is used to select extra assumptions for the problem to be solved which may include:

---
[1] MS-DOS® is a registered trademark of Microsoft Corporation, USA.
[2] Unix® is a registered trademark of The Open Group, USA.
[3] Windows® is a registered trademark of Microsoft Corporation, USA.

Fig. 8.1. System Design of LoLA

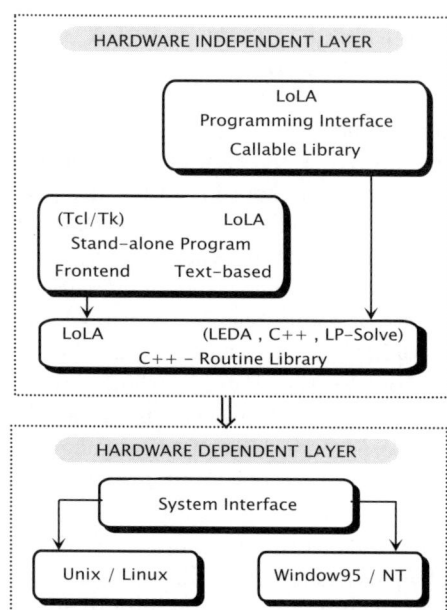

Fig. 8.2. LoLA *Frontend*

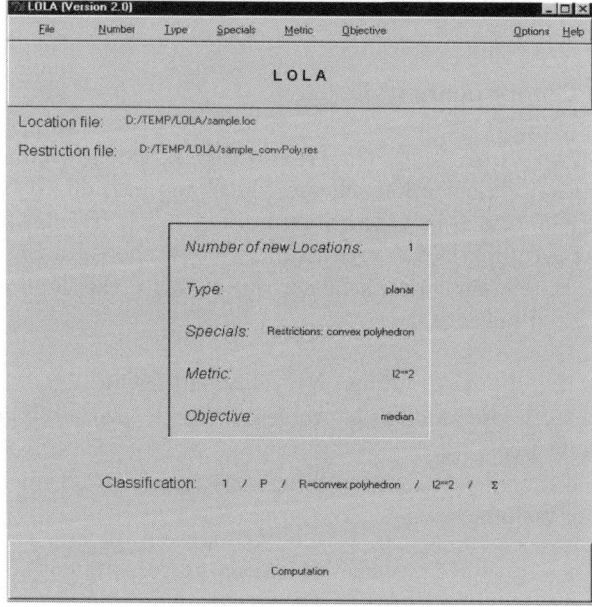

equal weights If the weights of the existing locations are all equal, fast procedures are available for some types of location problems, e.g. $1/T/w_i = 1/d(V,T)/\Sigma$.

restrictions With the option outside we can choose whether the forbidden region is inside or outside the given restrictions. Possible restriction types are: polyhedron, convex polyhedron, circle, rectangle and all (the restriction file must describe one or more of the above four possibilities in arbitrary order and number).

none (default option)

Metric is used to select the distance function. For planar problems the following norms are available: l_1, l_2, l_2^2, l_∞ and l_p, which are denoted in LoLA by l1, l2, l2**2, linf and lp, respectively. For the l_p–norm, the value of p can be selected in Options under Preferences. Self–created gauges (gauge) and block–norms (block) that is, symmetrical gauges, are also available. For Graph and Tree algorithms, we have the options d(V,V) and d(V,G) (d(V,T)), respectively, where the optimal points are searched only on the node set V or on the entire graph G (tree graph, T).

Objective to select the type of objective function to be minimized. median and center functions as well as their multicriteria extensions Q–median and Q–center are available.

In addition to the above options, the *Frontend* of LoLA contains the following menus:

File is used to select files providing data for a specific problem, e.g. Load Location, Load Restriction among others. The buttons View Location and View Restriction allow the display of the input data in a separate window (see Fig. 8.5). Graphical Edit and Create Example provide tools to create new data files.

Help is used to call the online help browser.

Options is used to set general preferences on the maximum number of iterations, the default metric for l_p, the choice of a heuristic for p–facility network problems or other problem dependent settings.

Computation is used to start the optimization.

Example 1 and 2 (cont.). The input data for the planar problem (Example 1) regarding existing locations and their weights are contained in the file "sample.loc", while "sample_convPoly.res" is the restriction file (available in every standard distribution of LoLA). The input data for the network problem (Example 2) are contained in the file "sample.gra". Input files can be generated with any ASCII–editor (see Sect. 8.2.4 for a description of the data format) or with the graphical editor that is available under Graphical Edit *in the* File *menu. In this editor, the location data can be entered using the mouse. To solve our two problems, three steps must be performed.*

a) *First, the problem type is specified by selecting the following options in the menu bar.*

	Options	Example 1	Example 2
	Number	1–facility	N–facility *and then enter* 4
	Type	planar	graph
	Specials	convex polyhedron	–
	Metric	$l2**2$	d(V,V)
	Objective	median	median
	Options	–	N/G/•/d(V,V)/• *and then select* Exchange–heuristic

Depending on the problem type, the classification string on the LoLA *screen now reads*

$$1/P/R = convex\,polyhedron/l2**2/\Sigma \qquad or \qquad 4/G/\bullet/d(V,V)/\Sigma\ .$$

b) *Next, the input files are loaded using the following menu entries that are available under* File.

	Example 1	Example 2
	Load Location	Load Graph
	Load Restriction	

c) *Clicking on* Computation, LoLA *will provide an output window with the solution (see Fig. 8.3 for the planar problem).*

A solution window, as the one shown in Fig. 8.3, contains buttons to Save Results of the current solution in a file, to Refresh the solution window and to Close the window. Pressing the button View Results will show the coordinates and the objective function value of the optimal solution (see Fig. 8.3).

Solution windows of planar problems additionally contain the buttons for showing or hiding, respectively, the Convex Hull of the set of existing facilities, for showing/removing the Weights of the existing facilities and to view the unit ball of a special gauge: View Gauge.

In solution windows of network problems the buttons Node_Weights and Edge_Weights are available for showing or hiding the weights of the nodes and the edges, respectively, of the network (see Fig. 8.4).

For discrete location problems, buttons for showing or hiding the Fixed Costs of the supply points, the Weights of the demand points or to View the Cost Matrix are additionally available in the solution window.

Text–based User Interface. All algorithms of LoLA can be executed using command–line options in the *text–based mode* for which the *Frontend*, and therefore Tcl/Tk, is not needed. Upon invocation, LoLA returns the solution in text or graphical format. However, the latter can be completely

Fig. 8.3. Windows with the optimal solution and the numerical result for the planar problem (Example 1)

suppressed, rendering LoLA capable of operating text–only. This mode of operation is well suited to perform automated or repeated tasks or calling algorithms from other applications.

The *text–based mode* is automatically accessed if command–line options are detected, and it is the only available mode if the LoLA executable has been built with the configuration --withtcltk=no. The available options are

```
lola -a <algorithm> [-l|-r|-g|-d|-m <file>] [-p <n>] [-n <n>]
     [--output=<arguments>]
```

Options:
```
-a : the algorithm to run with the data (see below)
-l : <file> containing data describing the existing facilities
-r : <file> containing data describing the restriction(s)
-g : <file> containing data describing a (directed or
     undirected) graph
-d : <file> containing data describing polygonal gauge
     definitions
-m : <file> containing data describing a cost matrix for
     n-facility problems
-p : <p> is the value of p for an lp-norm
```

```
-n : <n> is the number of new facilities for problems on graphs
```

```
--output:
  this option takes a comma-separated list of arguments:
    (no)windowed : (don't) present the solution graphically;
    [file=]<file>: write solution as text into <file>;
  only the right-most argument of each type takes effect
```

Example 1 (cont.). To solve our planar problem $1/P/\mathcal{R} = convpoly/l_2^2/\Sigma$ using the input files "sample.loc" and "sample_convPoly.res", we need to type

```
lola -a in_l2sqr_sum -l sample.loc -r sample_convPoly.res
     --output:file=results.txt
```

The results of the optimization will be saved in the file "results.txt".

Example 2 (cont.). Typing

```
lola -a N_median_exchange -g sample.gra -n 4
```

will yield the solution for our multi-facility network problem (see Fig. 8.4).

Fig. 8.4. Windows with the solutions Opt1–Opt4 and the numerical result for the network problem (Example 2)

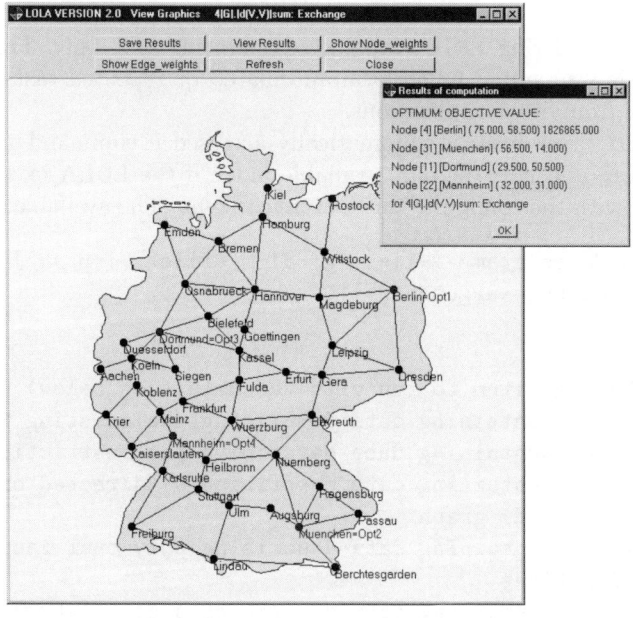

Programming Interface. All algorithms implemented in LoLA can be accessed via the *Callable Library* interface. Again using our examples, we will briefly explain how the LoLA–libraries can be used directly in a C++ program without making use of the LoLA *Frontend*.

Example 1 (cont.). First, we need to include the definitions of the routines required to handle location and restriction files (read, write) and the algorithms to solve the planar problem.

```
#include <LOLA/facs_util.h>
#include <LOLA/planealg.h>

void main() {
  // Variables to store the objective value, the
  // classification string and the names of the input files.
  double objval;
  string class_string;
  string loc_file="sample.loc";
  string restr_file="sample_convPoly.res";

  // List to save the solution points.
  list<location> PlanarSol;

  // Data structures of LoLA to handle locations,
  // restrictions and the available planar algorithms.
  facs_util FacsUtil;
  planealg PlaneAlg;
  restrictions Restr;

  // Read the input data of the planar problem describing
  // locations, i.e. x-coord, y-coord and their weight.
  ifstream locfile (loc_file);
  FacsUtil.ReadLoc(locfile);

  // Read a restriction file.
  ifstream restrfile (restr_file);
  Restr = ReadRestr(restrfile);

  // Solve the planar problem.
  objval = PlaneAlg.l2sqr_sum(FacsUtil,Restr);

  // The input data including locations, the restrictions
  // and the solution are graphically shown in a window.
  PlanarSol = PlaneAlg.alg_solution();
  class_string = "1/P/R = convex polyhedron/l2sqr/sum";
```

```
      FacsUtil.View(objval,class_string,PlanarSol,Restr);
}
```

Example 2 (cont.). For the network example, the following commands are required.

```
#include <LOLA/graph_util.h>
#include <LOLA/lgraphalg.h>

void main() {
  // Variables to store the objective value, the
  // classification string and the name of the input file.
  double objval;
  string class_string;
  string graph_file="sample.gra";

  // List to save the solution points.
  list<sol_typ> NetworkSol;

  // Data structures of LoLA to handle network problems;
  // the coordinates and names of the nodes are stored
  // separately in two data structures;
  // for N-facility problems additionally the assignments
  // of nodes to the solution points are computed.
  facilities GraphFacilities;
  list<string> LocTxt;
  graph_util GraphUtil;
  lgraphalg GraphAlg;
  list<int> BelongTo;

  // Load the input data for the network problem.
  ifstream graphfile (graph_file);
  GraphUtil.ReadGraph(graphfile,GraphFacilities,LocTxt);

  // Solve the network problem and show the solution.
  lolaundirected UGraph = GraphUtil;
  objval = GraphAlg.N_median_exchange(UGraph,4,BelongTo);

  NetworkSol = GraphAlg.alg_solution();
  class_string = "4/G/./d(V,V)/sum";
  GraphUtil.LGraphView(objval,class_string,NetworkSol,
                       GraphFacilities,LocTxt,BelongTo);
}
```

8.2.4 Format of the Input and Output Data

To solve facility location problems independent of a specific platform, LoLA reads input data and writes the results to ASCII files. LoLA interprets problem data for location problems using a descriptive language specifically designed for that task. There are two possibilities to generate input data files. One is to use any standard ASCII editor. The second is to use the graphical editor attached to LoLA. This tool allows a direct graphical input of data which can be converted into the LoLA data language afterwards.

In the following we describe the input and output formats for planar, restriction and network data files required for solving our example problems. For a complete description we refer to Hamacher et al. (1999). Note that the blanks in the environment specifications as given below cannot be omitted!

Input File Type: planar location data *loc*
begin {location} [d,Q]
$x_{11} \cdots x_{1d} \quad w_{11} \cdots w_{1Q}$ [*symbolic name of facility* 1]
$\vdots \qquad\qquad \vdots \qquad\qquad \vdots$
$x_{M1} \cdots x_{Md} \; w_{M1} \cdots w_{MQ}$ [*symbolic name of facility* M]
end {location}

d: Dimension of the facilities.
Q: For N–Facilities problems: Q is the dimension of the weights, which in this case is equal to the number N of new locations.
For Q–median or Q–center problems: The number of criteria (objectives) according to which the problem is to be solved.
x_{ij}: j–th coordinate of the i–th facility, $i = 1, \ldots, M$, $j = 1, \ldots, d$.
w_{ij}: For N–Facilities problems: For each new facility, there must be a weight representing the importance (demand) of this new facility with respect to the existing facilities. Hence, the value w_{ij} represents the weight of the j–th new facility with respect to the i–th existing facility. Note that in the case that $N = 1$, i.e. for 1–facility problems, only one weight w_i has to be specified for each existing facility.
For Q–median or Q–center problems: w_{ij} represents the weight (demand, importance) of the existing facility i with respect to the j–th criterion (objective).

Input File Type: restriction data *res*

In case of planar location problems, restrictions (forbidden regions or barriers for the new locations) can be specified using files of type *res*. Next, only the description of 2-dimensional convex polyhedra is given.

begin {restriction} [2]
begin {conpolyhedron} [m]

x_1 y_1 This environment should be alternatively used in case
x_2 y_2 of convex polyhedral restrictions.
$\vdots$

x_n y_n
end {conpolyhedron}
end {restriction}

m: This number indicates the type of restriction:
0: restriction is a forbidden region (default argument, can be omitted),
inf: restriction is a barrier.

Example 1 (cont.). The two files "sample.loc" and "sample_konvPoly.res" of the planar example problem are (partly) depicted in Fig. 8.5 using the buttons **View Location** *and* **View Restriction**.

Fig. 8.5. Location and restriction data files of the planar example

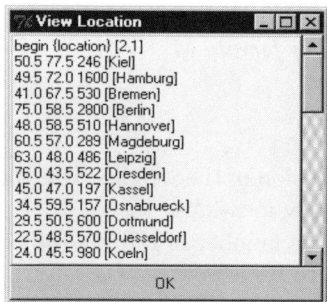
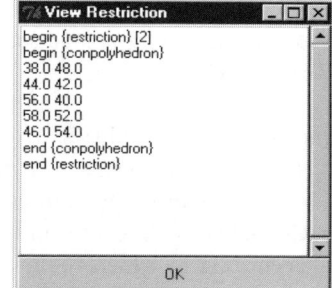

Input File Type: network data *gra*

The input format for network location problems consists of a *location* file and a file representing the adjacency matrix of the network (graph) in an *adjacencylist*. For this purpose, the format *adjlist_byname* is provided where the edges of the corresponding network can be specified using the symbolic names of their starting node (source node) and their end node (target node). The information of the existing facilities and the other nodes of the network is stored using the *location* format. Here the nodes can be specified by their (d – dimensional) coordinates which allows the use of location data from problems of planar type. They can be alternatively assigned the attribute NC, i.e. "no coordinates" have to be specified. If desired, the symbolic names of nodes in the adjacency list can be replaced by numbers corresponding to the sequence of the facilities in the location part of the network data file. Nodes of the network not representing an existing facility can be included in this list by setting their weights w_{ij} equal to zero.

```
begin {lolagraph}
begin {location} [d,Q]
```
$x_{11} \cdots x_{1d} \quad w_{11} \cdots w_{1Q}$ [symbolic name of facility 1]

$\qquad \vdots \qquad\qquad \vdots \qquad\qquad \vdots$

$x_{M1} \cdots x_{Md} \; w_{M1} \cdots w_{MQ}$ [symbolic name of facility M]
```
end {location}
begin {adjlist_byname}
```
$sourcename_1 \; targetname_1 \; ew_1$

$\qquad \vdots \qquad\qquad \vdots$

$sourcename_n \; targetname_n \; ew_n$
```
end {adjlist_byname}
end {lolagraph}
```

ew_i: Length of the i–th edge.

$sourcename_i$: Symbolic name of the starting node of the i–th edge in the location file.

$target_i$: Symbolic name of the end node of the i–th edge in the location file.

The information about the solution of a location problem is saved in files of type *sol*. Depending on the type of problem solved, this file may contain different information. In the first two environments, classification and objective value are self–explaining. In the environments polygonlist and graphpointset, one or several (as e.g. in the case of multicriteria problems) sets of optimal points/polyhedra can be given.

Output File Type: planar and network results *sol*
```
begin {result}
begin {classification}
```
classification
```
end {classification}
begin {objective value}
```
$z \; z \; z$
```
end {objective value}
```

For planar problems, the polygonlist contains the solution points/polyhedra:

```
begin {polygonlist}
begin {polygon}
```
$x_1 \; y_1$

$\vdots$

$x_n \; y_n$
```
end {polygon}
```
$\vdots$

```
begin {polygon}
x₁ y₁
⋮
xₙ yₙ
end {polygon}
end {polygonlist}
end {result}
```

whereas for the result of a network problem the solution points are given in the list `graphpointset`:

```
begin {graphpointset}
SolutionPoint 1
⋮
SolutionPoint n
end {graphpointset}
end {result}
```

The *SolutionPoint*s are described by

Node [i]: If the solution point is a node of the network.

Edge [i][j] t: If the solution point is on an edge of the underlying graph. i and j denote the source and the target node, respectively, of the edge and t, $0 \leq t \leq 1$, the relative distance of the solution point on the edge from i to j.

Example 2 (cont.). For our network problem the result file contains the following information:

begin {result}
begin {objective value}
1.99011e+006
end {objective value}
begin {graphpointset}
Node [13]
Node [39]
Node [33]
Node [23]
end {graphpointset}
end {result}

8.2.5 Function Reference

The algorithms available in LOLA for solving planar and network facility location problems are listed in Tables 8.2 and 8.3. A comma separated list at positions 2 or 3 of the classification scheme of Hamacher and Nickel (1998) indicates the available restrictions for a specific planar problem or on which type of networks the problem can be solved, respectively. Note that γ at position 4 denotes a general gauge. Finally, LoLA can solve discrete problems of the type $\#/D/\bullet/\bullet/\bullet$.

8.3 LoLA goes GIS

8.3.1 Introduction

In the preceding section, we introduced LoLA and explained how its input and output functionality is designed. An open question is how to get real world data easily into LoLA to solve real world problems.

Suppose we have a large set of facilities whose locations are saved in a certain database table or something similar. Although LoLA provides different ways of using these data, it cannot read any arbitrary input data file. On one hand, we can create the input file by hand, i.e. copy and paste; and on the other hand we can use the graphical editor of LoLA to create the file. Since both ways are very time consuming and ineffective, we developed a third option which consists of implementing a converter to create the location file from a database table.

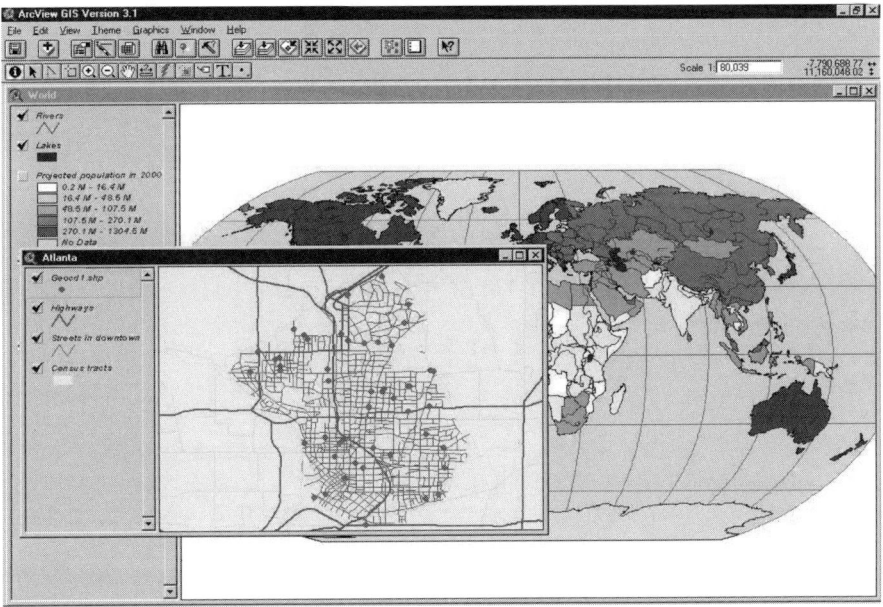

Fig. 8.6. ArcView GIS *Frontend*

Table 8.2. Algorithms which are available in LoLA for planar problems

Planar Problems
$1/P/\bullet, \mathcal{R}, \mathcal{R}^c/l_1/\Sigma$
$N/P/\bullet/l_1/\Sigma$
$1/P/\bullet/l_1/2 - \Sigma_{par}$
$1L/P/\bullet, \mathcal{R} = convpoly/l_1/\Sigma$
$1/P/\bullet, B, \mathcal{R} = convpoly/l_2/\Sigma$
$N/P/\bullet/l_2/\Sigma$
$1L/P/\bullet, w_i = 1, \mathcal{R} = convpoly/l_2/\Sigma$
$1/P/\bullet, \mathcal{R}, \mathcal{R}^c/l_2^2/\Sigma$
$N/P/\bullet/l_2^2/\Sigma$
$1/P/\bullet/l_2^2/Q - \Sigma_{par}$
$1/P/\bullet, \mathcal{R} = convpoly/l_p/\Sigma$
$N/P/\bullet/l_p/\Sigma$
$1L/P/\bullet, \mathcal{R} = convpoly/l_p/\Sigma$
$1/P/\bullet, \mathcal{R}, \mathcal{R}^c/l_\infty/\Sigma$
$N/P/\bullet/l_\infty/\Sigma$
$1/P/\bullet/l_\infty/2 - \Sigma_{par}$
$1L/P/\bullet, \mathcal{R} = convpoly/l_\infty/\Sigma$
$1/P/\bullet/\gamma/\Sigma$
$1/P/\bullet/\gamma/2 - \Sigma_{par}$
$1L/P/\bullet, \mathcal{R} = convpoly/\gamma_B/\Sigma$
$1/P/\bullet, w_i = 1, \mathcal{R} = convex/l_1/max$
$N/P/\bullet/l_1/max$
$1L/P/\bullet/l_1/max$
$1/P/\bullet, \mathcal{R}/l_2/max$
$1L/P/\bullet/l_2/max$
$1L/P/\bullet/l_p/max$
$1/P/\bullet, w_i = 1, \mathcal{R} = convex/l_\infty/max$
$N/P/\bullet/l_\infty/max$
$1L/P/\bullet/l_\infty/max$
$1/P/\bullet, \mathcal{R}/\gamma/\max$
$1L/P/\bullet/\gamma_B/\max$

Table 8.3. Algorithms which are available in LoLA for network problems

Network Problems
$1/G_D,G,T/ \bullet /d(V,V)/\Sigma$
$1/G/ \bullet /d(V,G)/\Sigma$
$1/T/ \bullet /d(V,T)/\Sigma$
$1/G/ \bullet /d(V,G)/2 - \Sigma_{par}$
$1/G_D,G/ \bullet /d(V,V)/Q - \Sigma_{par}$
$1/G_D,G/ \bullet /d(V,G)/Q - \Sigma_{par}$
$1/G_D,G/ \bullet /d(V,V)/Q - \Sigma_{lex}$
$1/G_D,G/ \bullet /d(V,G)/Q - \Sigma_{lex}$
$1/G_D,G,T/ \bullet /d(V,V)/\max$
$1/G_D,G/ \bullet /d(V,G)/\max$
$1/T/ \bullet /d(V,T)/\max$
$1/G_D/ \bullet /d(V,V)/Q - \max_{par}$
$1/G_D/ \bullet /d(V,G)/Q - \max_{par}$
$1/G/ \bullet /d(V,V)/Q - \max_{lex}$
$1/G_D/ \bullet /d(V,V)/Q - \max_{lex}$
$N/G/ \bullet /d(V,V)/\Sigma$
$N/G/ \bullet /d(V,V)/\max$

This is the point where Geographical Information Systems (GIS) come into play. These systems are designed for visualizing real world data of countries, states, cities, etc. with map data and additional information that allow decision makers to find solutions to problems such as locating facilities, routing, etc. An example of a *Frontend* of such a GIS is given in Fig. 8.6.

It seems to be a natural idea to link both systems, GIS and LoLA. LoLA needs real world data and an easy way to quickly input large data sets, while the GIS needs more sophisticated algorithms to be able to deal with locational decisions. One of the basic tasks of such a link is to transform data from GIS to LoLA. Additionally, the GIS should be able to call routines from LoLA, extract the solution of the location problem and visualize it in a map with the given data. Figure 8.7 depicts the scheme of the functionality the link should have.

An example of a Geographical Information System is ArcView GIS[4]. This GIS provides a large amount of location–relevant data, and new data is easy to add in the form of database tables, text files, etc. The large amount of data in ArcView GIS gives LoLA a large source of real world data. Second, new

---

[4] ArcView GIS® is a registered trademark of Environmental Systems Research Institute, Inc. (ESRI), USA.

Fig. 8.7. LoLA–GIS link

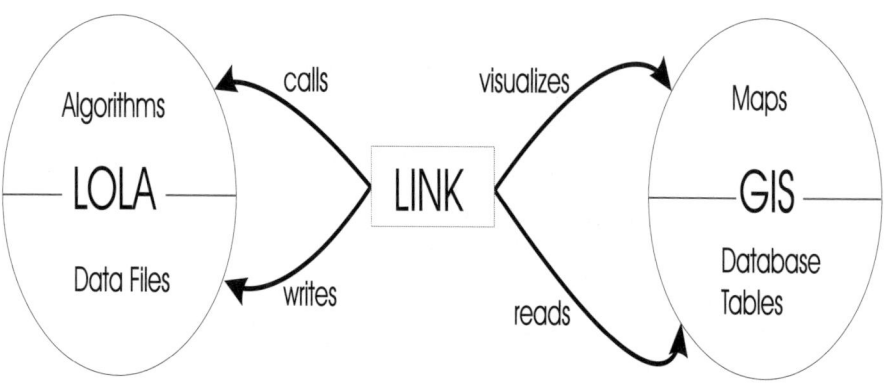

data can be easily added. This gives us a larger area of acting and solving problems.

A very useful tool of ArcView GIS is the script language which is called Avenue. It is easy to learn, easy to handle and has the ability to control every process of ArcView GIS. This language is what we need to combine ArcView GIS with LoLA.

We have implemented scripts that read data out of the ArcView GIS database, where all the locational data is stored, and convert the data into a LoLA input file. From another Avenue script, we can call LoLA and its routines as a console application, i.e. in the text–based mode; or we can call other executables with our own implementations which are based on LoLA. Since LoLA gives us the possibility to save the solution into a file we can get the solution back into ArcView GIS with another script, where we can visualize it. Hence, ArcView GIS together with Avenue fulfills the necessary conditions for linking it to LoLA; and, additionally, ArcView GIS meets the desired conditions that we mentioned above.

With the help of Avenue, we were also able to create and control menu options so that the user has an interface that allows him to communicate with ArcView GIS to select the algorithms of LoLA.

8.3.2 Implementation

First of all, we would like to export data from a GIS to LoLA in order to have real world data for our software. Since ArcView GIS was the selected GIS, we have to accept its conventions. The locational data concerning the facilities are stored in tables which we would like either to read from the database of ArcView GIS directly into LoLA or to write into an ASCII file which could be read by LoLA. Avenue gives us the means to accomplish this. Another advantage of Avenue is that it can directly access the coordinate data of the facilities which are hidden by ArcView GIS. With Avenue we can easily access the data base of ArcView GIS and write our own ASCII data files in the input format that LoLA uses. Below you will find the procedure for getting data of one record, for one facility, and writing the corresponding ASCII file:

```
lolaPoint = lolathemetab.GetLabelPoint(rec)
lolafile.WriteElt(lolaPoint.GetX.AsString + " " ++
    lolaPoint.GetY.AsString + " " ++
    lolathemetab.ReturnValue(lola3field, rec).AsString +
    " ["++ lolathemetab.returnValue(lolafield, rec) + "]").
```

The coordinates are written first, then a chosen value as weight and the name of the location. Running this procedure over each record we get the location data written in an ASCII file where only the head and the tail must be added in order to use it in LoLA.

Based on this script that we extended, the scripts for ArcView GIS increase the functionality of the ArcView–LoLA–link (ALL). We added our own control functions to have more control of ALL. In Fig. 8.8, the added LoLA button and the menu part for LoLA are shown.

Fig. 8.8. ArcView GIS–LoLA menu

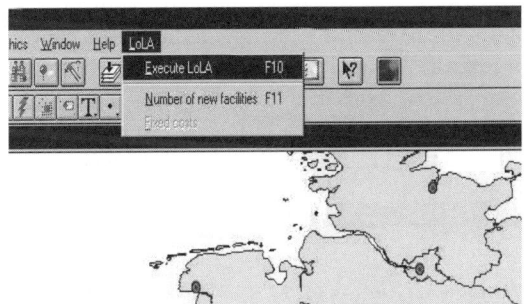

The menu option *number of new facilities* in Fig. 8.8 is used to specify the number of new facilities to build, i.e. a single facility, a specified number p for the p–median or center problem respectively, or to solve the Uncapacitated Facility Location Problem (UFLP) with an unspecified number. The dialog is shown in Fig. 8.9.

In the last option *fixed costs* we can define fixed costs for setting up new facilities (see Fig. 8.10). The option *Call* LoLA will start the process of writing the data file and calling LoLA. The LoLA button runs the same process. Upon starting the process, the dialog shown in Fig. 8.11 appears. Here the user has three fields to choose from. The first field is used to select a

Fig. 8.9. ArcView GIS–LoLA number of new facilities dialog

Fig. 8.10. ArcView GIS–LoLA fixed costs dialog

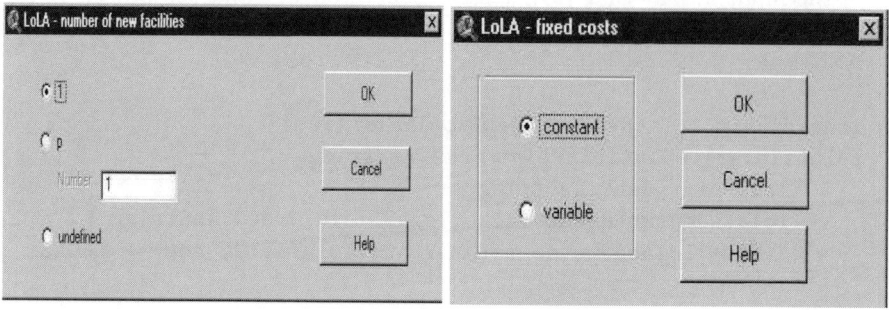

theme as a base for the locations. A theme is a visual representation of some geographic data. Within a theme the complete information of ArcView GIS with respect to a special data field, i.e. cities, countries, streets, etc., is stored. The data for the weights can be chosen in the second field. The dialog takes all numerical data that can can be found in the theme table and displays it. Then the user can choose one field from the list box which uses the data as weights. The last option allows the user to decide on the algorithm that runs with LoLA. This action depends on the number of facilities chosen in the *number* menu option. For single facility location problems there are in total six algorithms available for solving median and center problems with the distance functions l_1, l_2 and l_∞. For p–median/center problems and for the UFLP, there are three heuristics and an exact algorithm available.

Fig. 8.11. ArcView GIS–LoLA dialog

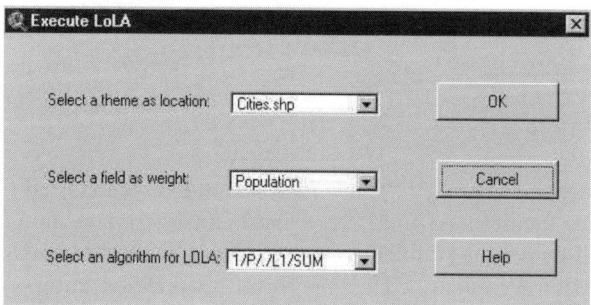

In Fig. 8.12 the visualization of the result of a p–median problem with 4 new facilities is shown. The new facilities are depicted with large points and the allocation of the existing facilities to the new ones is displayed with straight lines.

Fig. 8.12. ArcView GIS–LoLA solution

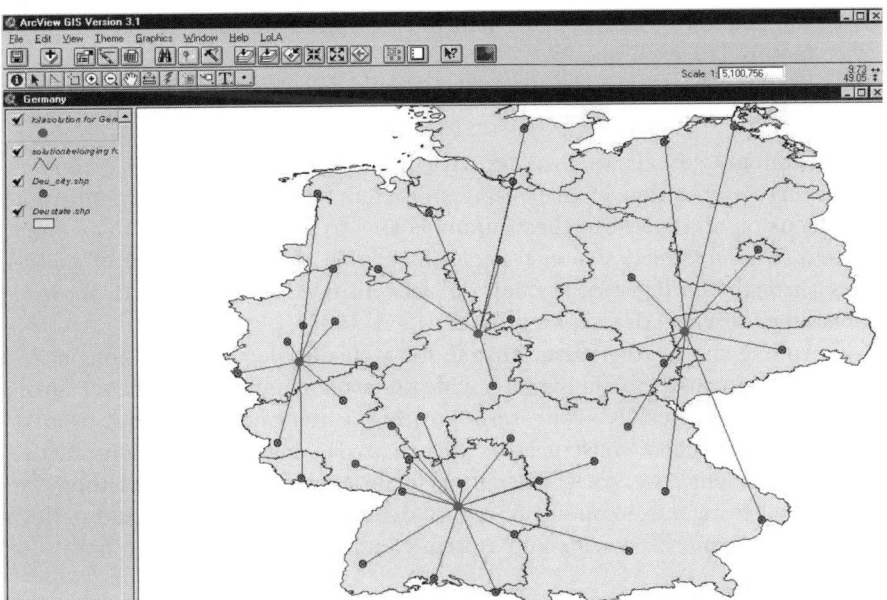

8.4 Supply Chain Management

8.4.1 The Role of Facility Location Planning in Supply Chain Management

A supply chain is a network of facilities (e.g. plants, distribution centers, warehouses) that performs a set of operations ranging from the acquisition of raw materials, the transformation of these materials into intermediate and finished products, to the distribution of the finished goods to the customers. The optimization of the complete supply chain is accomplished through efficient planning decisions. Three levels of planning can be distinguished depending on the time horizon. Strategic planning has a long–range scope and focus on the development of objectives and policies for the supply chain. Tactical planning is a medium–term activity concerned with the means by which the strategic objectives can be realized such as the effective use of existing resources. Finally, operational planning involves short–term decisions that deal with the efficient operation of daily activities.

The strategic design of a supply chain involves determining

- the number, location and capacity of manufacturing plants, distribution centers, warehouses, etc. that should be used;
- the number and location of existing service facilities that should be maintained;

- the set of suppliers that should be selected and the amount of raw materials that should be procured;
- the amount of intermediate and finished products that should be manufactured at each plant;
- the transportation channels that should be used to ship the products between the facilities.

Economy globalization along with rapid changes in technology and increasing competition in the business environment are forcing organizations to focus on and invest in their supply chains to quickly respond to customer needs. Well–planned strategic decisions enable the efficient flow of materials through the logistics system and lead to decreased costs and improved customer service (Bramel and Simchi–Levi, 1997).

Among the strategic issues listed above, facility location decisions have a key role in strategic planning for a wide range of organizations. The set up of a new facility is typically a long–term project involving high costs (e.g. property acquisition, facility construction) with a strong impact on various tactical and operational decisions. Therefore, facilities which are located today are expected to remain in operation for an extended period of time and perform well even as market trends and system conditions change.

8.4.2 Facility Location Models for Strategic Supply Chain Planning

As mentioned in the previous section, decisions concerning facility location have a major impact on the strategic design of supply chains. Although the term *supply chain management* is relatively new, the development of facility location models for the strategic design of production–distribution networks has received considerable attention in the last decades. In particular, emphasis has been put on Mixed Integer Programming (MIP) formulations. Aikens (1985) reviews some relevant MIP formulations for production–distribution systems. One of the earliest contributions in this area is dedicated to a multi–commodity problem comprising several plants with known capacities, distribution centers (DCs), and a number of customer zones (Geoffrion and Graves, 1974). Potential site locations for new DCs are known, but the particular sites to be used are selected based on minimizing total network costs. These costs consist of fixed charges for setting up a DC, variable operating costs (based on the amount shipped through a DC), and production and transportation costs for shipping products from a plant to a customer zone via a DC. The constraints in the model are related to capacity limits at the plants, customer demand satisfaction, single sourcing by customer zones and bounds on the throughput at each DC. A solution procedure based on Benders decomposition is proposed. A refined version of the model addressing real world problems was later developed by Geoffrion et al. (1978). The extended model considers single sourcing of customer zone by product type, nonlinear facility throughput constraints and trade–offs between distribution and customer service. The optimization procedure for solving this large scale MIP problem is again based on Benders decomposition.

In many practical situations, the selection of a potentially new facility is linked with the decision about the type of equipment that facility should use to handle a particular commodity. Hence, apart from the fixed charge for opening a facility, an additional fixed cost is incurred if an open facility is equipped to handle a given product or has a certain level of capacity installed (e.g. small, medium and large). Brown et al. (1987) considered such a generalization of the multi–product capacitated facility location problem. Their MIP formulation addresses the opening and closing of plants, the assignment of equipment to plants, and the commodities produced at each plant and delivered directly to customer zones. Variable production and transportation costs, fixed costs of equipment assignment, and fixed costs of plant operations are included in the objective function. The constraints considered in the model are comprised of customer demand satisfaction, maximum number of equipment assigned to each plant, single sourcing of equipment to plants, and upper bounds on commodities produced on each equipment at each plant. The problem is solved by applying a decomposition principle similar to that developed by Geoffrion and Graves (1974). Recently, Lee (1993), and Mazzola and Neebe (1999) addressed a similar problem in a 2–echelon context in which plant and customer locations were fixed and facility location decisions were restricted to a single echelon of DCs. Solution procedures developed for this problem use cross decomposition (Lee, 1993) and Lagrangian relaxation (Mazzola and Neebe, 1999).

Another type of problem that has practical relevance in industrial settings refers to the simultaneous selection of site locations for plants and DCs in a multi–commodity capacitated production–distribution system. Pirkul and Jayaraman (1996) apply Lagrangian relaxation to an MIP formulation of the problem and present a heuristic procedure for solving it. The objective function of the model minimizes the sum of the fixed cost of establishing and operating the plants and the DCs along with the variable cost of transporting products from the plants to the DCs and distributing the commodities from the DCs to the customers in order to satisfy the demands of the latter. Both plants and DCs have limited capacity.

The long–term nature of facility location decisions involves planning the operation of facilities in such a way that they can cope well with an uncertain future environment. Therefore, robust location decisions are of particular importance. For the case of incomplete information regarding future customer demands, Carrizosa and Nickel (1998) propose a model and a solution technique for solving a single facility continuous location problem. Robust decisions should also consider the dynamic aspect of facility location with respect to the timing of facility expansions and relocations during the whole planning horizon. While the literature on static facility location in production–distribution systems is quite extensive, the dynamic version of the problem has received considerably less attention. Owen and Daskin (1998) present a review on relevant MIP formulations in this area. A recent contribution dedicated to a 2–echelon multi–commodity capacitated facility location problem considers opening and closing both plants and DCs over a given number of time periods (Hinojosa et al., 2000). A Lagrangean relaxation scheme incorporating a heuristic procedure is proposed for solving the problem.

Due to the globalization of the economy and the emergence of global logistics, the development of models for the strategic design of international production–distribution systems has gained increasing importance. Such models address global features common to an international scenario in which the business activities of a firm are diversified among multiple countries. Critical issues in the strategic design of a global supply chain concern, for example, taxes and duties, exchange rates, trade barriers and government stability. Verter and Dincer (1995) present a literature review on analytical models for facility location in global supply chains. The authors observe that most existing models focus on the optimization of location and allocation decisions and neglect the interaction between financial and location decisions.

A number of case studies describing the application of facility location models to the strategic design of real life supply chains have been reported in the past decade, showing the growing awareness and importance that practitioners are devoting to this area. The rapid evolution of computer and communications technology has made possible the optimization of facility location decisions in real–world production–distribution systems. General–purpose mathematical programming software has become available to solve problems of realistic size. Examples have been reported by Breitman and Lucas (1987), Van Roy (1989), Martin et al. (1993), Pooley (1994), Camm et al. (1997), and Köksalan and Süral (1999) for various industry branches. In some cases, however, the problem size and complexity along with the management's wish to obtain "good" solutions in reasonable time have driven researchers to develop heuristic solution procedures. For example, this was the case for the p–median model proposed by Erkut et al. (2000) for an energy company. The model was solved using the greedy heuristic of Teitz and Bart (1968). To display the coverage areas of each selected facility, the solution generated by the heuristic is displayed in a geographical information system.

In practical situations, the analysis of a supply chain usually starts by examining how well the existing operations are being run with a view to optimizing the number and location of facilities under a given set of constraints, see e.g. Pooley (1994). Since data are usually not accurate enough, one may be interested in getting an indication of the approximate region where it would be sensible to site new facilities. Hence, instead of formulating a problem as an MIP, one may use a continuous approach (see Chap. 1 for a comprehensive review of continuous facility location models).

8.4.3 Supply Chain Planning with the SAP Advanced Planner and Optimizer

The SAP Advanced Planner and Optimizer[5] (SAP APO) is an integrated software application for supply chain planning (Bartsch and Teufel, 2000; SAP AG, Germany, 2000). It is part of the SAP Supply Chain Management Solution which combines the Enterprise Resource Planning (ERP) system SAP R/3 with an advanced supply chain planning tool offering integrated data and transactions for the management of the entire supply chain. It consists of a series of application tools for decision support ranging from the long–term strategic to the short–term operational planning. All applications use a consistent basis of master and transactional data. Data integration is offered to execution systems like SAP R/3 or Non–SAP systems. In addition, the connection to internet technologies enables enterprises the collaborative planning of their logistics activities. SAP APO comprises a series of advanced optimization techniques and algorithms combined with a high performance memory–resident object management technology.

Fig. 8.13. The SAP APO architecture

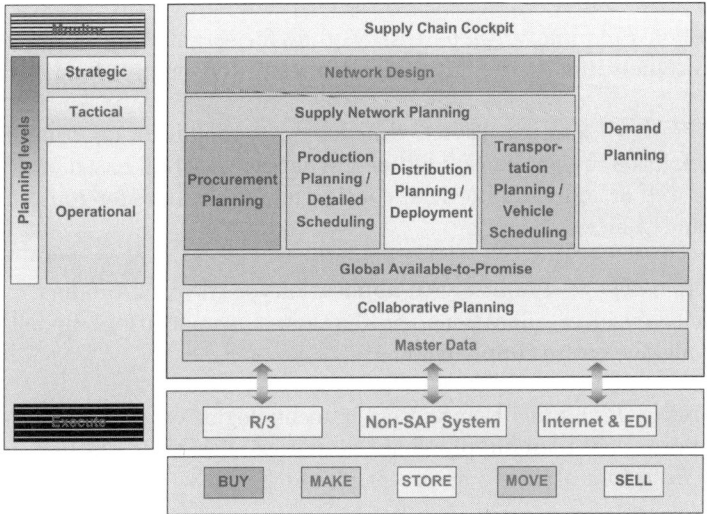

Figure 8.13 gives an overview of the main planning applications of APO. Next, a brief description of each application is presented.

---

[5] SAP Advanced Planner and Optimizer^{TM} is a registered trademark of SAP AG, Germany.

Supply Chain Cockpit: A graphical instrument panel for monitoring and controlling the supply chain.

Network Design: Strategic planning module to analyze and optimize the entire supply chain. Facility location decisions are supported as well as long-term sourcing, production and distribution decisions.

Demand Planning: A toolkit of statistical forecasting techniques and demand planning features for generating accurate forecasts.

Supply Network Planning: A planning module to create tactical production plans and sourcing decisions that considers the complete supply network and its constraints.

Procurement Planning: Workplace for the procurement of products, using optimization techniques and linked with collaboration processes.

Production Planning/Detailed Scheduling: Production planning tool using dynamic pegging and optimization techniques to generate executable plans. In addition, optimal production schedules are created for real–time scheduling for finite sequencing and final assignment of production resources.

Distribution Planning/Deployment: Module for establishing shipment plans for all products among the different facilities in the supply chain.

Transportation Planning/Vehicle Scheduling: A toolkit for determining optimal truck loads, freight consolidation and carrier selection based on shipment plans as well as individual orders, considering constraints for route determination and vehicle scheduling.

Global Available–to–Promise: An application for checking product availability which considers allocations, production, transportation capacities and costs along the entire supply chain.

Collaborative Planning: Using internet technologies enterprises are enabled to collaboratively plan supply chain activities with their business partners ranging from forecasting to shipment planning.

8.4.4 SAP APO Network Design

SAP APO Network Design[6] is a decision support tool which assists management in the analysis and redesign of supply chains. In particular, the tool supports strategic decision–making regarding the number, location and capacities of service facilities as well as the flow of products through the logistics

---

[6] SAP APO ® Network Design is a registered trademark of SAP AG, Germany.

network so as to minimize total costs. Both continuous and discrete mathematical models to be used under different levels of data availability are included in the tool. Typical decisions supported by SAP APO Network Design include those listed in Sect. 8.4.1 which arise in most production–distribution network systems.

SAP APO Network Design is closely linked with the Demand Planning and Supply Network Planning modules described in the previous section. Furthermore, it makes full use of the integrated basis of master data within SAP APO. Cost data required by Network Design can be obtained using the SAP Strategic Enterprise Management (SEM) module which allows the planner to create business models and to analyze financial plans, e.g. for the calculation of facility location costs taking taxes, interest rates and capital charges into account. Finally, an integration of the SAP Business Information Warehouse (BW) is available in which results obtained with Network Design can be explored more deeply using advanced query and reporting functions. Figure 8.14 illustrates the different components within the SAP APO system that are directly connected with the Network Design module. Furthermore, the integration with other SAP applications is shown.

Fig. 8.14. Integration of the SAP APO Network Design system with other SAP applications

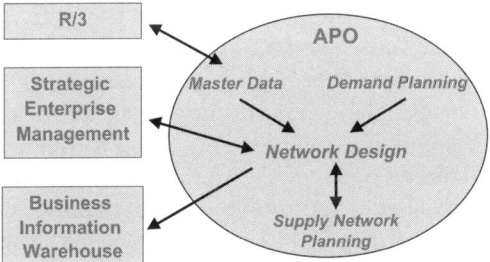

8.4.5 SAP APO Network Design Planning Algorithms

In industrial settings, the optimization of location and allocation decisions is often preceded by an evaluation of the structure of the already existing production–distribution network. Such an evaluation involves assessing the quality of the current locations of the service facilities, the degree to which their capacities are being well used and the allocation of customer demands to these facilities. SAP APO Network Design offers several tools for obtaining appropriate configurations of the supply chain. Both continuous and discrete models are integrated in SAP APO Network Design to support planning decisions under different levels of data availability (Kalcsics et al., 2000). If detailed data on the structure of the supply chain and its associated costs are known, then a discrete planning model should be employed, otherwise a continuous procedure is sufficient.

Upon exploring the configuration of the existing network, the decision maker can use additional tools to improve the network performance. These focus on the redesign of the network structure with respect to facility location, procurement, production and distribution decisions. Similar to the evaluation step, SAP APO Network Design offers tools for different levels of data aggregation in the redesign phase.

Analysis of a Production–Distribution Network System. To roughly assess the extent to which an operating supply chain is being well utilized with respect to shipping patterns and capacity utilization, SAP APO Network Design offers a quick diagnosis tool with low data requirements. The tool applies to the situation in which several facilities serve a set of geographically dispersed customer zones with known demands for a variety of products under the following simplifying assumptions: (i) all service facilities are identical; (ii) the price of a particular product is the same at every facility; (iii) the cost of acquiring the product is equal to the price plus the transportation cost from the facility; (iv) the transportation cost equals the product of distance traveled and a fixed price per unit distance; (v) each customer wishes to minimize the cost of acquiring the product.

These assumptions are often not completely satisfied in practice. However, our interest lies in the geometric interpretation of the assignment of customer zones to supplying facilities and, thus, a rough approximation of the catchment or trading areas of the facilities suffices. The above assumptions induce a subdivision of the total area under consideration into regions – the catchment areas of the facilities – such that all customers located in the same region are served by the same facility. This implies that the catchment area of a given site consists of all points for which that site is closer than any other site. For example, the catchment areas of M service facilities are determined by generating a *Voronoi diagram* which subdivides the plane into M polygonal regions, one for each facility. Customer locations within a Voronoi polygon are assigned to the facility in that cell. As a result, the size of a facility corresponds to the total amount of products shipped from that facility.

For a small example consisting of two manufacturing plants, two DCs, eleven customer zones and one product type, Fig. 8.15 shows the corresponding Voronoi assignment of customers to DCs and DCs to plants as displayed in the SAP APO Network Design application. The dottedd line separates the catchment area of the plant in Cheyenne from the catchment area of the plant in Chicago, while the solid line divides the areas served by the DCs located in Salt Lake City and Cincinnati.

Fig. 8.15. The Voronoi assignment of customers to DCs (solid line) and DCs to plants (dotted line) as displayed in the SAP APO Network Design application

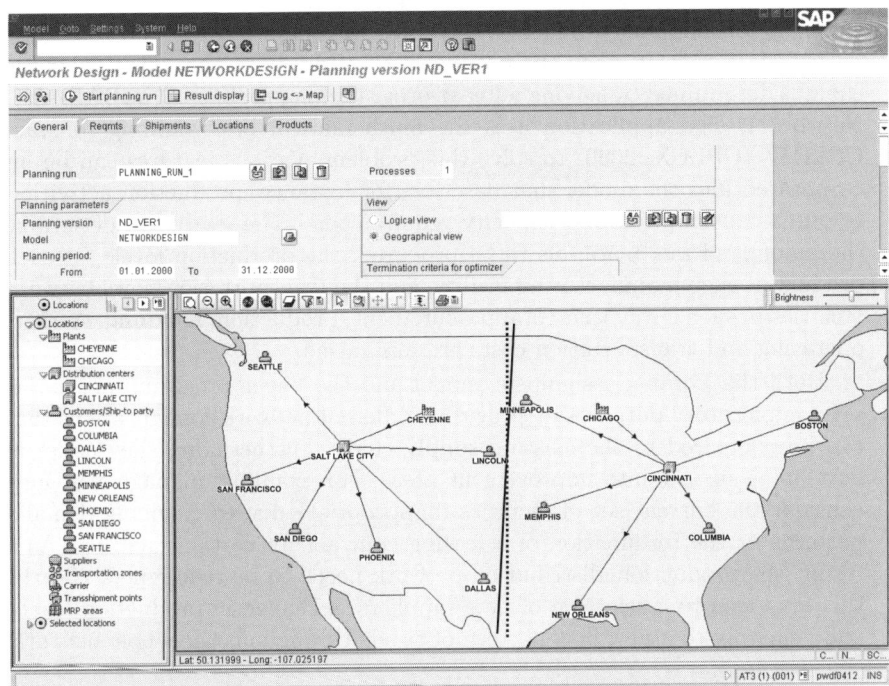

In mathematical terms, the Voronoi polygon associated with the ith facility is defined by

$$V_i = \bigcap_{1 \leq k \leq M : k \neq i} \left\{ p \in \mathbb{R}^2 : d(p, Ex_i) < d(p, Ex_k) \right\}, \qquad i = 1, \ldots, M$$

with Ex_i denoting the location in the plane of the ith facility and $d(p, Ex_i)$ denoting the distance between a point p and Ex_i. Distances in the plane are measured by the Euclidean norm and, as a result, the Voronoi diagram is computed with a sweep line algorithm which runs in $O(M \log M)$ time (Fortune, 1987). The Voronoi diagram has applications in many fields such as physics, robotics and facility location (Okabe et al., 2000).

The Voronoi assignment model provides a quick analysis of possible deviations to the current network configuration regarding demand allocation and capacity utilization at the existing facilities. In addition, the clustering of customers into regions such that each region is served by a single facility provides useful information for developing tactical and operational plans regarding the fleet of vehicles required to deliver the goods and the routes traveled by them. Another major advantage of the tool is that it helps to create a clear structure in the supply chain with very low data requirements.

If detailed data are available, for example, on the actual transportation channels for shipping products between facilities and on the corresponding transportation costs, then the optimal flow of products through the supply chain is determined by solving a linear programming problem. The SAP APO Network Design application uses the mathematical programming software CPLEX[7] (CPLEX, 2000) to solve this problem. Various features can be incorporated into the model that describe, for instance, production activities, resource consumption and capacity requirements. The goal is to determine the amount of raw materials to be procured, the production levels for each product at each manufacturing facility, and the transportation flows between facilities in such a way that total procurement, production, handling, storage, operating and transportation costs are minimized.

Both the Voronoi assignment model and the linear programming model serve exploratory purposes by providing different network configurations that can be compared to the existing supply chain. Furthermore, they assist in developing insight into improvement areas. For example, management may consider the current set of facilities inappropriate due to changing demand patterns or the termination of a leasing contract for certain facilities. As a result, the production–distribution network needs to be redesigned. Clearly, this may lead to a selection of new suppliers, a change in production levels, and generally to a new flow pattern of goods throughout the whole network.

Redesign of a Production–Distribution Network System. The redesign of a supply chain is supported by SAP APO Network Design through continuous and discrete models.

To solve a *conditional* location problem in which a number of facilities (say M) are already in place and the decision maker is considering the construction of some additional facilities (say N) without altering the positions of the existing facilities, SAP APO Network Design offers a geometric tool based on the Voronoi diagram technique. The major advantage of this approach is that it has very low data requirements which is a common situation in the strategic design of supply chains. Due to the long–term nature of facility location decisions, at the beginning of the planning horizon there is a considerable amount of uncertainty regarding relevant parameters that influence location decisions. Costs (e.g. facility, transportation) and demands (e.g. quantity, location) are examples of such parameters. Therefore, in such situations it is advisable to first use a continuous model and later apply a discrete model when detailed information becomes available.

The heuristic procedure developed for solving the conditional facility location problem considers the placement of the new facilities sequentially. Fig-

---

[7] $CPLEX^{TM}$ is a registered trademark of ILOG, Inc.

ure 8.16 summarizes the main steps required to locate the kth new facility given that $k-1$ new facilities were previously placed ($1 \leq k \leq N$).

Fig. 8.16. Heuristic procedure for locating a new facility

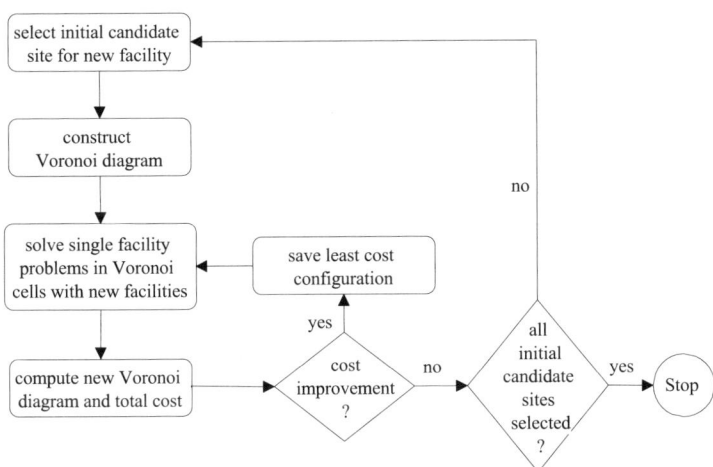

Based on the selection of an initial candidate site for the new facility, the plane is partitioned into sectors by constructing a Voronoi diagram using the $M+k$ facilities as the generator set. Next, the initial location of the kth facility is refined by determining the optimal site within its Voronoi cell. Moreover, improved positions for the first $k-1$ new facilities are also sought within their Voronoi cells. This is accomplished by solving a 1–median problem with the Weiszfeld algorithm (Love et al., 1988) in each cell containing a new facility. Upon obtaining new positions for all or some of the first k facilities, the network is reconfigured through the generation of a new Voronoi diagram. The entire process is repeated until the cost difference between two consecutive network configurations is less than a given tolerance. The cost of a solution is determined by summing up the system–wide costs which may range from procurement, production, handling, transportation and operating costs to fixed charges for opening new facilities. As soon as no cost improvement is obtained, another initial candidate site is selected for the kth new facility and the procedure is restarted. After exploring all initial candidate sites for the kth facility, the algorithm proceeds to find a suitable location for the $(k+1)$th facility using as a starting network structure the best configuration obtained so far. This process is repeated until the total desired number of new facilities has been located.

Based on a starting network configuration, six different initial candidate sites for a new facility are considered. The first set comprises Voronoi intersection nodes. Points are selected according to their weight which is measured by the sum of the demands of customers located in the cells adjacent to the points. The three points with the largest weights are selected. The second candidate set contains the Weber points of the three Voronoi cells with the largest customer demands.

For production–distribution systems with two echelons of facilities in which products are shipped from the first echelon (e.g. plants) to the customers via an intermediate second echelon (e.g. DCs), the procedure outlined in Fig. 8.16 can also be applied. In this case, it is desired to expand the second echelon by opening a given number of new facilities. From a transportation viewpoint, this raises the question of balancing the advantages of nearness to customers against nearness to plants. To deal with this problem, first a Voronoi diagram is constructed using the plant locations as the generator set. With the partition of the plane thus obtained, the assignment of DCs to plants is straightforward. Next, the algorithm shown in Fig. 8.16 is applied. While solving the 1–median problem in each Voronoi cell containing the kth new DC, the attraction effect incurred by the plant to which the DC is assigned is taken into account by including the distance from any point in the cell to the plant, weighed by

$$\max\left(\max_{m \in V_k} w_m, \frac{\sum_{m \in V_k} w_m}{2}\right)$$

with V_k denoting the Voronoi polygon associated with the kth new facility, and w_m denoting the total demand of the mth customer in V_k. ¿From the numerical tests performed with different weights, the above choice yielded on average the best results.

The above described algorithm can easily be adapted to the situation in which management desires to open new facilities but does not know the exact number. In this case, a site–independent fixed cost for locating a new facility is considered, and the most economical number of new facilities as well as their locations are determined in such a way that total costs are minimized. The decision maker can also specify a minimum and/or maximum number of new facilities to be located, and the heuristic procedure determines the extended network configuration with lowest total costs. For the small example depicted in Fig. 8.15 with an additional requisite for locating one new facility, Fig. 8.17 displays the network configuration and the associated customer–to–DC and DC–to–plant assignments obtained by SAP APO Network Design.

Often, data are not accurate enough to determine positions of new facilities exactly. Hence, the purpose of the continuous facility location algorithms implemented in SAP APO Network Design is to give the decision maker an indication of the approximate regions where it would be sensible to site new facilities. In the case that detailed data become available, facility location can be performed using a so–called discrete approach in which a finite set of alternative sites for the new facilities is defined beforehand, and the problem is to choose from these alternatives the best subset to satisfy demands.

Fig. 8.17. Location of the new facility as displayed in the SAP APO Network Design application

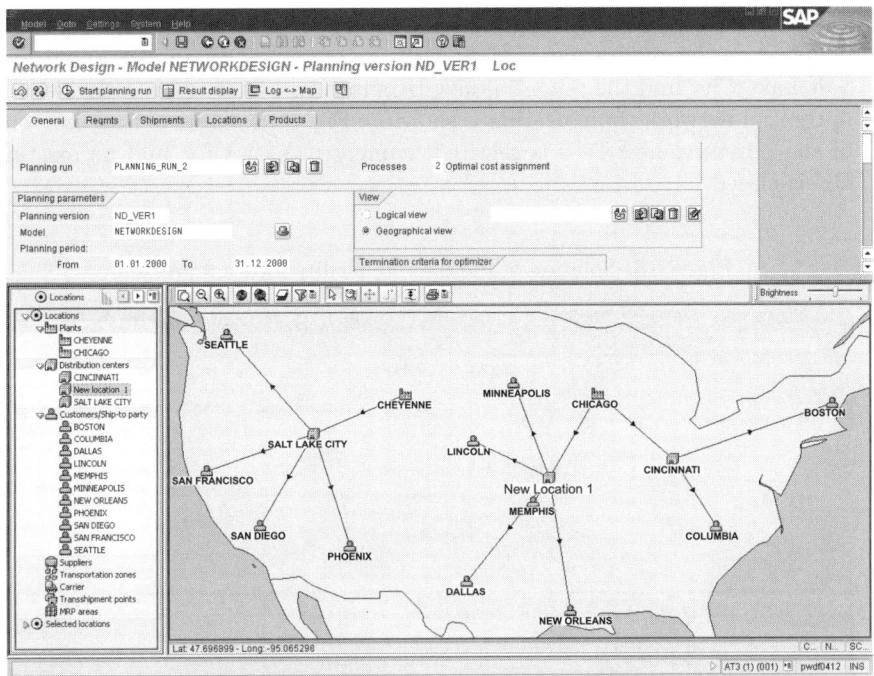

Furthermore, decisions regarding whether given existing facilities should continue to be operated or should be closed are also supported by the discrete model.

The advantage of this approach is that it can take explicit account of many items. These include fixed costs for setting up a new facility or closing an existing facility, procurement and production activities, actual transportation links for shipping intermediate and finished products throughout the network and the corresponding costs. In addition, production and handling resources with limited capacity can be considered, and the extension of the normally available amounts through, for example, overtime work at the expense of extra costs can also be modeled. The decision maker is also given the possibility to allow external demand not to be completely met. In the case that part or all of the demand of a customer for a given product is not satisfied, a penalty cost is charged. Observe that when resources are scarce, allowing demand to be partially satisfied ensures feasibility of the corresponding capacitated problem. The aim is to decide on the set of facilities to operate, the amount of products to procure, the amount of products to manufacture and the flow of products throughout the network so as to minimize system-wide costs ranging from procurement, production, distribution, handling, operating and transportation to fixed charges for opening or closing facilities. The resulting model is a large scale MIP problem which is solved with the commercial package CPLEX. Again using the small example of Fig. 8.15 and defining the set of alternative sites with the existing DCs in Cincinnati and Salt Lake City and the potential new DCs in Kansas City and Nashville, the optimized network configuration is shown in Fig. 8.18. Observe that only one of the potential new DCs is selected, namely Kansas City, and no existing DC is closed.

Fig. 8.18. Solution of the discrete facility location problem

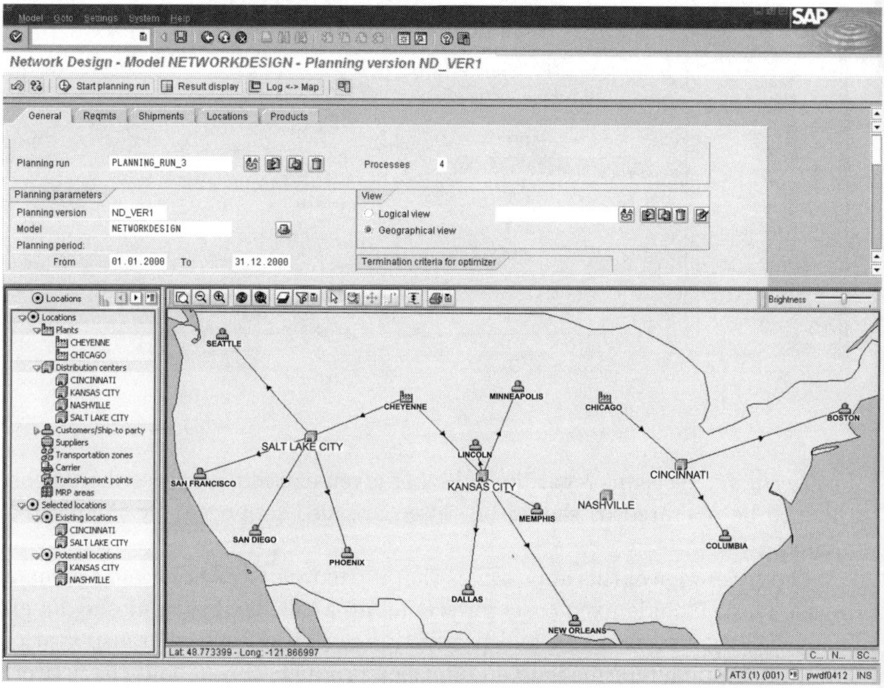

8.4.6 Planning with SAP APO Network Design

Before using SAP APO Network Design, the decision maker needs to define a model and a set of parameters. The model describes the supply chain and consists of master data objects such as locations, products, resources, production process models, transportation lanes and product demands. In general, the planning procedure consists of the following steps:

– setting up a planning scenario;

- performing the planning run;
- analyzing the planning results.

The decision maker can define several planning scenarios describing different conditions in the supply chain such as various trends in product demands. The results of the corresponding planning runs can be saved to be later compared.

Input data and parameters. To set up a planning scenario, the following data need to be specified.

Locations define facilities such as production plants, distribution centers and customers. Operating costs as well as fixed costs for opening a new facility or closing down an existing facility can be specified.

Products describe raw materials, intermediate and finished products. Specific product parameters can be defined in each facility where the product exists. These include variable storage costs and handling costs for the receipt and the issue of the product. Delivery time quotas can also be set to model service times for product delivery. Costs for late delivery apply in the case that transportation duration is longer than the required delivery time.

Transportation lanes define the transportation links that are available for shipping the products between the facilities in the supply chain. Variable transportation costs and times can be specified for different transportation modes (e.g. rail, truck).

Production process models (PPM) describe the bill of materials in a manufacturing facility. Each PPM comprises a list of material requirements for a given output product. Material consumption factors as well as variable production costs can be indicated.

Resources can be specified for production processes and handling activities. In addition, resource consumption factors per product and facility as well as total resource capacities are defined. The availability of a given resource can be extended beyond its normal capacity at the expense of some extra variable cost.

Demand can be specified for different products in various facilities. Demand plans can be obtained by using the SAP APO Demand Planning module which generates forecasts and aggregated demand quantities. To support different planning scenarios such as worst case and best case demand trends, demand quantities can be changed by using a factor which automatically increases or decreases all product demands. Variable penalty costs for non–delivery of customer demands can also be defined.

In addition to the above required data for describing a supply chain model, a planning period needs to be indicated. Furthermore, it is possible to create different planning scenarios by selecting only some of the objects of a given model. The cost structure of a scenario can also be easily modified by specifying several factors which allow to represent different cost situations.

Finally, in the discrete optimization models the decision maker can set the maximum desired runtime of an algorithm, and define the maximum desired deviation from the optimal solution. When a relative optimality tolerance is specified, the optimization terminates as soon as a feasible solution is found such that the difference between the value of that solution and that of the best lower bound, divided by the solution value, falls below the user defined tolerance. Both parameters, time limit and relative gap tolerance, enable the planner to obtain good solutions when the available runtime is restricted and/or the problem to be solved is very complex.

Performing Planning Runs. All algorithms can be started interactively from the Network Design application. The planning runs can also be executed as batch jobs, thus enabling the user to schedule a set of operations in advance. Upon termination of a planning run, the results are saved and can be retrieved later for comparison with other scenarios. Hence, the planner can analyze situations modeled with different data, e.g. worst and best case demand scenarios, various cost structures and a selection of different facilities.

Planning Results. The results of a planning run are displayed on a geographical map and also in tabular format. The results include the total costs of the planning scenario with all costs being differentiated by cost type (for example, overall production costs, costs for operating the selected facilities), and the number of selected and non–selected facilities. Moreover, the total costs per facility are specified along with the corresponding new geographical position in the case that one of the continuous facility location algorithms was run. Information regarding transportation lanes indicates transportation quantities and transportation costs as well as the transportation duration and the resulting penalty costs for late delivery. Product specific data in each facility include production, handling, storage and procurement quantities, and the corresponding costs. Data concerning the products can be aggregated by product type or by facility to enable the analysis, for example, of overall quantities of products at all facilities. Results regarding production processes give information on production quantities and the capacity utilization of the corresponding resources. In this way, the decision maker is able to see long–term capacity requirements.

For the solution displayed in Fig. 8.18, the different levels of detail that are available in the SAP APO Network Design application are shown in Fig. 8.19.

Fig. 8.19. Details of the solution shown in Fig. 8.18

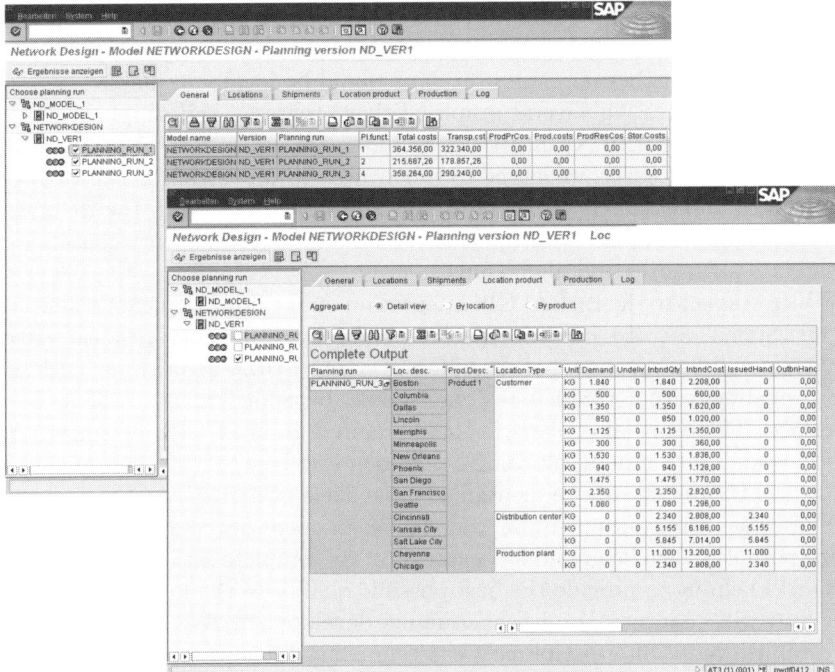

To compare different planning scenarios, the decision maker carries out various planning runs. This supports the analysis of what–if scenarios or best case and worst case calculations in order to find an overall best planning decision. All results can be downloaded to spreadsheet software like Microsoft Excel[8]. Results obtained with SAP APO Network Design can also be explored more deeply using the SAP Business Information Warehouse. It is possible to use maps to display a planning decision and show the selected locations, product quantities and costs assigned to them.

8.5 Outlook

In this chapter, we have shown how location software can successfully be developed by taking into account the different needs of the various users of such software. However, developing software is an ongoing process; and, therefore, a number of extensions are planned for the future. In the LoLA project, we intend to replace some of the software components used so far by more modern components. Until now, the LoLA *Frontend* has been completely

---

[8] Microsoft Excel® is a registered trademark of Microsoft Corporation, USA.

programmed in Tcl/Tk. Replacing this by a *Frontend* developed with JAVA[9] would allow us to bring the LoLA *Frontend* to the World Wide Web. Consequently, it would be possible to run LoLA inside a web browser, and the user could test (and use) LoLA without having to install the program on his local machine. Furthermore, all user data would be locally available.

We are also testing several graph editing programs and graph drawing tools to link with LoLA, and these would replace the currently available simple editor. Moreover, we intend to substitute the data structures of LoLA, which were implemented using the shareware tool LEDA, with the freeware C++ Standard Template Library (STL). STL is a powerful library containing basic data structures and algorithms.

With respect to the LoLA–GIS integration, many improvements still need to be carried out. To increase speed, a completely memory–resident data exchange will be implemented. In addition, more LoLA algorithms, such as those dealing with forbidden regions, should be added to the LoLA–GIS interface. There are also plans for a tight integration of GIS, optimization routines and data mining by using a common database.

In the future, the above mentioned extensions of LoLA and the LoLA–GIS interface will continue to be available as an open source software. Hence, every potential developer will have access to the complete set of source codes. This will help us to provide the best possible quality to the location community and will guarantee an ever continuing development. We would also like to establish more interdisciplinary projects using the current available code basis of LoLA to speed up the information exchange within the location community.

For the commercial applications described in Sect. 8.4, we are mainly concerned with adding relevant functionality to the software. This includes developing heuristic procedures for the discrete facility location problems already covered by the software. In addition, location–routing models, multi–period and multi–criteria facility location models will also be subject of attention.

References

Aikens, C. H. (1985) Facility location models for distribution planning. *European Journal of Operational Research* **22**, 263–279

Bartsch, H., Teufel, Th. (2000) *Supply Chain Management mit SAP APO: Supply Chain-Modelle mit dem Advanced Planner & Optimizer*. Bonn: Galileo Press (in German)

Berkelaar, M. (1995) *Mixed Integer Linear Program Solver, Version 2.3*. ftp://ftp.es.ele.tue.nl/pub/lp_solve/

Bramel, J., Simchi-Levi, D. (1997) *The logic of logistics: Theory, algorithms and applications for logistics management*. New York: Springer-Verlag.

Brandeau, M. L., Chiu, S. S. (1989) An overview of representative problems in location research. *Management Science* **35**, 645–674

---

[9] JAVA® is a registered trademark of Sun Microsystems, Inc., USA.

Breitman, R. L., Lucas, J. M. (1987) PLANETS: A modeling system for business planning. *Interfaces* **17**, 94–106

Brown, G. G., Graves, G. W., Honczarenko, M. D. (1987) Design and operation of a multicommodity production/distribution system using primal goal decomposition. *Management Science* **33**, 1469–1480

Camm, J. D., Chorman, T. E., Dill, F. A., Evans, J. R., Sweeney, D. J., Wegryn, G. W. (1997) Blending OR/MS, Judgment, and GIS: Restructuring P&G's supply chain. *Interfaces* **27**, 128–142

Carrizosa, E. J., Conde, E., Muñoz, M., Puerto, J. (1995) The generalized Weber problem with expected distances. *RAIRO* **29**, 35–57

Carrizosa, E., Nickel, S. (1998) Locating a robust facility. *Operations Research Proceedings*, 532–540

CPLEX (2000) *Reference Manual, Version 6.5.* ILOG, Inc., Incline Village, Nevada. http://www.cplex.com

Eiselt, H. A., Laporte, G., Thisse, J.-F. (1993) Competitive location models: A framework and bibliography. *Transportation Science* **27**, 44–54

Erkut, E., Myroon, T., Strangway, K. (2000) TransAlta redesigns its service-delivery network. *Interfaces* **30**, 54–69

EWGLA (2000) Homepage of the European Working Group on Locational Analysis. http://www.vub.ac.be/EWGLA

Fortune, S. J. (1987) A sweepline algorithm for Voronoi diagrams. *Algorithmica* **2**, 153–174

Geoffrion, A. M., Graves, G. W. (1974) Multicommodity distribution system design by Bender's decomposition. *Management Science* **20**, 822–844

Geoffrion, A. M., Graves, G. W., Lee, S. J. (1978) Strategic distribution system planning: A status report. In *Studies in Operations Management.* Hax, A. C. (ed.). Amsterdam: North-Holland, 179–204

Hamacher, H. W., Hennes, H., Nickel, S. (1999) *LoLA – Library of Location Algorithms, Version 2.0*, University of Kaiserslautern. http://www.mathematik.uni-kl.de/~lola

Hamacher, H. W., Nickel, S. (1998) Classification of location models. *Location Science* **6**, 229–242

Handler, G. Y., Mirchandani, P. B. (1979) *Location on networks: Theory and algorithms.* MIT Press, Cambridge.

Hinojosa, Y., Puerto, J., Fernandez, F. R. (2000) A multiperiod two-echelon multicommodity capacitated plant location problem. *European Journal of Operational Research* **123**, 271–291

Kalcsics, J., Melo, T., Nickel, S., Schmid-Lutz, V. (2000) Facility location decisions in supply chain management. In *Operations Research Proceedings 1999.* Inderfurth, K., Schwödiauer, G., Domschke, W., Juhnke, F., Kleinschmidt, P., Wäscher, G. (ed.). Berlin: Springer-Verlag, 467–472

Köksalan, M., Süral, H. (1999) Efes beverage group makes location and distribution decisions for its malt plants. *Interfaces* **29**, 89–103

Lee, C. Y. (1993) A cross decomposition algorithm for a multiproduct-multitype facility location problem. *Computers and Operations Research* **20**, 527–540

Love, R. F., Morris, J. G., Wesolowsky, G. O. (1988) *Facilities location: Models & methods.* New York: North-Holland.

Martin, C. H., Dent, D. C., Eckhart, J. C. (1993) Integrated production, distribution, and inventory planning at Libbey-Owens-Ford. *Interfaces* **23**, 68–78

Mazzola, J. B., Neebe, A. W. (1999) Lagrangian-relaxation-based solution procedures for a multiproduct capacitated facility location problem with choice of facility type. *European Journal of Operational Research* **115**, 285–299

Mehlhorn, K., Näher, S., Uhrig, C. (2000) *LEDA – Library of Efficient Data types and Algorithms, Version 4.1.* Max-Planck Institut Saarbrücken. http://www.mpi-sb.mpg.de/LEDA/

Okabe, A., Boots, B., Sugihara, K., Chiu, S. N. (2000) *Spatial tessellations: Concepts and applications of Voronoi diagrams.* Second edn. Chichester: Wiley Series in Probability and Mathematical Statistics.

Ousterhout, J. (2000) *Tcl/Tk Tool Command Language, Version 8.3.* http://dev.scriptics.com/

Owen, S. H., Daskin, M. S. (1998) Strategic facility location: A review. *European Journal of Operational Research* **111**, 423–447

Pirkul, H., Jayaraman, V. (1996) Production, transportation and distribution planning in a multicommodity tri-echelon system. *Transportation Science* **30**, 291–302

Pooley, J. (1994) Integrated production and distribution facility planning at Ault Foods. *Interfaces* **24**, 113–121

SAP AG, Germany (2000) *SAP Advanced Planner and Optimizer.* http://www.sap-ag.de/solutions/scm/apo/apo_over.htm

SOLA (2000) Homepage of the Section on Location Analysis. http://www.uscolo.edu/sola/sola.html

Teitz, M. B., Bart, P. (1968) Heuristic methods for estimating the generalized vertex median of a weighted graph. *Operations Research* **16**, 955–961

Van Roy, T. J. (1989) Multi-level production and distribution planning with transportation fleet optimization. *Management Science* **35**, 1443–1453

Verter, V., Dincer, M. C. (1995) Global manufacturing strategy. In *Facility location: A Survey of applications and methods.* Drezner, Z. (ed.). New York: Springer-Verlag, 263–282

9 Telecommunication and Location

Eric Gourdin[1], Martine Labbé[2], and Hande Yaman[2]

[1] France Telecom R&D, DAC/OAT, 38-40, rue du General Leclerc,
Issy-les-Moulineaux Cedex 9, 92794, France. e-mail:
eric.gourdin@francetelecom.fr
[2] Universite Libre de Bruxelles, Service de Mathematiques de la Gestion,
Boulevard du Triomphe CP 210/01, 1050 Bruxelles,Belgium. e-mail:
mlabbe@smg.ulb.ac.be; hyaman@smg.ulb.ac.be

9.1 Introduction

The usage of telecommunication networks has considerably changed in the last decade. The traditional concept of telecommunication networks exclusively dedicated to the telephony is now completely outdated. The new telecommunication networks will carry without distinction (or, more likely, with differentiated treatment according to the service) voice, video or data traffic. Most of the future telecommunications will be carried over the Internet. This revolution, already started a few years ago, is characterized by a spectacular growth in traffic (the traffic doubles every ten months). Another crucial factor in the evolution of networks, especially in Europe, is the liberalization of the telecommunication market, forcing all historical operators to adapt very quickly to the new competitive environment. The context for the development of the next generation networks has changed considerably giving rise to many new problems, among which the optimal choice of locations for concentrator nodes remains a crucial one, especially when combined with other features of the new telecommunication networks. This explains the interest of many researchers for such problems. The aim of this survey is to gather some significant examples of location problems that arise in the design of modern telecommunication networks. For earlier surveys, one can refer to Boffey (1989) and Klincewicz (1998).

In a typical telecommunication network, the traffic is gathered from many sources, progressively combined in order to fill links of increasing capacity and finally forwarded to its destination. Hence, most telecommunication networks are naturally structured in a multi-layer hierarchical architecture. At a lower level of such an architecture, the traffic is collected to be sent across an upper level. Although the real telecommunication (voice, video or data, nation wide or international) networks are usually structured into many such levels, most design problems considered by the telecommunication companies concern only a part of the overall network. A generic telecommunication network consists of access networks which connect the terminals (user nodes) to concentrators (switches or multiplexers) and a backbone network which interconnects these concentrators or connects them to a central unit (root).

Facility Location: Applications and Theory.
Edited by Z. Drezner and H.W. Hamacher
© 2002 Springer-Verlag, ISBN 3-540-42172-6

Telecommunication networks differ in the designs of these access networks and the backbone network. The terminals can be connected to concentrators by point-to-point links which results in a star topology or they may be connected by multidrop links which can have a tree, path or ring structure. Similarly, there are different topologies for the backbone network. This network can be fully connected or meshed. If there exists a central unit, then the backbone network can be a star, a tree or a path. There are also some models where terminals can be directly connected to the central unit. To specify the structure of a given network, we use the notation by Klincewicz (1998) which is "backbone structure/access network structure". For example, a tree/star network means that the backbone network is a tree and the access networks are star networks. We present examples of some telecommunication networks in Figures 9.1 to 9.3.

The study of the location problems for telecommunication network design dates back to the 1960's when Hakimi (1964), (1965) introduced the 1-median and p-median problems to locate switching centers in communication networks. Since then many models considering different design issues have been developed. Before discussing these models, we list the basic questions related with the design of a telecommunication network:

- how many concentrators are needed,
- where should these concentrators be located,
- what should be the capacities of these concentrators,
- how the terminals should be assigned to the concentrators,
- how the terminals assigned to a concentrator should be connected to this concentrator and what should be the capacities of the links,
- how the concentrators should be connected to each other or to a central node and what should be the capacities of the links used.

It is hard to develop tractable models that can come up with answers to all questions above simultaneously. So, usually the design is done in an iterative manner as follows:

- Decide about the number and locations of concentrators and the assignment of the terminals to these concentrators
- Design the access networks
- Design the backbone network.

The idea of this iterative method is that once the location of concentrators and the assignment of terminals to these concentrators are done, the design of the access networks and the design of the backbone network become independent and can be handled separately. Some models we encountered in the literature deal with the location phase only, while some others consider the location and design phases simultaneously assuming some topologies for the access networks and the backbone network.

Fig. 9.1. A fully connected/tree network

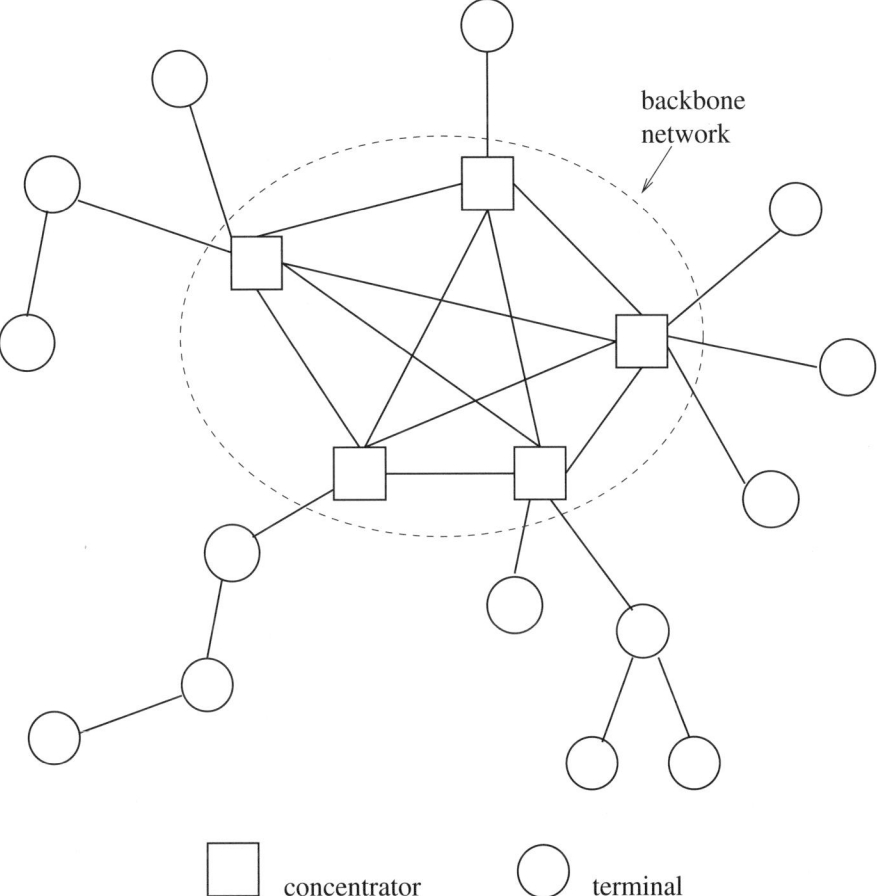

Telecommunication network design models also differ in the way the possible networks are evaluated. When evaluating a given network design, one is concerned with issues such as the cost of establishing the network, the cost of operating the network, its reliability and capability to supply the demands of terminals in a reasonable time. While some models are only interested in minimizing the cost of the network, some others consider both the cost of the network and the average delay cost. The reliability issue is usually handled by connecting the terminals to several concentrators rather than a single one.

There are also models that consider the expansion of the telecommunication networks over time. One may be willing to optimize the expansion of link capacities or one may be also allowed to install new concentrators. These models are called dynamic.

Fig. 9.2. A star/star network

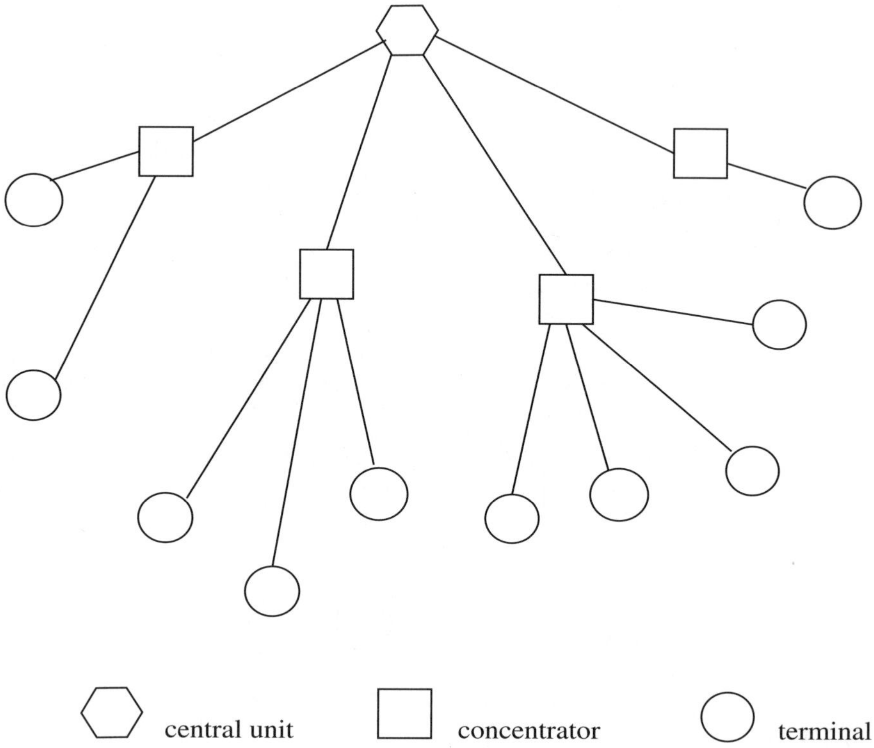

◇ central unit   □ concentrator   ○ terminal

We classified the models into three classes as uncapacitated, capacitated and dynamic models. So this survey is organized as follows: In Section 9.2, we consider the uncapacitated models. Section 9.3 covers the capacitated concentrator location problem. In Section 9.4, we discuss the generalizations of the capacitated concentrator location problem. Finally, in Section 9.5, we present some examples of dynamic models.

9.2 Uncapacitated Models

The uncapacitated location models for telecommunication networks are concerned with the location of concentrators and the assignment of terminals to these concentrators assuming that there is no capacity constraint for the concentrators. Some models also determine the links of the backbone network and the access networks. Usually these models assume an a priori structure for these networks.

The uncapacitated facility location problem (UFLP) is a core problem for the design of uncapacitated telecommunication networks. These models are generalizations of the UFLP or they include the UFLP as a subproblem.

Fig. 9.3. A path/star network

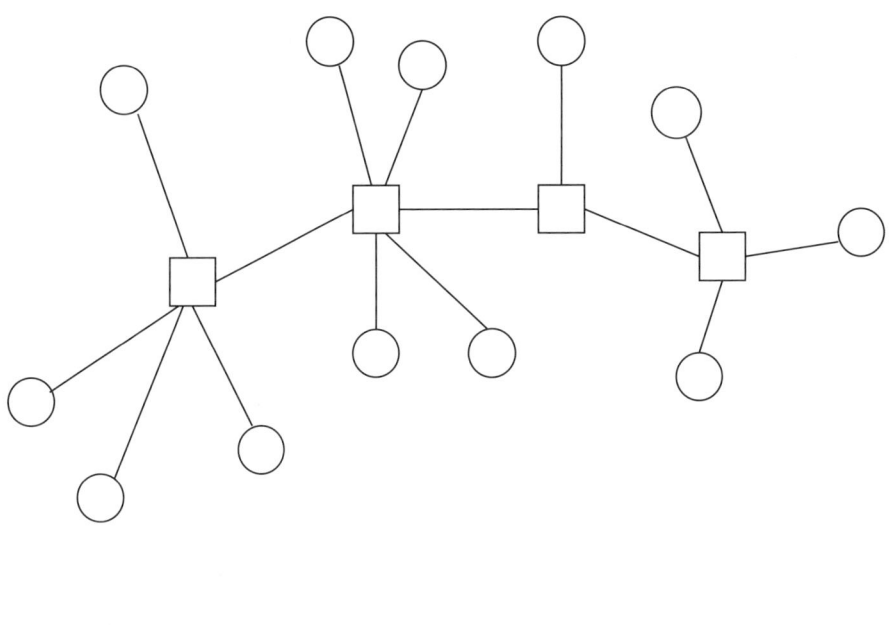

□ concentrator   ○ terminal

The UFLP can be stated as follows: given the set of terminals N and the set of possible locations for the concentrators M, determine the number and the location of concentrators and assign the terminals to these concentrators. The goal is to minimize the sum of the cost of installing concentrators and of the cost of serving terminals via the installed concentrators. The formulation of the UFLP is as follows:

$$\min \sum_{i \in N} \sum_{j \in M} C_{ij} x_{ij} + \sum_{j \in M} F_j y_j \tag{9.1}$$

$$\text{subject to: } \sum_{j \in M} x_{ij} = 1 \quad \forall i \in N \tag{9.2}$$

$$x_{ij} \leq y_j \quad \forall i \in N, j \in M \tag{9.3}$$

$$x_{ij} \in \{0, 1\} \quad \forall i \in N, j \in M \tag{9.4}$$

$$y_j \in \{0, 1\} \quad \forall j \in M \tag{9.5}$$

where C_{ij} denotes the cost of assigning terminal i to concentrator at location j, F_j denotes the cost of installing a concentrator at location j,

$$y_j = \begin{cases} 1 & \text{if a concentrator is installed at location } j \\ 0 & \text{otherwise,} \end{cases}$$

for all $j \in M$ and

$$x_{ij} = \begin{cases} 1 \text{ if terminal } i \text{ is assigned to the concentrator at location } j \\ 0 \text{ otherwise,} \end{cases}$$

for all $i \in N$ and $j \in M$. Constraints (9.2) and (9.4) state that each terminal should be assigned to exactly one concentrator and constraints (9.3) state that a terminal can be assigned to a concentrator only if this concentrator is installed.

The UFLP can be solved as the first phase of the iterative design procedure. Then the backbone and access networks can be designed separately. There are also models which consider the location of concentrators and the design of backbone network and the access networks simultaneously. These models vary in the topologies chosen for the network. Some of the networks have a central unit. If the concentrators can be directly connected to this central unit, then the backbone network is a star. But if the costs of establishing these links are high, then the backbone network can be a tree or a path. If there is no central unit, then concentrators are connected to each other. If the response time and reliability are critical, then the network can be a complete network. But again if the costs of establishing links are high, the network can be optimized. The same argument applies for the access networks. The terminals can be connected to concentrators directly forming a star network around each concentrator. Otherwise, it may be a tree or a path. In fact, if we define C_{ij} to be the cost of establishing a link between terminal i and concentrator j and F_j to be the cost of installing a concentrator at location j and connecting it to the central unit, then the UFLP corresponds to the design of a star/star network.

For more information about the UFLP, one may refer to Krarup and Pruzan (1983), Cornuéjols et al. (1990) and Labbé et al. (1995). Below, we discuss some variants of this problem.

Marianov and Rios (1996) consider the uncapacitated facility location problem with an additional constraint that imposes a threshold on the probability of finding all concentrators busy. This way they incorporate the quality of service into the model. In order to write this constraint in a tractable form, they assume an M/M/1 queueing system. They also impose the same constraint in the p-median formulation and report computational experiments using CPLEX 3.0 with this model.

Helme and Magnanti (1989) consider the design of a satellite communications network. Each terminal is directly assigned to a concentrator and concentrators are directly assigned to a central unit forming a star/star network. The difference of this model with respect to the UFLP is that it includes the operating costs for the concentrators. If two terminals are assigned to the same concentrator, they use the capacity of the local switch at the concentrator. On the other hand, if they are assigned to two different concentrators, they use the capacity of these two concentrators. To compute the cost of operating the concentrators, it is necessary to differentiate the traffic between terminals assigned to the same concentrator and the traffic between terminals assigned to different concentrators. Let C be the unit cost per capacity of the concentrator and S be the unit cost per capacity of the local switch at the concentrator. We denote by t_{ik} the traffic between two terminals i and k and by t_i all the traffic incident at terminal i, i.e.

$$t_i = \sum_{k \in N} t_{ik} + \sum_{k \in N} t_{ki}.$$

The set of concentrators to which node i can be assigned is denoted by M_i. We also define $M_{ik} = M_i \cap M_k$. Then the cost function in this variant of the UFLP becomes:

$$\sum_{i \in N} t_i \sum_{j \in M_i} C_{ij} x_{ij} - (2C - S) \sum_{i \in N} \sum_{k \in N} t_{ik} \sum_{j \in M_{ik}} x_{ij} x_{kj} + \sum_{j \in M} F_j y_j$$

which is quadratic. The constraints and the variables are the same as in the UFLP. The authors present a linearization for the problem and a branch and bound algorithm to solve it. They provide computational results and discuss the improvement obtained by using this exact approach rather than some heuristic approaches.

Chung et al. (1992) develop a model for designing a fully connected/star type network, i.e. all the concentrators are connected to each other and the terminals assigned to a concentrator are directly connected to this concentrator. The authors develop a formulation where the cost function includes the cost of installing concentrators, cost of assigning terminals to concentrators and the cost of interconnecting concentrators. So the model they have is the same as the UFLP except that the cost function also includes the term $\sum_{j \in J} \sum_{l \in J} B_{jl} y_j y_l$ where B_{jl} is the cost of connecting concentrators j and l. Thus the total cost function is quadratic. In the remaining part of the paper, this formulation is linearized and a dual-based solution procedure and computational results are presented.

Mateus et al. (1994) design a tree/tree network. They decompose the design problem into three separate phases. The location phase consists of an uncapacitated facility location problem where the number of concentrators to be installed is bounded from above, i.e. the constraint $\sum_{j \in M} y_j \leq k$ is added to the UFLP. Once this problem is solved, the concentrators are located and the terminals are assigned to concentrators. Each terminal i has a known demand t_i. Now, the goal is to establish the access network topology and the dimensioning. For each concentrator and the set of terminals assigned to it, the problem is to find a minimum cost tree network with the concentrator as the root. We have the constraint that the installed capacity on each link should satisfy the demand. Suppose that $G = (N, A)$ is the subgraph for concentrator p. Let T^p be the set of terminals assigned to p. The formulation is as follows:

$$\min \sum_{(i,j) \in A} V_{ij} f_{ij} + \sum_{(i,j) \in A} B_{ij} z_{ij} \qquad (9.6)$$

$$\text{subject to:} \sum_{(p,i) \in A} f_{pi} = \sum_{i \in T^p} t_i \qquad (9.7)$$

$$\sum_{(j,i) \in A} f_{ji} - \sum_{(i,j) \in A} f_{ij} = t_i \quad \forall i \in T^p \qquad (9.8)$$

$$\sum_{(j,i) \in A} f_{ji} - \sum_{(i,j) \in A} f_{ij} = 0 \quad \forall i \in N \backslash (T^p \cup \{p\}) \qquad (9.9)$$

$$f_{ij} \leq \overline{M} z_{ij} \quad \forall (i,j) \in A \qquad (9.10)$$

$$f_{ij} \geq 0 \quad \forall (i,j) \in A \qquad (9.11)$$

$$z_{ij} \in \{0,1\} \quad \forall (i,j) \in A \qquad (9.12)$$

where f_{ij} is the flow on arc (i,j),
$$z_{ij} = \begin{cases} 1 \text{ if arc } (i,j) \text{ is used} \\ 0 \text{ otherwise} \end{cases}$$
and $\overline{M}$ is a big enough constant. For each arc (i,j), we have a variable cost V_{ij} and a fixed cost B_{ij}.

In this formulation constraints (9.7), (9.8) and (9.9) are flow conservation constraints. Constraints (9.10) say that there can be a flow on arc (i,j) only if this arc is used in the network.

Rather than solving this model directly, the authors reduce the graph to an approximate graph $G' = (N', A')$ where $N' = T^p \cup \{p\}$. Graph G' is a complete graph where each arc length corresponds to the shortest path distance between its head and tail. Then they solve the model for this reduced graph. The flow conservation constraints become

$$\sum_{(p,i) \in A'} f_{pi} = \sum_{i \in T^p} t_i$$

$$\sum_{(j,i) \in A'} f_{ji} - \sum_{(i,j) \in A} f_{ij} = t_i \quad \forall i \in T^p$$

and we also have the following constraints

$$\sum_{j \in N' \backslash \{i\}} z_{ji} = 1 \quad \forall i \in T^p$$

which state that the number of arcs entering each node is 1. The authors present a greedy type heuristic and an arc substitution heuristic to solve this model. Then they convert the solution on the reduced graph to a solution for the original graph. The third phase of the problem which involves the design of the backbone network is modeled and solved the same way as the second phase.

Pirkul and Nagarajan (1992) design a tree/star network using a two-phase algorithm. In the first phase, they use a sweep algorithm to divide the set of nodes into regions. In the second phase, they develop for each region a formulation to determine a path from the furthest node of the region to the central node such that all nodes on this path are concentrator locations and the terminals in the region are connected to these concentrators via point-to-point links. The model is as follows:

$$\min \sum_{i \in N} \sum_{j \in M} C_{ij} x_{ij} + \sum_{j \in M, j \neq s,t} F_j y_j + \sum_{j \in M} \sum_{l \in M} B_{jl} z_{jl} \quad (9.13)$$

$$\text{subject to: } \sum_{j \in M} x_{ij} = 1 \quad \forall i \in N \quad (9.14)$$

$$x_{ij} \leq y_j \quad \forall i \in N, j \in M, j \neq s,t \quad (9.15)$$

$$\sum_{j \in M} z_{sj} = 1 \quad (9.16)$$

$$\sum_{j \in M} z_{jt} = 1 \quad (9.17)$$

$$\sum_{j \in M} z_{lj} - \sum_{j \in M} z_{jl} = 0 \quad \forall l \in M, l \neq s,t \quad (9.18)$$

$$\sum_{j \in Q} \sum_{l \in Q} z_{jl} \leq |Q| - 1 \quad \forall Q \subseteq M \quad (9.19)$$

$$y_j = \sum_{l \in J} z_{jl} \quad \forall j \in M, j \neq s,t \quad (9.20)$$

$$x_{ij} \in \{0,1\} \quad \forall i \in N, j \in M \quad (9.21)$$

$$y_j \in \{0,1\} \quad \forall j \in M \quad (9.22)$$

$$z_{jl} \in \{0,1\} \quad \forall j, l \in M \quad (9.23)$$

where
$$z_{jl} = \begin{cases} 1 \text{ if nodes } j \text{ and } l \text{ are connected by a backbone path link} \\ 0 \text{ otherwise,} \end{cases}$$

N is the set of terminals and M is the set of concentrators in the region. The cost of establishing link (j,l) is B_{jl}. A vector z satisfying constraints (9.16), (9.17), (9.18) and (9.23) defines a path from a starting node s to the central node t. Nodes s and t are not candidate location to receive concentrators. Constraints (9.19) eliminate subtours on this path. Constraints (9.20) state that a node is on the backbone path if and only if a concentrator is installed at this node.

This model is solved for each region separately. The cost function consists of the cost of concentrators, cost of assigning terminals to concentrators and the cost of installing the links of the backbone path. Since this formulation has features of both the shortest path problem and the UFLP, the authors apply Lagrangian Relaxation to decompose the problem into these two problems. They solve the Lagrangian dual problem by subgradient optimization, discuss ways of generating feasible solutions and present computational results.

Current and Pirkul (1991) consider a close problem where the concentrators are connected to each other by a path of primary arcs and the terminals are connected to the concentrators by paths of secondary arcs. The aim is to design such a topology to minimize the cost of concentrators and the cost of primary and secondary arcs used to establish the links. We assume $M = N$. The model is as follows:

$$\min \sum_{i \in M} \sum_{j \in M} C_{ij} x_{ij} + \sum_{j \in M} F_j y_j + \sum_{j \in M} \sum_{l \in M} B_{jl} z_{jl} \quad (9.24)$$

subject to:
$$\sum_{j \in M} x_{ij} + y_j = 1 \quad \forall i \in M \quad (9.25)$$

$$\sum_{i \in Q} \sum_{j \in Q} x_{ij} \leq |Q| - 1 \quad \forall Q \subseteq M \quad (9.26)$$

$$y_j \leq \sum_{l \in M} z_{jl} \quad \forall j \in M \quad (9.27)$$

$(9.16) - (9.19)$ and $(9.21) - (9.23)$.

The vector z defines the primary path and constraints (9.25) and (9.26) ensure that each node is either a concentrator or it is connected to a concentrator by a path of secondary arcs. Constraint (9.27) states that each concentrator should be on the primary path. Different from constraint (9.20), this constraint allows terminals to be on the primary path, too. To solve this problem, the authors present two heuristics based on Lagrangian Relaxation and provide computational results.

Lee et al. (1996) consider the design of a network where the access networks are stars and the concentrators are connected to each other by a minimum spanning tree. They present a formulation to determine a tree/star network which minimizes the cost of installing concentrators, the cost of assigning terminals to these concentrators and the cost of establishing the links of the spanning tree. This model is similar to the model given by Pirkul and Nagarajan (1992) if we replace constraints (9.16)-(9.18) that state that z defines a path that passes through all concentrators by the constraint

$$\sum_{j \in M} \sum_{l \in M} z_{jl} = \sum_{j \in M} y_j - 1$$

which implies that z is the incidence vector of a tree that spans all concentrators. In fact, the current model is a generalization of the other. The authors apply Lagrangian Relaxation to this model. They relax in a Lagrangian manner all constraints except the one stating that z defines a spanning tree so that the resulting problem is a minimum spanning tree problem. They solve the Lagrangian Dual by subgradient optimization, present a heuristic to convert the Lagrangian solution to a primal feasible solution and provide computational results.

Gavish (1982) formulates a problem where the terminals are connected to the concentrators via multidrop links which are capacitated, and concentrators are connected to a central unit by point-to-point links. This results in a star/tree network. The objective function involves the cost of establishing the links and installing the concentrators. There are different types of links with different costs and capacities. Let L denote the set of different kinds of links. Below we present this formulation:

$$\min \sum_{i \in N} \sum_{j \in M} \sum_{l \in L} C_{ijl} x_{ijl} + \sum_{j \in M} \sum_{l \in L} (F_j + C_{j1l}) y_{jl} \qquad (9.28)$$

$$\text{subject to:} \sum_{j \in M \cup N} \sum_{l \in L} x_{ijl} = 1 \quad \forall i \in N \qquad (9.29)$$

$$\sum_{l \in L} y_{jl} \leq 1 \quad \forall j \in M \qquad (9.30)$$

$$\sum_{j \in M \cup N} f_{ij} - \sum_{j \in N} f_{ji} = t_i \quad \forall i \in N \qquad (9.31)$$

$$\sum_{i \in N} f_{ij} \leq \sum_{l \in L} Q_{j1l} y_{jl} \quad \forall j \in M \qquad (9.32)$$

$$f_{ij} \leq \sum_{l \in L} Q_{ijl} x_{ijl} \quad \forall i \in N, j \in M \cup N \qquad (9.33)$$

$$x_{ijl} \in \{0,1\} \quad \forall i \in N, j \in M, l \in L \qquad (9.34)$$

$$y_{jl} \in \{0,1\} \quad \forall j \in M, l \in L \qquad (9.35)$$

$$f_{ij} \geq 0 \quad \forall i \in N, j \in M \qquad (9.36)$$

where node $1 \in M$ is the central unit, C_{ijl} is the cost of installing a link of type l between nodes i and j. The traffic volume generated by terminal i is t_i. The maximum traffic that a line type l can carry between these nodes is denoted by Q_{ijl}. The variables are defined as follows:

$$x_{ijl} = \begin{cases} 1 \text{ if terminal } i \text{ is connected to concentrator } j \text{ by link type } l \\ 0 \text{ otherwise} \end{cases}$$

$$y_{jl} = \begin{cases} 1 \text{ if concentrator } j \text{ is connected to the central unit by link type } l \\ 0 \text{ otherwise} \end{cases}$$

and f_{ij} is the flow on arc (i,j). Constraints (9.29) state that each terminal should be linked to a concentrator or to another terminal using a single type of link. Similarly, constraints (9.30) state that each concentrator can be connected to the central unit using a single type of link. Constraints (9.31) are flow conservation constraints. Constraints (9.32) state that the flow into a concentrator should not be larger than what the links that connect this concentrator to the central unit can support. Constraints (9.33) are the capacity constraints for the links between the terminals and concentrators and for the links among terminals.

Gavish (1992) discusses the evolution of network topologies and network design process. Then he formulates a model to design a network with no a priori given network topology. Let P denote the set of communicating pairs. For each communicating pair p of nodes, the author defines a set of routes R_p that can support the traffic between these nodes. We define $R = \cup_{p \in P} R_p$. We denote by L_r all the links on a given route r. The problem is to choose the routes and install concentrators and the links on these routes to minimize the total cost of installing concentrators and links, the cost of traffic transfer and the cost of delay encountered by messages, so that the traffic between these communicating pairs can flow. It is assumed that messages have exponentially distributed lengths. The problem is defined over the graph $G = (M, A)$. The model is as follows:

$$\min \sum_{j \in M} F_j y_j + \sum_{(i,j) \in A} C_{ij} z_{ij} + \sum_{(i,j) \in A} T_{ij} f_{ij}(x_r) +$$

$$+ D \sum_{(i,j) \in A} f_{ij}(x_r)/(f_{ij}(x_r) - Q_{ij}) \qquad (9.37)$$

subject to:
$$\sum_{r \in R_p} x_r = 1 \quad \forall p \in P \qquad (9.38)$$

$$x_r \leq z_{ij} \quad \forall (i,j) \in L_r, r \in R \qquad (9.39)$$

$$z_{ij} \leq y_i \quad \forall (i,j) \in A \qquad (9.40)$$

$$z_{ij} \leq y_j \quad \forall (i,j) \in A \qquad (9.41)$$

$$x_r \in \{0,1\} \quad \forall r \in R \qquad (9.42)$$

$$y_j \in \{0,1\} \quad \forall j \in M \qquad (9.43)$$

$$z_{ij} \in \{0,1\} \quad \forall (i,j) \in A \qquad (9.44)$$

where
$$x_r = \begin{cases} 1 \text{ if route } r \text{ is chosen} \\ 0 \text{ otherwise} \end{cases}$$

$$y_j = \begin{cases} 1 \text{ if a concentrator is placed at location } j \\ 0 \text{ otherwise} \end{cases}$$

$$z_{ij} = \begin{cases} 1 \text{ if link } (i,j) \text{ is installed} \\ 0 \text{ otherwise} \end{cases}$$

and $f_{ij}(x_r)$ is the traffic on arc (i,j) which is a linear function of x_r and $f_{ij}(x_r)/(f_{ij}(x_r) - Q_{ij})$ is the average delay on arc (i,j) where Q_{ij} is the capacity of link (i,j). The cost per traffic unit on arc (i,j) is T_{ij} and the cost per delay unit is D.

Constraints (9.38) state that for each communicating pair of nodes, a route should be chosen. Constraints (9.39) state that if route r is chosen then all links on this route should be installed and by constraints (9.40) and

(9.41) we have that if a node is the head or the tail of an installed link, then it should receive a concentrator.

The author presents a Lagrangian Relaxation based solution procedure to solve this nonlinear formulation.

Finally, we present a model due to Chardaire et al. (1999) for designing a two level star/star network. The authors consider the design of a network with two levels of concentrators, i.e. each terminal is connected to a first level concentrator which is connected to a second level concentrator. All second level concentrators are connected to a central unit. They present two integer programming formulations and show that the linear relaxations of both formulations lead to the same bound. Let N be the set of terminals, M the set of possible sites for the first level concentrators and K the set of possible sites for the second level concentrators. Below, we present the first formulation due to Chardaire et al. (1999):

$$\min \sum_{i \in N} \sum_{j \in M} C_{ij} x_{ij} + \sum_{j \in M} \sum_{k \in K} B_{jk} y_{jk} + \sum_{k \in K} F_k z_k \quad (9.45)$$

$$\text{subject to: } \sum_{j \in M} x_{ij} = 1 \quad \forall i \in N \quad (9.46)$$

$$x_{ij} \leq \sum_{k \in K} y_{jk} \quad \forall i \in N \ \forall j \in M \quad (9.47)$$

$$y_{jk} \leq z_k \quad \forall j \in M \ \forall k \in K \quad (9.48)$$

$$\sum_{k \in K} y_{jk} \leq 1 \quad \forall j \in M \quad (9.49)$$

$$x_{ij} \in \{0,1\} \quad \forall i \in N \ \forall j \in M \quad (9.50)$$

$$y_{jk} \in \{0,1\} \quad \forall j \in M \ \forall k \in K \quad (9.51)$$

$$z_k \in \{0,1\} \quad \forall k \in K \quad (9.52)$$

where C_{ij} is the cost of assigning terminal i to first level concentrator j, B_{jk} is the sum of the cost of setting a first level concentrator at location j and of connecting this first level concentrator to the second level concentrator k and F_k denotes the cost of installing a second level concentrator at location k. The variables are defined as follows:

$$y_{jk} = \begin{cases} 1 & \text{if first level concentrator } j \text{ is connected to second level concentrator } k \\ 0 & \text{otherwise} \end{cases}$$

for all $j \in M$ and $k \in K$ and

$$z_k = \begin{cases} 1 & \text{if a second level concentrator is installed at location k} \\ 0 & \text{otherwise} \end{cases}$$

for all $k \in K$.

A simulated annealing algorithm is given to find good feasible solutions to this formulation. For the second formulation, the authors present a family of structural cuts that can strengthen the linear programming relaxation. They also show that the Lagrangian Relaxation for the second formulation with or without the cuts give the same bound as the linear programming relaxation.

9.3 Capacitated Concentrator Location Problem

Capacitated models are developed to decide about the locations of concentrators and the assignment of terminals without violating the capacity constraints of the concentrators. This capacity can be defined in terms of the number of the terminals assigned to a concentrator or it may be in terms of the demands of the terminals. These problems are the same as the capacitated facility location problem with single sourcing, i.e. each demand point can be assigned to a single facility and it is also called the capacitated concentrator location problem (CCLP) (see Mirzaian (1985), Klincewicz and Luss (1986)).

In this section, we give the formulation of the capacitated concentrator location problem and discuss some polyhedral results and solution methods.

The CCLP is defined as follows: Given a set of terminals N with known demands d_i for each terminal $i \in N$, and a set of possible locations for concentrators M with capacities Q_j for each $j \in M$, the aim is to install concentrators and assign each terminal to exactly one concentrator so that the total cost of installing concentrators and the cost of assigning terminals to concentrators is minimized and the capacities of the concentrators are sufficient to supply the demand of terminals assigned to them. The other parameters and variables are defined as in the UFLP. The CCLP can be formulated as follows:

$$\min \sum_{i \in N} \sum_{j \in M} C_{ij} x_{ij} + \sum_{j \in M} F_j y_j \tag{9.53}$$

$$\text{subject to:} \sum_{j \in M} x_{ij} = 1 \quad \forall i \in N \tag{9.54}$$

$$\sum_{i \in N} d_i x_{ij} \leq Q_j y_j \quad \forall j \in M \tag{9.55}$$

$$x_{ij} \in \{0, 1\} \quad \forall i \in N, j \in M \tag{9.56}$$

$$y_j \in \{0, 1\} \quad \forall j \in M \tag{9.57}$$

Constraints (9.54) and (9.56) ensure that each terminal is assigned to exactly one concentrator and constraints (9.55) ensure that terminal i is assigned to concentrator j only if a concentrator is installed at location j and the capacity of this concentrator is sufficient to supply the demands of all terminals assigned to it. The cost function (9.53) is the sum of the cost of assigning terminals to concentrators and the cost of installing concentrators.

An alternative formulation can be obtained by replacing constraints (9.55) with the following set of constraints:

$$\sum_{i \in N} d_i x_{ij} \leq Q_j \quad \forall j \in M \tag{9.58}$$

$$x_{ij} \leq y_j \quad \forall i \in N, j \in M \tag{9.59}$$

or simply by adding constraints (9.59) to the above formulation.

Similar to the UFLP, the CCLP can be solved as the first phase of the iterative procedure to design telecommunication networks. Moreover, it also corresponds to the design of a star/star topology if the costs are defined appropriately.

9.3.1 Polyhedral Properties of the CCLP and Exact Solution Methods

We will discuss the valid and facet defining inequalities for the CCLP. For basic definitions, one may refer to Nemhauser and Wolsey (1988).

If the integrality constraints (9.56) on variables x_{ij} are replaced by the constraints $0 \leq x_{ij} \leq 1$ for all i, j in the above formulation of the CCLP, the resulting problem is the capacitated facility location problem (CFLP). The difference is that, in the CFLP we are allowed to supply the demand of terminals by more than one concentrator. As the CFLP is a relaxation of the CCLP, the valid inequalities for the CFLP are also valid for the CCLP. Moreover, under some conditions, the facet defining inequalities for CFLP are also facet defining for the CCLP. So we start with the polyhedral structure of the CFLP.

Aardal et al. (1995) studies the valid inequalities and facets of the polytope associated to the CFLP. She defines the feasible domain X as:

$$\sum_{j \in M} v_{ij} = d_i \quad \forall i \in N$$

$$v_j = \sum_{i \in N} v_{ij} \quad \forall j \in M$$

$$\sum_{j \in M} v_j = d(N)$$

$$v_j \leq Q_j y_j \quad \forall j \in M$$

$$v_{ij} \geq 0 \quad \forall i \in N \ \forall j \in M$$

$$0 \leq y_j \leq 1 \quad \forall j \in M$$

$$y_j \text{ integer} \quad \forall j \in M$$

where v_{ij} is the flow between terminal i and concentrator j and $d(S) = \sum_{i \in S} d_i$. Let polytope PX be the convex hull of X. It is assumed that $\sum_{k \in M} Q_k - Q_j \geq d(N)$ for all $j \in M$. The author proves that the dimension of PX is $mn + m - n$, lists trivial facets, discusses valid inequalities and introduces two new classes of valid inequalities, called the effective capacity inequalities and the submodular inequalities. Below we list the results due to Aardal et al. (1995):

Knapsack Facets: Let $J \subset M$ be such that $\sum_{j \in J} Q_j > \sum_{j \in M} Q_j - d(N)$. Set J is called a cover with respect to N and M and a minimal cover if, in addition, for all $S \subset J$, we have $\sum_{j \in S} Q_j \leq \sum_{j \in M} Q_j - d(N)$. Since the capacity of concentrators in $M\backslash J$ is not sufficient to supply all the demands of terminals, at least one of the concentrators in J has to be installed. So we have that $\sum_{j \in J} y_j \geq 1$ is valid. In fact under some conditions, this inequality induces a facet of PX.

Theorem 1. *If J is a minimal cover with respect to M and N, $Q_{min} = \min_{j \in J} Q_j$ and $\sum_{j \in M \backslash J} Q_j + Q_{min} > d(N)$, then $\sum_{j \in J} y_j \geq 1$ defines a facet of $conv(X \cap \{y \in \{0,1\}^m : y_j = 1 \ \forall j \in M \backslash J\})$.*

These knapsack cover inequalities can be generalized by picking a $M' \subset M$ and J' which is a minimal cover with respect to N and M' and applying sequential lifting to variables $y_j = 0$ in $M \backslash M'$ and $y_j = 1$ in $M' \backslash J'$. The resulting inequality which defines a facet of PX is of the form:

$$\sum_{j \in M' \backslash J'} \beta_j y_j + \sum_{j \in M \backslash M'} \alpha_j y_j + \sum_{j \in J'} y_j \geq 1 + \sum_{j \in M' \backslash J'} \beta_j.$$

Flow Cover Inequalities: A subset $J \subset M$ such that $\lambda = \sum_{j \in J} Q_j - d(N) > 0$ is called a flow cover set with respect to M and N. This means the maximum value that the sum of the flow to the concentrators in set J can take is $d(N)$. If we close one of the concentrators in set J, say j, then this maximum will not change if $Q_j \leq \lambda$ and will decrease be $Q_j - \lambda$ otherwise. This observation leads to the following result:

Theorem 2. *Let J be a flow cover with respect to M and N such that $max_{j \in J} Q_j > \lambda$ and $\sum_{j \in M} Q_j > d(N) + Q_k$ for all $k \in J$. Then the flow cover inequality*

$$\sum_{j \in J} v_j + \sum_{j \in J}(Q_j - \lambda)^+ (1 - y_j) \leq d(N)$$

defines a facet of PX.

The author generalizes this idea and introduces another family of inequalities called the effective capacity inequalities and proves the following theorem.

Effective Capacity Inequalities: Let $J \subseteq M$ and $K \subseteq N$. For each $j \in J$ choose $K_j \subseteq K$. Define $\overline{Q}_j = min\{Q_j, d(K_j)\}$. J is a flow cover with respect to M and K if $\lambda = \sum_{j \in J} \overline{Q}_j - d(K) > 0$.

Theorem 3. *Let $J \subset M$ be a flow cover with respect to M and K. Let $Q \subset J$ be the set of concentrators for which $\overline{Q}_j < Q_j$. Assume $\sum_{k \in M} Q_k > Q_j + d(N)$ for all $j \in M$. The inequality*

$$\sum_{j \in J} \sum_{i \in K_j} v_{ij} + \sum_{j \in J}(\overline{Q}_j - \lambda)^+ (1 - y_j) \leq d(K)$$

defines a facet of PX if and only if

$$\forall j_1, j_2 \in Q, K_{j_1} \cap K_{j_2} = \emptyset$$
$$K_j = K \ \forall j \in J \backslash Q$$
$$\cup_{j \in Q} K_j \subset K$$
$$\overline{Q}_j > \lambda \ \forall j \in Q$$
$$|K| \leq 1 \Rightarrow \exists j \in J \backslash Q \ \text{with} \ Q_j > \lambda.$$

Another family of valid inequalities by Aardal et al. (1995) is the family of submodular inequalities.

Submodular Inequalities: A set function f on $N = \{1,..n\}$ is submodular if $f(A) + f(B) \geq f(A \cup B) + f(A \cap B)$ for all $A, B \subseteq N$. Define $\phi_j(A) = f(A \cup \{j\}) - f(A)$ for $j \in N \backslash A$ as the incremental function. Let $K \subseteq N$, $J \subseteq M$ and $K_j \subseteq K$ for all $j \in J$. The function

$$f(J) = \max \sum_{j \in J} \sum_{i \in K_j} v_{ij}$$

$$\text{subject to:} \quad \sum_{i \in K_j} v_{ij} \leq \overline{Q}_j y_j \quad \forall j \in J$$

$$\sum_{j \in J, i \in K_j} v_{ij} \leq d_i \quad \forall i \in K$$

$$v_{ij} \geq 0 \quad \forall i \in K, j \in J$$

$$y_j = 1 \quad \forall j \in J$$

is submodular on M.

Theorem 4. *Let $K \subseteq N$, $J \subseteq M$ and $K_j \subseteq K$ for all $j \in J$. The submodular inequality*

$$\sum_{j \in J} \sum_{i \in K_j} v_{ij} + \sum_{j \in J} \phi_j(J \backslash \{j\})(1 - y_j) \leq f(J)$$

is valid for PX.

The author shows that submodular inequalities are at least as strong as the effective capacity inequalities and gives the conditions when these inequalities define facets. She also discusses combinatorial and lot sizing inequalities.

Aardal (1998) gives heuristics for the separation of the above inequalities and presents computational results.

Leung and Magnanti (1989) study the valid inequalities and facets of the CFLP where all concentrators have the same capacity. They consider the following feasible domain F:

$$x_{ij} \leq y_j \quad \forall i \in N, j \in M$$
$$\sum_{j \in M} x_{ij} \leq 1 \quad \forall i \in N$$
$$\sum_{i \in N} d_i x_{ij} \leq Q y_j \quad \forall j \in M$$
$$x_{ij} \geq 0 \quad \forall i \in N, j \in M$$
$$y_j \leq 1 \quad \forall j \in M$$
$$y_j \text{ integer} \quad \forall j \in M$$

and the convex hull PF of F. They prove that PF is full-dimensional and give the trivial facets. Then they define a new family of facet defining inequalities called the residual capacity inequalities:

$$\sum_{i \in N'} \sum_{j \in M'} d_i x_{ij} - r \sum_{j \in M'} y_j \leq D - r \lceil D/K \rceil \tag{9.60}$$

where $N' \subset N$, $M' \subset M$, $D = \sum_{i \in N'} d_i$ and $r = D \pmod{K}$ with $r = K$ if D is a multiple of K. Then r can be interpreted as the residual demand to be satisfied by the last concentrator when $\lceil D/K \rceil - 1$ concentrators supply the demands with full capacity. Inequality (9.60) is valid and defines a facet of PF when $1 \leq r \leq K-1$ and $|M'| \geq \lceil D/K \rceil$.

Leung and Magnanti (1989) also consider the convex hull of F with the additional constraint that x_{ij}'s are integer variables, i.e. $\text{conv}(F \cap \{x_{ij} \in \{0,1\} \ \forall i \in N, \ \forall j \in M\})$. The residual capacity inequalities are also valid for this polyhedron but not facet defining in the general case. In case, the demands of the terminals are equal, we can assume that each terminal has unit demand and the capacity is in terms of the number of terminals assigned. Then the inequalities become:

$$\sum_{i \in N'} \sum_{j \in M'} x_{ij} - r' \sum_{j \in M'} y_j \leq |N'| - r' \lceil |N'|/K \rceil \tag{9.61}$$

Theorem 5. *The valid inequalities (9.61) are facet defining for the CCLP when the demands of all terminals are equal.*

For the CCLP with unit demands, Leung and Magnanti (1989) develop an alternative formulation, where the capacity of a concentrator is viewed as a collection of unit capacity concentrators. The following variables are defined:
$$x_{ijk} = \begin{cases} 1 \text{ if terminal } i \text{ is assigned to unit } k \text{ of concentrator } j \\ 0 \text{ otherwise} \end{cases}$$
for all $i \in N$, $j \in M$ and $k = 1,..,Q_j$.
$$y_j = \begin{cases} 1 \text{ if a concentrator is installed at location } j \\ 0 \text{ otherwise} \end{cases}$$

and $\overline{y}_j = 1 - y_j$ for all $j \in M$. Then the following constraints define the problem:

$$\sum_{j \in M} \sum_{k=1}^{Q_j} x_{ijk} \leq 1 \quad \forall i \in N \tag{9.62}$$

$$\sum_{i \in N} x_{ijk} + \overline{y}_j \leq 1 \quad \forall j \in M, k = 1, .., Q_j \tag{9.63}$$

$$\sum_{k=1}^{Q_j} x_{ijk} + \overline{y}_j \leq 1 \quad \forall i \in N, j \in M \tag{9.64}$$

$$x_{ijk} \in \{0,1\} \quad \forall i \in N, j \in M, k = 1, .., Q_j \tag{9.65}$$

$$\overline{y}_j \in \{0,1\} \quad \forall j \in M \tag{9.66}$$

The above set of constraints defines a set packing problem. It is well known that solutions of such a problem are the independent sets of vertices of an intersection graph, G. This intersection graph is an undirected graph in which there exists a node for each variable of the problem and there exists an edge between two nodes if and only if the corresponding variables appear in the same constraint. The cliques of this graph give rise to families of facets which are the same as the constraints (9.62), (9.63) and (9.64) of the problem.

The odd holes of the intersection graph G can be associated with a matrix A whose rows correspond to $N' \subset N$ and columns correspond to $\overline{M} = \{(j,k)|j \in M', k = 1, .., Q_j\}$ with $M' \subset M$, $|M'| = |N'| - 1$. Each row of A has exactly two 1's and each column has either one or two 1's. If $\overline{M}'$ is the set of columns with two 1's, then $|M'| - |\overline{M}'|$ is odd. The set of vertices of G corresponding to the set of variables $\{x_{ijk}|i \in N', (j,k) \in \overline{M}, a_{ijk} = 1\} \cup \{\overline{y}_j | j \in M' \backslash \overline{M}'\}$ form an odd hole and the corresponding odd hole inequality is:

$$\sum_{i \in N', (j,k) \in \overline{M}} a_{ijk} x_{ijk} + \sum_{j \in M' \backslash \overline{M}'} \overline{y}_j \leq |N'| + \frac{|M' \backslash \overline{M}'| - 1}{2} \tag{9.67}$$

which is valid. The authors mention that the graph G does not contain any other facet defining subgraph. Finally, they compare the two formulations and say that the linear relaxations of both formulations are equivalent.

Neebe and Rao (1983) formulate the CCLP as a set partitioning problem and solve it by a branch and price algorithm. The formulation is as follows: Let an activity $a_j = (a_{1,j}, .., a_{n,j})$ for concentrator j be a binary vector defined as:
$$a_{i,j} = \begin{cases} 1 \text{ if terminal } i \text{ is assigned to the concentrator at location } j \\ 0 \text{ otherwise} \end{cases}$$
and $\sum_{i \in N} d_i a_{i,j} \leq Q_j$. Let M_j denote the set of indices k of all activities $a_{i,j}^k$ for concentrator j.

The cost of a non-null activity $a_{i,j}^k$ is $c_j^k = \sum_{i \in N} C_{ij} a_{i,j}^k + F_j$ and the cost of a null activity is 0. The decision variables x_j^k are defined as:
$$x_j^k = \begin{cases} 1 \text{ if activity } a_j^k \text{ is active} \\ 0 \text{ otherwise} \end{cases}$$
for all $j \in M$ and $k \in M_j$.

Then the formulation follows:

$$\min \sum_{j \in M} \sum_{k \in M_j} c_j^k x_j^k \qquad (9.68)$$

$$\text{subject to:} \sum_{j \in M} \sum_{k \in M_j} a_{i,j}^k x_j^k = 1 \quad \forall i \in N \qquad (9.69)$$

$$\sum_{k \in M_j} x_j^k \leq 1 \quad \forall j \in M \qquad (9.70)$$

$$x_j^k \in \{0,1\} \quad \forall j \in M, k \in M_j \qquad (9.71)$$

The linear programming relaxation of the above formulation is solved using column generation, since it is burdensome to enumerate all columns. Let B be a feasible basis for the linear relaxation of (9.69)-(9.71) with corresponding dual variables $\sigma = (\sigma_1, .., \sigma_{n+m})$. The next column to enter the basis is the one with the largest positive reduced cost. This column is found by solving a problem which decomposes into m knapsack problems. For each $j \in M$, the following problem is solved:

$$\max \theta_j = \sum_{i \in N} (\sigma_i - C_{ij}) t_{ij} + \sigma_{n+j} - F_j \delta_j \qquad (9.72)$$

$$\text{subject to:} \sum_{i \in N} d_i t_{ij} \leq Q_j \delta_j \qquad (9.73)$$

$$t_{ij} \in \{0,1\} \quad \forall i \in N \qquad (9.74)$$

$$\delta_j \in \{0,1\} \qquad (9.75)$$

Let t_j^* be the optimal solution with an objective value of θ_j^*. The value $\theta = \max_{j \in M} \theta_j^* = \theta_p^*$ is computed. If $\theta \leq 0$, then the linear relaxation is solved. Otherwise, the column (t_p^*, e_p) is introduced into the basis, where e_p is a unit vector with 1 at the pth position.

If the solution to the relaxed problem is all integer, then $x_{ij} = \sum_{k \in M_j} a_{i,j}^k x_j^k$ for all $i \in N$ and $j \in M$, which is also integer. If the solution is not integer, the algorithm proceeds by branching on the fractional variables. The computational results with this algorithm are given in the paper.

9.3.2 Lagrangian Relaxation Based Heuristics

We encounter four Lagrangian Relaxations for the CCLP. The first one is obtained by dualizing the capacity constraints (9.58). Let $v \geq 0$ denote the

vector of Lagrangian multipliers. For a given vector $v \geq 0$, the relaxed problem is as follows:
$LR(v)$

$$\min \sum_{i \in N} \sum_{j \in M} (C_{ij} + v_j d_i) x_{ij} + \sum_{j \in J} F_j y_j - \sum_{j \in M} v_j Q_j$$

subject to: (9.54), (9.56), (9.57), (9.59)

which is an UFLP. The Lagrangian Dual is

$$LD(v) = \max_{v \geq 0} LR(v)$$

Mirzaian (1985) formulates the problem where all concentrators have the same capacity and each terminal places a unit demand so that the capacity constraints (9.55) become $\sum_{i \in N} x_{ij} \leq Q y_j$ for all $j \in M$. In this formulation, terminals are allowed to be directly assigned to the central unit. For this special case, the author proves that the Lagrangian Relaxation obtained by dualizing capacity constraints (9.55) in the CCLP gives the same bound as the linear programming relaxation of the problem.

Klincewicz and Luss (1986) develop a Lagrangian Relaxation based heuristic for the CCLP. They apply Lagrangian Relaxation by dualizing capacity constraints and solve the UFLP subproblems by the dual ascent and dual adjustment algorithm of Erlenkotter (1978) without the branch and bound phase. An "add" procedure which adds facilities one at a time is used to generate an initial feasible solution and an upper bound for the problem. Then the Lagrangian Relaxation heuristic starts with the dual variables set to 0. If the resulting UFLP gives a feasible solution for the original problem, the procedure ends. Otherwise, the dual variables for the violated constraints are increased to the minimum value so that some terminals are assigned to other concentrators. Then the algorithm switches to a version of subgradient optimization and proceeds till the ratio between the upper and the lower bound is smaller than a given value or the current solution is worse than a previous one for the original problem or the maximum number of iterations allowed is reached. When the subgradient optimization phase terminates, if a feasible solution is found then an adjustment heuristic which may update the terminal assignments is applied to improve this solution.

The Lagrangian Relaxation Heuristic failed to find a feasible solution for only one of the 24 problems Klincewicz and Luss (1986) used to test their algorithm. Darby-Dowman and Lewis (1988) define a class of problems for which the Lagrangian Relaxation where the capacity constraints are dualized will result in an infeasible solution regardless of the values of the multipliers. A lower bound L on the number of concentrators to be installed can be computed as follows: Let $Q_{(j)}$ denote the jth largest capacity and $D = \sum_{i \in N} d_i$ denote the total demand. Then L may be determined using the following relationship:

$$\sum_{j=1}^{L-1} Q_{(j)} < D \le \sum_{j=1}^{L} Q_{(j)}$$

Next, let $\sum^{+(n-1)}$ denote the summation of the $(n-1)$ most positive entities. If there are less than $(n-1)$ positive entities, then the operator denotes the sum of all positive entities. Consider the following inequality:

$$F_q > \sum_{i \in N}^{+(n-1)} (C_{ip} - C_{iq}) \qquad (9.76)$$

If inequality (9.76) is satisfied for all $p, q \in M$, then in the solution of the relaxed problem, only one concentrator is installed. The authors generalize the condition to the case in which the solution of the relaxed problem will involve at most $(L-1)$ concentrators to be installed:

$$F_q > \sum_{i \in N}^{+(n-L+1)} (C_{ip} - C_{iq}).$$

If the above inequality is satisfied for all $p, q \in M$, then the relaxed problem will have a solution with at most $(L-1)$ concentrators installed.

The second Lagrangian Relaxation for the CCLP is obtained by dualizing the assignment constraints (9.54) with Lagrangian multipliers u. This leads to the following problem for a given vector u:
$LR(u)$

$$\min \sum_{i \in N} \sum_{j \in M} (C_{ij} - u_i) x_{ij} + \sum_{j \in J} F_j y_j + \sum_{i \in N} u_i$$

subject to: (9.55), (9.56), (9.57), (9.59)

and the Lagrangian Dual is $LD(u) = max_u LR(u)$.

Pirkul (1987) considers this Lagrangian Relaxation and develops a heuristic based on this relaxation. The problem $LR(u)$ is decomposed into m knapsack problems one for each concentrator as follows:

$$LR_j(u) = \min \sum_{i \in N} (C_{ij} - u_i) x_{ij} + F_j y_j$$

$$\text{subject to:} \sum_{i \in N} d_i x_{ij} \le Q_j y_j$$

$$x_{ij} \le y_j \quad i \in N$$

$$x_{ij} \in \{0, 1\} \quad \forall i \in N, \ y_j \in \{0, 1\}$$

If the optimal solution to this knapsack problem is positive, then concentrator j is installed. Otherwise, y_j is set to 0. The Lagrangian Dual is solved using subgradient optimization.

Sridharan (1993) considers the same Lagrangian Relaxation and uses an aggregate capacity constraint,

$$\sum_{j \in M} Q_j y_j \geq \sum_{i \in N} d_i \tag{9.77}$$

to strengthen the formulation. The relaxation $LR(u)$ with the additional constraint (9.77) decomposes into m independent knapsack problems for each j. Specifically,

$$LR(u) = \min \sum_{j \in M} LR_j(u) y_j + \sum_{i \in N} u_i$$
$$\text{subject to: } (9.57), (9.77)$$

where

$$LR_j(u) = \min \sum_{i \in N} (C_{ij} - u_i) x_{ij} + F_j$$
$$\text{subject to: } \sum_{i \in N} d_i x_{ij} \leq Q_j$$
$$x_{ij} \in \{0, 1\} \quad \forall i \in N$$

Let x^* be the optimal solution for $LR_j(u)$ and y^* be the optimal solution for $LR(u)$ for a fixed u. Then an optimal solution for $LR(u)$ is given by $y_j = y_j^*$ and
$$x_{ij} = \begin{cases} x_{ij}^* & \text{if } y_j^* = 1 \\ 0 & \text{otherwise} \end{cases}$$

This Lagrangian Relaxation may lead to a better bound compared to bound of the LP relaxation unless $d_i = d$ for all $i \in N$ and $Q_j (\bmod d) = 0$ for all $j \in M$. A heuristic based on this relaxation is presented in the paper. This heuristic uses the fact that $LR(u)$ not only returns a lower bound for the problem but also fixes a set of variables y_j to one, which then leads to a single source transportation problem defined on the set of concentrators for which $y_j = 1$. The solution of this problem yields an upper bound. Computational results with this algorithm are presented in the remaining part of the paper.

Hindi and Pienkosz (1999) consider the same problem but propose an algorithm that combines a greedy constructive heuristic with a restricted neighborhood search to find feasible solutions. The greedy heuristic proceeds as follows: For each terminal and concentrator pair define a penalty π_{ij} as:

$$\pi_{ij} = \begin{cases} C_{ij} & \text{if concentrator } j \text{ is already installed,} \\ C_{ij} + \frac{F_j d_i}{Q_j} & \text{otherwise.} \end{cases}$$

Given a set of assigned terminals, compute the regret value for an unassigned terminal as the difference between the two lowest penalties for concentrators whose remaining capacity is sufficient to supply the demand of this terminal. So the aim is to assign all the terminals iteratively, by assigning the terminal with the maximum regret to the concentrator that gives the smallest penalty and that has sufficient capacity and updating penalties and remaining capacities at each iteration.

If the greedy heuristic fails to find a feasible solution, then the second phase starts with the set of installed concentrators maintained. The second phase tries to shift or exchange assignments of terminals to reach feasibility. Finally, the third phase of the algorithm which is a restricted neighborhood search, again shifts and exchanges assignments of terminals but it also considers the effect of these moves on the total cost. This heuristic is used at each iteration of the subgradient optimization to find feasible solutions and to compute upper bounds.

Barcelo and Casanovas (1984) consider the same Lagrangian Relaxation and give a two stage heuristic to obtain feasible solutions. The first stage involves deciding about a subset of concentrators to be installed based on the information from $LR(u)$ and the second phase concerns assigning terminals to these concentrators.

Holmberg et al. (1999) develop a Lagrangian Relaxation heuristic which utilizes a primal heuristic based on a repeated matching algorithm. They also present a branch and bound algorithm and compare the results with CPLEX.

Ronnqvist et al. (1999) use the repeated matching algorithm as a heuristic approach by itself. A packing is defined to be a subset of assignments of terminals to concentrators satisfying the capacity constraints and where a terminal cannot be assigned to more than one concentrator. The heuristic uses repeated matchings to form feasible packings from the previously determined ones with decreasing cost.

Barcelo et al. (1991) reformulate the problem by introducing auxiliary variables w_{ij}'s for x_{ij}'s. They add the constraint that $x_{ij} = w_{ij}$ for all $i \in N, j \in M$, modify the cost function as

$$\min \alpha \sum_{i \in N} \sum_{j \in M} C_{ij} x_{ij} + \beta \sum_{i \in N} \sum_{j \in M} C_{ij} w_{ij} + \sum_{j \in M} F_j y_j$$

where α and β are fixed values and $\alpha + \beta = 1$. They replace the assignment constraints by

$$\sum_{j \in M} w_{ij} = 1 \quad \forall i \in N.$$

Then the constraints $x_{ij} = w_{ij}$ are relaxed and the problem decomposes into two parts, one part for the variables x_{ij}'s without the assignment constraints, which further decomposes into knapsack problems and the second part involving the variables w_{ij}'s only without the capacity constraints which is a semi-assignment problem. An algorithm based on this relaxation is presented in the paper.

Mirzaian (1985) considers the same relaxation for the formulation without constraints (9.59) and shows that this can lead to a better bound than the bound of the linear programming relaxation since this Lagrangian dual is the same as the linear relaxation of the original problem with the additional constraints (9.59). So what follows is another Lagrangian Relaxation for the problem with the additional set of constraints (9.59) dualized with multipliers $\lambda \geq 0$, and this leads to the following problem for a fixed vector $\lambda \geq 0$:
$LR(\lambda)$

$$\min \sum_{i \in N} \sum_{j \in M} (C_{ij} + \lambda_{ij}) x_{ij} + \sum_{j \in M} F_j y_j - \sum_{j \in M} \sum_{i \in N} \lambda_{ij} y_j$$

subject to: (9.54), (9.55), (9.56), (9.57)

and the Lagrangian Dual is $LD(\lambda) = max_{\lambda \geq 0} LR(\lambda)$. The author develops an approximation algorithm based on this relaxation. This algorithm uses subgradient optimization to solve the Lagrangian Dual and includes post optimization heuristics to improve the upper bounds at each iteration and at the end of the algorithm.

Beasley (1993) relaxes both the capacity and assignment constraints of the CCLP. The dual variables corresponding to the capacity constraints (9.55) are $v_j \geq 0$ and the ones corresponding to the assignment (9.54) constraints are u_i. The resulting problem is as follows:
$LR(u, v)$

$$\min \sum_{i \in N} \sum_{j \in M} C_{ij} x_{ij} + \sum_{j \in M} (F_j - v_j) y_j + \sum_{i \in N} u_i$$

subject to: (9.59), (9.56), (9.57)

with $C_{ij} = C_{ij} - u_i + v_j(d_i/Q_j)$.

The Lagrangian Dual problem is $max_{v \geq 0, u} LR(u, v)$. The author gives a heuristic based on this relaxation. Agar and Salhi (1998) suggest some improvements for this procedure.

9.4 Capacitated Models

The CCLP is the core problem for designing a capacitated telecommunication network. The capacitated models in the literature are generalizations of the CCLP. For example, there are models where the decision maker can decide about the capacity of concentrators to be installed. This problem is called multitype capacitated concentrator location problem (see Lee (1993), Amiri (1997)).

For some networks, reliability is a critical issue. To increase reliability, terminals can be assigned to more than one concentrator. This problem is called the CCLP with multiple homing (see Tang et al. (1978), Pirkul et al. (1988)).

Next, we discuss these models that are generalizations of the CCLP.

9.4.1 CCLP with Multitype Concentrators

The CCLP with multitype concentrators is a generalization of the CCLP in the sense that for each candidate location for concentrators there is a given set of concentrator types each with a different capacity and cost. So, the decisions involved are to determine the location and type of concentrators and to assign the terminals to the concentrators subject to capacity constraints. We can model this problem as follows:

$$\min \sum_{i \in N} \sum_{j \in M} C_{ij} x_{ij} + \sum_{j \in M} \sum_{k \in K} F_{jk} y_{jk}$$

$$\text{subject to:} \sum_{j \in M} x_{ij} = 1 \quad \forall i \in N$$

$$\sum_{k \in K} y_{jk} \leq 1 \quad \forall j \in M$$

$$\sum_{i \in N} d_i x_{ij} \leq Q_{jk} y_{jk} \quad \forall j \in M, k \in K$$

$$x_{ij} \in \{0,1\} \quad \forall i \in N, j \in M$$

$$y_{jk} \in \{0,1\} \quad \forall j \in M, k \in K$$

where K is the set of concentrator types, F_{jk} is the cost of installing a concentrator of type k at location j, Q_{jk} is the capacity of a concentrator of type k at location j and

$$y_{jk} = \begin{cases} 1 \text{ if a concentrator of type } k \text{ is installed at location } j \\ 0 \text{ otherwise,} \end{cases}$$

for all $j \in M$ and $k \in K$.

For this problem, Lee (1993) proposes an algorithm based on cross decomposition, which combines Benders Decomposition and Lagrangian Relaxation and presents computational results.

Amiri (1997) studies the multitype capacitated concentrator location problem with delay costs. He models the system as a M/M/1 queueing system and gives two Lagrangian Relaxation based heuristics to solve this problem.

9.4.2 CCLP with Multiple Homing

In the CCLP, each terminal is assigned to a single concentrator. However, to increase the reliability of the network, the decision makers may prefer to assign terminals to more than one concentrator.

Tang et al. (1978) consider the problem where it is necessary to connect terminal i to exactly β_i concentrators. The model is the same as the CCLP, except the assignment constraint (9.54) which is replaced by

$$\sum_{j \in M} x_{ij} = \beta_i \quad \forall i \in N.$$

This model assumes that the capacity of concentrators is in terms of the number of terminals assigned to it. The authors first present an algorithm to solve the assignment problem once the locations of concentrators are given. Then they develop an add heuristic to solve the CCLP with multiple homing.

Pirkul et al. (1988) study the problem where each terminal is assigned to two concentrators, one for the primary coverage and the second for the secondary or backup coverage. They also consider the case where each terminal is required to be assigned to k concentrators. They present a Lagrangian Relaxation with primary and secondary coverage (assignment) constraints dualized. Then the problem reduces into an independent knapsack problem for each concentrator. The Lagrangian Dual is solved by subgradient optimization and a heuristic to construct a feasible solution at each iteration of the subgradient optimization is given.

9.4.3 Other Capacitated Models

Monticone and Funk (1994) consider the problem where a set of communicating pairs of terminals are given that can be connected to each other directly or using exactly two concentrators. The aim is to minimize the cost of establishing these links and the cost of installing concentrators. It is assumed that each communicating pair places a unit demand. Some partitioning and reduction techniques are applied to reduce the computation times.

9.5 Dynamic Models

When the demand of terminals increases over time, it may be necessary to expand the sizes of the concentrators and the capacities of the links in a network. The problem of optimally expanding a network over a given planning horizon to meet the increasing demand is called Dynamic Facility Location Problem or Multiperiod Facility Location Problem. Some examples of this problem have application in telecommunication.

Shulman (1991) considers the multiperiod capacitated concentrator location problem where there are concentrators of different types. The author also imposes an upper bound on the number of concentrators that can be installed at a given location. The aim is to minimize the cost of installing concentrators, the cost of delivering demand of terminals and the operation cost. It is assumed that the demand of a terminal may be supplied by several terminals. Two cases of the problem are considered. In the first case, the mix of different types of concentrators at a given location is allowed, while in the second case, it is not allowed. Two algorithms are proposed for these two cases.

Balakrishnan, Magnanti and Wong (1995) consider the expansion of a telecommunication network and they assume that there are no concentrators in the network at the beginning of the planning horizon. The problem is to install concentrators and expand the sizes of the links with minimum cost and meet the projected demand. They develop a decomposition method based on Lagrangian Relaxation and dynamic programming.

Chardaire, Sutter and Costa (1996) develop a model to determine when and where to install additional concentrators in an environment where the demand of terminals increases. The aim is to minimize the cost of installing new concentrators, operating them and the cost of assigning terminals. The authors prove that the Lagrangian Relaxation gives the same bound as the linear relaxation of the problem and they present a heuristic approach to solve the problem. They report computational results.

Acknowledgement

The research of the third author was supported by France Telecom R&D under contract no. 99 1B 774.

References

Aardal, K., Y. Pochet and L.A. Wolsey (1995) "Capacitated Facility Location: Valid Inequalities and Facets", *Mathematics of Operations Research*, 20, 562-582.

Aardal, K. (1998) "Capacitated Facility Location: Separation Algorithms and Computational Experience", *Mathematical Programming*, 81, 149-175.

Agar, M.C. and S. Salhi (1998) "Lagrangian Heuristics Applied to a Variety of Large Capacitated Plant Location Problems", *Journal of Operational Research Society*, 49, 1072-1084.

Amiri, A. (1997) "Solution Procedures for the Service System Design Problem", *Computers and Operations Research*, 24, 49-60.

Balakrishnan, A., T.L. Magnanti and R.T. Wong (1995) "A Decomposition Algorithm for Local Access Telecommunications Network Expansion Planning", *Operations Research*, 43, 58-76.

Barcelo, J. and J. Casanovas (1984) "A Heuristic Lagrangian Algorithm for the Capacitated Plant Location Problem", *European Journal of Operations Research*, 15, 212-226.

Barcelo, J., E. Fernandez and K.O. Jornsten (1991) "Computational Results from a New Lagrangian Relaxation Algorithm for the Capacitated Plant Location Problem", *European Journal of Operations Research*, 53, 38-45.

Beasley, J.E. (1993) "Lagrangean Heuristics for Location Problems", *European Journal of Operations Research*, 65, 383-399.

Boffey, T.B. (1989) "Location Problems Arising in Computer Networks", *Journal of Operational Research Society*, 40, 347-354.

Chardaire, P., J.L. Lutton and A. Sutter (1999) "Upper and lower bounds for the two-level simple plant location problem", *Annals of Operations Research*, 86, 117-140.

Chardaire, P., A. Sutter and M.C. Costa (1996) "Solving the Dynamic Facility Location Problem", *Networks*, 28, 117-124.

Chung, S., Y. Myung and D. Tcha (1992) "Optimal Design of a Distributed Network with a Two-Level Hierarchical Structure", *European Journal of Operations Research*, 62, 105-115.

Cornuéjols, G., G.L. Nemhauser and L.A. Wolsey (1990) "The Uncapacitated Facility Location Problem" in *Discrete Location Theory*. P.B. Mirchandani and R.L. Francis (eds), Wiley, New York, 119-171.

Current, J. and H. Pirkul (1991) "The Hierarchical Network Design Problem with Transshipment Facilities", *European Journal of Operations Research*, 52, 338-347.

Darby-Dowman, K. and H.S. Lewis (1988) "Lagrangian Relaxation and the Single Source Capacitated Facility Location Problem", *Journal of Operational Research Society*, 39, 1035-1040.

Erlenkotter, D. (1978) "A Dual Based Procedure for Uncapacitated Facility Location", *Operations Research*, 26, 992-1009.

Gavish, B. (1982) "Topological Design of Centralized Computer Networks: Formulations and Algorithms", *Networks*, 12, 355-377.

Gavish, B. (1992) "Topological Design of Computer Communication Networks- The Overall Design Problem", *European Journal of Operations Research*, 58, 149-172.

Hakimi, S.L. (1964) "Optimum Locations of Switching Centers and the Absolute Centers and Medians of a Graph", *Operations Research*, 12, 450-459.

Hakimi, S.L. (1965) "Optimum Distribution of Switching Centers in a Communication Network and Some Related Graph Theoretic Problems", *Operations Research*, 13, 462-475.

Helme, M.P. and T.L.Magnanti (1989) "Designing Satellite Communication Networks by Zero-One Quadratic Programming", *Networks*, 19, 427-450.

Hindi, K.S. and K. Pienkosz (1999) "Efficient Solution of the Large Scale, Single Source Capacitated Plant Location Problems", *Journal of Operational Research Society*, 50, 268-274.

Holmberg, K., M. Ronnqvist and D. Yuan (1999) "An Exact Algorithm for the Capacitated Facility Location Problem with Single Sourcing", *European Journal of Operations Research*, 113, 544-559.

Klincewicz, J.G. (1998) "Hub Location in Bacbone/Tributary Network Design: a Review", *Location Science*, 6, 307-335.

Klincewicz, J.G. and H. Luss (1986) "A Lagrangian Relaxation Heuristic for Capacitated Facility Location with Single Source Constraints", *Journal of Operational Research Society*, 37, 495-500.

Krarup, J. and P.M. Pruzan (1983) "The Simple Plant Location Problem: Survey and Synthesis", *European Journal of Operations Research*, 12, 36-81.

Labbé, M., D. Peeters and J.F. Thisse (1995) "Location on Networks" in *Network Routing*, Handbooks in Operations Research and Management Sciences, Vol. 8, M.O. Ball, T.L. Magnanti, C.L. Monma and G.L. Nemhauser (eds), North-Holland, Amsterdam, 551-624.

Lee, C.Y. (1993) "An Algorithm for the Design of Multitype Concentrator Networks", *Journal of Operational Research Society*, 44, 471-482.

Lee Y., B.H. Lim and J.S. Park (1996) "A Hub Location Problem in Designing Digital Data Service Networks: Lagrangian Relaxation Approach", *Location Science*, 4, 185-194.

Leung, J.M.Y. and T.L. Magnanti (1989) "Valid Inequalities and Facets of the Capacitated Plant Location Problem", *Mathematical Programming*, 44, 271-291.

Marianov, V. and M. Rios (1996) "Queueing-Plant Location and Queueing-p-pmedian Models for the Design of Circuit Switched Telecommunications Networks with Constrained Buffer(Cell Queue) Length at Switches", *Proceedings ALIO-EURO Conference*, Valparaiso, Chile, 115-127.

Mateus, G.R., F.R.B. Cruz and H.P.L. Luna (1994) "An Algorithm for Hierarchical Network Design", *Location Science*, 2, 149-164.

Mirzaian, A. (1985), "Lagrangian Relaxation for the Star-Star Concentrator Location Problem: Approximation Algorithm and Bounds", *Networks*, 15, 1-20.

Monticone, L.C. and G. Funk (1994) "Application of the facility location problem to the problem of locating concentrators on an FAA microwave system", *Annals of Operations Research*, 50, 437-454.

Neebe, A.W. and M.R. Rao (1983) "An Algorithm for the Fixed-Charge Assigning Users to Sources Problem", *Journal of Operational Research Society*, 34, 1107-1113.

Nemhauser, G.L. and L.A. Wolsey (1988) "Integer and Combinatorial Optimization" , Wiley, New York.

Pirkul, H. (1987), "Efficient Algorithms for the Capacitated Concentrator Location Problem", *Computers and Operations Research*, 14, 197-208.

Pirkul, H. and V. Nagarajan (1992) "Locating Concentrators in Centralized Computer Networks", *Annals of Operations Research*, 36, 247-262.

Pirkul, H., S. Narasimhan and P. De (1988) "Locating Concentrators for Primary and Secondary Coverage in a Computer Communications Network", *IEEE Transactions on Communications*, 36, 450-458.

Ronnqvist, M., S. Tragantalerngsak and J. Holt (1999) "A Repeated Matching Heuristic for the Single Source Capacitated Facility Location Problem", *European Journal of Operations Research*, 116, 51-68.

Shulman, A. (1991) "An Algorithm for Solving Dynamic Capacitated Plant Location Problems with Discrete Expansion Sizes", *Operations Research*, 39, 423-436.

Sridharan, R. (1993) "A Lagrangian Heuristic for the Capacitated Plant Location Problem with Single Source Constraints", *European Journal of Operations Research*, 66, 305-312.

Tang, D.T., L.S. Woo and L.R. Bahl (1978) "Optimization of Teleprocessing Networks with Concentrators and Multiconnected Terminals", *IEEE Transactions on Computers*, 27, 594-604.

10 Reserve Design and Facility Siting

Charles ReVelle[1] and Justin C. Williams[1]

Department of Geography and Environmental Engineering, Johns Hopkins University, Baltimore, Maryland. e-mail: revelle@jhu.edu; jcwjr@jhunix.hcf.jhu.edu

10.1 Introduction

In the past 15 years, a new field of spatial inquiry, supported by the mathematics of combinatorial optimization, has burst on the scene. This field, the design/selection of nature reserves for the preservation of species, landscapes, and ecological processes, happens to intersect sharply with an older line of spatial study, discrete location science. The parallels in formulation of problems and in methods of solution are quite surprising. The student of discrete location science will find immediate familiarity with the concepts of nature reserve design that focus on selection of parcels for species and for landscape preservation/representation.

The purpose of this chapter is to lay out the central problems in these two fields, side by side, and compare their core formulations, finding similarities between them where they occur and pointing to the differences that may separate them. We will also compare the methods for solution of these problems that have been used up to the present and recommend, for certain cases, new approaches that may efficiently resolve problems that are now potentially difficult to resolve. ReVelle et al. (2001) also review the twin problems of reserve design for species protection and location science, but emphasize reserve design and offer an economic perspective and interpretation to reserve design problems.

At least five fundamental problem classes in discrete location science either have or are likely to have significant counterparts in the design and selection of nature reserves. These are location set covering, maximal covering, backup/ redundant covering, probabilistic covering, and expected covering. Our plan for the paper is to lay out formulations in each of these five classes of discrete location problems and then to present, for each formulation, the comparable problem in the design/ selection of nature reserves. The focus in the development of reserve networks will be the identification of parcels for species preservation and protection. In addition, we comment on applicable solution methods and their anticipated properties.

For each location problem, a network of demand nodes and connecting arcs is assumed to exist. Each arc has attached to it a value that represents the distance or travel time between the two nodes the arc connects. The shortest distance or time between any two nodes in the network is presumed

---

Facility Location: Applications and Theory.
Edited by Z. Drezner and H.W. Hamacher
© 2002 Springer-Verlag, ISBN 3-540-42172-6

to have been calculated. Facilities that are to serve the demand nodes may be located at a subset of the demand nodes or sites.

Similarly, for each problem in the design/selection of nature reserves, a set of parcels exists (the parcels are not necessarily connected), and a number of species exists in each parcel. Each species may reasonably be expected to survive in the subset of parcels that contains habitat suitable for survival of the species.

10.2 Set Covering Problems

10.2.1 The Location Set Covering Problem

This problem seeks to locate the least number of facilities such that each demand node has at least one facility sited at a node within a specified maximum distance or time (Toregas et al., 1970; Toregas and ReVelle, 1973; ReVelle et al., 1975). The notation used to state the problem is:

i, I = the index and set of demand points or nodes;
j, J = the index and set of eligible facility sites or nodes;
d_{ij} = the shortest distance or time between points or nodes i and j;
S = a distance or time standard; a facility sited at some node j within the standard of a demand node i is eligible to serve the demand node;
$N_i = \{j | d_{ij} \leq S\}$ is the set of nodes j within distance or time S of node i; these are the nodes eligible to house facilities which "cover" node i;
$x_j \in \{0, 1\}$. it is 1 if a facility is sited at j, and 0 otherwise.

The Location Set Covering Problem (LSCP) is stated as a linear zero-one program as:

$$\text{Min} \left\{ \sum_{j \in J} x_j \right\} \quad (10.1)$$

s.t.

$$\sum_{j \in N_i} x_j \geq 1 \ \forall i \in I \quad (10.2)$$

$$x_j \in \{0, 1\} \ \forall j \in J$$

The objective (10.1) minimizes the number of facilities needed to cover each and every demand node by at least one facility. The constraints (10.2) insure that all demand nodes i are "covered," i.e., each has at least one facility within S distance or time units of it. The variables are required to assume values of zero or one.

The LSCP was originally stated by Toregas, et al. (1970), who utilized relaxed linear programming (LP) to solve the problem. In the approximately 5% of cases in which the relaxed LP solution was not all zero-one, the LP was supplemented by a single cutting plane constraint which resolved the problem in integers. Toregas and ReVelle (1973) also showed how a repeated sequence of logical reductions of the constraint matrix (row and column elimination) and subsequent selection of essential variables could be employed to solve most LSCP's without linear programming optimization. The network base of the LSCP seems to contain a relational structure that makes this problem easier to solve by LP or by reductions when compared to the solution of other set covering problems which do not have a network base. This special relational structure (discussed in Toregas and ReVelle, 1973), if it exists, may be found in the covering sets (N_i) and their relationships to each other.

Networks and other spatial systems for which the triangle inequality holds seem more likely to possess this relational structure. Recent computational experience by Schilling et al. (2000) with a related location problem, the p-median problem, suggests that the greater the extent of satisfaction of the triangle inequality, the more likely that exact methods such as linear programming will resolve the problem without resort to branch and bound. Additionally, greater satisfaction of the triangle inequality appears to lessen the amount of branching and bounding when it is needed. Finally, heuristic methods are more often successful in finding the global optimum in the p-median problem when the triangle inequality holds.

10.2.2 The Species Set Covering Problem

This problem operates on a set of eligible parcels, each of which contains a subset of species (i.e., individuals or populations of a species) from a list of species which have been designated as important for protection/ preservation. For a particular species, one can identify a set of parcels whose habitat is such that if the species currently exists in that parcel, it is likely that the species could continue to survive in that parcel if the parcel were given some form of protected status. The Species Set Covering Problem (SSCP) is to choose the least number of parcels (or perhaps the least-cost set of parcels) in such a way that each species is represented (is likely to survive) in at least one parcel. The SSCP was first articulated by Margules et al. (1988) and was soon after recognized as the set covering problem of integer programming/ operations research by Possingham et al.(1993) and Underhill (1994). Needed notation for the set covering formulation is:

i, I = the index and set of all species;
j, J = the index and set of all parcels;
M_i = the set of parcels j which contain species i and in which species i is likely to survive if the parcel is protected; and
$x_j \in \{0, 1\}$. it is 1 if parcel j is selected for inclusion in the reserve set, and 0 otherwise.

The problem is stated as:

$$\text{Min} \left\{ \sum_{j \in J} x_j \right\} \qquad (10.3)$$

s.t.

$$\sum_{j \in M_i} x_j \geq 1 \ \forall i \in I \qquad (10.4)$$

$$x_j \in \{0, 1\} \ \forall j \in J$$

The objective (10.3) seeks to minimize the number of parcels required to represent all species at least once in the reserve. The constraints (10.4) insure that each and every species has at least one parcel chosen which will preserve it.

Of course, the reader sees that objectives (10.1) and (10.3) are identical in format and that constraints (10.2) and (10.4) are likewise identical in format. Although the formats of the two problems match exactly, an important difference may exist. The covering sets M_i of the SSCP may not have the special relational structure that frequently exists in the covering sets N_i of the LSCP. Whether or not particular instances of the SSCP have this type of relational structure may depend on the data, that is, on the level of spatial correlation in the distribution of species within parcels. Strong correlation or groupings of species would suggest the existence of a relational structure, and vice versa. In comparison to the LSCP, a lack of such structure in the SSCP would be expected to yield the following solution properties for the SSCP: a decrease in the effectiveness of an LP/IP approach in the sense that an LP cut would less frequently resolve the solution in integers; likewise, the application of LP plus branch and bound would not be as efficient in resolving the solution in integers; and furthermore, the matrix reduction approach would be more likely to encounter "cyclic" matrices and hence terminate without reaching a solution.

In fact, our experience has been that data structures for LSCP problems for which the linear programming solutions terminate with fractions will also cause the reduction approach to terminate with cyclic matrices i.e., without finding an answer. On the other hand, if the species can be divided into two or more non-overlapping sets, the set covering problem can be decomposed into two or more smaller sub-problems, each corresponding to one of the non-overlapping sets.

The central question to which these observations lead is, "What methods might be most successfully employed to find globally optimal solutions to the SSCP given that the methods successfully applied to the LSCP may not fare as well when applied to the SSCP?" One possible answer would be to attempt combinations of methods such as LP followed by branch and bound followed by a reductions approach to find solutions at the nodes of the branch and bound tree. Variations of cutting planes and reductions, used individually or in combination, could well be effective. A number of methods for the general set covering problem exist, but fundamentally, the question is not yet answered.

10.3 Maximal Covering Problems

10.3.1 The Maximal Covering Location Problem

Maximal covering problems are well known in the operations research literature (see Chung, 1986; Kolesar, 1980; Dwyer and Evans, 1981), but had their origin in location problems (Church and ReVelle, 1974.) The Maximal Covering Location Problem (MCLP) derives from a consideration of the location set covering problem. Whereas the LSCP sites the minimum number of facilities needed to cover all demand nodes within S distance units, the MCLP admits that resources may be insufficient to cover all of the demand nodes. With limited facility resources, that is, with just p facilities available to be sited, and with different populations at each of the demand nodes, the MCLP seeks to locate the facilities in such a way that the largest possible demand or population is covered. The concept of coverage is the same as before: a demand node is considered covered if there is at least one facility sited within the distance standard S. New needed notation is:

a_i = the population at demand node i, and
$y_i \in \{0, 1\}$. it is 1 if demand node i has one or more facilities sited within S distance units, and 0 otherwise.

The Maximal Covering Location Problem is stated in non-standard form as:

$$\text{Max} \quad \left\{ \sum_{i \in I} a_i y_i \right\} \quad (10.5)$$

s.t.

$$y_i \leq \sum_{j \in N_i} x_j \quad \forall i \in I \quad (10.6)$$

$$\sum_{j \in J} x_j = p \quad (10.7)$$

$$y_i \text{ and } x_j \in \{0, 1\} \quad \forall i \in I, \ j \in J$$

The objective (10.5) seeks to maximize the total covered population. Constraints (10.6) allow a demand node i to be covered only if one or more facilities are sited at nodes j in the set N_i (i.e., within S units of i). Constraint (10.7) requires exactly p facilities to be sited.

The MCLP was originally stated and solved by Church and ReVelle, 1974, who offered three initial approaches to the problem. The first approach, dubbed greedy adding (GA), began with the single site that provided the maximal possible population coverage and successively added those sites, one at a time, which incrementally increased the coverage values the most. The second approach, called greedy adding with substitution (GAS), utilized the GA algorithm as the backbone. However, each stage (each value of p), begins with the solution found by the best facility addition to the previous solution and tries to improve the solution by one-at-a-time exchanges of facility nodes for non-facility nodes. Besides these heuristic approaches, the authors showed that linear programming was effective in producing 0,1 solution for problems up to 55 nodes in about 80% of cases. When linear programming produced fractional solutions, it was only necessary to supplement it by modest amounts of branch and bound. Hillsman (1984) as well as Church and ReVelle (1976) noted the equivalence of a data-modified p-median and the maximal covering problem and pointed out that any heuristic (such as vertex substitution) useful for the p- median could also be utilized for the maximal covering problem.

The observation that any algorithm developed for the p-median problem can be modified and applied to the maximal covering location problem suggests that p-median algorithms more recent than one-opt vertex substitution (Teitz and Bart, 1967) might be useful for larger versions of the MCLP. These newer algorithms include those of Densham and Rushton (1996) and Rosing and ReVelle (1998). We return to this point in our discussion of the maximal covering species problem, which we formulate next.

10.3.2 The Maximal Covering Species Problem (MCSP)

As in the SSCP, species (i.e., individuals and populations) are distributed among the parcels in the system. Of interest are those parcels which may harbor a species in sufficient numbers and with suitable habitat that fosters long term survival of the species in that parcel. The MCSP seeks to choose p parcels for a reserve in such a way that the largest number of species in need of protection is represented in the reserve (Church et al., 1996). The MCSP uses the same notation as the MCLP with the exception that:

i, I = the index and set of species;

j, J = the index and set of parcels; and

$y_i \in \{0, 1\}$. it is 1 if species i is represented by at least one parcel in M_i, the set of parcels j in which i is likely to survive, and 0 otherwise.

The MCSP is stated in a non-standard form as a zero-one linear program:

$$\text{Max} \quad \left\{ \sum_{i \in I} y_i \right\} \tag{10.8}$$

s.t.

$$y_i \leq \sum_{j \in M_i} x_j \quad \forall i \in I \tag{10.9}$$

$$\sum_{j \in J} x_j = p \tag{10.10}$$

$$y_i \text{ and } x_j \in \{0, 1\} \quad \forall i \in I, \; j \in J$$

If the parcels have different costs c_j, the problem could be stated as a multiple objective problem, in which the number of species represented is maximized and total cost of the parcels selected is minimized:

Objective 1: $\max \left\{ \sum_{i \in I} y_i \right\}$

Objective 2: $\min \left\{ \sum_{j \in J} c_j x_j \right\}$

The multi-objective problem statement indicates that a tradeoff exists between the number of species "protected" and the investment in the reserve network.

The solution of the MCSP, (10.8), (10.9), (10.10) can be approached by the same general means that were applied to (10.5), (10.6), (10.7), the MCLP. However, as pointed out in the discussion of the LSCP and SSCP, the performance of the algorithms would likely depend on whether the covering sets M_i have the special relational structure discussed above. If this structure turned out to be lacking, solutions for the MCSP derived using the methods discussed above (a greedy adding procedure, a vertex exchange method, or LP plus branch and bound) could fall short of the optimum more frequently, fall further from the optimum, or require more branching and bounding than solutions for the MCLP derived by the same methods. Thus, although the MCLP and the MCSP look alike superficially, the ability of an analyst to solve large instances of the two problems may be somewhat different. The newer p-median algorithms, mentioned above, may be helpful in solving large or difficult MCSP's.

10.4 Redundant/Backup Coverage Problems

10.4.1 Facility Siting Problems

The application of redundant and backup coverage concepts to the siting of facilities is due to Berlin (1973), Daskin and Stern (1981), Hogan and ReVelle (1983), and Eaton et al. (1986). The notion of redundant coverage of demands consists of simply counting the number of coverers of each demand node in excess of first coverage. Redundant coverage of a demand node may be weighted by the population at the node or by the frequency of the need for service at the node; we continue to represent the population or frequency of service required by a_i. The redundant coverage facility siting problem may be stated as:

$$\text{Max} \quad \left\{ \sum_{i \in I} a_i s_i \right\} \quad (10.11)$$

s.t.

$$\sum_{j \in N_i} x_j - s_i = 1 \quad \forall i \in I \quad (10.12)$$

$$\sum_{j \in j} x_j = p \quad (10.13)$$

$$x_j \text{ integer} \quad \forall j \in J$$

where s_i is the number of coverers of node i in excess of one; it is, in essence, the surplus variable in the ith covering equation. Note that multiplication of the excess coverage variable by a_i increases the importance of redundant covering node i in proportion to its need for service.

In backup covering, in contrast to the redundant covering model, coverage beyond the second coverer is not counted or valued in any way in the objective function. Only the second coverer has value placed on it in the objective (the distance standard for backup coverage need not be the same as the distance standard for primary coverage). Two backup covering location models may be stated. In the first model, the population with backup coverage is maximized, given p facilities, and every demand node is covered at least once; this is the same requirement as the constraints of the LSCP. Of course the number of facilities to be sited, p, must be greater than or equal to the minimal number required from the LSCP. In mathematics, the problem is stated as:

$$\text{Max} \quad \left\{ \sum_{i \in I} a_i u_i \right\} \quad (10.14)$$

s.t.

$$u_i \leq \sum_{j \in N_i} x_j - 1 \quad \forall i \in I \quad (10.15)$$

$$\sum_{j \in j} x_j = p \quad (10.16)$$

$$u_i \in \{0, 1\} \text{ and } x_j \text{ integer} \quad \forall i \in I, \; j \in J$$

where the only new notation is: $u_i \in \{0, 1\}$; it is 1 if node i has one or more backup coverers (i.e., has two or more coverers in total), and 0 otherwise. Constraint (10.15) insures that u_i will be zero unless two or more coverers of node i have first been achieved.

The second backup covering model does not require primary coverage, but treats both primary coverage and backup coverage as objectives to be maximized, given the availability of p facilities to place on the network. This bi-objective model may be stated as:

$$\text{Obj 1: } Max \quad \left\{ \sum_{i \in I} a_i y_i \right\} \tag{10.17}$$

$$\text{Obj 2: } Max \quad \left\{ \sum_{i \in I} a_i u_i \right\}$$

s.t.

$$u_i + y_i \leq \sum_{j \in N_i} x_j \quad \forall i \in I \tag{10.18}$$

$$u_i \leq y_i \quad \forall i \in I \tag{10.19}$$

$$\sum_{j \in j} x_j = p \tag{10.20}$$

$$u_i, y_i \in \{0,1\}, \text{ and } x_j \text{ integer} \quad \forall i \in I, \ j \in J$$

No new notation is introduced in this model. The first objective is maximum primary coverage and the second objective is maximum backup coverage. Constraints (10.18) and (10.19) together say that first and second coverage cannot both be achieved unless two facilities are within S of node i (10.18) and second coverage cannot be achieved unless primary coverage is first achieved (10.19).

10.4.2 Redundant and Backup Species Covering Problems

The models in the preceding category have been extended to the coverage of species by Malcolm and ReVelle (2001). Again, there are two basic model types, a model that focuses on redundant coverage of species and models that focus on backup species coverage. The translation is relatively immediate with the single exception that no population weights are attached to the coverage of each species.

However, weights reflecting the rarity of a species may be attached to the primary coverage or backup coverage variables. The greater the rarity, the larger the weights which would be attached to primary coverage or backup coverage for a particular species. In addition, some species may be classified into a category called "rare", while others are simply candidates or targets for preservation. Primary coverage for rare species might well be required. Given such a requirement, two objectives which could be traded off against one another are first coverage of the "target" species (which are not rare) and backup or redundant coverage for those species which are classified as rare. Details of these variations as well as possible weighting formulae are described in Malcolm and ReVelle (2001).

10.5 Chance Constrained Covering Models

10.5.1 Chance Constrained Location Covering Models

Siting models with constraints on the reliability of performance are most often applied in settings where servers are dispatched from their bases to demand nodes on an emergency basis. These siting models enforce or try to achieve a desired level of reliability (or probability) that a server or supplier will be available (i.e., free to be dispatched) somewhere in the system within the distance or time standard to respond to a call for service. An example is the siting of ambulance dispatch stations. Ambulances may be located at these stations and the number of vehicles and their positions may be set in such a way that points of demands have a server free within a stated time with a specific level of reliability.

We discuss two types of siting models incorporating the reliability of service provision. In the first model (ReVelle and Hogan, 1989a) the reliability constraint is enforced for every demand node and the least number of servers is sought. This model is called the probabilistic location set covering problem (PLSCP). In this and the subsequent model we assume that each "point" of demand is a sub-region of the area being served and that supply points (points eligible to house a server) are available at the centroid of each of the demand areas; i.e., demand nodes and candidate supply nodes coincide. Also, in this model the server variable can take any positive integer variable, that is, a supply node may contain more than one server, a condition that cannot be considered in reserve design problems where set-aside can occur just once.

If we knew in advance the probability p_j of an individual server being busy at each supply node j, the reliability constraints for each demand node could be written:

$$1 - \prod_{j \in N_i} (p_j^{x_j}) \geq \alpha \quad \forall i \in I \qquad (10.21)$$

where $x_j = 0, 1, 2 \ldots$ is the number of servers at site j. This constraint says that the probability that one or more servers in the set N_i (i.e., within the standard S) is free to be dispatched is greater than or equal to α. Unfortunately, such probabilities cannot be divined without prior knowledge of location, so the constraint is a concept rather than an enforceable statement. In the setting of species preservation, however, such a constraint has more meaning, as we shall see.

As an alternative to (10.21), we can develop the idea of an average busy fraction for servers (e.g., vehicles) in the same sector of the planning region for which the reliability constraint is being written, i.e., within the set N_i. Said another way, if the reliability constraint is being written for demand node i, the servers (vehicles) for which the average busy fraction is calculated are those in N_i, the eligible coverers for demand node i. This average busy fraction is

$$q_i = \frac{t' \times \sum_{k \in M_i} f_k}{24 \times \sum_{j \in N_i} x_j} \qquad (10.22)$$

where:
q_i = the average busy fraction in the region around i;
t' = the average duration of a call in hours;
f_k = the frequency of calls at demand k (calls/day); and
M_i = the set of demand nodes within S of node i.

The numerator of (10.22) is the total time (in hours) spent servicing calls each day in N_i. The denominator is the total service time (in hours) available daily within N_i. The terms in (10.22) that do not involve number of servers (x_j variables) can be lumped together by the following definition.

$$F_i = \frac{t' \times \sum_{k \in M_i} f_k}{24}$$

The chance constraint (10.21) can now be replaced by a different constraint.

$$1 - \left(\frac{F_i}{\sum_{j \in N_i} x_j}\right)^{\sum_{j \in N_i} x_j} \geq \alpha \qquad (10.23)$$

Constraint (10.23) has no exact inverse, i.e., no analytical solution for the number of vehicles required in N_i. However, it does have a numerical solution. For constraint (10.23) to be met, the number of vehicles in N_i must be greater than or equal to b_i, where b_i is defined as the smallest integer satisfying (10.23). The linear equivalent of (10.23) is

$$\sum_{j \in N_i} x_j \geq b_i \qquad (10.24)$$

where b_i is the smallest integer satisfying

$$1 - \left(\frac{F_i}{b_i}\right)^{b_i} \geq \alpha \qquad (10.25)$$

and where, again, x_j is not 0 or 1 but the integer number of vehicles at j.

The probabilistic location set covering problem (PLSCP) seeks the least number of servers in the system subject to coverage being available within S with α reliability. It is stated simply as

$$\text{Min} \left\{ \sum_{j \in J} x_j \right\} \quad (10.26)$$

s.t.

$$\sum_{j \in N_i} x_j \geq b_i \; \forall i \in I \quad (10.27)$$

$$x_j \text{ integer} \quad \forall j \in J$$

where b_i is the smallest integer satisfying (10.25) and x_j is the integer number of servers at j. An interesting tradeoff curve can be developed between the number of servers needed in the system and the reliability level of coverage, α. A comparable tradeoff curve in the species coverings models will be informative as well.

The second type of chance constrained siting model places a limit on the number of servers and seeks, within that limit, to maximize the number of calls for service that can be responded to with α reliability (ReVelle and Hogan, 1989b). That is, the reliability statement is converted from a requirement to a goal which is sought for each demand node. Demand node i needs b_i servers sited within N_i in order to have one or more servers available with α reliability. Hence, the problem becomes one of maximizing the total number of calls, over all demand nodes i, which have b_i servers sited within N_i, where b_i servers in N_i achieve the desired level of reliability α, according to equation (10.25).

The problem is known as the maximum availability location problem (MALP). Two versions of the maximum availability location problem can be offered. A first formulation given in ReVelle (1993) for explanatory purposes is more straightforward, but it is less likely to have an LP formulation terminate in all integers. In non-standard form, it is:

$$\text{Max} \left\{ \sum_{i \in I} f_i v_i \right\} \quad (10.28)$$

s.t.

$$\sum_{j \in N_i} x_j \geq b_i v_i \quad \forall i \in I \quad (10.29)$$

$$\sum_{j \in J} x_j = p \quad (10.30)$$

$$x_j \text{ integer and } v_i \in \{0, 1\} \quad \forall i \in I, \; j \in J$$

where:

x_j = the integer number of servers at j;
f_i = calls per day at demand node i; and
$v_i \in \{0,1\}$ it is 1 if b_i or more servers are within N_i, and 0 otherwise.

Constraint (10.29) states for each i that if the number of servers within N_i is b_i or more, v_i can be one or more. However, because v_i can only be zero or one, it will be one. If the number of servers is less than b_i, then v_i must be a number less than one and hence must be zero. While this formulation is easy to understand, integer termination of the LP is unlikely and branch and bound needs could be extensive. The next formulation is more complex but branching and bounding should proceed more swiftly.

The MALP described in Hogan and ReVelle (1989b) also utilizes the b_i from (10.24) and (10.25) but in a more integer friendly way. The problem statement is:

$$Max \quad \left\{ \sum_{i \in I} f_i y_{ib_i} \right\} \tag{10.31}$$

s.t.

$$\sum_{k=1}^{b_i} y_{ik} \leq \sum_{j \in N_i} x_j \quad \forall i \in I \tag{10.32}$$

$$y_{ik} \leq y_{ik-1} \quad \forall i \in I, \text{ and } \forall k = 2, \ldots, b_i \tag{10.33}$$

$$\sum_{j \in j} x_j = p \tag{10.34}$$

x_j integer and $y_{ik} \in \{0,1\}$

where $y_{ik} \in \{0,1\}$; it is 1 if demand area i has at least k servers in N_i, and 0 otherwise.

Constraint (10.32) says that the number of times that node i is covered within S (the sum of the y_{ik}) must be less than or equal to the number of servers within S (the sum of the x_j). The left hand side of (10.32) can go no larger than b_i even if the number of servers in N_i (the right hand side) exceeds b_i, a reasonable situation. Constraint (10.33) says that the formulation cannot count y_{ik} coverers for i unless y_{ik-1} coverers have already been achieved. The y_{ik} variables are descriptively referred to as "counting variables." (A linear programming solution to this problem should upper bound the y_{ik} at 1). The objective maximizes the sum of the products of calls at i and the variables y_{i,b_i} which indicates whether or not b_i coverers have been achieved within N_i that is, whether or not demand area i is covered within S with α reliability. The formulation described by (10.28) - (10.30) and (10.31) - (10.34) are fully equivalent with the latter being the more integer-friendly.

The model for MALP can be criticized from the standpoint of the independence assumption inherent in the calculation of sector busy fractions. In fact, the sector busy fractions are likely dependent variables. Marianov and ReVelle (1996), show how to use queuing theory to get at server busy fractions, thus correcting the problem of the independence assumption in the original MALP.

10.5.2 Chance Constrained Species Covering Models.

The probabilistic location siting models have interesting counterparts in the species covering models. Although we do not know, at the time of this writing, of anyone who has previously formulated the counterpart of the probabilistic location set covering problem, it is not hard to state. Indeed the constraints of the model we will set up appear in a larger model due to Haight et al. (2000), a model with a quite different objective. We need to define the following:

p_{ij} = probability that species i will survive in parcel j where the term survival means that a desired population size of the species will exist in the parcel at some specified future time; and

$q_{ij} = (1 - p_{ij})$ = probability species i will not survive (be represented) in parcel j.

Alternatively, p_{ij} may be interpreted as the probability that species i will be found in parcel j. The value of parameter p_{ij} may be based on the ecological characteristics and other feature of the parcel. Here, we assume the p_{ij} is independent of the survival of the species in other parcels, or the survival of other species in the same or separate parcels. The probability that species i survives in none of the parcels is:

$$\prod_{j \in J} q_{ij}^{x_j}$$

where x_j is once again restricted to 0 or 1 and is 1 if the parcel is chosen for protection status. If species i is not present in parcel j and is not expected to be (accidentally) introduced or capable of survival in j if it were introduced, then q_{ij} takes a value of 1. The probability that species i survives in one or more parcels is:

$$1 - \prod_{j \in J} q_{ij}^{x_j}$$

If the level of reliability with which species i is desired to survive somewhere in the system of reserves is to be at least α, we can write the chance constraint as:

$$1 - \prod_{j \in J} q_{ij}^{x_j} \geq \alpha \qquad (10.35)$$

or

$$\prod_{j \in J} q_{ij}^{x_j} \leq 1 - \alpha \qquad (10.36)$$

which is the same as saying that the probability of no survival anywhere in the system is less than or equal to (1- α). The reader will easily see that (10.35) is an equivalent form to the chance constraint given by (10.21). We did not pursue (10.21) because of the extreme difficulty of estimating the p_j. Because the constraint (10.36) is straightforward to linearize by taking logarithms, the full probabilistic species covering problem (PSCP) can be written:

$$\text{Min} \quad \left\{ \sum_{j \in J} x_j \right\} \qquad (10.37)$$

s.t.

$$\sum_{j \in J} (\log q_{ij}) x_j \leq \log(1 - \alpha) \; \forall i \in I \qquad (10.38)$$

$$x_j \in \{0, 1\} \qquad \forall j \in J$$

This model seeks the minimum number of parcels such that each of the target species survive with α reliability in at least one parcel in the system. The solution of the 0-1 program represented by (10.37) and (10.38) is not expected to be easy because no special structure exists in the constraints which would favor 0-1 termination. That is, the formulation is not integer friendly in the sense of ReVelle (1993). Extensive branching and bounding could well be needed. As a consequence, heuristic procedures would need to be developed for rapid solution of reasonably sized problems of this form.

Just as the probabilistic species covering problem parallels the probabilistic location set covering problem (PLSCP), a parallel model to the maximum availability location problem (MALP, (10.31) - (10.34)) has been stated. Haight et al. (2000) maximized the number of species that are expected to be represented in a reserve system with α reliability given a limit on the total area of parcels which can be included in the reserve. For simplicity, our binding constraint will be a limit on the number of parcels, p, which can be included in the reserve system rather than a limit on the total area.

In the maximum species covering problem, (10.8), (10.9), (10.10), we use y_i as a 0-1 variable indicating whether or not a species was represented in one or more selected parcels. Here we use y_i to indicate whether or not species i survives in the system with at least α reliability. Haight et al. (2000) used a different level of α for each species. For simplicity we use just one level of α, but the mathematics are the same. No other new notation is needed; the x_j

and y_i are 0-1 variables and the problem may be stated as:

$$\text{Max} \quad \left\{ \sum_{i \in I} y_i \right\} \tag{10.39}$$

s.t.

$$\prod_{j \in J} q_{ij}^{x_j} \leq (1-\alpha)^{y_i} \quad \forall i \in I \tag{10.40}$$

$$\sum_{j \in J} x_j = p \tag{10.41}$$

$$x_j, y_i \in \{0, 1\} \quad \forall i \in I, \quad j \in J$$

The objective (10.39) simply counts the number of species "preserved" with α reliability. Constraint (10.41) limits the number of parcels to p. Constraint (10.40) says that a species is not counted as preserved ($y_i = 1$) unless the probability of it not surviving anywhere in the system is less than or equal to $(1 - \alpha)$. To put numbers in these equations, if the desired probability of survival is 0.95, the probability of not surviving is less than or equal to 0.05.

However, different levels of reliability may well reflect the degree of importance or rarity of the species with higher required levels of reliability possibly applied to those species most endangered. More abundant species may achieve their desired levels of reliability of survival simply as a consequence of the constraints on rare species survival. To see that (10.40) models the desired relation, we take the logarithm of both sides of (10.40), giving

$$y_i \leq \frac{\sum_{j \in J} (\log q_{ij}) x_j}{\log(1-\alpha)} \tag{10.42}$$

We can see that unless the numerator of (10.42) is greater than or equal to the denominator (Constraint (10.36) is satisfied), y_i will be forced to zero, as a 0-1 variable. Of course to solve this problem as a linear integer program, constraint (10.42) would be substituted for (10.40), and the set (10.39), (10.42), (10.41) would be submitted as objective and constraints.

Unlike other maximal covering problems, this one is not particularly integer friendly, requiring extensive branching and bounding to resolve in integers. We note here that if the same α level is applied to all species, that the model will generally choose to "protect" first, those species with the highest likelihoods of survival not necessarily those species which are priorities because of endangerment. In fact, the endangered species could well be ignored because of their decreased likelihoods of survival. Species-specific reliabilities would seem to be appropriate for such a situation. That is, in addition to constraint (10.42), the analyst would append constraint (10.38) with an appropriate species-specific reliability α_i for each species i. This same potential weakness of making the easy choices first will be seen in the next set of species covering models to be discussed shortly in the next section.

There is another aspect of these models that is imperfect and deserves attention in future research. The survival probability of species i in parcel j will likely depend on the protection status of other parcels containing species i as well as the preservation of parcels which contain species that influence the survival of i.

10.6 Expected Covering Models

10.6.1 Expected Covering Location Models

Daskin (1983) originated the maximal expected covering location problem (MEXCLP). Daskin built the model utilizing an estimate of a single system-wide busy fraction. The estimate was based on the system call frequency, on the duration of a call and on the number of vehicles in the system. That is, equation (10.22), expanded to encompass the entire system, was used as the means to calculate the system busy fraction. In fact, equation (10.22) is a particularization of Daskin's original system-wide busy fraction equation.

Daskin's model sought the maximum expected value of population coverage within the time standard, given p facilities to be sited on the network. The model uses the same variables utilized earlier in section 10.5.1. In the model of section 10.5.1, y_{ik} takes a value of 1 if demand i is covered at least k times and is 0 otherwise. Using the system-wide busy fraction, Daskin calculated, for each demand i, the increase in the expected coverage when the number of servers in N_i increases from $k-1$ to k. That increase is calculated in the following way.

The probability of at least one server being free within N_i (coverage achieved) if there are k servers in N_i is equal to $1 - q^k$. The probability of at least one server being free in N_i if there are $k-1$ servers is $1 - q^{k-1}$. The increase in expected coverage in moving from $k-1$ to k servers is $(1-q^k) - (1-q^{k-1}) = (1-q)q^{k-1}$. Hence, the call weighted expected coverage is

$$\sum_{k=1}^{n_i} f_i(1-q)q^{k-1} y_{ik}$$

where n_i is the largest number of servers in N_i. Using this term, we can construct the formulation as

$$\text{Max} \left\{ \sum_{i \in I} \sum_{k=1}^{n_i} f_i(1-q)q^{k-1} y_{ik} \right\} \quad (10.43)$$

s.t.

$$\sum_{k=1}^{n_i} y_{ik} \leq \sum_{j \in N_i} x_j \quad \forall i \in I \quad (10.44)$$

$$\sum_{j \in J} x_j = p \quad (10.45)$$

$$x_j, y_{ik} \in \{0, 1\}$$

The objective (10.43) is the call-weighted expected coverage of all demands in the system. Constraints (10.44) say that the coverage counting variables for demand i are limited by the number of coverers sited in N_i. The constraints (10.33) from the model in section 10.5.1 are not needed to make the y_{ik} enter in the correct order in MEXCLP because the weights $[(1-q)q^{k-1}]$ decline as k increases, causing y_{1k} to enter before y_{2k}, y_{2k} before y_{3k}, etc. Although Daskin utilized a heuristic for solution to MEXCLP, experience with MALP (10.31) - (10.34), a strikingly similar model, suggests that this model should be integer friendly to an LP/ Branch and Bound approach. Although this model uses only a single system busy fraction, it remains an elegant representation of an interesting problem. The model could be solved iteratively with new solutions that utilize q_i's calculated by equation (10.22) using the number of servers in N_i from the prior solution.

10.6.2 Species Expected Covering Models

Once again, a species model exists which parallels the location model, in this case paralleling MEXCLP. This model, originated by Polasky et al. (2000), maximizes the expected number of species "covered". Here, coverage of an individual species may be measured by: (1) the probability of representation in the system given probabilities of the species being found in each parcel (Polasky et al., 2000); or (2) the probability of the species surviving at one more parcels in the system. The reserve network may have either (1) an upper limit on the number of parcels that can be selected, or (2) an upper limit on the investment that can be made in the network, given parcels with different costs, the latter situation being the one described by Polasky et al. (2000).

The objective, which uses the p_{ij} defined earlier, may be written in two different but equivalent forms. The first uses the expression for species survival developed in section 10.5.2:

$$\text{Max}\left\{\sum_{i \in I}(1 - \prod_{j \in J}(1-p_{ij})^{x_j})\right\}$$

which can be written (equivalently) as:

$$\text{Min}\left\{\sum_{i \in I}(\prod_{j \in J}(1-p_{ij})^{x_j})\right\} \qquad (10.46)$$

The second form, which is the one used by Polasky et al. (2000), represents the probability of species i survival or representation as

$$1 - \prod_{j \in J}(1 - p_{ij}x_j)$$

where the product term is the probability of no survival of i anywhere in the system. The objective would be written as:

$$Max\left\{\sum_{i\in I}\left(1-\prod_{j\in J}(1-p_{ij}x_j)\right)\right\}$$

or

$$Min\left\{\sum_{i\in I}\prod_{j\in J}(1-p_{ij}x_j)\right\} \quad (10.47)$$

Both (10.46) and (10.47) involve the sum of product terms and do not appear to be linearizable. However, either (10.46) or (10.47) can be optimized by heuristic methods. The complete problem would have either (10.46) or (10.47) optimized subject to either

$$\sum_{j\in J}x_j = p \quad (10.48)$$

or

$$\sum_{j\in J}c_j x_j \leq B \quad (10.49)$$

where:

$c_j =$ the cost of parcel j, and
$B =$ an overall budget limit.

Polasky et al. (2000) briefly describe a greedy algorithm which adds a site to the previous set of sites in a greedy way. Other more potent heuristic procedures based on the metaheuristics of combinatorial optimization, such as tabu search, genetic algorithms, etc., are likely to be described in the future as researchers come to grips with this fundamental problem. These procedures are anticipated to lead to better objective values than those found by greedy techniques.

Polasky et al. (2000) note that it is possible to weight species by taxonomic uniqueness or contribution to the preservation of genetic diversity. We would suggest weighting by rarity as well, since, if we were able to optimize exactly the above problem without weights and without parcel costs, the solution would choose first those species most likely to survive and last those species with low likelihood of survival.

Perhaps more appropriate than weighting to reflect rarity would be either of two extensions. In the first extension, constraints (10.38), requiring a stated probability of survival or representation, could be added to the problem for the endangered species. Solution becomes problematic, however, as the greedy heuristic no longer applies. A second extension would meld the approach of Haight et al. with that of Polasky et al. (2000) to create a two-objective problem. The first of the two objectives would be objective (10.46) or objective (10.47) (Polasky et al., 2000) which maximizes the expected number of species represented or expected to survive. This objective would be optimized over all species. The second objective would be to maximize the number of species preserved with α reliability, using (10.39) and (10.42), where the species are those in the endangered category. Thus, the trade-off in this bi-objective problem is between the expected number of all species preserved and the number of endangered species preserved with α reliability. This problem does not appear to be beyond solution by heuristic means.

As in section 10.5.2, the survival probability of one species is assumed to be independent of the survival probability of other species as determined by the parcels chosen for the reserve network. And the probability of a particular species surviving in some parcel is assumed independent of its survival elsewhere in the system.

10.7 Conclusion

We have described five fundamental classes of problems from location science and their counterparts from the literature of reserve design. Parallels in formulation and problem statement of these combinatorial optimization problems are striking, but solution methods, it is noted, may vary in effectiveness in the two areas. Ideas from the more mature field of location science may offer possibilities for both new formulations and new methods of solution. We also note that no spatial relations of parcels were considered in this review. For instance, the potential of protecting a preserved parcel with a buffer of preserved parcels, or insisting on contiguity, proximity, or desired aggregate shape characteristics of preserved parcels was not examined.

References

Berlin, G., (1972) "Facility Location and Vehicle Allocation for Provision of an Emergency Service," Doctoral Dissertation, Johns Hopkins University, Baltimore, MD.

Chung, C-H., (1986) "Recent applications of the maximal covering location planning (MCLP) model," *Journal of the Operational Research Society*, 37, 735-746.

Church, R., D. Stoms, and F. Davis (1996) "Reserve selection as a maximal covering location problem," *Biological Conservation*, 76, 105-112.

Church, R. and C. ReVelle, (1974) "The maximal covering location problem," *Papers of the Regional Science Association*, 32, 101-118.

Church, R. and C. ReVelle, (1976) "Theoretical and computational links between the p-median, location set covering, and the maximal covering location problem," *Geographical Analysis*, 8, 406-15.

Daskin, M. and E. Stern, (1981) "A hierarchical objective set covering model for emergency medical service deployment," *Transportation Science*, 15, 137-52.

Daskin, M., (1983) "A maximum expected covering location model: formulation, properties, and heuristic solution," *Transportation Science*, 17, 48-70.

Densham, P. and G. Rushton, (1992) "A more efficient heuristic for solving large p-median problems," *Papers in Regional Science*, 71, 307-329.

Dwyer, F. and J. Evans, (1981) "A branch and bound algorithm for the list selection problem in direct mail advertising," *Management Science*, 27, 658-67.

Eaton, D., Sanchez, U., Mi, H., Lantigua, R. and J. Morgan, (1986) "Determining ambulance deployment in Santo Domingo, Dominican Republic," *Journal of the Operational Research Society*, 37, 113-26.

Haight, R., ReVelle, C. and Snyder, S., (2000) "An integer optimization approach to a probabilistic reserve site selection problem," *Operations Research*, forthcoming.

Hillsman, E., (1984) "The p-median structure as a unified linear model for location-allocation analysis," *Environment and Planning A*, 16, 305-318.

Hogan, K. and C. ReVelle, (1985) "Concepts and applications of backup coverage," *Management Science*, 32, 1434-1444.

Kolesar, P., (1980) "Testing for vision loss in glaucoma suspects," *Management Science*, 26, 439-448.

Malcolm S. and C. ReVelle (2001) "Models for Preserving Species Diversity with Backup and Redundant Coverage," in submission.

Margules, C., Nichols, A. and R. Pressey, (1988) "Selecting networks of reserves to maximize biological diversity," *Biological Conservation*, 43, 63-76.

Marianov, V. and C. ReVelle, (1996) "The queueing maximum availability location problem: a model for the siting of emergency vehicles," *European Journal of Operational Research*, 93, 110-120.

Polasky, S., Camm, J., Solow, A., Csuti, B., White, D. and R. Ding, (2000) "Choosing reserve networks with incomplete species information," *Biological Conservation*, 94, 1-10.

Possingham, H., Day, J., Goldfinch, M., and F. Salzborn, (1993) "The mathematics of designing a network of protected areas for conservation," in *Decision Sciences, Tools for Today*, Proceedings of the 12th Australian Operations Research Conference, eds. D. Sutton, E. Cousins, and C. Pierce (ASOR, Adelaide, 1993).

ReVelle, C. and K. Hogan, (1989a) "The maximum reliability location problem and alpha reliable p- center problem: derivatives of the probabilistic location set covering problem," *Annals of Operations Research*, 18, 155-174.

ReVelle, C. and K. Hogan, (1989b) "The maximum availability location problem," *Transportation Science*, 23, 192-200.

ReVelle, C., Williams, J. C., and Boland, J. (2001) "Counterpart models in location science and reserve selection science," *Environmental Modeling and Assessment*, forthcoming.

ReVelle, C., (1993) "Facility Siting and Integer-Friendly Programming," *European Journal of Operational Research*, 65, 147-158.

ReVelle, C., Toregas, C., and L. Falkson, (1976) "Applications of the location set covering problem," *Geographical Analysis*, 8, 65-74.

Rosing, K. and C. ReVelle, (1997) "Heuristic concentration: two stage solution construction," *European Journal of Operational Research*, 97, 75-86.

Schilling, D., Rosing, R., and C. ReVelle, (2000) "Network distance characteristics that affect computational effort in location problems," *European Journal of Operational Research*, 127, 525-536.

Teitz, M. and P. Bart, (1968) "Heuristic methods for estimating the generalized vertex median of a weighted graph," *Operations Research*, 16, 955-961.

Toregas, C. and C. ReVelle, (1973) "Binary logic solutions to a class of location problems," *Geographical Analysis*, 5, 145-155.

Toregas, C., Swain, R., ReVelle, C. and L. Bergman, (1971) "The Location of Emergency Service Facilities," *Operations Research*, 19, 363.

Underhill, L., (1994) "Optimal and suboptimal reserve selection algorithms," *Biological Conservation*, 35, 85-87.

11 Facility Location Problems with Stochastic Demands and Congestion

Oded Berman[1] and Dmitry Krass[1]

Joseph L. Rotman School of Management, University of Toronto, 105 St. George Street, Toronto, Ontario, Canada M5S 3E6. e-mail: berman@rotman.utoronto.ca; krass@mgmt.utoronto.ca

11.1 Introduction

In this chapter we address the problem of finding optimal locations for a set of facilities in the presence of stochastic customer demand and potential congestion at the facilities. We refer to this class of problems as "Location Problems with Stochastic Demand and Congestion" (LPSDC). For the most part, we will restrict the discussion to problems on networks, even though the models can often be extended to discrete location setting (by introducing an alternative distance measure) without much difficulty.

The obvious importance of addressing facility location problems in the presence of various uncertainties has lead to a large body of literature on the subject. LPSDC models primarily concentrate on two sources of uncertainly: (1) the actual amount and timing of the demand generated by each customer location, and (2) a possible loss of demand (or a monetary penalty) due to facility's inability to provide "adequate" service to (some of the) customers due to congestion at the facility.

The problem is to find the best locations for a set of facilities, as well as to determining the service capacity k_j ("number of severs") to be stationed at facility j. The resulting system can thus be viewed as a spatially distributed queuing system with M queues, and $\sum_{j=1}^{M} k_j$ (possibly distinguishable) servers. Even the descriptive analysis of such systems (i.e., assuming the locational decisions have already been made) can stretch the current capabilities of queuing theory. The "exact" original problem combines the complications of "classical" location problems (most of which are known to be NP-complete) with the complicated dynamics of queuing systems - resulting (nearly always) in "intractable" models. Thus, in building a 'practical' LPSDC model one invariably has to make simplifying assumptions and approximations to render the model tractable. We will attempt to review the main modelling approaches that have been developed for this problem, the structure of assumptions underlying these models, and the implications of these assumptions.

One important area of applications for LPSDC models is that of location of emergency service facilities - such as hospitals, police stations, fire stations and ambulances. The ability to respond to a call for service within adequate time is particularly pertinent in such problems, often recognized in legislature governing operation of emergency service facilities (for example, a common North American standard for ambulances is to be able to respond to top-priority calls within 3 minutes). The unpredictability of the number and arrival times of calls for service, and the effect on system performance of congestion that results when some of the facilities receive too many calls for service during a certain time period are also basic features of the underlying problem. Indeed, historically, the problem of siting emergency service facilities has provided the main impetus for most of the research in this area.

The other important, but less well analyzed, area of applications is location of retail outlets or other service facilities, where the amount of total business (customer demand) at a facility may be adversely affected when the service rate slows down due to congestion. While some of the models developed for location of emergency facilities can be applied to non-emergency settings as well, these two classes of applications do induce different features in the underlying problems, resulting in different modeling requirements. We will discuss these differences in greater detail in the sequel.

We should note that throughout this chapter we have chosen to focus primarily on multiple facility/service unit models - i.e., the ones capable of handling several facilities, each with, possibly, several servers. While models developed for locating only a single service unit often serve as building blocks for these more general models, we chose not to focus on the results that do not easily translate to more general settings (we refer the reader to Berman *et al* (1990) for a review of some of the key results when the number of servers equals to one). We stress that the objective of this chapter is not to provide a comprehensive review of all work in the area, but to select a sample of what we believe to be the main directions of research.

In the following sections we give a more detailed introduction to the underlying general model, discuss some the main types of modeling approaches, and outline some of the differences and similarities between models for emergency and non- emergency services.

11.1.1 Components of the General Model

LPSDC models are primarily concerned with interactions between four sets of elements:

 – customers - who generate calls for service,
 – facilities - which house resources ("servers") needed to provide the service,
 – servers - that actually provide the required service, and
 – call for service - that originate with the customers and are handled by matching a customer with an available server.

The other components necessary to describe an LPSDC model are: types of service provision (do customers travel to facilities to obtain servers, or do mobile servers travel to customer locations?), nature and consequences of congestion (what happens when a facility receives too many request for service?), customer behavior assumptions (do customers decide which facility to obtain service from, or is there a "central authority" that matches customers with facilities), type of objectives, and other special requirements (typically expressed as constraints) such as "coverage standards". We provide a brief description of each of these components.

Throughout we assume a given network $G = (N, A)$, where N is the node set and A is the arc set. For $x, y \in N$ we use $d(x, y)$ to denote the shortest-path distance from x to y.

1. **Customers** Customers are assumed to be located at the nodes of the network. We assume that proportion h_i of all calls for service originates from node $i \in N$, with $\sum_{i \in N} h_i = 1$. It is typically assumed that the total customer demand for service is a time-homogeneous Poisson process with rate λ. It follows that the service call process for each node i is a Poisson process with rate $\lambda_i = h_i \lambda$. While most models use the customer demand structure described above, there have been some attempts to incorporate possible loss of demand due to congestion. This can be accomplished by re-defining the demand rate at node i to be $\bar{\lambda}_i = \lambda_i h(C)_i$, where C is some measure of the congestion cost incurred by the customer, and $h(\cdot)$ is a non-increasing function. Unless specifically noted otherwise, throughout the remainder of this chapter we will generally assume that λ_i is not affected by congestion.

2. **Facilities** We will assume that a maximum of M facilities are to be located. We assume that a discrete set of potential facility locations X has been identified (with $M < |X|$), and that $X \subset N$. This assumption is made without loss of generality: following arguments presented in Berman, Larson and Chiu (1985), it can be shown that if it is allowed to locate facilities anywhere along the arcs of the network, an optimal solution can be found within a discrete set of locations consisting of nodes of the network augmented by some interior points along the arcs. Thus, by augmenting the original node set with some additional "dummy" nodes, we can assume that X is nodal.

3. **Servers** Each facility j can house between 1 and K servers. Depending on the nature of service provided by the facility, the servers are either be *fixed* - i.e., permanently housed at the facility, or *mobile* - traveling to customer's location to provide service. The number of servers $k_j \in 1, \ldots, K$ housed at the facility j is a decision variable in the model.

4. **Calls for Service** A call for service generally requires a "match up" between the originating customer and one of the available servers on the system. This is normally accomplished as follows:
   – First, it is necessary to determine whether location i is "covered" by the system. It is usually assumed that in order for a customer to be "covered", certain *coverage standards* have to be met (e.g., sufficient

number of servers must be located in the vicinity of the customer, etc.). These coverage standards often arise from legislative or administrative rules. If customer location i is not "covered", all service calls originating from i are automatically rejected by the system (irrespective of whether the system is currently congested or not). There is typically a penalty for failing to cover location i. An alternative interpretation of not extending coverage to a customer is that this customer is covered by some "reserve" or "alternative" service (e.g., a private ambulance service); the "no coverage" penalty can then be interpreted as a subcontracting fee.

- Once it has been determined that the call for service came from one of the "covered" customers, an evaluation is made on whether the current state of the system allows it to process the call. This evaluation often proceeds in two stages: first, *districting rules* and customer's location are used to determine customer's "subsystem" - i.e., which facilities and servers could potentially process this call (this may involve all servers on the network, or only servers based at the facilities located within certain travel radius of customer's location, etc.). Next, the number of outstanding calls in this subsystem is evaluated, and the decision is made on whether to accept or reject the call. This decision is usually based on the queuing capacity of the subsystem (e.g., for a "loss" queue a rejection might occur if no servers are currently available; in other cases there might be a limit to how many calls can be queued at any given time). There is normally a penalty associated with rejecting a call. We underscore that the rejection of a call from a "covered" customer is based on the state of the system - as opposed to rejection of calls from "non-covered" customers, which is automatic. Note that districting rules determine the degree of co-operation between various facilities and servers in the system.

- Next, the accepted call is *routed* (i.e., assigned) to one of the facilities. This assignment may depend on certain *routing rules*, as well as on the current state of the system (e.g., a call might always be routed to the closest facility, or the call might be routed to the closest facility with at least one free server, etc.). The routing rules also depend on customer behavior assumptions - i.e., whether the choice of which facility should process the call is up to a customer, or up to some central authority. We refer to the case where the customer decides which facility should process the call as *user choice*, and to the case a central authority makes this decision as *directed choice*.

- An accepted call at a certain facility is generally placed in queue until a server becomes available. Once this happens, the server and the customer are "matched up". In case of mobile servers, this requires the server to travel from their current position to the customer location (incurring *travel cost*), followed by on-scene service (incurring "on-scene" cost). Upon completion of on-scene service, the server might either return to its "home" facility (if no calls are in queue, or if certain "of-scene" service is required as part of servicing a call), or travel directly to the location of next call. Customer's total response time thus consists of two components: the *waiting time* - incurred while the call is waiting in queue (if the call has to wait at all), and the *travel time* between customer's and server's locations (thus the response time is counted from the moment the server or customer begin their travel until the start of the on-scene service period). The *waiting cost* is charged on the total response time of each call. Note that in case of mobile servers, while the travel time back to the facility and off-scene service time are not charged directly as part of the cost structure, they certainly affect server availability and thus - indirectly - system costs.

11.1.2 Formulation of the General LPSDC

Based on the description of model components above, we now provide a general formulation for the LPSDC problem. The primary goal here is to provide a formulation of sufficient generality so that all main LPSDC models described in subsequent sections can be viewed as special cases. This also allows us to introduce some of the required notation and formalize some of the key concepts discussed above. On the other hand, we will not worry about the tractability of this general formulation - its main purpose being illustrative, rather than practical.

The primary decision variables in the model are:

$$x_j = \begin{cases} k \text{ if there is a facility at location } j \text{ and } k \text{ servers have} \\ \quad \text{been allocated to this facility, where } k \text{ is a positive integer} \\ 0 \text{ if no facility is located at } j \end{cases} \quad j \in X$$

and

$$y_i = \begin{cases} 1 \text{ if customer i is covered by one of the facilities} \\ 0 \text{ otherwise} \end{cases} \quad i \in N$$

We use **x** to represent the facility location vector, and **y** to represent the coverage vector.

Next we define the components of the objective function:

- Let NC be the cost of not providing coverage to customer at location i (assessed on per service call basis). Then the total "no-coverage" cost is given by

$$TC_{NC} = \sum_{i \in N} NC \lambda_i (1 - y_i)$$

(recall that $\lambda_i = \lambda h_i$ represents the rate of calls for service originating from location i).

- Let RC be the cost of rejecting a call from customer i assuming this customer is covered. Suppose that given a location vector $\mathbf{x}$ and a coverage vector $\mathbf{y}$, the probability of rejecting a call from customer i is $P_R^i(\mathbf{x}, \mathbf{y})$ (note that this quantity depends on both the location of the facilities, and on the utilization of the facilities - thus it is a function of both $\mathbf{x}$ and $\mathbf{y}$). Then the total "call rejection" cost is

$$TC_{RC} = \sum_{i \in N} RC P_R^i(\mathbf{x}, \mathbf{y}) \lambda_i y_i$$

Obviously, the probability $P_R^i(\mathbf{x}, \mathbf{y})$ also depends on the districting rules and other operational characteristics of the system (such as on-scene and off-scene service time distributions, whether servers are mobile or fixed, etc.) This quantity is usually non-linear in both the location vector and in the coverage vector. When rejected calls are handled by "reserve units" that are also handling calls from non-covered customers, $RC = NC$ can be assumed.

- Let WC be the waiting cost per call per unit time. The waiting time consists of two components: the time call spends waiting in queue ("queue waiting time") and the expected travel time between customer's and server's location. Let $T_W^i(\mathbf{x}, \mathbf{y})$ be the expected queue waiting time for call originating from i given facility location vector $\mathbf{x}$ and coverage vector $\mathbf{y}$ (where the expectation is conditional on the call not being rejected by the system). Also, let $P_D^{ij}(\mathbf{x}, \mathbf{y})$ be the probability that the call originating at i is served by a server with home location at j, given that this call is not rejected by the system. We note that both of the above quantities depend on districting and routing rules, as well as on the operational characteristics of the system. Assuming average travel speed v and recalling that $d(i, j)$ is the shortest-path distance between locations i and j, the total waiting cost is given by

$$TC_{WC} = \sum_{i \in N} WC(1 - P_R^i(\mathbf{x}, \mathbf{y})) \lambda_i y_i [T_W^i(\mathbf{x}, \mathbf{y}) + \frac{1}{v} \sum_{j \in X} P_D^{ij}(\mathbf{x}, \mathbf{y}) d(i, j)]$$

- Let LC^{jk} be cost of locating k service units at location j. Then the total facility location cost is given by

$$TC_{LC} = \sum_{j \in X} \sum_{k=1}^{K} LC^{jk} 1\{x_j = k\},$$

where, as mentioned earlier, K is the maximum number of servers that can be located and $1\{\cdot\}$ is the indicator function. This cost component is often replaced by a budgetary constraint.

The objective function of the "generic" LPSDC model can now be written as

$$TC(x, y) = TC_{NC} + TC_{RC} + TC_{WC} + TC_{LC}$$

We next turn our attention to constraints of the problem. Typically these include:

- An upper limit M on the total number of facilities that can be located:

$$\sum_{j \in X} 1\{x_j > 0\} \leq M \qquad (11.1)$$

- An upper limit K on the total number of servers that can be located:

$$\sum_{j \in X} x_j \leq K \qquad (11.2)$$

- *Coverage standards.* These can take a variety of forms, depending on the coverage requirements being used. Perhaps the simplest (and oldest) form of these constraints simply requires that a certain minimum number of servers b_i must be positioned within certain maximum distance of each customer location i. Let N_i be the subset of potential facility location sites within the required distance of i. Then the constraint can be expressed as:

$$\sum_{j \in N_i} x_j \geq b_i y_i. \quad i \in N \qquad (11.3)$$

A more sophisticated form of the coverage constraint may impose probabilistic requirements on response times. For example, consider a typical three-minute response time requirement for top-priority ambulance calls. The constraint might require that this three-minute response time limit be met at least 95% of the time for all calls. Another form of constraint might impose an upper limit on the proportion of the rejected calls P_R^i. Schematically, we can represent a general constraint as follows. Let $SL_i(\mathbf{x}, \mathbf{y})$ be a random variable represented the "service level" delivered by the system to customer demand point i, where the "service level" is tailored to a particular coverage standard constraint (e.g., response time). We assume that the stochastic process describing our system is sufficiently well-behaved that a steady-state limiting distribution of this service level measure exists, and that $SL_i(\mathbf{x}, \mathbf{y})$ is a random variable with this distribution. Let u_i be the desired bound on the service level (e.g., 3 minutes) - that is we want to insure that the event $SL_i(\mathbf{x}, \mathbf{y}) \leq u_i$ occurs sufficiently frequently. Let α_i represent the minimum desired frequency of this event (e.g., 95% of the time). Then a general service level constraint can be represented by

$$P[SL^i(\mathbf{x}, \mathbf{y}) \leq u_i)] \geq \alpha_i y_i \quad i \in N \qquad (11.4)$$

While this form of the constraint looks forbidding, it is often possible to determine the "certainty equivalent" of this constraint - which, under some simplifying assumptions on the system's operations, frequently turns out to be of the form (11.3) (where the parameter b_i would be computed during the pre-processing). This approach will be discussed in more detail in the sequel.

The general LPSDC problem can now be formulated as:

$$min\ TC(\mathbf{x}, \mathbf{y}) \tag{11.5}$$

Subject to
Constraints (11.1), (11.2), and (11.4)

$$x_j \in 0, \ldots, K,\ y_i \in 0, 1.$$

Obviously, in order to make the formulation above tractable, we require some way to represent the various system performance parameters used in developing the objective function and the constraints (i.e., rejection probabilities $P_R^i(\mathbf{x}, \mathbf{y})$, queue waiting times $T_W^i(\mathbf{x}, \mathbf{y})$, etc.). Unfortunately, general analytic expressions for these quantities are typically unavailable. This leads to two possible approaches: the first approach requires making certain simplifying assumptions on the system operations (such as simple districting rules, negligible travel times, etc) - under which the analytical forms of these expressions (or, at least, approximations of these expressions) can be derived. The second approach involves using descriptive-based techniques (e.g., simulation) to compute required system performance measures for particular values of the location vector $\mathbf{x}$, coupled with some heuristic technique to select new "trial" values of $\mathbf{x}$; the process is repeated until some acceptable solution is achieved. Both approaches are described in subsequent sections.

11.1.3 Classes of Models: Coverage vs. Median-Type Models

Two popular classes of models in the deterministic location literature are Coverage-Type and Median-Type problems. These two classes of models have direct parallels in LPSDC literature as well.

Coverage-Type problems are generally concerned with providing adequate coverage to a customer, rather than with minimizing travel-related costs. In deterministic models, "adequate coverage" typically means that there is at least one facility within a specified maximum distance of the customer. Using N_i to represent the number of all facility locations that "adequately cover" customer location i, the "adequate coverage" requirement translates into constraint (11.3) above with $b_i = 1$ for all $i \in N$.

There are two subclasses of coverage problems: Set Cover-type problems and Maximal Weight Cover-type problems. In the deterministic set cover location problem, every customer location must be covered and the objective is to minimize the total number of facilities. In the LPSDC (i.e., stochastic) environment this translates into a model where the objective function terms

TC_{NC}, TC_{RC}, and TC_{WC} are dropped (the term TC_{RC} is dropped because it is assumed that adequate coverage is assured by the constraints). In principle, variables **y** can be dropped from the formulation as well, even though in practice it is sometimes more convenient to retain them. It can be seen that the set cover framework in the stochastic context significantly simplifies the objective function of the LPSDC, which becomes linear. The non-linearity of the feasible region induced by constraint (11.4) remains. However, as mentioned earlier, under some additional assumptions it is possible to linearize these as well, yielding a linear IP formulation very similar in structure to its counterpart for the deterministic set cover location problem. Unfortunately, the assumptions required to achieve this linearization are often quite strong.

The second sub-class of deterministic coverage-type problems is Maximal Weight Cover models. Here there is no requirement that every customer must be covered, but the set of facilities is fixed. The objective is to maximize the total weight (demand) of covered customers. In the equivalent LPSDC (stochastic) formulation, terms TC_{RC}, TC_{WC}, and TC_{LC} are dropped from the objective function, and constraints (11.1), (11.2) are converted to equalities. Once again, it can be observed that the objective function becomes linear, and the LPSDC can be formulated as a linear IP, provided we succeed in linearizing constraints (11.4). Both classes of coverage-type LPSDC models are discussed in Section 11.2 below.

Median-type deterministic location problems seek to minimize the average (or total) travel cost of all customers. It is usually assumed that the number of facilities is fixed, all customers are "covered" automatically (i.e., there are no explicit "adequate coverage" constraints), and each customer is served by the closest facility. Translated to the stochastic context, this implies dropping terms TC_{NC} and TC_{LC} from the objective function of the LPSDC model, converting constraints (11.1) and (11.2) to equalities, and dropping variables **y** (along with constraints 11.4)). The resulting formulation has linear constraints, but highly non-linear objective function. Consequently, only heuristic solution approaches are available for this class of models. Median-type LPSDC problems are described in Section 11.3 below.

One interesting sub-class of median-type LPSDC models involves models with "elastic" demand. In these models the demand originating from customer location i is a decreasing function of the total "congestion cost" experienced by customers from i, where "congestion cost" is defined as the sum of rejection cost TC_{RC} and waiting cost TC_{WC}. These models (which are especially appropriate in non-emergency settings) were discussed extensively by Brendau et al (1998). Unfortunately, these models (which also include competitive aspects) lead to very difficult formulations, making them impractical for realistic-size problems.

11.1.4 Mobile vs. Fixed Servers

As mentioned earlier, one of the key design parameters in LPSDC models is the nature of the servers - i.e., whether the servers are mobile or fixed. Our general formulation of the LPSDC model applies to both cases. However, the performance characteristics of the system required for computing the objective function and the left-hand side of the constraint (11.4) change drastically depending on the nature of servers.

To appreciate this difference, consider a system with one facility and K servers. Suppose the system has queue capacity of 1 (i.e., at most one call may wait in queue at any given time). Assume the on-scene service has exponential distribution. Consider first the fixed server case. Here the facility operates as a $M/M/K/1$ queue, and all performance characteristics (such as expected wait, proportion of lost calls, etc.) can be readily computed from the standard queuing formulas. Thus, it is possible to write down an analytical formulation of the exact LPSDC model, which will, of course, have non-linearities in both the objective function and the constraints.

Suppose now that the servers are mobile. Assume that whenever a call for service arrives, the server travels from the facility to the customer's location, provides on-scene service (exponentially distributed) and then returns back to the facility before it can be dispatched to the next call. Assume further that travel times are deterministic. In this case, the total service time is a sum of the exponential on-scene service time, travel time from the facility to the customer, and the travel time back to the facility (the latter two components might be different if the outbound and inbound travel speeds are different). It can be seen that the distribution of travel times is no longer exponential - therefore the facility operates as a $M/G/K/1$ system. We note that no analytical formulas are available for such systems, and thus the analytical formulation of LPSDC model cannot be obtained (however, there are several approximations available that can be used to obtain an approximate formulation - see Section 11.3 for more details on this approach).

Now suppose that there are two different facilities, one housing $k < K$ servers, and the other $K - k$ servers. As before, we assume that a server always returns to its "home" facility after completing the call (thus the travel always starts from the facility). Assume that the closest available server is always dispatched to the call. Now the distribution of service times depends on which server is dispatched to the call (or, more precisely, where the server is dispatched from). We are thus dealing with an $M/G/K/1$ system with *distinguishable servers* - for which even approximate analytical results are unavailable. Thus, solution of the LPSDC model requires a combination of a numerical (e.g., simulation) technique to evaluate system performance coupled with a search procedure to identify the best location.

We note that the situation is complicated still further if the assumption that travel always starts from the facility is dropped (i.e., if servers are allowed to travel directly from one call location to another). In this case, in addition to having distinguishable servers, we must also take into account that service times for successive calls are no longer independent.

¿From the preceding discussion, it should be clear that the nature of LPSDC model changes drastically depending on whether the servers are fixed or mobile. Fixed-server case tends to be easier (even though the difficulties it presents are still formidable). The mobile-server case nearly always requires strong simplifying assumptions and/or heuristic solution approaches. The approaches described in Sections 11.2 and 11.3 below include both fixed and mobile server models.

The directed choice (i.e., centralized dispatching) vs. user choice (customers selects the facility that will process their call) customer behavior assumptions is also related to the types of servers used in the model. In case of mobile servers, directed choice assumption is appropriate in most cases (and employed in all published models). In case of fixed servers, the directed choice assumption is less realistic, but is still made in some of the literature (primarily because it simplifies the analysis).

Finally, we note that the two principle classes of applications of LPSDC models - emergency vs. non-emergency servers are also related to the server type and customer behaviour assumptions. Models developed for emergency services applications (locations of fire, police and ambulance stations) typically have mobile servers, directed choice, and well defined (usually probabilistic) coverage standards. Non-emergency service systems typically involve fixed servers, and user choice. Coverage standards may be absent entirely (i.e., all customers are automatically covered) or take deterministic form (all customers within a certain radius of the facility are covered).

The plan for the rest of the Chapter is as follows. After discussing Coverage-Type and Median-Type LPSDC models in sections 11.2 and 11.3, respectively, we present some concluding remarks and discussion of future research in Section 11.4.

11.2 Coverage Problems with Stochastic Demand and Congestion

In this section we discuss Coverage-type LPSDC problems. As noted in the Introduction, the models discussed in this section are based on two classical deterministic location problems: the Set Coverage location problem and the Maximal Weight Cover problem. We refer the reader to Chapters 2 and 3 of the current volume for further discussion of these models. These two classes of models are closely related in terms of both, their formulations and the solution approaches. This relationship carries over to the stochastic versions of these models as well - conceptually, the transformation from a Set Coverage LPSDC model to a corresponding Maximal Weight Cover LPSDC model and vice versa are usually straightforward.

A more fundamental difference between various coverage-type LPSDC models is in the nature of service provided - i.e., the difference between fixed and mobile servers. As explained earlier, the underlying system performance characteristics change drastically depending on whether servers are mobile or fixed. On the other hand, many of the models described in the sequel use only a rough approximation of the actual system performance, thus obscuring this difference somewhat. However, one key difference that always remains between models with fixed and mobile servers is in the assumption on service provision: in mobile server models, different calls from the same customer location might be served by units housed at different facilities. Thus, for a call from location i, all service units stationed in N_i are potentially able to provide service (recall that N_i is the set of facility locations within a specified maximum distance of i). In other words, the assignment of servers to customers is dynamic - it depends on the conditions (e.g., which servers are available) at the time when the call arrives. In the case of fixed servers, it is commonly assumed that each customer at i selects (either of their own accord, or as a result of some centralized allocation system) a service facility in N_i, which then handles all calls from this customer. Thus, the assignment of customers to facilities is static - made once and remaining in force from that point on (or, at least, over a significant time interval). While one can certainly think of scenarios where servers are fixed but customer assignments are dynamic (e.g., if customers have to call some centralized number every time they want to obtain service, and are then directed to the least congested facility), these situations occur infrequently in practice. Models with mobile servers (or, more precisely, with dynamic customer assignments) are described in Section 11.2.1 below, while those with fixed servers (static assignments) are covered in Section 11.2.2.

Another difference between various coverage-type LPSDC models is the underlying probabilistic mechanism used to describe congestion. Here, the models can be broadly divided into two groups: models with "explicit" versus "implicit" congestion. The former group applies queuing-based techniques to explicitly represent congestion at service facilities. The second group is somewhat less ambitious, using only approximate techniques to describe the congestion (which often requires additional parameters - such as estimates of utilization rates of the servers - that the "explicit" models generate internally). The demarcation line between these two model classes is somewhat blurry, with the more advanced "implicit" models approaching the "explicit" description of congestion.

11.2.1 Models With Dynamic Server Assignments (Mobile Servers)

As discussed above, the models described in this section assume dynamic assignment of servers to calls for service: that is, any server which is available and which is located within certain maximum distance of customer's location can provide service. Moreover, all models in this section employ the same coverage criterion: the probability that a service call finds an available server must be sufficiently high.

Most of the models in this section can trace their origins to the Maximum Expected Covering Location Problem (MEXCLP) developed by Daskin (1983), which provides a natural starting point for our discussion.

Consider a Coverage-type LPSDC model where the service level requirements are based on the proportion of calls for service to which a server can be dispatched immediately (i.e., for which the waiting time in queue is zero). Assume that for customer location i, service calls can only be processed by units stationed within a certain maximum distance from i. As in (11.3), let N_i be the potential facility locations within required distance of i. To consider the probability that a server may not be available when a demand call occurs, Daskin (1983) makes 3 simplifying assumptions: (i) severs operate independently from one another—thus the probability that server j is busy, is independent of the probability that server k is busy for $k \neq j$; (ii) the probability that a server is busy is identical for all servers; (iii) the server busy probabilities are invariant with respect to server location *or server workload*. These assumptions imply that the probability of finding a particular server busy at any given time is equal to p, which is treated as the external parameter to the model.

Assume for the moment that all location decisions have been made, and that the number of servers located within N_i equals to k. The number of busy servers can now be modeled using the binomial probability distribution: the probability that a service call from i finds an available server is $1 - p^k$.

Now suppose the number of servers within N_i is increased to $k+1$. Then the probability of finding an available server is increased by

$$r_{k+1} = p^k(1-p).$$

Let y_{ik} be the decision variable taking on a value of 1 if customer location i is covered by k or more servers, and 0 otherwise. The probability that a call from i finds an available server can now be written as

$$\sum_{k=1}^{K} r_k y_{ik}.$$

Let α be the lower bound on the expected number of calls from each location that are to be "covered" (i.e., that should immediately find an available server). Then the service level constraint can be written as:

$$\sum_{k=1}^{K} \lambda_i r_k y_{ik} \geq \alpha, \quad i \in N \tag{11.6}$$

where, as mentioned earlier, λ_i is the rate of arrival of calls from customer location i. As in Section 11.1 above, let LC^{jk} be the cost of stationing k

servers at location j. Assuming this cost to be proportional to the number of servers k, and letting $LC^j = LC^{jk}/k$ for $j \in X$, we now obtain the following set coverage problem:

$$\min \sum_{j \in X} LC^j x_j \tag{11.7}$$

s.t.

$$\sum_{j \in N_i} x_j \geq \sum_{k=1}^{K} y_{ik} \quad i \in N$$

$$\sum_{k=1}^{K} \lambda_i r_k y_{iK} \geq \alpha, \quad i \in N$$

$$y_{ik} = 0, 1 \quad i \in N, k \in \{1, \ldots, K\}, \quad x_j = 0, 1, \ldots, K$$

Here x_j is a decision variable representing the number of servers to locate at site j and K is the maximum number of servers that can be located.

Alternatively, we can fix the total number of servers in the system to K and maximize the total expected number of "covered" calls (i.e., the service level delivered by the system):

$$\max \sum_{i \in N} \sum_{k=1}^{K} \lambda_i r_k y_{ik} \tag{11.8}$$

s.t.

$$\sum_{j \in N_i} x_j \geq \sum_{k=1}^{K} y_{ik} \quad i \in N$$

$$\sum_{j \in X} x_j = K$$

$$y_{ik} = 0, 1 \quad i \in N, k \in \{1, \ldots, K\}, \quad x_j = 0, 1, \ldots, K$$

Daskin (1983) uses the second formulation and suggests a node substitution heuristic for solving (11.8) parametrically in server availability parameter p. We note that since both formulations (11.7) and (11.8) are linear Integer Programs, standard IP solvers (e.g., CLPEX) could also be applied. Note also that since $r_k \leq r_{k+1}$, $y_{ik} \geq y_{i(k+1)}$ holds automatically (in both versions of the formulation).

It is clear that the main simplification in (MEXCLP) formulations above - the linearization of the "coverage standards" constraint - is achieved by assuming that the primary indicator of system performance (probability of finding an available server) is external to the system. Strictly speaking, (MEXCLP) is not a "true" LPSDC model, since the congestion is not captured in the model. However, (MEXCLP) has led to several subsequent models that have attempted to integrate congestion effects.

Batta et al. (1989) attempted to relax the assumption that each server's availability is independent of that of the other servers. Following Larson (1975), the assumption on servers busy probability is relaxed by the following adjustment of the objective function of (11.8): each term $r_k = (1-p)p^{k-1}$ is multiplied by a correction factor $Q(K, p, k-1)$ which attempts to account for interdependence between the servers. The main idea is that if location i is covered by k servers, than in order for the $k-th$ serve to respond to the call for service from i, the other $k-1$ servers must be occupied elsewhere. Assuming that for each call the servers are chosen at random without replacement, the correction factor can be computed to be:

$$Q(K,p,k) = \frac{\sum_{j=k}^{K-1}\left\{\frac{(K-k-1)!(K-j)}{(j-k)!}(\frac{K^j}{K!})p^{j-k}\right\}}{[(1-p)\sum_{j=0}^{K-1}(\frac{K^j}{j!})p^j + (\frac{K^K p^K}{K!})]} \quad k=0,\ldots,K-1$$

Therefore the new objective function of the MEXCLP is

$$\max \sum_{i \in N} \sum_{k \in N_i} r_k Q(K, p, k-1) \lambda_i y_{ik}.$$

Alternatively, the same adjustment can be made in the service level constraint of the formulation (11.7). According to computational results in Batta et al. (1989), the adjustment factor can result in significant changes in both the objective function and optimal solution of (MEXCLP) when server busy probability p is large (over .5). We note that Batta $et\ al$ (1989) also proposed an alternative model - based on the Hypercube queuing model of Larson (1974). This approach (which is fundamentally different from the MEXCLP-based approach above) is discussed in the sequel.

The adjustment factor $Q(K,p,k)$ developed above, while accounting for some interdependence between servers, still treats server busy probability p as an external parameter. An alternative approach was pursued by ReVelle and Hogan (1989). As in the MEXCLP, the number of busy servers is assumed to be binomially distributed. However the assumption that the server busy probability is identical to all severs is relaxed - instead, the parameter p is replaced by a "local estimate" p_i, which is different for different customer locations. This estimate is based on the following two-part assumption, which will also be seen in other LPSDC models:

Districting Assumption: Recall that $N_i \in X$ represents the potential facility sites that are within the specified maximum distance of customer location i. Let $M_i \in N$ be the set of other customer locations that lie within the maximum distance of i (when $N = X$, $N_i = M_i$). Then,

(i) *All* service calls originating from locations in M_i are served by facilities in N_i.
(ii) Facilities in N_i *only* service calls originating from M_i.

The "districting assumption" is not realistic for most real-life systems: customer locations on the periphery of M_i will likely direct some (or most) of their calls to facilities outside of N_i; at the same time, facilities on the periphery of N_i are likely to devote a significant proportion of their resources to serving calls outside of M_i. Thus, the "districting assumption" might result in over-estimation of the required number of facilities and service units in some parts of the system, while under-estimating resource requirements in other parts. Nevertheless, this assumption greatly simplifies the underlying stochastics of the system, making the problem much more tractable, as illustrated below.

Let p_i be the probability (fraction of time) that a server in N_i is busy. A local estimate for p_i is given by estimating ρ_i, defined as the ratio of the utilization rate of a server in N_i divided by the number of servers in N_i. The number of servers in N_i is given by:

$$\sum_{j \in N_i} x_j.$$

Under the districting assumption, an estimate of ρ_i is given by a product of an estimate of the average service time t (in hours) of a call (assumed to be identical for all calls) and an estimate of the arrival rate of calls in N_i:

$$\sum_{n \in M_i} \lambda_n.$$

Therefore,

$$\rho_i = \frac{24t \sum_{n \in M_i} \lambda_n}{T \sum_{j \in N_i} x_j}, \qquad (11.9)$$

where T is an estimate of the total number of hours during the day that each server is available (if service units are staffed around the clock, $T = 24$ can be used). We can now use ρ_i in place of p in the MEXCLP service level constraint (11.6) corresponding to location i. However, this presents a problem, since the constraint is now highly non-linear. Fortunately, we can linearize the constraint as follows.

Using the binomial distribution as before, following the approach suggested by Chapman and White (1974), the service level constraint requiring that the probability that at least one server in N_i is available is greater than or equal to α can be stated as:

$$1 - (\rho_i)^{\sum_{j \in N_i} x_j} \geq \alpha.$$

This non-linear constraint is equivalent to the linear constraint

$$\sum_{j \in N_i} x_j \geq b_i \qquad (11.10)$$

where b_i is the smallest integer satisfying

$$1 - \left(\frac{t \sum_{n \in M_i} \lambda_n}{T b_i}\right)^{b_i} \geq \alpha$$

(the b_i values can be computed during the pre-processing stage). We can now state the following set cover-type model:

$$\min \sum_{j \in X} LC^j x_j \qquad (11.11)$$

s.t.

$$\sum_{j \in N_i} x_j \geq b_i \quad i \in N$$

$$x_j = 0, 1, \ldots, K$$

(here the y_i variables have been dropped since we are assuming that every customer must receive adequate coverage). Alternatively, we can write the following Maximal Weight Cover-type formulation:

$$\max \sum_{i \in I} \lambda_i y_i \qquad (11.12)$$

$$\sum_{j \in N_i} x_j \geq b_i y_i \quad i \in N$$

$$\sum_{j \in X} x_j = K$$

$$y_i = 0, 1, \quad i \in N, \quad x_j = 0, 1, \ldots, K, \quad j \in X.$$

Note that the first constraint - representing the linearization of the "coverage standards" - is exactly of the form (11.3), as discussed in the Introduction. We also note that the use of binomial distribution in formulations (11.11) and (11.12) implicitly assumes that all servers in N_i function independently.

Formulation (11.12) is essentially equivalent to ReVelle and Hogan's Maximum Availability Location Problem (MALP) of ReVelle and Hogan (1989). The only difference is that the latter assumes that each facility can contain only 1 server - thus variables x_j are binary and the RHS of the last constraint represents the total number of facilities, rather than the total number of servers. We also note that another (earlier) formulation of MALP was given by ReVelle and Hogan (1989a).

Note however that a subtle "error" was made in moving from the set cover formulation (11.11) to the maximal weight cover-type formulation (11.12). The ability to pre-compute values of b_i, $i \in N$ depends on being able to estimate the utilization factor ρ_i in advance. However, when not all customers in M_i are covered (as is the case in (11.12)), the total call arrival rate in N_i is given by

$$\sum_{n \in M_i} \lambda_n y_n,$$

and thus ρ_i is itself a function of the decision variables and cannot be computed in advance. Of course, we can still use the value of ρ_i computed by (11.9) as an upper bound on the "true" utilization rate (this bound ignores the fact that not all calls from M_i will need to be processed). This allows us to derive formulation (11.12) as before. However, we should realize that the values of parameters b_i should be regarded as conservative upper bounds - they likely over-estimate the actual number of servers required in N_i for $i \in N$. Note that the previous statement assumes that the districting assumption holds - as discussed earlier, the failure of that assumption may by itself lead to over- or under-estimation of the required system resources.

All the models described so far use "implicit" representation of congestion in the system (by relying on the binomial distribution, they essentially ignore the fact that some of the servers might be less available than others).

Marianov and ReVelle (1996) extended the formulation (11.12) above by introducing a more explicit, queuing-based representation of the system. The "districting assumption", which is essential to their development, is stated in an equivalent form in Marianov and ReVelle (1996): for any neighborhood (set) M_i the number of calls for service that originate outside M_i and are attended by servers stationed in N_i is roughly the same as the number of calls that originate in M_i and are served by service units outside N_i. As a result, each neighborhood M_i can be treated as a separate queuing system. The second strong assumption made in Marianov and ReVelle (1996) is that travel times of service units to customers are much smaller than service times (and can thus be ignored when estimating server availability). Assuming exponential distribution of service times, with service rate μ_i for service units stationed in N_i, and supposing for the moment that a total of k service units are stationed in N_i, the service units stationed in N_i can be viewed as an $M/M/k$ queuing system. The probability that all service units are busy is given by

$$P_k^i = \frac{\frac{1}{k!}\rho_i^k}{\sum_{n=0}^{k} \frac{\rho_i^n}{n!}} \qquad (11.13)$$

where $\rho_i = \frac{\lambda^i}{\mu_i}$ and $\lambda^i = \sum_{n \in M_i} \lambda_n$. The server availability requirement can be written as:

$$1 - P_k^i \geq \alpha, \quad i \in N.$$

Since the LHS is clearly an increasing function of k, this requirement is satisfied, provided the number of servers in N_i is at least b_i, where b_i be the smallest integer that satisfies

$$P_{b_i}^i \leq 1 - \alpha.$$

Therefore, the "coverage standard" (server availability) constraint can be replaced by the constraint

$$\sum_{j \in N_i} x_j \geq b_i y_i, \quad i \in N \qquad (11.14)$$

(where x_j and y_i are defined as in (11.12) above). Thus, the linearized form of the "coverage standard" constraint is exactly the same as in (11.10), leading to an MALP-like formulation - the only difference is in the way parameters b_i are estimated. Marianov and ReVelle (1996) refer to this formulation as "Queuing Maximum Availability Problem" (Q-MALP). They also formulate the set-cover version of this model, which they call the "Queuing Location Set Covering Problem" (Q-LSCP) - see Marianov and ReVelle (1994). In the latter, all y_i are set to 1 and the total number of service units is minimized (actually, in marianov and ReVelle (1994, 1996) it is assumed that each location houses exactly one service unit - and thus the decision variables x_j are binary, we relaxed this assumption in the formulation above).

Note that the statements made above with regards to the (MALP) formulation (11.12) apply to (Q-MALP) formulation as well: the derivation of parameters b_i above ignores the fact that not all customers in M_i might be covered, and thus overestimates the total call arrival rate (and, consequently, the required number of servers) in M_i.

Ball and Lin in (1993) derive a similar constraint to (11.10) using completely different techniques, allowing them to partially relax the "districting assumption". Let $\bar{N}_i$ represent the set of actual facility locations in N_i (assuming, for the moment, that location decisions have already been made). A "failure of an individual system at node i" is said to occur when there is no available server in $\bar{N}_i$ to respond to a demand call from node i. Let q_i be the probability that an individual system at node i fails. First, a necessary condition for the failure of an individual system at node i is derived. Let $PT(j)$ be the set of all demand points covered by a station at node j, and k_j the number of vehicles at station j. Then a demand from node i cannot be served if $\sum_{j \in \bar{N}_i} k_j$ vehicles have been assigned to uncompleted calls in $\bigcup_{j \in \bar{N}_i} PT(j)$ - the combined coverage area of all facilities in $\bar{N}_i$ at the time that the call occurs. For a set S of facility nodes, let $A(S)$ be the number of demand calls that arose in $\bigcup_{j \in S} PT(j)$ and have not been completed at some time t. The following necessary condition can now be stated: "if there is no server available to service a call from node i, then $A(S) \geq \sum_{j \in S} k_j \quad \forall S \subseteq \bar{N}_i$." Let $\hat{T}$ be an upper bound on a service time duration of a call. Then $A(S) \leq D(S)$, where $D(S)$ is the total number of demand calls that arose in $\bigcup_{j \in S} PT(j)$ during the interval $(t - \hat{T}, t)$.

Let $\text{INT}(i)$ be the event $\left\{ D(S) \geq \sum_{j \in S} k_j \quad \forall S \subseteq \bar{N}_i \right\}$. It follows that $q_i \leq P[\text{INT}(i)]$. Assuming that demand calls are generated according to a Poisson distribution it is further shown in Ball and Lin (1993) that

$$q_i \leq Pr[\text{INT}(i)] \leq \text{PROD}(i)$$

where,

$$\text{PROD}(i) = \prod_{j \in \bar{N}_i} P(D(j) \geq k_j)$$

Let x_{jk} denote the binary decision variable that is equal to 1 if k vehicles are stationed at location j and 0 otherwise, $j \in X$ and $k = 1, 2, \ldots, K$. Requiring that

$$\text{PROD}(i) \leq 1 - \alpha$$

is equivalent to requiring

$$\prod_{j \in \bar{N}_i} \prod_{1 \leq k \leq K} P[D(j) \geq k]^{x_{jk}} \leq 1 - \alpha \qquad (11.15)$$

It is shown in Ball and Lin (1993) that (11.15) is equivalent to the linear constraint

$$\sum_{j \in N_i} \sum_{1 \leq k \leq K} a_{jk} x_{jk} \geq b_i \quad \forall i \in N$$

where,

$$a_{jk} = -\log p[D[j] \geq k],$$
$$b_{jk} = -\log [1 - \alpha]$$

(these quantities can be computed during the pre-processing stage). The (set cover-type) problem considered in Ball and Lin (1993) is formulated as follows:

$$\min \sum_{j \in X} \sum_{1 \leq k \leq K} LC^{jk} x_{jk} \quad (11.16)$$

s.t.

$$\sum_{j \in N_i} \sum_{1 \leq k \leq K} a_{jk} x_{jk} \geq b_i \quad \forall i \in N$$

$$\sum_{1 \leq k \leq K} x_{jk} \leq 1 \quad \forall j \in X$$

$$x_{jk} = 0, 1 \quad \forall j, k$$

where LC^{jk} is the cost of housing k servers at station j (as in Section 11.1).

It can be seen that the derivation above retains part (i) of the districting assumption (all calls originating from M_i are still serviced only by servers in N_i), but removes part (ii), since servers in N_i are now responsible for calls in a much larger set $\bigcup_{j \in N_i} PT(j)$. In fact, it assumes that only servers from N_i are available for covering calls originating from this set. Since, in reality, the latter statement is likely not true (some of this call traffic will be handled by servers from other neighborhoods), the formulation (11.16) will always over-estimate the number of servers required to meet the "availability standard". It should also be recognized that this formulation makes another strong assumption: the derivation assumes a deterministic service time $\hat{T}$ for each call - this value is then used as an upper bound on actual service times. We find this assumption particularly troublesome: it is well known that the on-scene service time distribution in many emergency systems can be closely approximated by an exponential distribution. For example, while a typical police call for service might only take a few minutes to process, a few calls will require many hours of processing time. Thus, it is not clear what value of $\hat{T}$ should be used - if we use the actual upper bound on the service time (i.e., several hours in the example above) - the required number of servers will be wildly over-estimated by the model. If we set $\hat{T}$ equal to the average service time instead, an underestimation may result.

Nevertheless, we note that model (11.16) can be used to obtain an upper bound on the required number of servers - i.e., the numbers computed by the model guarantee that coverage standards at each customer location are met. This guarantee cannot be made by any of the previously reviewed models, which can lead to both over-estimation and under-estimation of the coverage standards at different customer locations.

We close this section by discussing the second model presented in Batta *et al* (1989). This model uses the Hypercube model (Larson, 1974) to describe system behavior for a fixed set of facility locations. The Hypercube model - which is discussed further in Section 11.3 below, is a model of $M/M/k$ queuing system with distinguishable servers. As discussed in Section 11.1 above, this is the most realistic representation of the system out of all the models discussed in this section (even though the "true" representation requires a model of $M/G/k$ queue with distinguishable servers - for which no analytic description is available). Unfortunately, the Hypercube model is a descriptive one - no closed-form solutions of system performance measures are available; this quantities can only be obtained by a numerical solution of a system of equations. Thus, no reduction to a closed-form mathematical formulation (such as the ones we have seen earlier) is possible for Hypercube-based models. Batta *et al* (1989) employ a node substitution heuristic - which starts with a "trial" set of facility locations, uses the Hypercube model to estimate system performance characteristics, and then evaluates the effects of shifting one of the facilities to a new location. They report that the optimal solution to the (MEXCLP) model (with or without the adjustment factor $Q(K,p,k)$ developed above) provides a good starting point for their Hypercube-based procedure.

11.2.2 Models with Static Server Assignments (Fixed Servers)

In this section we describe models that assume "static" assignments of customers to service facilities: each customer is allocated (assigned) to a particular facility, and only servers stationed at that facility may process calls for service from this customer. As discussed in Section 11.1 above, this situation is typical in case of fixed server locations (such as hospitals) - where customers travel to facilities to obtain service. We emphasize that the assignment of a customer to a facility might be centralized (the "directed choice environment") or made by each individual customer by employing some decision rule - e.g., always get service from the closest facility (the "user choice" environment). We will revisit this issue at the end of the current section.

Two models for fixed servers were described by Marianov and Serra (1998). These models are stated in Maximal Weight Cover form: the problem is to locate M *fixed* facilities (i.e., users travel to facilities to obtain service) and allocate the users to them to maximize coverage. A user is considered covered if: (i) she is within a time (distance) standard from the facility she is allocated to, and (ii) the probability that she will enter a queue at that center with at most c waiting users is at least α. An alternative version of requirement (ii) is (ii'): the probability that the user will be served within time τ upon arrival at the facility is at least α.

Using our usual notation with binary decision variables x_j, $j \in X$ ($x_j = 1$ if a facility is opened at j, 0 otherwise) and y_{ij}, $i \in N, j \in X$ (where $y_{ij} = 1$ if customer at i is assigned to a facility at j), the following general formulation can be stated:

$$\max \sum_{i \in N} \lambda_i \sum_{j \in X} y_{ij} \quad (11.17)$$

$$s.t. \quad y_{ij} \leq x_j \quad \forall i \in N \quad j \in N_i$$

$$\sum_{j \in N_i} y_{ij} \leq 1 \quad \forall i \in N$$

$$\sum_{j \in X} x_j = M$$

$$P \text{ (facility } j \text{ has at most } c \text{ users in queue)} \geq \alpha \quad \forall j \in X \quad (11.18)$$

or
$$P \text{ (waiting time at facility } j \leq \tau) \geq \alpha \quad \forall j \in X \quad (11.19)$$

$$y_{ij}, x_j = 0, 1 \quad i \in N, \quad j \in X$$

The formulation above is, for the most part, a standard formulation of the Maximal Weight Cover Problem: the first constraint guarantees that a user cannot be allocated to a node unless there is a facility at that node, the second constraint ensures that each user is allocated to at most one facility and the third constraint limits the number of facilities to M. The only outstanding issue is how to formulate constraints (11.18) and (11.19), which correspond to requirements (ii) and (ii'), respectively. Marianov and Serra (1998) propose two alternative formulations of these constraints under different sets of assumptions.

First, assume that each facility houses exactly one service unit and that the average on-scene service time for a facility located at node j is $1/\mu_j$ for $j \in X$. Since each customer is allocated to exactly one facility (which is a form of the "districting assumption" seen earlier), each facility functions as an $M/M/1$ queuing system. A facility at location j has a total arrival rate of

$$\lambda_j = \sum_{i \in N} \lambda_i y_{ij}. \tag{11.20}$$

We can now utilize the standard results for the $M/M/1$ queuing (e.g., see Kleinrock (1975)) to express constraints (11.18) and (11.19). For a facility located at node j, the probability that the length of queue is at most c is given by $1 - \rho_j^{c+2}$, where $\rho_j = \lambda_j/\mu_j$. Thus constraint (11.18) is equivalent to:

$$1 - \rho_j^{c+2} \geq \alpha. \tag{11.21}$$

After simple algebraic manipulation, this can be re-written as:

$$\sum_{i \in N} \lambda_i y_{ij} \leq \mu_j \sqrt[c+2]{1-\alpha} \quad \forall j \in X, \tag{11.22}$$

which is linear, and can be used in place of (11.18) in the formulation above.

Similarly, the requirement that a user arriving to the facility located at node j will wait at most τ units of time with probability of at most α can be written as:

$$1 - e^{-(\mu_j - \lambda_j)\tau} \geq \alpha, \quad j \in X,$$

which, after taking logarithms and using (11.20) can be re-written as

$$\sum_{i \in N} \lambda_i x_{ij} \leq \mu_j + \frac{1}{\tau} \ln(1-\alpha) \quad \forall j \in X. \tag{11.23}$$

Thus, constraints (11.19) can be replaced by the linear constraints (11.23), leading to a linear integer programming formulation.

The second model considered in marianov and Serra (1998) assumes that there are exactly k servers at each of the facilities. Here the results for the $M/M/k$ queuing system Kleinrock (1975) can be utilized. Constraint (11.18) is equivalent to requiring that the probability of at least $c+1$ users are waiting in the queue is less than or equal to $1 - \alpha$, and can be expressed as follows (see Kleinrock (1975)):

$$\sum_{r=k+c+1}^{\infty} \frac{P_0 k^k}{k!} \left(\frac{\rho_j}{k}\right)^r \leq 1 - \alpha \quad \forall j \in X, \tag{11.24}$$

where

$$P_0 = \frac{\rho_j^k}{(1-\frac{\rho_j}{k})k!} + \sum_{r=0}^{k-1} \frac{\rho_j^r}{r!} \quad \forall j \in X$$

and $\rho_j = \lambda_j/\mu_j$ as before. Following some algebraic manipulations, (11.24) can be rewritten as

$$\sum_{r=0}^{k-1} \frac{(k-r)k!k^c}{r!} \frac{1}{\rho_j^{k+c+1-r}} \geq \frac{1}{1-\alpha} \quad \forall j \in X \quad (11.25)$$

In this form, it is clear that the Left-hand Side of the expression above is a decreasing function of ρ_j. Thus, there must exist some critical value ρ_α such that (11.25) holds whenever $\rho_j \leq \rho_\alpha$. Moreover, this critical value of ρ_α can be computed in advance. It follows that constraint (11.18) can again be re-written as a linear constraint:

$$\sum_{i \in N} \lambda_i y_{ij} \leq \mu_j \rho_\alpha \quad \forall j \in X.$$

The resulting integer program is solved in Marianov and Serra (1998) using a standard Branch and Bound solver (CLPEX). A simple heuristic is also shown to provide good results.

In Marianov and Rios (2000), they apply this methodology to the problem of finding optimal location of switches in ATM communication networks. In this application, the requirement (ii) is linearized as above, but - unlike in the formulation (11.17) above, the total number of facilities is left as a decision variable (with the total facility location cost term TC_{LC} added to the objective function).

While the total waiting time requirement (11.19) is not treated for the k-server case in Marianov and Serra (1998), a linearized version of this constraint can be obtained using similar technique as above, together with the waiting-time distribution results for $M/M/k$ queuing systems. For completeness we sketch out this derivation below.

Let

$$f(\rho) = \frac{\rho^k}{(k-1)!(k-\rho)} p_0,$$

where $\rho = \lambda/\mu$ and

$$p_0 = \left[\sum_{n=0}^{k-1} \frac{\rho^n}{n!} + \frac{\rho^k}{(k-1)!(k-\rho)}\right]^{-1}.$$

Then the waiting-time distribution function for $M/M/k$ queue is given by

$$W_q(t) = \begin{cases} 1 - f(\rho) & \text{if } t = 0 \\ 1 - f(\rho)exp[-\mu(k-\rho)t] & \text{if } t > 0 \end{cases} \quad (11.26)$$

It is easy to see that

$$\frac{1}{f(\rho)} = 1 + \sum_{n=0}^{k-1} \frac{(k-1)!(k-\rho)}{n!\rho^{k-n}}$$

is a decreasing function of ρ. From (11.26) it now follows that $W_q(t)$ is also a decreasing function of ρ. The constraint (11.19) specifies that

$$W_q^j(\tau) \geq \alpha, \quad j \in X,$$

where $W_q^j(t)$ is the waiting time distribution for a facility located at j. By the preceding arguments, this constraint is equivalent to

$$\rho_j \leq \rho_{j\alpha},$$

where the quantity on the RHS is a constant that can be computed in advance (note that since $W_q^j(t)$ depends explicitly on μ_j, this constant will be different for different j). Thus, constraint (11.19) can be replaced by a linear constraint

$$\sum_{i \in N} \lambda_i y_{ij} \leq \mu_j \rho_{j\alpha} \quad \forall j \in X. \quad (11.27)$$

It is useful to observe an important difference between the final version of the formulation (11.17) above - that is, the formulation we would obtain after linearizing constraints (11.18) or (11.19) - and the formulations discussed in Section 11.2.1 above. In the latter, the linearization of the standard of coverage constraints led to a new constraint of the form (11.3) - that required that a certain number of servers be positioned around each customer location i, but did not restrict the number of customers utilizing a particular facility. In (11.17), the linearized standard of coverage constraints take the form of (11.27), which has the effect of capping the capacity of each facility. This has the effect of turning this formulation into a *capacitated* facility location problem, whereas formulations in Section 11.2.1 still have the uncapacitated form. This difference is due to the fact the total number of servers in Marianov and Serra's (1998) model is fixed in advance (and identical at all facilities), where it was flexible and part of the decision in the previous models.

The capacitated nature of formulation (11.17) has important consequences with regards to potential applications. As mentioned earlier, this model was designed for location of fixed servers (medical centers, distribution centers, and banks are mentioned as potential applications in Marianov and Serra (1998)). However, this model clearly makes the "directed choice" assumption (customers may only utilize the facility they are allocated to - see Section 11.1 for more discussion). In uncapacitated coverage-type facility location models, the difference between directed choice and user choice assumptions is irrelevant - since the capacity of each facility is unlimited, customers may, in fact, utilize any facility they wish, instead of the one they were allocated to by the model. In capacitated models this is not the case, and unless customer allocations are strictly enforced, some of the facilities may well be overwhelmed. This is especially important since capacitated models are known to lead to "counter-intuitive" customer allocations (where customers in the immediate vicinity of one facility are allocated to a more distant one, or - worse yet - left uncovered by the model). This occurs because the model tries to solve a knapsack-like problem for each facility - cramming as many customers as it can under the capacity cap. It is not clear how the directed choice system can be effectively enforced for facilities such as hospitals or banks. We note that the model (11.17) above can be converted to a user choice model by assuming that customers will normally utilize the closest facility, and adding constraints to enforce this assumption.

It is also interesting to note that the models in Section 11.2.1 allowed for flexible number of servers at each facility, whereas the models presented in the current section specify the number of servers at each facility in advance (in fact, they assume same number of servers at each facility - but this assumption can be easily relaxed). This is due to the fact that most of the models in Section 11.2.1 rely on some form of the "districting assumption" - which allows them to estimate the total arrival rate in each neighborhood N_i in advance (or, at least, approximate it in the case of maximal weight cover-type models).

As we have seen, using queuing techniques, it is possible to express the "coverage standard" constraints through performance characteristics of the respective queuing system (e.g., as in (11.24) above). The LHS of these expressions is typically a decreasing function of both, the utilization rate ρ and the number of servers k. In Section 11.2.1, using the "districting assumption" we can estimate ρ in advance, allowing us to linearize the constraint with respect to the number of servers k (by finding the smallest value of k under which the constraint is met) - thus allowing for a flexible number of servers. For models presented in the current section, allocation of customers to facilities is part of the decision problem - thus ρ cannot be estimated in advance. Therefore, to linearize the constraint, we need to fix the number of servers at each facility in advance.

11.3 Problems with Median-Type Objective: The Stochastic Queue Model

11.3.1 Introduction

In this section we discuss LPSDC models where the main objective is to minimise the average response time in the system (where the response time is measured from the initial arrival of the call to the start of on-scene service). As discussed in the Introduction, these models - which we view as stochastic extensions of median-type problems - generally share the following features:

- The total number of locations - M - is assumed to be fixed in advance. All models discussed in the current section assume that the servers are mobile, and that each server has a "home" location to which it must return after each call for service. We use the binary decision vector $\mathbf{x} = (x_j)$, $j \in X$ to indicate whether a facility has been located at j. We also use k_j to represent the number of servers located at j (where $k_j = 0$ when $x_j = 0$, and $k_j > 0$ when $x_j = 1$).
- When a call for service occurs, the closest available service unit is dispatched, assuming that at least one is available. Note that all customer locations are assumed to be "covered" (in the sense discussed in Section 11.1) - thus there are no "coverage standards".
- The service time of a typical service unit includes several components:
    1. Travel to the scene of the call from the service station.
    2. Providing on-scene service time.
    3. Travel back to the station.
    4. Possibly providing off-scene service.

    The mean service time for a service unit located at j is:

    $$S(j) = \sum_{i \in N} h_i^j (w_{ij} + \frac{\beta_j}{v} d(i,j)) \qquad j \in X \qquad (11.28)$$

    where w_{ij} is the sum of mean on and off-scene service times, $\beta_j > 1$ is a constant that allows different travel speed to and from the scene of the call (if those speeds are identical, $\beta_j = 2$ can be used) and v is the average travel speed. The quantity h_i^j is the probability that a service unit from facility at j is dispatched to customer location i, given that it is dispatched to a call for service (note the difference from the quantity P_D^{ij} defined in earlier in Section 11.1). Calls that find all service units busy enter a FCFS queue.
- The queue of calls waiting to be served is assumed to have a fixed capacity $c \geq 0$ (however, the case of $c = \infty$ will also be considered). Calls that occur when the length of the queue is c are lost and incur a penalty $RC > 0$ to the system (as mentioned in Section 11.1, RC can be the cost of dispatching a reserve unit or a unit from another network).

- We use $T_R(\mathbf{x})$ to represent expected response time to a random call for service that is not rejected by the system. This quantity is the sum of two components: (i) $T_W(\mathbf{x})$ – the expected waiting time in the queue, and (ii) $t(\mathbf{x})$ – the expected travel time to a call. Note that T_W is related in an obvious way to the quantities T_W^i (the expected waiting time for a call from location i) defined in Section 11.1 earlier. We can write the following expression for T_R:

$$T_R(\mathbf{x}) = T_W(\mathbf{x}) + t(\mathbf{x}). \tag{11.29}$$

- Let P_R be the probability that a random call is rejected by the system (i.e., that it occurs while there are already c calls waiting in the queue). The objective function minimizes the weighted sum of the mean response time and the cost of rejecting a customer from the system. This quantity - which we called TC_{WC} earlier can now be written as

$$TC_{WC}(\mathbf{x}) = (1 - P_R)T_R(\mathbf{x}) + P_R RC.$$

Note that we set the waiting cost $WC = 1$ (an alternative interpretation is that RC now represents the rejection cost relative to the waiting cost).

In view of the discussion above, we now obtain the following formulation, that underlies all of the models discussed in the current section:

$$\min TC_{WC}(\mathbf{x}) = (1 - P_R)T_R(\mathbf{x}) + P_R RC \tag{11.30}$$

$$\text{S. t.} \sum_{j \in X} x_j = M$$

$$k_j \leq K x_j, \quad j \in X$$

$$\sum_{j \in X} k_j = K$$

$$x_j \in \{0, 1\}, k_j \in \{0, \ldots, K\}, j \in X,$$

where K represents the maximum number of servers in the system. Here the first constraint sets the total number of facilities to M, the second one ensures that no service units are located at j unless a facility is opened there, and the last one limits the total number of servers in the system.

The stochastic behaviour of the underlying system can be described as follows (as always, we assume for now that all locational decisions have been made). Let us number the facilities from 1 to M and let $k(m)$ be the number of servers located at facility m for $m \in \{1, \ldots, M\}$ (of course, $k(m) = k_j$ for some $j \in X$). The state of the system at a particular point in time can be represented as an $M + 1$-dimensional vector, where the m-th component for $m \leq M$ represents the number of servers located at facility m that are currently busy, and the $M + 1$-st component represents the number of calls waiting in the buffer. Let s_m be the value of the m-th component. Note that $s_m \in \{0, 1, \ldots, k(m)\}$ for $m = 1, \ldots, M$ and $s_{M+1} \in \{0, \ldots, c\}$.

We can subdivide all possible states into two types: "F" (for "free") and "B" (for "busy"). States of type F include at least one component m such that $s_m < k(m)$ - that is, there is at least one available server somewhere in the system. It follows that $s_{M+1} = 0$ for any state of type F. States of type B have $s_m = k(m)$ for $m = 1, \ldots, M$ - that is, all the servers are busy.

It is convenient to number the states of type B by the number of calls waiting in the buffer (that is, by the value of the last component). Let B_b be the state of type B with $b \geq 0$ calls waiting in the buffer. There is no "natural" numbering system for states of type F. Fortunately, there are only $Q = \prod_{m=1}^{M}(k(m) + 1) - 1$ states of this type. Thus, we can number states of type F from 1 to Q arbitrarily.

The state-space description of the system developed above is complete, given the assumption that a server can only be dispatched from the facility (i.e., a server must return to its "home" facility after each call) - this makes it unnecessary to record the location of the call as part of the state space. Indeed, it can be seen that if a system is currently in a state B_b with $b > 0$, then the next state will either be B_{b+1} (if the next arrival happens before a service completion) or B_{b-1}. Similarly, if the current state is $F_{q'}$ then the next state will be either some $F_{q''}$ or B_0. The system just described is an instance of an $M/G/K/c$ queuing system with distinguishable servers. If we also assume that the total service time (including inbound and outbound travel times, as well as on- and of-scene service times) is exponential, we obtain an $M/M/K/c$ queuing system with distinguishable servers. Larson (1974) used the state-space description above to obtain steady-state probabilities for such a system (obtained as solutions to a large system of equations) - this is the basis of the Hypercube model. However, the exponential service assumption is unlikely to be realistic: while the on-scene (and, possibly off-scene) service times do appear to follow the exponential distribution in many applications, the travel times often do not - hence the total service times might not be exponential. Unfortunately no analytic techniques are currently available to obtain steady-state probabilities for an $M/G/K/c$ queue with distinguishable servers.

In the following sections we consider several versions of problem (11.30) under some simplifying assumptions. In spite of these assumptions, the models covered in the sequel tend to have a much more detailed representation of the stochastics of the underlying system compared to the models discussed in

Section 11.2 above. On the other hand, most of the models discussed below lead to highly non-linear formulations - heuristic techniques are often the only viable solution approaches.

11.3.2 One Facility Case: Stochastic Loss Median and Stochastic Queue Median Problems

In this section we review the relevant results for the one-facility case. For the most part, our discussion is limited to the results that will be necessary for subsequent discussion of multi-facility models. We start with the single server case - i.e., both M and K are set to 1. The problem of locating a single service unit on G is solved for the two cases: $c = 0$ and $c = \infty$ by Berman, Larson and Chiu (1985).

When $c = 0$, a call will either find the service unit available at its home location x or will be rejected. Therefore Problem (11.30) is reduced to :

$$\min_{x \in X} z(x) = (1 - P_R(x))t(x) + P_R(x)RC \qquad (11.31)$$

where $P_R(x) = \rho(x) = \lambda S(x)$ is the average fraction of time the server is busy given that it is located at x when available, and

$$t(x) = \sum_{i \in N} h_i d(x, i)/v. \qquad (11.32)$$

An optimal solution of the problem is called the "Stochastic Loss Median" (SLM) of the network G. It is shown in berman et al. (1985) that the 1-median node of G is also a SLM (recall that a 1–median is a node which minimizes $t(x)$).

When $c = \infty$, no customer is rejected and therefore problem (11.30)) reduces to

$$\min_{x \in X} T_R(x) = T_W(x) + t(x), \qquad (11.33)$$

where $t(x)$ is defined in (11.32). Since this is a $M/G/1$ queuing model, the expression for the waiting time is given by (see Kleinrock (1975)):

$$T_W(x) = \frac{\lambda S^2(x)}{2(1 - \lambda S(x))} \qquad \text{when } \lambda S(x) < 1, \qquad (11.34)$$

where $S(x)$ is the expected service time and $S^2(x)$ is the second moment of the service time. The expression for $S(x)$ is given in (11.28), which in this case simplifies to:

$$S(x) = \sum_{i \in N} h_i \left(w_{ix} + \frac{\beta}{v} d(x, i) \right). \qquad (11.35)$$

The expression for the second moment of service time is:

$$\bar{S}^2(x) = \sum_{i \in N} h_i E\left[\tilde{w}_{ix} + \frac{\beta}{v}d(x,i)\right]^2, \qquad (11.36)$$

where $\tilde{w}_{ix}$ is the random variable representing total on- and off-scene service time associated with a service call from i served by a service unit from x, and the expectation $E[\cdot]$ is taken with respect to the distribution of $\tilde{w}_{ix}$.

Note that the objective function is now non-linear. However, it is relatively easy to evaluate numerically for any fixed facility location $x \in X$, and since we have assumed that $X \in N$, the optimal location can be found by complete enumeration. The optimal location in the $c = \infty$ case is called the Stochastic Queue Median (SQM) of the network G. The SQM of the network generally does not coincide with the 1-median.

The assumption that X is a subset of nodes, or even that X is finite, is not necessary. In fact, it is assumed in berman et al. (1985) that a server can be located at a node or anywhere along any link in the network. Suppose the server is located on a link connecting nodes a and b at a point d' which is at a distance of d' from node a and $l - d'$ from node b, where l is the length of link (a,b). The set of nodes N can be divided into two disjoint sets A and B as follows:

$$A = \{i \in N; d' + d(a,i) \leq (l - d') + d(b,i)\}, \quad B = N - A.$$

When d' varies from 0 to l on link (a,b), the sets A and B may change. Breakpoints are defined as points on (a,b) where the sets A and B change. We call the range $[d_1, d_2]$ where d_1 and d_2 are two consecutive breakpoints on (a,b) a "primary region".

Note that the sets A and B are invariant within any primary region. It was shown in Berman et al. (1985) that, when finite, $T_R(x)$ is a continuous convex function on $[d_1, d_2]$. Therefore, the (local) minimum of $T_R(x)$ within any primary region can be computed analytically (Berman et al., 1985). Thus, the search for the optimal location can be reduced to a finite set, consisting of all the nodes plus local minima within the primary regions along the links - we refer the reader to berman et al. (1985) for details. In fact, this result (along with its generalizations in later works) motivated our initial assumption that X is finite and nodal. We also note that many extensions of the model with one service unit are presented in Berman et al. (1990).

The results presented above can be generalized to the multiple server $(K > 1)$ case by employing formulas for $M/G/K$ queuing systems. Assuming $c = 0$, Chiu and Larson (1985) extend the results obtained for the single service unit case discussed above. The formulation (11.31) can be applied directly to the $K > 1$ case by using a different expression for P_R. For the $M/G/K/0$ queue (where 0 is the capacity of the queue), P_R is given by the

Erlang loss formula Kleinrock (1975):

$$P_R = \frac{\rho(x)^K/K!}{\sum_{k=0}^{K} \rho(x)^k/k!}. \tag{11.37}$$

It is shown in Chiu and Larson (1985) that, as in the $K = 1$ case, the optimal location of the service station is a 1-median even when $K > 1$.

When $c = \infty$ and $K > 1$, a closed form expression like (11.34) does not exist for the expected waiting time of a random call. In Batta and Berman (1989), Berman and Batta use the following approximate expression for $T_W(x)$ developed by Nozaki and Ross (1978):

$$T_W(x) = \frac{\lambda^K S^2(x)(S(x))^{K-1}}{2(K-1)![K-\lambda S(x)] \sum_{k=0}^{K-1}(K-i)}, \quad \text{when } \lambda S(x) < K.$$

(11.38)

The problem is now identical to the one formulated in (11.33) except that (11.38) is used for $T_W(x)$. Some questions were raised about the adequacy of the approximation provided in (11.38). According to the computational results in Batta and Berman (1989), this approximation does not provide accurate values of the objective function of (11.33). However, it does appear to result in the optimal location of the service facility. We note that an alternative approach would be to use a simulation-based technique to evaluate $T_W(x)$ directly in this case. This approach was not used in Batta and Berman (1989) because they did not restrict X to be nodal (see Batta and Berman (1989) for a thorough discussion of the case where the facility can be located anywhere in the network).

11.3.3 M Stations, Single Service Unit at a Station, No-cooperation

We have seen that when only one facility is to be located, it is possible to obtain expressions for the required system performance measures by employing results for queuing systems with exponential arrivals and general service distribution. In the current section we extend these results to the case of multiple service facilities, but under the assumption that servers do not co-operate.

In the general problem (11.30), the facilities are implicitly assumed to co-operate with one another - this co-operation is imposed by the assumption that the closest available unit is always dispatched to a call. Thus, when all service units at a particular station are busy, and a call that is normally serviced by a unit from that station arrives, an available service unit from another station will be dispatched. In the current section we assume that there is no cooperation between the stations, i.e., in the scenario above, the call would have to wait until one of the service units at its "normal" station frees up. The assumption of "no co-operation" is similar to the "districting assumption" we saw in Section 11.2 earlier. Under this assumption the problem takes on the following form: we want to simultaneously determine service territories (by partitioning the customer demand nodes into areas to be served by one facility), and to find a location for the facility within each territory, so that the overall response time to a random call for service in the network is minimised.

Note that in the $c = 0$ case, the SLM within each territory coincides with the 1-median, so the overall problem is similar to the standard (i.e., deterministic) M-median problem.

For the rest of the current section we focus on the $c = \infty$ case. Even if we assume that all facility locations have already been picked, the problem of finding districts for pre-located facilities is, by itself, quite challenging. In Berman and Larson (1985), they considered the districting problem for two stations, each housing a single service unit, that are located at known locations x^1 and x^2 in G. Given disjoint districts N^1 and N^2 $(N^1 \cup N^2 = N)$ the expected response time in each districts, $T_R(x^j)$ $j = 1, 2$ can be written using expressions (11.32 – 11.36), where the summations in expressions (11.32, 11.35, 11.36) pertain to the nodes in districts j, $j = 1, 2$ and the fraction of demand from any node $i \in N^j$ is $h_i^j = \frac{h_i}{h^j}$ where $h^j = \sum_{i \in N^j} h_i$ $j = 1, 2$.

The problem now can be expressed as follows:

$$\min_{N^1 \cup N^2 = N,\ N^1 \cap N^2 = \emptyset} T_R^{1,2} = h^1 T_R(x^1) + h^2 T_R(x^2). \tag{11.39}$$

It is shown in Berman and Larson (1985) that the optimal districting policy can be categorised by ranges of the total call rate λ as follows:

Region A: $0 \leq \lambda \leq \lambda_A$
Region B: $\lambda_A < \lambda < \lambda_C$
Region C: $\lambda_C \leq \lambda < \lambda_D$
Region D: $\lambda_D \leq \lambda$
where λ_A, λ_C and λ_D are constants.

Region A includes λ values small enough so that $h^1 t(x^1) + h^2 t(x^2)$ dominates $T_R^{1,2}$, and therefore the optimal policy is based on the proximity of each station to the demand nodes. The problem thus turns into a standard 2-median problem.

Region D includes all λ values for which a feasible solution does not exist (the queues in at least one of the districts build up to infinity). It is not difficult to observe from (11.34) that

$$\lambda_D = \min\{(h^1 S(x^1))^{-1}, (h^2 S(x^2))^{-1}\}. \tag{11.40}$$

The optimal policy for region C can be found due to the observation that when λ is close enough to $\lambda_D (\lambda < \lambda_D)$ either the denominator of $T_W(x^1)$ or that of $T_W(x^2)$ dominates the value of $T_R^{1,2}$ (since one of the two facilities is operating close to its maximum utilization). Therefore the optimal policy is one that minimizes the function

$$\max[h^1 S(x^1), h^2 S(x^2)].$$

The problem can be formulated as the following linear integer programming:

$$\min V \qquad (11.41)$$

S.t. $\quad \sum_{i \in N} y_{ij}[h_i(\beta d(x^j, i)/v + w_{ix^j}] \leq V \qquad j = 1, 2$

$$\sum_{j=1}^{2} y_{ij} = 1 \qquad i \in N$$

$$y_{ij} = 0, 1 \qquad i \in N, \qquad j = 1, 2.$$

where

$$y_{ij} = \begin{cases} 1 & \text{if customer node } i \text{ is assigned to service facility } j \\ 0 & \text{otherwise.} \end{cases}$$

Here the first set of constraints represents the objective function. The second set ensures that each customer node is assigned to some facility.

If the value of λ falls in region B, neither the linear term $t(x)$, nor the expected service time term dominate, and the full non-linearities of the problem must be considered. Moreover, there may be several optimal policies for different λ values. The problem (using the decision variables y_{ij} defined above) can be formulated as follows:

$$\sum_{j=1}^{2} \left\{ \frac{(\sum_{i \in N} y_{ij} h_i) \lambda \sum_{i \in N} y_{ij} h_i E[\beta d(x^j, i)/v + \tilde{w}_{ix^j}]^2}{2\left[1 - \lambda \sum_{i \in N} y_{ij} h_i(\beta d(x^j, i)/v + w_{ix^j})\right]} + \sum_{i \in N} y_{ij} h_i d(x^j, i)/v \right\}$$

(11.42)

s.t. $\quad \sum_{j=1}^{2} y_{ij} = 1 \qquad i \in N$

$$\sum_{i \in N} (y_{ij} h_i(\beta d(x^j, i)/v + w_{ix^j}) < \frac{1}{\lambda} \qquad j = 1, 2$$

$$y_{ij} \in \{0,1\} \quad i \in N \quad j = 1,2.$$

The second set of constraints above guarantee that an infinite queue does not form in any district.

Due to the non-linearities in the formulation above, an exact optimization approach is not practical. Instead, the following heuristic algorithm was developed. The heuristic to solve the districting problem starts with an initial solution (which may, for example, be the optimal solution for regions A or C). Then the method of convex combination Wagner (1975) is applied. Essentially in this method a continuous version of the problem is considered where at each iteration the solution is improved by finding a direction in which the objective function is decreased (this direction is found by using a linear approximation of the objective function in (11.42)). The solution obtained by this method may involve the splitting of at most one node between the two servers. In such a case the solution is expressed as a convex combination of two integer solutions that are identical except for the split node. A sequence of node switches is then performed to find an improved solution.

To find simultaneously the optimal location and districts for the two stations, the following method was developed in Berman and Mandowsky (1986). Given an initial solution (x_0^1, x_0^2), "optimal" districts (N_1^1, N_1^2) can be obtained using the method shown above. Now, given (N_1^1, N_1^2), and using $\lambda^j = \lambda h^j$ (h^j is defined using N_1^j for $j = 1, 2$), the best location of the server in each district is found using the method to find an SQM as described in Section 11.3.2. Using the new solution (x_1^1, x_1^2), a new districting policy (N_2^1, N_2^2) is obtained by again applying the districting heuristic. The algorithm continues in this fashion (alternating allocation and location steps), until there is no further improvement in the objective function value.

To obtain the optimal location $(x^1, x^2, \ldots, x^M)$ of M stations and corresponding optimal districts $(N^1, N^2, \ldots, N^M)$ the method for $p = 2$ is modified in Berman and Mandowsky (1986). Since we do not know *a priori* if the value of λ lies in regions A, B or C, the optimal solutions for regions A and C are both used as initial solutions for the iterative sequence of solution improvements. Given an initial solution $(N_0^1, N_0^2, \ldots, N_0^M)$, an optimal location vector $(x_1^1, x_1^2, \ldots, x_1^M)$ is found by solving the SQM problem for each N_0^i. Now, for each pair of stations the best districting policy is found while keeping the other districts and stations unchanged. A cycle is defined as an application of the districting problem to all $\binom{M}{2}$ pairs. At the end of a cycle the best districting solution is taken as the initial solution of the next cycle. To start the next cycle, we first find new optimal locations for the districts that have changed, and then apply the districting cycle. This process continues until no further improvement is made.

11.3.4 M Stations, Single Unit at a Station, Server Co-operation

Here we consider the general problem (11.30) of Section 11.3.1 when the number of service units in each station is one, and when the capacity of the queue $c = \infty$. The problem when co-operation between stations is allowed is considered by Berman, Larson and Parkan (1987).

Assume, for now, that all the locational decisions have been made. Recall the state-space description of the system developed at the end of Section 11.3.1, where each state is described by an $M+1$-dimensional vector with the first M dimensions representing the number of busy servers at each facility, and the last dimension representing the number of calls waiting in the buffer. Because we assume that there is only one service unit at each facility, a type-F state (i.e., a state with at least one free server) is described by an M-dimensional binary vector with at least one component not equal to 1, while a type-B state (a state where all servers are busy) is described by vector of M ones followed by a non-negative integer.

The key idea here is to use the Hypercube model of Larson (1974) to approximate our system as an $M/M/M$ queue with distinguishable servers. This allows us to compute the proportion of each customer node's demand that is handled by a particular facility - which is roughly equivalent to the "allocation step" of the algorithm described in the previous section. We can then employ a heuristic search procedure to perform the "location" step, and repeat the process until no further improvements to the objective function can be made.

Assume for the time being that the travel time components of the service times are negligible and the on-scene times are exponentially distributed. Thus, the service times are exponentially distributed, and the Hypercube model can be used to compute steady-state behaviour of the system. This model explicitly takes into account our dispatching rules (e.g., always dispatching the closest available server to the call). Using the steady-state probabilities $P(F_s)$ and $P(B_s)$ for each state, the fraction of all dispatches that send server with home location at j to demand node i, f_{ij}, can be obtained ($\sum_{i,j} f_{ij} = 1$). These fractions take into account dispatches that do and do not incur any waiting time delay. A crucial step in obtaining accurate results from the Hypercube model is using the "Mean Time Calibration" procedure, which allows us to (partially) relax the assumption that travel times are negligible. The procedure works as follows. Given an initial location vector $\mathbf{x}_0$ the Hypercube model is run with service rates μ^j for each service unit j that ignore the travel time components. After running the Hypercube model μ^j are modified by including in $1/\mu^j$ the travel time components $\sum_{i \in N} h_i^j (d(x^j, i)\frac{\beta_j}{v} + w_{ij})$ derived from the Hypercube model. The Hypercube model is then run again using the new value of μ^j, and so on until μ^j values do not change appreciably.

Using the estimated f_{ij} $i \in N$, $j = 1, \ldots, M$, the parameter h_i^j (representing the proportion of calls from i that are handled by server at j) can be approximated by $\hat{h}_i^j$:

$$\hat{h}_i^j = \frac{f_{ij}}{\sum_{n \in N} f_{nj}} \qquad i \in N. \tag{11.43}$$

At this point the "allocation" step is complete and we have all the information required to perform the "location" step. Two heuristics are proposed for this purpose in Berman et al. (1987). Heuristic 1, which is a modification of a heuristic suggested by Jarvis (1976), is based on the 1-median problem, whereas heuristic 2 utilizes the SQM model described earlier.

The "1-median heuristic" concentrates on minimizing the average travel time (11.32). Using $\hat{h}_i^j$ in place of h_i, we can solve the 1-median problem to find the best location for station j. This procedure is repeated for all $j = 1, \ldots, M$.

The "SQM heuristic" works similarly, except that the $\hat{h}_i^j$ is also used in place of h_i in the expressions for first and second moments of the service time (11.35,11.36). In addition, we also require an estimate of λ_j - the total arrival rate faced by facility j. This estimate is given by

$$\hat{\lambda}_j = 2 / \left[2S(x^j) + \frac{S^2(x^j)}{T_W(\mathbf{x})} \right] \qquad j = 1, \ldots, M.$$

Here $\hat{\lambda}_j$ is the arrival rate that gives the same expected response time using the $M/G/1$ queuing system and the Hypercube model.

Once the new optimal location for each facility is computed, the Hypercube model is ran again to re-estimate the traffic proportions $\hat{h}_i^j$. This process is repeated until the objective function stabilizes.

Computational results in Berman et al. (1987) show that except for intermediate values of λ, the two heuristics described above provide very similar results. For intermediate values of λ, the SQM heuristic appears to be superior.

Finally, we note that the Hypercube model describes the behaviour of individual servers, not service facilities. Thus, multi-server facilities can be modelled in the "allocation" step by simply specifying the same home locations for some of the servers (in a way, it is best to think of the state-space description above as $K + 1$-dimensional, rather than $M + 1$-dimensional). In the "location" step, it is entirely possible that the optimal locations of some of the service units will coincide. Thus, the method described above might lead to several service units being located at one facility.

If the number of service units at each station is fixed in advance, then the K-unit 1-facility SQM model can be used as the basis of the locational heuristic. However, to solve a "true" problem with multi-unit facilities, the number of service units to locate at each station must itself be a part of the decision problem. In addition, the costs must be taken into account, as it is likely cheaper to house an extra unit at an existing facility than to open a new facility dedicated to that unit. None of the approaches described above account for these additional features.

11.3.5 General Number of Stations with Call-to-Call Travel

Until now we have assumed that the service units always return to the home location following the completion of a call for service. This assumption is not very realistic in some applications: for example, police units usually do not return to their station house before the end of their shift. The assumption is more sound in case of firefighting units, but only tenuous in case of ambulances. For mobile non-emergency servers, call-to-call travel also appears to be the norm, rather than the exception.

In a recent work, Berman and Vasudeva (2000) considered the problem of locating general number of service units for the case $c = \infty$ when service units return to their home location only if no calls are waiting for service; otherwise units will travel from call to call to provide the service required. The resulting problem is quite complicated even in the 1-facility, 1-service unit case since the service times of different calls for service are no longer independent. We briefly sketch out the approach of Berman and Vasudeva (2000) below.

As usual, we start by dealing with a 1-facility, 1-service unit problem. The following approximation is used to estimate the expected response time: it is assumed that the system functions as a modified $M/G/1$ with different service time distributions for calls that initiate a busy period and calls that occur during the busy period. This assumption is motivated by the fact that call-to- call travel occurs during the busy periods, while "regular" facility-to-call travel prevails during the slack periods. Based on results in Odoni (1969), the following expression for $T_W(x)$ is used:

$$T_W(x) = \frac{\lambda}{1 - \lambda(S' - S(x))} \left[\frac{A}{2(1 - \lambda S')} \right] \quad \text{if } 1 - \lambda S' > 0 \qquad (11.44)$$

where

$$A = \sigma_1^2(x) + (S(x))^2 + \lambda[S(x)\{\sigma_2^2 + (S')^2\} - S'\{\sigma_1^2(x) + S(x)^2\}] \qquad (11.45)$$

and $S(x)$, S', $\sigma_1^2(x)$ and σ_2^2 represent the means and variances of the service time for calls initiating a busy period and for calls that occur during the busy period, respectively. The model described in Berman and Vasudeva (2000) estimates ρ – the fraction of time that a service unit is busy, $S(x), S', \sigma_1^2(x)$ and σ_2^2. This approach is then generalized to the 1-facility, K-service units case by employing the approximations of Nozaki and Ross (1978) for $T_W(x)$ as in Section 11.2.1 above.

Finally, to extend the model to the multiple facility case, the general location-allocation approach of the preceding section is employed. A generalisation of the Hypercube model is used to estimate the fraction of calls from each customer node handled by each facility. These estimates are then used in the 1-facility model above (either the 1-server or the K-server version can be employed) to find a (possibly) new location for each facility, and the process is repeated until no further improvements in the objective function are made. We refer the reader to Berman and Vasudeva (2000) for full details. Of course, this approach can only treat the case where the number of service units at each station is pre-determined.

We also note that the 1-facility, 1-service unit problem with the call-to-call regime and $c < \infty$ was analysed in Berman and Vasudeva (2000a) and Jamil et al. (1994). However, no multi-facility extensions of these approaches have been proposed to date.

11.4 Conclusions and Open Problems

In this Chapter we have attempted to provide an overview of facility location models with stochastic demand and congestion - which we called "LPSDC models". After defining a general LPSDC model in Section 11.1, we have overviewed the two main streams of research in LPSDC models: the coverage-type models in Section 11.2, and median-type models in Section 11.3. We have tried to overview both, the stochastics of the underlying system, as well as the computational approaches employed by various models.

It is fair to say that, in spite of the large body of literature on LPSDC models, overall the field remains quite open. It suffices to point out that none of the models attempt to capture all of the relevant features specified in our general formulation in Section 11.1. This, in large part, is due to the difficulties posed by the underlying model, and the ever-present struggle between modelling realism and solvability.

The coverage-type models of Section 11.2 tend to make very strong assumptions about the stochastics of the underlying system. In many cases, the effects of such crucial features as dispatching rules, degree of server cooperation, etc. - that certainly have strong effects on the performance of underlying system - are not captured by the models. However, by using a simplified representation of the system behaviour, these models do lead to quite tractable formulations that can often be solved to optimality for realistic-size systems using standard integer programming software or efficient heuristic algorithms. It should be kept in mind, however, that not much is known about the quality of the underlying solutions. Basic questions remain, such as: do the systems designed as a result of these models indeed provide "adequate coverage" (as expressed by the coverage standards embedded in the models) under realistic conditions? Does a particular model tend to over- or under-estimate the required resources, and under what conditions does over- or under-estimation tend to occur? While these questions might prove too difficult for rigorous theoretical analysis, much insight could be gained from intensive computational testing using emergency system simulators, that have gained a wide measure of acceptance in practice. In particular, implementations of the Hypercube model are used by many services as a test bed for various operational decisions. These implementations could be used to test system configurations designed by various LPSDC models as well. While it is obvious that many of the underlying assumptions made by the models described in Section 11.2 are not realistic, it is difficult to predict to what extent this affects the quality of the resulting solutions. However, we feel that following the testing suggested above, it might be possible to recommend the better-performing models for practical design of emergency systems

The median-type models overviewed in Section 11.3 tend to use more sophisticated stochastic representations of the system - pushing, in many cases, against the very boundaries of the state of the art in queuing theory. The price paid for this realism is the difficulty of the resulting models, for which, in many cases, only heuristic algorithms are available. In addition, more theoretical work is required to evaluate the performance of these algorithms - basic properties like convergence guarantees or error bounds are generally not available (even though the methods do seem to perform well in computational experiments). While the assumptions made by median-type models are, in general, milder than for their cover-type counterparts, some of them (e.g., the "facility-to-call travel assumption") are nevertheless quite strong and could potentially have serious impact on the resulting solutions. Again, a thorough testing of the performance of the resulting solutions using system simulators would seem to be in order.

Perhaps the most promising field of research in this area are models for non-emergency fixed service facilities. In spite of the obvious importance of this subject, only the first steps have been made in modeling these systems (some of the models are discussed in Section 11.2.2). On the other hand, as discussed in Sections 11.1 and 11.2 above, the stochastic behaviour of the underlying system is substantially simpler than for the case of mobile servers, and can be described using the available results from queuing theory. While these problems are by no means easy, we think further research in this area could well lead to very fruitful results.

Acknowledgement This research was supported by NSERC. We wish to thank the referees for their very useful suggestions on the initial version of this paper.

References

Ball, M. and F. Lin (1993) "A Reliability Model Applied to Emergency Service Vehicle Location," *Operations Research*, 41,18-36.

Batta, R. and O. Berman (1989) "A Location Model for a Facility Operating as an $M/G/k$ Queue," *Networks*, 19, 717-729.

Batta, R.,J. Dolan and N. Krishnamurthy (1989) "The Maximal Expected Covering Location Problem: Revisited," *Transportation Science*, 23, 277-287.

Berman O., R. Larson, A. Odoni, S. Chiu and R. Batta (1990) "Locations of Mobile Units in a Stochastic Environment," *Discrete Location Theory*, P. Mirchandani and R. Francis, Editors, John Wiley and Sons, NY.

Berman, O. and R.C. Larson (1985) "Optimal 2-Faciltiy Network Districting in the Presence of Queuing," *Transportation Science*, 19, 261- 277.

Berman, O., R.C. Larson and S.S. Chiu (1985) "Optimal Server Location on a Network Operating as an $M/G/1$ Queue," *Operations Research*, 33, 746-771.

Berman, O., R. Larson and C. Parkan (1987) "The Stochastic Queue p-Median Problem," *Transportation Science*, 21, 207-216.

Berman, O. and R.R. Mandowsky (1986) "Location-Allocation on Congested Networks," *European Journal of Operational Research*, 26, 238- 250.

Berman, O. and S. Vasudeva (2000) "Approximating Performance Measures for Public Services," working paper, Joseph L. Rotman School of Management, University of Toronto.

Berman, O. and S. Vasudeva (2000a) "Approximating Performance Measures for a Network of Unreliable Machines," working paper, Joseph L. Rotman School of Management, University of Toronto.

Brendau, M.L., Chiu, S.S., Kumar, S., and T.A. Grossman (1998), "Location With Markte Externalities", in *Facility Location: A Survey of Appilcations and Methods*, Z. Drezner, ed., Springer, 121-150.

Chapman, S.C. and J.A. White (1974) "Probabilistic Formulations of Emergency Service Facilities Location Problems," ORSA/TIMS Conference, San Juan, Puerto Rico.

Chiu, S.S. and R.C. Larson (1985) "Locating an n-Server Facility in a Stochastic Environment," *Computers and Operations Research*, 12, 509-516.

Daskin, M.S. (1983) "A Maximum Expected Covering Location Model: Formulation, Properties and Heuristic Solution," *Transportation Science*, 17, 48-70.

Jamil, M., R. Batta and D.M. Malon (1994) "The Travelling Repairpersons Home Base Location Problem," *Transportation Science*, 28, 2, 150-161.

Jarvis, J.P. (1976) "A Location Model for Spatially Distributed Queueing systems," *Proceedings of the 1976 IEEE-SMC International Conference on Cybernetics and Society*, 32-35.

Kleinrock, L. (1975) *Queueing Systems*, John Wiley and Sons, New York, NY.

Larson, R.C. (1974) "A Hypercube Queuing Model for Facility Location and Redistricting in Urban Emergency Services," *Computers and Operations Research*, 1, 67-95.

Larson, R.C. (1975) "Approximating the Performance of Urban Emergency Service Systems," *Operations Research*, 845-868.

Marianov V. and C. ReVelle (1994) "The Queueing Probabilistic Location Set Covering Problem and Some Extension'" *Socio-Economic Planning Sciences*, 28, 3, 167-178.

Marianov V. and C. ReVelle (1996) "The Queueing Maximal Availability Location Problem: a Model for the siting of Emergency Vehicles," *European Journal of Operations Research*, 93, 110-120.

Marianov V. and D. Serra (1998) "Probabilistic Maximal covering Location-Allocation for Congested System." *Journal of Regional Science*, 38, 3, 401-424.

Marianov V. and M. Rios (2000) "A Probabilistic Quality of Service Constraint for a Location Model of Switches in ATM Communications Networks," *Annals of Operations Research*, accepted for publication.

Odoni, A.R. (1969) "An Analytical Investigation of Air Traffic in the Vicinity of Terminal Areas," Ph.D. dissertation, Massachusetts Institute of Technology.

Nozak, A. and S.M. Ross (1978) "Approximating In-Finite Capacity Multi-Service Queues with Poisson Arrivals," *Journal of Applied Probability*, 15, 826-834.

ReVelle, C. and K. Hogan (1989) "The Maximum Reliability Location Problem and α-Reliable p-center Problem: Derivatives of the Probabilistic Location Set Covering Problem" *Annals of Operations Research*, 155-174.

ReVelle, C. and K. Hogan (1989a) "The Maximum Availability Location Problem" *Transportation Science*, 23(3), 192-200.

Wagner, H.M. (1975) *Principles of Operations Research*, Prentice-Hall, New York, Chapter 5.

12 Hub Location Problems

James F. Campbell[1], Andreas T. Ernst[2], and Mohan Krishnamoorthy[2]

[1] College of Business Administration, University of Missouri - St Louis, 8001 Natural Bridge Road, St Louis, Missouri 63121-4499, USA.
email: campbell@umsl.edu

[2] CSIRO Mathematical and Information Sciences, Private Bag 10, Clayton South MDC, Vic 3169, Australia.
email: {andreas.ernst,mohan.krishnamoorthy}@cmis.csiro.au

12.1 Introduction

Hub location research has become an important area of location theory over the past two decades. This is due in large part to the use of hub networks in modern transportation and telecommunication systems. These systems serve demand for travel or communication between many origins and many destinations, where economies of scale exist in the cost for such travel or communication. Rather than serving every origin-destination demand with a direct link, a hub network provides service via a smaller set of links between origins/destinations and hubs, and between pairs of hubs. Such a network allows a large set of origins and destinations to be connected with relatively few links, via central hub facilities. The use of fewer links in the network concentrates flows and allows economies of scale to be exploited. Hub location problems involve locating hub facilities and designing hub networks.

Hub location problems differ from classical facility location problems in several key ways. In a classical discrete facility location problem, demand for service occurs at discrete points, facilities are located at discrete points and the objective is generally related to the distance or cost between the facilities and the demand points. In hub location problems demand is specified as flows between many origins and many destinations, and hub facilities serve as switching (or connection) and consolidation (or concentration) points for the origin-destination flows. As a switching point, a hub allows flows to be redirected. In addition, a hub may perform a consolidation or concentration function to combine many small separate flows (for example, from different origins) into larger flows. Hubs may also perform the opposite function to split a large flow into separate smaller flows for different destinations. Thus, hubs are intermediate points along the paths followed by origin-destination flows. (When a hub corresponds to an origin/destination for a particular flow, then it is also the endpoint of the flow.)

Transportation applications of hub location models include air passenger travel, air freight travel, express shipments (for example, overnight delivery systems), large trucking systems, postal operations and rapid transit systems. Demand is usually specified as flows of passengers or goods between city pairs,

---

Facility Location: Applications and Theory.
Edited by Z. Drezner and H.W. Hamacher
© 2002 Springer-Verlag, ISBN 3-540-42172-6

and these flows are transported in vehicles of some type (aircraft, motor vehicles, trains, etc.). Hub facilities are transportation terminals or sorting centers. The geographic scope of transportation hub location problems may be local (for example, within a city) or global. The strong economies of scale in transportation encourage the consolidation of flows.

Telecommunications applications of hub location models include a wide variety of distributed data networks in areas such as computer communication, telephone networks, video teleconferences, distributed computer processing, etc. Demand is for transmission of information (data, voice, video, etc.), and this occurs over a variety of physical media (for example, telephone lines, fiber optic cables, or co-axial cables) or through the air, as with satellite channels and microwave links. Hub facilities are generally electronic devices such as multiplexors, concentrators, switches, gates, etc. The scope of telecommunications hub location systems ranges from local networks (for example, with a single building) to global networks utilizing satellites. The large fixed costs for communication links, and the economies of scale in transmission and utilization, encourage the development of hub networks.

The focus of this chapter is on new discrete or network hub location models in which location of the hubs is a key decision. Research that addresses hub network design, without determining hub locations is (generally) not considered. Thus, the considerable research on design of networks for a given set of hub (or backbone) nodes is not covered here. (See Gendron et al. (1999) and Klincewicz (1998) for references.) Further, our main focus is on new work, especially that not covered in previous reviews and surveys of hub location research.

12.2 Background

This section provides background on hub location research and highlights previous hub location review papers. It also indicates links to related literature and summarizes the early work on hub location models.

12.2.1 Previous Reviews

The two earliest reviews of hub location research are from 1994 by Campbell (1994a) and O'Kelly and Miller (1994). Campbell provided a survey of the growing body of network hub location research and presented a classification scheme for the different models and problems considered. O'Kelly and Miller focussed more on the topological alternatives available in hub networks. These two works summarized the early research on mathematical approaches to hub location problems, which began with the seminal work of O'Kelly for continuous (O'Kelly, 1986b) and discrete location (O'Kelly, 1987). (Perhaps the earliest hub location model is in Goldman (1969) which extends the node optimality property of Hakimi (1964, 1965) to what is essentially the hub median problem.)

With the burst of activity in hub location research through the 1990's two more recent review papers have appeared. Klincewicz (1998) reviewed work in the telecommunications area that included the design of hub networks *and* the location of hub nodes. Bryan and O'Kelly (1999) surveyed work primarily in the context of air transportation and identified directions for future research.

The purposes and foci of these review papers have differed somewhat, in part reflecting the authors' orientations and interests, as well as an indication of the different disciplines involved. Hub location has been studied from various perspectives including geography, regional science, location theory, operations research, transportation, telecommunications and computer science. Research efforts are currently underway in all these areas by researchers around the globe.

12.2.2 Fundamental Hub Location Models

One result of the early research in hub location was exploration of the linkages between hub location models and classical facility location models. For each of the fundamental classical discrete facility location problems, (p-median problem, uncapacitated facility location problem, p-center problem, covering problems), analogous hub location problems have been formulated and studied: p-hub median problem, uncapacitated hub location problem, p-hub center problem, and hub covering problems (Campbell, 1994b).

One important difference between classical (non-hub) facility location and hub location, is that multiple and single allocation versions exist for hub location problems. In a single allocation hub location problem, each demand point must be allocated to (communicate with) exactly one hub node. Thus, all flows to and from each demand point travel via the same hub node. In a multiple allocation hub location problem each demand point may be allocated to (communicate with) more than one hub. The greater flexibility with multiple allocation allows lower cost solutions, and simplifies solution, since for a given set of hub nodes, each origin-destination flow can be routed separately from all others via the least cost path. Thus, each type of hub location problem has a single and multiple allocation variant. Note that most classical uncapacitated facility location problems involve single allocation, since each demand point is allocated to the nearest (or least costly) facility.

The hub location literature also includes capacitated problems, just as in the classical facility location literature. However, capacities in hub location problems differ from those in classical facility location problems because of the different nature of demand. In a hub location problem there can be capacities on the hub nodes as well as on the flows between hubs or between hubs and non-hubs. Capacities at the hub nodes could be for the total flow through the node, or just for the flow into the node, as in the postal sorting application discussed in Ebery at al. (2000), Ernst and Krishnamoorthy (1999).

The hub location problems that have received the most attention from researchers so far are the p-hub median problem, and the capacitated and uncapacitated hub location problem. We now present basic mixed integer linear programming formulations for these problems. Each problem can be formulated in a variety of ways (see Campbell (1994b) for a discussion of alternate formulation approaches), but the most effective general approach seems to be that introduced in Ernst and Krishnamurthy (1996, 1998b) based on tracking *flows* on arcs for each specific origin. Before presenting the MILP formulations we provide the necessary notation.

Consider a complete graph $G = (V, E)$ with a node set $V = \{v_1, \ldots, v_M\}$, where nodes correspond to origins/destinations and potential hub locations. The given demand for flow between node i and node j is W_{ij} and the distance from node i to node j is d_{ij}, where these distances satisfy the triangle inequality. Denote the number of hubs to locate by p. Each origin-destination path can be viewed as consisting of three components: collection from an origin to the first hub, transfer between the first and last hub, and distribution from the last hub to the destination. Paths involving only a single hub are also possible and can be thought of as a special case in which the transfer is a null step. Parameters χ, α and δ reflect the unit costs for collection (origin-hub), transfer (hub-hub), and distribution (hub-destination), respectively. Thus, the origin-destination path from non-hub origin i to non-hub destination j via hubs k and l ($i \to k \to l \to j$), incurs a cost $\chi d_{ik} + \alpha d_{kl} + \delta d_{lj}$ per unit flow. Generally α is used as a discount factor to provide reduced unit costs on arcs between hubs to reflect economies of scale, so $\alpha < \chi$ and $\alpha < \delta$. Thus, the basic hub location model can be viewed as a two level network where the access level includes arcs connecting non-hub origins and destinations to hubs, and the hub level includes transfer arcs connecting hubs.

Basic hub location models assume that every origin-destination path includes at least one hub node (i.e. all flows are routed via at least one hub), and that the cost per unit flow is discounted between *all* hub pairs using α. The effect of this, along with the triangle inequality, is that the access level network consists of single arcs connecting non-hubs to hubs, and the hub level network is a complete graph on the hubs. Basic hub location models have also focused primarily on costs for flows in the network and for fixed costs of hubs.

To model the basic hub location problems we define three sets of decision variables corresponding to the three components of an origin-destination path. The collection and distribution components each involve a single access arc, which might be from a node to itself, if it is a hub. In the basic models the transfer component involves a single arc, though in more complex models the transfer component may involve several arcs. Note also that the transfer component does not exist if the origin-destination path is

$origin \to hub \to destination$. The decision variables are:
Z_{ik} = flow from origin i to hub k,
Y_{kl}^i = flow from hub k to hub l that originates at origin i, and
X_{lj}^i = flow from hub l to destination j that originates at origin i.

In addition to the decision variables for flows on arcs, we have binary variables for locating hubs: $H_k = 1$ if node k is a hub, and 0 otherwise.

The uncapacitated multiple allocation p-hub median model (sometimes denoted UMApHMP in the literature) is to locate p hub facilities to minimize the total flow cost. This problem will be denoted as p-hub/D/MA/•/$\sum_{flow}$ using the standard classification scheme for location problems (more details on the way hub location problems are classified in this scheme are given in Sect. 12.4). Using the above variables this problem can be written as:

Problem p-hub/D/MA/•/$\sum_{flow}$

$$\text{Min.} \sum_{i \in V} \left[\sum_{k \in V} \chi d_{ik} Z_{ik} + \sum_{k \in V} \sum_{l \in V} \alpha d_{kl} Y_{kl}^i + \sum_{l \in V} \sum_{j \in V} \delta d_{lj} X_{lj}^i \right] \quad (12.1)$$

$$\text{Subject to} \quad \sum_{k \in V} H_k = p, \quad (12.2)$$

$$\sum_{k \in V} Z_{ik} = O_i, \quad \forall \, i \in V, \quad (12.3)$$

$$\sum_{l \in V} X_{lj}^i = W_{ij}, \quad \forall \, i, j \in V, \quad (12.4)$$

$$Z_{ik} + \sum_{l \in V} Y_{lk}^i = \sum_{l \in V} Y_{kl}^i + \sum_{j \in V} X_{kj}^i, \quad \forall \, i, k \in V, \quad (12.5)$$

$$X_{lj}^i \leq W_{ij} H_l, \quad \forall \, i, j, l \in V, \quad (12.6)$$

$$Z_{ik} \leq O_i H_k, \quad \forall \, i, k \in V, \quad (12.7)$$

$$Z_{ik}, Y_{kl}^i, X_{lj}^i \geq 0, \quad \forall \, i, j, k, l \in V, \quad (12.8)$$

$$H_k \in \{0, 1\}, \quad \forall \, k \in V. \quad (12.9)$$

where:
O_i = total flow originating at origin i:

$$O_i = \sum_{j \in V} W_{ij}, \quad \forall \, i \in V. \quad (12.10)$$

The objective (12.1) sums the cost for collection, transfer and distribution. Constraint (12.2) ensures that the appropriate number of hubs are selected. Constraint (12.3) ensures that all flow from each origin leaves the origin. Constraint (12.4) ensures that all flow for each origin-destination pair arrives at the proper destination. Constraint (12.5) is the flow conservation equation at the hubs. Constraints (12.6) and (12.7) ensure that hub nodes are established for every distribution and collection movement, respectively.

The uncapacitated single allocation p-hub median problem is similar to the above problem, but each non-hub node is restricted to be allocated to a single node. This can be formulated similarly, but we can restrict the Z_{ik} variables to be binary and eliminate the X_{lj}^i variables. We can also replace the binary H_k variables for locating hubs with Z_{kk}. Now the decision variables are:

$Z_{ik} = 1$ if node i is allocated to a hub at node j, and 0 otherwise, and
$Y_{kl}^i = $ flow from hub k to hub l that originates at origin i.

The uncapacitated single allocation p-hub median problem (sometimes denoted USApHMP in the literature) can then be formulated:

Problem p-hub/D/SA/•/$\sum flow$

$$\text{Min.} \sum_{i \in V} \sum_{k \in V} d_{ik} Z_{ik} \left(\chi O_i + \delta D_i \right) + \sum_{i \in V} \sum_{k \in V} \sum_{l \in V} \alpha d_{kl} Y_{kl}^i \quad (12.11)$$

$$\text{Subject to} \sum_{k \in V} Z_{kk} = p, \quad (12.12)$$

$$\sum_{k \in V} Z_{ik} = 1, \quad \forall\, i \in V, \quad (12.13)$$

$$\sum_{j \in V} W_{ij} Z_{jk} + \sum_{l \in V} Y_{kl}^i = \sum_{l \in V} Y_{lk}^i + O_i Z_{ik}, \quad \forall\, i, k \in V, \quad (12.14)$$

$$Z_{ik} \leq Z_{kk}, \quad \forall\, i, k \in V, \quad (12.15)$$

$$Y_{kl}^i \geq 0, \quad \forall\, i, k, l \in V, \quad (12.16)$$

$$Z_{ik} \in \{0,1\}, \quad \forall\, i, k \in V. \quad (12.17)$$

where:
$D_i = $ total flow destined for destination i:

$$D_i = \sum_{j \in V} W_{ji}, \quad \forall\, i \in V. \quad (12.18)$$

The objective (12.11) sums the cost for collection, distribution and transfer. Constraint (12.12) ensures that the appropriate number of hubs are selected. Constraint (12.13) ensures that each non-hub is allocated to a single hub. Constraint (12.14) is the flow conservation equation at the hubs. Constraint (12.15) ensures that hub nodes are established for every distribution and collection movement.

The multiple and single allocation uncapacitated hub location problems are similar to the multiple and single allocation hub median problems, except that the number of hubs is not specified explicitly (as in constraint (12.2) or (12.12)), but is determined implicitly by the optimization. Thus the uncapacitated multiple allocation hub location problem (sometimes denoted UMAHLP in the literature) can be formulated:

Problem hub/D/MA/•/$\sum_{flow} + \sum_{hub}$

$$\text{Min.} \sum_{i \in V} \left[\sum_{k \in V} \chi d_{ik} Z_{ik} + \sum_{k \in V} \sum_{l \in V} \alpha d_{kl} Y_{kl}^i + \sum_{l \in V} \sum_{j \in V} \delta d_{lj} X_{lj}^i \right] \quad (12.19)$$
$$+ \sum_{k \in V} F_k H_k$$

Subject to (12.3) – (12.9)

where F_k is the fixed cost of locating a hub at node k.

The uncapacitated single allocation hub location problem (sometimes denoted USAHLP in the literature) can be formulated similarly:

Problem hub/D/SA/•/$\sum_{flow} + \sum_{hub}$

$$\text{Min.} \sum_{i \in V} \sum_{k \in V} d_{ik} Z_{ik} (\chi O_i + \delta D_i) + \sum_{i \in V} \sum_{k \in V} \sum_{l \in V} \alpha d_{kl} Y_{kl}^i + \sum_{k \in V} F_k Z_{kk} \quad (12.20)$$

Subject to (12.13) – (12.17)

The capacitated versions of the above hub problems can be formulated by adding a constraint to restrict the total flow at a hub. To restrict the total flow through hub node k to a maximum of Γ_k add:

$$\sum_{i \in V} \left[Z_{ik} + \sum_{l \in V} Y_{lk}^i \right] \leq \Gamma_k H_k \quad \forall k \in V \quad (12.21)$$

To restrict only the flow collected at a hub, as in the postal application described in Ebery at al. (2000), Ernst and Krishnamoorthy (1999) where the capacity is on the amount that can be sorted at a hub and a hub sorts only the inbound flow, add:

$$\sum_{i \in V} Z_{ik} \leq \Gamma_k H_k \quad \forall k \in V \quad (12.22)$$

12.2.3 Related Literature

There is considerable literature on a variety of problems closely related to discrete hub location problems. This includes research on continuous space hub location problems, where the hub locations are allowed to be located anywhere in a continuous region. See Aykin (1995b), Aykin and Brown (1992), O'Kelly (1986b), O'Kelly (1992a), O'Kelly and Miller (1991), Suzuki and

Drezner (1997) for examples. Detailed discussion of this literature is beyond the scope of this chapter.

One large area of related literature is the research on designing hub networks, but without the hub location component. The relevant models generally are those with two (or more) levels where the different levels form a hierarchy. For example in two level hierarchical networks the lower level may be a local access network connecting non-hub demand points to hubs, and the upper level may be a "backbone" network connecting all the hub nodes. Often a set of hub nodes, demand points, flows, and arc costs is given, and the problem is to design a minimum cost network that meets customer requirements.

There is a considerable literature on network design problems in which the location of the hub (backbone) nodes is specified. These are very difficult problems in general and there are many variants depending on the specific applications, the objective(s), and the constraints considered. See Crainic et al. (2000), Gendron et al. (1999), Klincewicz (1998), Lederer and Nambimadon (1998) for overviews of this area of research, including the design of local access networks. Note that the demand in much of this research is specified at nodes, rather than between node pairs as in hub location research. For example, local access network design problems may involve locating facilities to concentrate or consolidate flows (often called concentrators, multiplexors, or hubs) to exploit economies of sale, in addition to designing the network. In these problems all flows may have a common origin (or destination) corresponding to the node in the hub level network linked to the local access network. Constraints on the design of the local access network may include particular topologies, such as rings or trees, or require specified levels of reliability.

Another area of related research is multicommodity network flows with concave costs (Minoux, 1989). In these problems the demand for each origin-destination pair is treated as a separate commodity and the flow cost along an arc is a nondecreasing concave function of the flow on the arc. While hub location models have a discrete set of cost rates for flows, these models treat economies of scale more continuously with a concave cost function. Although the problem environment is similar to that for hub location models, multicommodity network flow models are generally concerned with designing a network and routing flows, not with locating facilities at nodes. These models are appropriate when the network topology is essentially given and the main aim is to determine capacities for the arcs.

A final area of related research is that on continuous demand many-to-many distribution problems with transshipments. In these works, the demand is treated as a continuous density over a geographic region, and analytical expressions can be derived for the optimal locations of hubs (transshipment facilities) and optimal average distance and cost. Models may include transportation, inventory and facility costs. Campbell (1993) derives analytical solutions under the assumption that the demand is uniformly distributed and using rectilinear distances. Daganzo (1999) provides a summary of this literature, and provides analytical formulae to evaluate the benefits of one or more transshipments.

12.2.4 "First Generation" Hub Location Models

The "first generation" of hub location research, summarized in Campbell (1994a) and O'Kelly and Miller (1994), produced important advances in understanding hub systems and developed basic models focused primarily on minimizing flow cost and fixed facility costs, for rather constrained situations where a single cost discount is used for *all* flows between hubs. One important consequence of discounting all flows between hubs was that the optimal hub level network was a complete graph on the hubs. Thus, the design of the hub level network was determined by the location of the hub nodes. Some researchers extended the basic models to include features such as direct origin-destination flows (not via a hub) (Aykin, 1994), capacities (Aikin, 1994), and additional objectives, including mode choice (O'Kelly and Lao, 1991) and congestion (Gavish, 1992; O'Kelly, 1986a).

In spite of the rather simple nature of the first generation hub location models, these have turned out to be quite difficult to solve; more difficult than the corresponding classical facility location problems. Much early research involved applying standard heuristic and optimal solution approaches of operations research to the basic hub location problems. While problems of moderate size (for example, 50 origins/destinations and 5 hubs) can be solved to optimality in reasonable times, larger more realistic problems are too computationally intensive. Thus, in spite of the apparent simplicity of the early hub location models, efficient optimal solution techniques have remained elusive and researchers appreciate well the complexity and difficulty of these problems.

12.3 Recent Trends

Following the successes achieved and challenges identified in the first generation of hub location research, a second generation of research has produced substantial progress. This research sought to address some shortcomings of the early hub location models, as well as to solve larger problems to optimality.

One area of focus in this new hub location research has been to extend the first generation models to be more realistic by integrating hub location and network design decisions. The basic hub location models are primarily location models, not network design models, since the hub locations defined both the hub level network (as a complete graph on the hubs) and the access network.

One approach to integrate network design and hub location was to consider a more complex and accurate hub level network. This includes models with flow dependent cost discounts (Bryan, 1998; Bryan and O'Kelly, 1999; O'Kelly, 1998; O'Kelly and Brian, 1998), models that relax the requirement that all flows between hub nodes are discounted Campbell et al. (2000a, 2000b), and models with minimum flow thresholds (Aykin, 1996; Podnar and Skorin-Kapov, 1999; Podnar et al., 1999).

Another approach to integrate network design and hub location was to consider a more complex and accurate access network. This includes models that allow direct origin-destination paths (not via a hub node) (Aykin, 1995a), multi-stop access paths (for example, *origin 1* → *origin 2* → *hub*) (Chou, 1990; Klincewicz, 1998), and collection and distribution routes (Nagy and Salhi, 1998), as in the location routing problem (Laporte, 1989).

Another area of research for enhancing hub location modeling was the level of service. The level of service may be viewed in terms of (maximum) path lengths, numbers of stops at hubs, or travel/transmission times. Note that in the basic hub location models the maximum path length is three arcs with a single transfer arc between hubs. No paths included multiple transfer arcs, since the hub level network was a complete graph on the hubs. To address alternate levels of service, models have restricted paths to a specific number of hub stops (or transfers) or to a specific number of arcs (O'Kelly, 1998a; Sasaki et al., 1997, 1999). Some research also addressed scheduling issues (Kara and Tansel, 1999b).

Researchers have also sought to extend hub location models by considering additional costs. Transportation oriented hub location resesarch had focussed mainly on flow costs, along with fixed costs for hubs when the number of hubs is not prescribed. Telecommunications oriented research, and network design research in general, have generally included fixed costs for arcs and hubs, along with other costs, possibly including flow costs and delay costs. Developing hub location models that include more of the relevant costs is important to produce more realistic and useful results.

Finally, several researchers extended hub location models to include new objectives. These works include minimizing the maximum cost or the latest arrival (Kara and Tansel, 1999b 2000), hub covering models (Kara and Tansel, 1999a) and hub location models with competition (Marianov et al., 1999).

While many researchers have concentrated on new models, others have sought better solution methods for the existing (first generation) hub location models. The goal is to solve larger problems to optimality. Many realistic hub location problems involve hundreds, or possibly a few thousand origins-destinations. Research in this area has involved both developing new solution algorithms, and adapting existing successful algorithms or approaches for non-hub location problems to hub location problems (for example Boland et al., 2000; Hamacher et al., 2000; Klincewicz, 2000; Mayer and Wagner, 1998).

12.4 Models and Taxonomy

In this section we provide an overview over the different variants of the basic hub location models that have been studied in the literature. In keeping with the remainder of this book, the problems have been classified according to the classification scheme for location problems by Hamacher and Nickel (1998).

The classification scheme has five positions. The first will always contain the word 'hub' to indicate that the new facilities to be located are hubs which enable communication between the existing facilities (nodes). In addition the first position may indicate the number of hubs to be located when this is prescribed as part of the problem.

The second position in the classification scheme is usually 'D', indicating that the location of hubs is discrete – they have to be located at one of the given nodes. Note that some authors restrict the possible location of hubs to be a subset of the nodes. We will not distinguish this in the classification as it does not substantially alter the problem; the restriction just makes it easier to solve.

Most of the additional constraints and variations in hub location models are indicated in the third position. The most common of these are multiple allocation (MA), single allocation (SA) and capacity constraints on hubs (cap), which have already been introduced above, but there are also several other options.

The fourth position, relating to the relationship between new and existing facilities is usually •, which in the hub location context denotes the usual distance function based on given inter-node distances d_{ij} with multipliers χ, α, δ. This could also be written more verbosely as $d(\mathcal{V}, \mathcal{V})$.

Finally the type of objective function is indicated in the last position. Normally this involves $\sum_{flow}$, denoting the minimization of the flow costs between all pairs of nodes. In hub location problems there is often an additional term $\sum_{hub}$ denoting the fixed costs of establishing hubs.

In the remainder of this section we provide an overview of the different variants of hub location problems that have been considered in the literature, in terms of their objectives, network components and constraints. In each case we provide an indication of how each variant fits into the classification scheme.

12.4.1 Objective

The objective in vast majority of hub location research is minimization of costs, where the particular types of cost(s) included depend on the application and context. Transportation oriented hub location research focuses on flow based transportation costs. Telecommunications oriented hub location research generally includes fixed cost for constructing or acquiring the network (for example, fixed costs for arcs and hubs). Some hub location researchers have also addressed non-cost objectives involving travel times, coverage measures, competition and congestion.

Hub median problems: These problems have the standard objective function as given in (12.1) and (12.11) minimizing the flow costs between all pairs of nodes for a given number of hubs. This corresponds to the median problem in (non-hub) facility location. In the classification scheme these are indicated by the objective function $\sum_{flow}$ and they usually have the number of hubs prescribed.

Hub location problems: In these problems the optimal number of hubs is determined as part of the problem, usually by incorporating the fixed cost of establishing hubs in the objective along with other costs, such as flow costs. The objective captures the tradeoff of increasing fixed costs from more hubs, with reduced flow costs from less circuitous routings. This option has been considered by a number of authors (e.g. Ernst and Krishnamoorthy, 1999; Klincewicz, 1996; Mayer and Wagner, 1998; O'Kelly, 1992b). O'Kelly et al., (1996) provide graphs showing the interaction between the fixed cost, the cost discount for inter-hub travel (α), and the number of hubs in the optimal solution.

Hub center problems: The p-hub center problem is to locate p hubs to minimize the maximum distance or cost between any pair of nodes. This is denoted by 'max' in the fifth position of the classification. The single allocation variant (p-hub/D/SA/•/max) of this problem has been considered by Kara and Tansel (2000). A variation of this is introduced in Kara and Tansel (1999b) where the objective is to minimize the maximum travel time between any pair of nodes including transit time at the hubs (p-hub/D/SA,transit/•/max). Another variation is to impose the maximum travel time as a constraint giving the required performance, and then to minimize the cost of setting up such a network (the number of hubs required), resulting in a covering problem which can be classified as hub/D/SA,max.time/•/$\sum_{hub}$ (Kara and Tansel, 1999a).

Fixed costs on arcs: Fixed costs for arcs are important in many applications, particularly in telecommunications networks, where the cost of establishing arcs is high compared to the cost of operating them. Note that arcs may represent links that are privately owned, such as telecommunication lines or railways, or publicly owned, such as airways in air transportation. Fixed costs for arcs are represented by a $\sum_{arc}$ term in the last position of the classification.

Flow (transportation) costs: Some models have adopted flow cost functions that are concave to reflect economies of scale (the cost per unit flow decreases as the flow increases) (Bryan, 1998; O'Kelly and Bryan, 1998, 2000). In particular these authors consider a piecewise linear, concave cost function for the flows between hubs that can be classified as (hub/D/MA/$f_{transfer}$: $convex/\sum_{flow}$). Note that this is equivalent to considering multiple possible discount factors α, but with corresponding fixed costs which increase for greater discounts. Modeling this problem as an integer program requires additional integer variables making it significantly harder to solve.

Other objectives: Some research has begun to address some hub location problems with different non-cost objectives. Marianov et al. (1999) consider hub location in a competitive model to maximize the flow captured. This can be classified as (hub/D/MA/•/$\sum_{flowcap}$).

12.4.2 Network Components

A hub network consists of two types of nodes connected by one or more types of arcs. The two types of nodes are non-hub origins/destinations and hub nodes.

Hub nodes: Discrete hub location models select hubs from a set of potential hub nodes. The set of potential hub nodes usually coincides with the demand points (origins/destinations) or is a subset of the demand points. However, potential hub nodes can also be points other than demand points.

Hub nodes generally provide a switching or sorting function and a consolidation or concentration function. Hubs that perform only the switching/sorting function, but not the consolidation/concentration function have been termed *isolated* hubs (Campbell et al., 1999). A number of authors have considered using two classes of hub nodes (Hall, 1989; O'Kelly and Lao, 1991; O'Kelly98a): mini-hubs that are used to serve flow within a region only and one or more major hub nodes that are used for flow between nodes in different regions.

Demand (origin-destination flows): Demand in hub location problems is specified as flows between origins and destinations. Generally this is provided as input data for each problem instance. The two most commonly used data sets for transportation oriented research are the CAB data sets based on air passenger travel in the USA in 1970 as collected by the Civil Aeronautics Board (see Fotheringham (1983) for details), and the AP data sets based on metropolitan operations of Australia Post (Ernst and Krishnamoorthy, 1996, 1999). The CAB data sets are based on continental scale flows of air passengers and the hubs are airport facilities where passengers can change aircraft. The AP data sets are based on flows of mail between post offices in a major metropolitan area and the hubs are mail sorting centers. One difference between these data sets is that in the AP data sets there are flows from a node to itself (which much visit a hub for sorting), while this does not occur in the CAB data sets.

In order to make the models more realistic it is also possible to consider the effects of competition rather than simply assuming a fixed given demand. Marianov et al. (1999) propose a model in which the amount of flow between a pair of nodes depends on the cost of travel between these nodes compared to a given price that represents the market value.

Arcs: Arcs form the network connecting non-hub nodes and hub nodes. Arcs between two hub nodes generally have a discounted cost rate per unit flow to reflect economies of scale from consolidation/concentration. These arcs with discounted cost rates are referred to as hub arcs (Campbell, 1998). While basic hub location models assumed that the hubs were fully connected by hub arcs, a more general hub arc location problem is to locate a fixed number of q hub arcs. In this model hub nodes are defined by the end points of the hub arcs. This type of model can be classified as q-hub-arc/D/MA/•/$\sum_{flow}$. Campbell et al. (2000a, 2000b) examine optimal solutions for several variants of the hub arc location model.

Some models allow a non-hub node to communicate directly with other non-hub nodes (not via a hub) (Aykin, 1994, 1995b). The possibility that the hub nodes are not used at all in this type of model can be prevented by either limiting the node pairs that can communicate directly or imposing a significantly higher cost per unit flow on these direct links than for collection and distribution links (or by both of these methods). In the classification scheme this type of model is indicated with the word 'direct' in the third position to show that as well as being allocated to a hub, a node can also have direct connections to other nodes.

Network topology: The basic single allocation hub location problems have their network topology defined by the assumption that the hub nodes form a complete graph, and that the allocation of non-hub nodes to hubs is in the form of a star network. Several other topologies are possible. For example if the hub nodes are connected by a tree the overall network structure is that of a spanning tree. This type of topology has been considered by several authors under slightly different names including hierarchical hub model (Chou, 1990), hierarchical tree-star network (Kim and Tcha, 1992) and digital data service network (Lee et al., 2000). In the classification scheme this restriction can be indicated by including a $\mathcal{T}$ in the third position.

Another option, in which the hub network is a ring (cycle), can be indicated by "ring" in the third position. This topology is more important in telecommunications than transportation. Lee et al. (1993) considers such a hub location and network design problem with ring topology and single allocation (hub/D/SA,ring/•/$\sum_{hub} + \sum_{arc}$). Klincewicz (1998) describes other instances with ring topologies.

While the telecommunications literature considers a variety of alternative topologies for the access level network (see Klincewicz (1998)), very little transportation oriented work has been published that does not include direct links from each origin/destination to a hub. One recent paper by Nagy and Salhi (1998) proposes a network in which the allocation part consists of a multiple vehicle routing problem using the hub as a depot (p-hub/D/veh.routing/•/$\sum_{flow} + \sum_{hub}$). This gives rise to a type of ring structure access network, though with some additional complications due to constraints on vehicle capacity and maximum trip length. Another area that has received little attention to date is the effect of different network topologies on the flow between origin and destination nodes.

12.4.3 Constraints

Capacities on nodes: Two types of capacities on hub nodes have been introduced above. In the airline application, where capacity is determined by the number of runways and terminal size, it is reasonable to limit all through traffic. On the other hand, in postal applications the capacity constraint arises from a limitation of the amount of incoming mail that can be sorted at the hub, so that only the flow arriving from non-hub nodes needs to be constrained. However this difference has very little influence on the solution technique, so we will denote both of these variants by the word 'cap' in the third position of the classification. It is interesting to note that Ebery et al, (2000a) showed that the solutions of capacitated problems may have the rather surprising property that flow from a hub node to itself is routed via another hub, unless this is specifically prohibited.

In much of the literature the constraint of using exactly p hub nodes is dropped when capacities are imposed on the hubs, so that the optimal number of hubs is determined endogenously by the model as part of the optimization. Aykin (1994) on the other hand restricts the number of hubs to be exactly p, while imposing capacities on throughput and considering fixed costs on the hubs (p-hub/D/SA,cap,direct/•/$\sum_{flow} + \sum_{hub}$).

Capacities on arcs: Capacity on inter-hub arcs are proposed by Bryan (1998), who explores the relationship between capacities and the piecewise linear cost function for inter-hub arcs proposed in O'Kelly and Brian (1998a) for a given set of hub nodes. On the whole the impact of imposing capacities appears to be relatively small when the set of hub nodes is fixed.

Thresholds on arcs: In examining solutions of the basic hub location models, several researchers noted that the flow volumes across arcs in optimal solutions do not necessarily match the assumption that inter-hub transfers can be discounted due to greater flow volumes. Aykin (1996) was one of the first to suggest the use of flow thresholds to improve the basic hub location problems in this regard. Aykin suggests various types of thresholds based on the amount of flow going through a hub or the inter-hub flows. Flows on arcs in the access network also might have unrealistically low volumes, especially in multiple allocation problems, and Campbell (1994b) presented models with flow thresholds on spokes.

Bryan (1998) proposes a thresholding scheme where inter-hub arcs carrying less than the minimum amount of flow are simply not allowed. One method is to modify the multiple allocation p-hub median problem by dropping the assumption that all pairs of hub nodes are directly connected. Alternatively, flow thresholds can be used to determine the number of hubs. Dropping the requirement that there are exactly p hubs would produce a solution where every node is a hub but with very low flow volumes. Adding a threshold constraint ensures that only a small number of hubs is chosen. A drawback of this method is that the threshold constraint will also affect the solution of the allocation problem so that not all flow will travel along the shortest path through the hub network.

A more detailed discussion of a particular thresholding scheme is provided by Podnar et al. (1999a, 1999b) who consider a scheme where the per-unit cost of flow across an arc is either discounted or incurs the full cost depending on whether the flow across the arc exceeds some given threshold or not. While motivated by the hub location literature, the models proposed in Podnar et al. (1999b) strictly speaking fall out of the area of location problems as they do not explicitly include hubs but simply select a set of arcs in the network to be discounted. Furthermore it seems advisable to modify the model in Podnar et al. (1999b) to include a fixed cost for establishing a discount arc, so that the cost as a function of flow volume is piecewise linear, rather than having a significant discontinuity when the threshold is reached. Skorin-Kapov (2000b) builds on this model with a game-theoretic approach that looks at the interaction between the different users of the network.

Performance constraints: A variety of performance constraints may be included in hub location models to ensure that the hub network can effectively handle the traffic. These constraints are most common in telecommunications-based research and may be of various forms, including limits on the percentage of calls blocked due to insufficient capacity, limits on the transmission time or limits on queue lengths or delays. See Klincewicz (1998) for details. In transportation-oriented research Marianov and Serra (2000) model hub location subject to a constraint on the length of the queue of aircraft waiting for a runway at a hub (p-hub/D/MA,queue/•/$\sum_{flow} + \sum_{hub}$).

12.5 Applications

12.5.1 Air Transportation

A large number of authors have considered hub location in hub-and-spoke networks for passenger airlines. For an overview see Bryan and O'Kelly (1999). Some of the significant factors are the flow economies of scale that these networks can achieve, and the ability to provide good connections between a large number of locations without having to connect each pair of destinations. Some of the differences between air passenger hub networks and air freight hub networks are discussed by O'Kelly (1998b). These arise mainly from the importance of convenience for passengers, whereas the overall performance and cost of the network is paramount for freight.

Some authors have considered the special case where airlines only offer one-stop flights (Drezner and Drezner, 2001; Sasaki et al., 1997, 1999). In this case only one hub is used in any path and hence the hub location problem can be turned into the simpler p-median problem. The p-hub median problem arises when two-stop services are allowed. Finally it is possible to allow any number of stops leading to a more general network design problem. These three cases are compared by Jaillet et al. (1996). If multiple allocation is allowed, a hybrid model is possible in which isolated hubs serve small regional areas with interconnected hubs used for travel over larger distances (Campbell et al., 1999).

Marianov and Serra (2000) consider the effect of the hub network design on the congestion at the airports chosen as hubs. To this end they model the hubs as M/D/c queues. They then solve hub location problems with additional constraints to ensure that an appropriate level of service can be achieved at each of the hubs. Additionally they consider optimization of the number of runways to be built at each airport. Earlier work on congestion at airports includes the simulation model of Grove and O'Kelly (1986).

An important issue in air passenger transport is the effects of pricing and competition. However so far relatively little work has been done on the relationship between pricing and hub network design. Skorin-Kapov (1998) uses cooperative games with the aim of determining a 'fair' distribution of the cost of providing the service among the users. An alternative approach is given by Marianov et al. (1999) who consider a model in which the demand between any pair of nodes i and j depends on the cost of traveling from i to j via the hub network (discretised to a number of distinct demand levels).

For airline networks the discount for large flow volumes is achieved via the use of larger aircraft, which in principle could occur anywhere in the network. This has led several authors to suggest alternative models to the original hub location models (in which all inter-hub links are discounted). Campbell et al. (2000a, 2000b) consider locating a fixed number of discounted hub-arcs, while Jaillet et al. (1996) opt for a more general network design formulation in which a number of aircraft of different types can be allocated to any link, each with their own capacity and fixed cost.

12.5.2 Rapid Transit

Nickel et al. (2000) consider hub location problems in urban public transport networks. In this application it is unrealistic to assume that all hubs are directly connected and that the costs are proportional to Euclidean distances and hence satisfy the triangle inequality. For these reasons they develop a new formulation based on the uncapacitated multiple allocation hub location problem, but with the location of hub-arcs (offering discounted movement between hubs) a separate decision variable. This type of model provides greater flexibility in network design, but is also significantly harder to solve. Similar models are proposed by Campbell et al. (2000a, 2000b) with a view to ensuring that the inter-hub movements actually attract as much flow as possible.

12.5.3 Postal Networks

The postal application has been discussed by Ernst and Krishnamoorthy (1996, 1999) based on the study of a metropolitan postal delivery system for Australia Post. Here the nodes represent post code districts, flow corresponds to mail volume, and the hub nodes are not only consolidation points but also used to sort mail. Some of the significant differences to the more common airline applications include:

- Collection and distribution costs are not the same as these involve quite different processes and transportation modes.
- A node usually sends flow to itself. This involves sending the mail to the hub node where it is sorted and then returning it to its originating location.
- Capacities on the hub nodes, if present, are imposed only on the total collection of flow (which has to be sorted) and not on the transfers.

A standard data set for testing hub location algorithms has been made available based on data provided by Australia Post. This data set contains 200 nodes – smaller data sets with 10, 20, 25, 50 or 100 nodes can be derived through aggregation of the nodes. The data set can be obtained from the the OR Library (Beasley, 1990) or from the authors.

A different hub location model for postal operations is described by Donaldson et al. (1999). They provide an integer programming model that allows routes via at most one hub, and use the model to evaluate the location of cross-docking sorting centers (hubs) for mail distribution in the USA. Their models include time constraints and allow direct origin-destination shipments.

12.5.4 Service Network Design

An application related to postal network design is design of overnight or express parcel delivery networks. This is a part of what is often called service network design in the transportation literature (Crainic, 1999). These networks typically use planes for transportation between hubs, and planes or trucks for transportation between origins/destinations and hubs. When trucks are predominantly used for all links, (as in Europe with its shorter distances, as opposed to the USA), the transportation cost discount on the inter-hub arcs may be quite small or non-existent. However even in this situation the use of hubs can still be advantageous (Vahrenkamp, 1998). The transportation literature on service network design is quite large (see Crainic, 1999)), but it has been primarily focussed on routing and scheduling for a given set of hub nodes.

The critical feature in express parcel delivery systems is the need to meet performance standards in terms of delivery time. An early description of this problem is given by Hall (1989) who provides some qualitative results. Kara and Tansel (1999b) produced an integer programming formulation and solution algorithm for this problem. Each origin node has a ready time when the cargo for all other nodes is available and can be sent to the hub. However transfers from the hub can only occur once all of the cargo has been collected from the allocated origin nodes. Similarly, distribution from a hub to destination nodes can only occur once all of the transfers to the hub have been completed. Several different objectives are possible: minimizing the maximum delivery time, minimizing the number of hubs required to achieve a given service standard or minimizing the weighted travel times. This model could also be used for multi-modal systems in which the hub nodes serve as transshipment points. Furthermore it can be trivially extended to include delays at the hubs for the time taken to transfer the cargo between aircrafts or trucks. In some cases it makes sense to optimize the choice of transportation mode (see O'Kelly and Lao (1991a)), rather than assuming that all inter-hub transfers are performed by plane with collection and distribution occurring by truck. This leads to an interesting trade off between the time and cost required to make the delivery.

If there are no time windows for the availability of the mail, then the problem of designing a network for guaranteed time distribution becomes simply a p-hub center problem. This case is considered by Iyer and Ratcliff (1990) when the underlying transportation network is a tree. An interesting variant introduced in Iyer and Ratcliff (1990) is to add a central sorting system to be located anywhere on the graph. All mail must be routed via the hubs, which act as local consolidation and distribution points, and the central sorting system. This leads to an interesting hybrid problem consisting of both a p-hub center problem, combined with a 1-median problem.

12.5.5 Trucking

Large scale trucking networks are very similar to airline and postal networks in providing transportation between many origins and destinations. Not surprisingly, many large trucking networks employ hubs for transferring and consolidating shipments. However, most of the studies on hub and spoke networks in the trucking industry address the operation and design of such networks without optimizing the location of hubs. See for example, the load planning models for less-than-truckload (LTL) trucking (Powell, 1986; Powel and Sheffi, 1983, 1989).

A number of works have specifically addressed hub location in truckload and LTL trucking. In truckload trucking hubs may be used not for the consolidation benefits (since truckloads need no consolidation), but to reduce driver turnover by providing transfer points that can reduce the driver's trip length. This allows drivers to return home more frequently. Hunt (1998) describes truckload trucking operations and presents heuristic algorithms for locating relay points (hubs). A number of simulation studies have been carried out to investigate hub location and the performance of hub and spoke networks in the truckload trucking industry (Taha and Taylor, 1994; Taha et al., 1996; Taylor et al. 1995, 1999). Taylor et al. (1999) show that using a number of regional hubs is better than using just a single central hub, or than sending all shipments direct.

LTL applications are more similar to the airline and postal applications described earlier, as LTL carriers rely on consolidation of small shipments to provide access to the strong economies of scale in trucking. For some LTL applications the collection and distribution operations can be treated as vehicle routing problems. This means that multiple nodes can be serviced by the same vehicle as part of a round trip starting at the hub which serves as a depot. In this context a number of additional constraints can be considered:

- Capacities on the vehicles.
- Maximum distances or durations for the round trips.
- A fixed fleet size, either per hub or to be distributed between the hubs.
- If pickup and delivery are performed separately then two distinct allocation problems need to be solved, one for collection and one for distribution.

Nagy and Salhi (1998) introduce this type of hub location problem and discuss the relationship with other areas including terminal location, vehicle routing and location-routing.

12.5.6 Telecommunications

Most research on hub networks in the telecommunications area concentrates on the cost of establishing the network rather than the cost for flow volumes between origin-destination pairs. In other words the optimal solution is determined by minimizing the fixed costs of establishing the network, including costs for hub nodes, hub arcs and allocation arcs, without considering the cost of satisfying the demand using this network. This is a reasonable approach in telecommunications applications where the operating cost is very small compared to the establishment cost of a network (unlike the airline or postal applications). Klincewicz (1998) surveys hub location research in the telecommunications area, so we focus below on new references not included in Klincewicz (1998) and on some key earlier works.

An example of hub location in telecommunications is given by Chung et al. (1992) which use problem hub/D/SA/$\sum_{hub} + \sum_{arc}$ to model the design of large-scale data communication networks. A somewhat more general network structure for both the access and the hub level networks is permitted in the models by Yoon et al. (1998b, 1998c). In these models a set of hub nodes and a set of edges spanning the graph must be chosen (each incurring a fixed cost), such that the cost of communication between a node and its hub, as well as communication between the hubs via the sub-graph is minimised. A significant difference to standard hub location problems is that the cost of communication (volume of flow exchanged) between two hub nodes is independent of the allocation of nodes to hubs. This is because the cost of communication between hubs is measured by the cost of installing cables (which are assumed to have an infinite capacity), with the fixed cost for arcs being used to establish the conduit network.

In order to increase the robustness of the hub network, Kim et al. (1995) and Yoon and Tcha (1996) consider slightly different models where each non-hub node is allocated to two different hubs. This gives rise to an uncapacitated dual allocation hub location problem, though again only the fixed costs for establishing the network are considered.

A similar problem is considered by Garfinkel (1996) but with emphasis on determining the arc capacities for a given network containing two hubs, where all flow can be sent either via the hubs or directly between the origin and destination node. They show that this relatively simple problem, in which the location of the hubs and the allocation of nodes to hubs are known, is already NP-Hard.

Another application related to telecommunications is the configuration of distributed computer systems described by Gavish (1987). Here each node produces transactions that need to be recorded in a data base and reports need to be generated from this data. A number of computers are to be located at different locations to handle these transactions and generate reports. This can be modeled as a single allocation hub median problem in which the hubs represent computers and non-hub nodes can be either origins (sources of transactions) or destinations (reports). Transfers occur when the computer which has been assigned to generate a report requires data stored at one of the other hubs. The number of hubs is limited by a fixed cost of opening hubs in the model proposed by Gavish. An extension of this model in which the cost of congestion on the network is also considered is presented in Gavish (1992b).

12.6 Solving Hub Location Problems

12.6.1 Complexity Results

Only a limited amount of work has been done to analyze the complexity of hub location problems, mostly for the single allocation p-hub median problem. Problem p-hub/D/SA/•/$\sum_{flow}$ is known to be NP-Hard. In fact, even for a given set of hubs, the assignment problem of optimally allocating the non-hub nodes to hubs is already NP-Hard (Kara and Tansel, 1998). This problem is equivalent to the Multi-Processor Assignment Problem which is known to be NP-Hard even for $p = 3$. Polynomially solvable special cases of the allocation problem exist when the matrix of flows W_{ij} is sparse (for example if the non-zero entries form a tree or a k-tree over the set of nodes). Another special case is $p = 2$. Sohn and Park (1997) showed that problem 2-hub/D/SA/•/$\sum_{flow}$ can be transformed into a minimum cut problem and hence solved in polynomial time.

The single allocation hub center problem (p-hub/D/SA/•/max) is also NP-Complete as shown by Kara and Tansel (1999a). Again it can be shown that even the allocation part of the problem for a given set of hubs is NP-Complete.

The multiple allocation problem for fixed hubs, on the other hand, can be solved in polynomial time using an all pairs shortest path algorithm provided the hubs are uncapacitated (Ernst and Krishnamoorthy, 1998a; Sohn and Park, 1998). This is true for both the hub median problem ($\sum_{flow}$ objective) as well as for the center problem. Let $\mathcal{H} \subset V$ be a given set of hubs then the minimum cost multiple allocation can be calculated as follows:

$$c'_{kj} = \min_{l \in \mathcal{H}} \{\alpha\, d_{kl} + \delta\, d_{lj}\} \text{ for all } k \in \mathcal{H},\ j \in V.$$

$$c_{ij} = \min_{k \in \mathcal{H}} \{\chi\, d_{ik} + c'_{kj}\} \text{ for all } i, j \in V.$$

$$C = \sum_{i,j \in V} W_{ij}\, c_{ij}.$$

The above algorithm gives the minimum cost C for distributing the flows W_{ij} via the given set of hubs. The c_{ij} values give the lowest cost path from i to j. The allocation of node i to hubs can be obtained by storing the hubs $k, l \in \mathcal{H}$ for which the minimum in the definition of c_{ij} and c'_{ki} is attained. Hence the optimal value for the multiple allocation hub center problem for given hubs can be obtained by simply taking the maximum of all of the c_{ij}. In general $c_{ij} \neq c_{ji}$, however if $\chi = \delta$ then the solution is symmetrical and the number of computations can be almost halved by only calculating c_{ij} for $i \leq j$. The complexity of the above algorithm is $\mathcal{O}(n^2 p)$ where $p = |\mathcal{H}|$.

12.6.2 Pre-Processing

As with many other combinatorial optimization problems, pre-processing of the problem data is a useful technique to improve the efficiency of solving the problems in practice, though it does not change the theoretical complexity of the algorithms. There are a number of methods that can be used for most of the hub location variants to reduce the size of MILP formulations and hence to make the problem more tractable.

The most important pre-processing technique is to observe that in most cases the flow from i to j takes the same path in the optimal solution as the flow from j to i. The main exception to this is where multiple allocation is allowed and capacity constraints or different collection and distribution costs can lead to non-symmetrical solutions. However provided the symmetry between $i-j$ and $j-i$ paths exists, this can be used to modify the W_{ij} matrix to simply set $W'_{ij} = W_{ij} + W_{ji}$ and $W'_{ji} = 0$ for all $i < j$. Thus all variables that track the $j - i$ flow can be eliminated from consideration.

While it is generally not possible to rule out specific hub locations based on the structure of the problem data, the geometric nature of the locations and costs can often be used to eliminate some of the allocation choices. For example consider problem p-hub/D/SA/•/$\sum_{flow}$. Intuitively a node i would never be allocated to some hub k in the optimal solution if node k is too far from i. If there is another hub l much closer than k, it would be preferable to allocate i to l rather than k. In particular if

$$\max\{\chi,\delta\}\,d_{ik} \geq \max\{\chi,\delta\}\,d_{il} + \alpha\,d_{lk},$$

then any optimal solution that has both k and l as hubs will not allocate i to k. This observation can be used in a number of ways. For example it can be used to eliminate the Y^i_{kl} variable in the formulation (12.11)–(12.17). Alternatively if there are a large number of different nodes l none of which can be hubs in the optimal solution if i is allocated to k, then this can be used in calculating a lower bound with the aim of eliminating the Z_{ik} variable altogether. Similar arguments based on costs and the structure of the optimal solution can also be used to great effect for multiple allocation problems (see for example Boland et al. (2000)). This type of preprocessing only requires that the hubs are uncapacitated and that the distances between nodes satisfy the triangle inequality.

12.6.3 Linear Programming Based Approaches

The most common approach to solving hub location problems is using linear programming. The formulations used fall into two broad categories. The first uses $\mathcal{O}(n^4)$ variables in order to track the flow between every origin-destination pair (eg the formulation for the uncapacitated multiple allocation hub location problem given in (12.23)–(12.26) below). This type of formulation is very tight but also very large as it not only has a large number of variables but also requires $\mathcal{O}(n^3)$ constraints. The fact that both the number of rows and the number of columns increases rapidly with problem size means that these formulations are not amenable to branch-and-price or branch-and-cut techniques. Hence it is very difficult to solve the LP relaxation for problems involving more than 25 nodes. However these types of formulations still provide a good starting point for algorithms using dual ascent or lagrangean relaxation since they are quite tight.

In order to make hub location problems more tractable for standard LP solvers, it is important to reduce the problem size. Hence the formulations given in Sect. 12.2.2 aggregate all flow from a single origin (equivalently flow could be aggregated based on its destination). This reduces both the number of variables and the number of constraints by roughly a factor of n. This means that the LP relaxations are not as tight as for the larger formulations but can be obtained significantly more quickly. The result is an increase in the size of the branch and bound tree but the total computational time is reduced. Thus not only can the hub location problems be solved more quickly, but also larger problems be solved.

At least for single allocation problems, the number of variables can be reduced further to $\mathcal{O}(n^2)$ continuous variables with the same number of binary variables (e.g. Ebery et al. (2000)). Unfortunately preliminary investigations by the authors showed that this does not bring significant benefits for two reasons. Firstly the LP relaxations are too weak, significantly increasing the number of branch and bound nodes that have to be evaluated. Secondly it is not possible to reduce the number of constraints as well, and hence there is no significant reduction in the computational time required to solve each LP relaxation. The only non-trivial special case for which a smaller LP formulation can be made to work reasonably well is when $p = 3$ (see Ebery (2001)).

For the hub center problem p-hub/D/SA/•/max, Kara and Tansel (1999a) provide several MILP formulations. Based on computational experiments they conclude that the most effective of these is:

$$\min X$$

Subject to $\quad X \geq \sum_{k}(\chi\, d_{ik} + \alpha\, d_{kl})\, Z_{ik} + \delta\, d_{lj}\, Z_{lj} \quad \forall\, i,j,l \in V$

(12.12), (12.13), (12.15) & (12.17).

Polyhedral Properties: Very little is known about the polyhedral properties of any of the hub location problems described in this chapter. Recently Hamacher et al. (2000) have developed some results for the uncapacitated multiple allocation hub location problem. They establish that the dimension of the polytope is $n^2 q + n - q$, where q is the number of origin-destination pairs exchanging flow (or $n^4 - n^2 + n$ in the case where the flow matrix has no zero entries). The studies of the polytope leads them to formulate this problem as:

Problem hub/D/MA/•/$\sum_{flow} + \sum_{hub}$

$$\text{Min.} \sum_{i,j,k,l:k \leq l} C_{ijkl} X_{ijkl} + \sum_k F_k H_k \quad (12.23)$$

$$\text{Subject to} \sum_{k,l:k \leq l} X_{ijkl} = 1 \quad \forall\, i,j \in V \quad (12.24)$$

$$\sum_{l:l<k} X_{ijlk} + \sum_{l:l \geq k} X_{ijkl} \leq H_k \quad \forall\, i,j,k \in V \quad (12.25)$$

$$X_{ijkl} \geq 0 \quad \forall i,j,k \leq l \in V \quad (12.26)$$

$$\text{and} \quad H_k \in \{0,1\}$$

where

$$C_{ijkl} = \min\{W_{ij}\,(\chi\,d_{ik} + \alpha\,d_{kl} + \delta\,d_{lj}),\ W_{ij}\,(\chi\,d_{il} + \alpha\,d_{lk} + \delta\,d_{kj})\}.$$

The X_{ijkl} variables are also binary but constraint (12.26) is sufficient as the flow for any origin-destination pair will always take the shortest path through the hub network. This formulation is tighter than the tightest previously known formulation by Klincewicz (1996). Constraints (12.25) and (12.26) as well as the upper bound on H_k have been shown to be facets of the integer polytope in Hamacher et al. (2000).

For 3-hub/D/SA,fixed hubs/•/$\sum_{flow}$, that is the case where there are exactly 3 hubs and these hubs are given, Sohn and Park (2000) present a novel MILP formulation for the allocation problem and show that the inequality constraints in their formulation are facet defining.

Dual ascent: Mayer and Wagner (1998) developed a dual ascent method for the uncapacitated multiple allocation hub location problem. They take as their starting point the dual problem of the formulation by Campbell (1994b), which is similar to (12.23)–(12.26) but with a weaker form of constraint (12.25). This dual problem can be solved efficiently as a special transportation problem. Unfortunately the lower bounds obtained in this way are weak, hence they tighten the primal formulation, which destroys the special nature of the dual structure. Nevertheless by solving the weaker formulation optimally and then using this as a starting point for performing dual ascent in the tighter formulation, Mayer and Wagner achieve tight enough lower bounds to allow their branch and bound algorithm to find optimal solutions. Computationally this method appears to be superior to the dual ascent method by Klincewicz (1996) and also outperforms the MILP approach using the formulation given in Sect. 12.2.2 for large problems.

Dual ascent is also frequently used for hub location problem variants in telecommunications where there are only fixed costs and no flow costs to be minimized (e.g. Kim et al. 1995; Kim and Tcha, 1992; Lee et al., 1993; Yoon and Current, 1998).

Lagrangean Relaxation algorithms: Lagrangean relaxation has been used by Aykin (1994) to solve a capacitated multiple allocation p-hub median problem (p-hub/D/MA,cap/•/$\sum_{flow} + \sum_{hub}$). It has also been used in the telecommunications area by Lee et al. (2000) and for uncapacitated single allocation problems by Gavish (1987, 1992). In the latter the technique has been used to deal with a non-linear objective function that penalizes congestion in the network.

Lagrangean relaxation has also been used to good effect by Pirkul and Schilling (1998) to solve the standard problem p-hub/D/SA/•/$\sum_{flow}$. Consider the following formulation for the single allocation p-hub median problem (Skorin-Kapov et al., 1996):

$$\text{Min.} \sum_{i \in V}\sum_{j \in V}\sum_{k \in V}\sum_{l \in V} (\chi\, d_{ik} + \alpha\, d_{kl} + \delta\, d_{lj}) X_{ijkl} \tag{12.27}$$

$$\text{Subject to} \quad \sum_{k \in V} Z_{kk} = p, \tag{12.28}$$

$$\sum_{k \in V} Z_{ik} = 1 \quad \forall\, i \in V, \tag{12.29}$$

$$\sum_{l \in V} X_{ijkl} = Z_{ik} \quad \forall\, i,j,k \in V, \tag{12.30}$$

$$\sum_{k \in V} X_{ijkl} = Z_{jl} \quad \forall\, i,j,l \in V, \tag{12.31}$$

$$\sum_{k \in V}\sum_{l \in V} X_{ijkl} = 1 \quad \forall\, i,j \in V, \tag{12.32}$$

$$Z_{ik} \leq Z_{kk} \quad \forall\, i,k \in V, \tag{12.33}$$

$$X_{ijkl} \geq 0 \quad \forall\, i,j,k,l \in V, \tag{12.34}$$

$$Z_{ik} \in \{0,1\} \quad \forall\, i,k \in V. \tag{12.35}$$

By relaxing constraints (12.29)–(12.31) the problem decomposes into two sub-problems that can each be solved very quickly. Note that (12.32) acts only as a strengthening constraint for the relaxation and that some of the ideas in Sect. 12.6.2 need to be applied to obtain an effective algorithm.

12.6.4 Enumerative Algorithms

Since the multiple allocation problem can be solved very quickly for a given set of hubs, and the number of integer variables (hub locations) in multiple allocation is relatively small, this has suggested the use of enumerative algorithms as a reasonable approach for solving hub location problems (Aykin, 1995a; Ernst and Krishnamoorthy, 1998b).

The simplest way to make use of the shortest path algorithm for the allocation problem is to enumerate all possible hub locations for a problem such as the multiple allocation p-hub median problem. While this algorithm is exponential in p, it is polynomial in n and for most problem instances the number of hubs is comparatively small. As each hub combination can be evaluated very quickly, this gives a viable solution approach for multiple allocation problems provided the number of hubs to be enumerated is comparatively small (up to about 5) (Campbell et al. 2000b; Ernst and Krishnamoorthy, 1998b).

For larger instances the shortest path method for the allocation problem can be combined with a branch and bound algorithm to determine the location of the hubs (Ernst and Krishnamoorthy, 1998a). The method starts by dividing the set of nodes into a number of k clusters and enumerating all possible allocations of the p hubs to the k clusters. For each such placement of the hubs into the clusters, a lower bound can be calculated by solving the multiple allocation problem where any node is treated as a hub if it is in a cluster that contains at least one hub. Branching is then performed by choosing a cluster containing one or more hubs and sub-dividing this cluster to resolve the position of the hubs. This approach can also be used to solve the single allocation p-hub median problem since the multiple allocation problem provides a lower bound to the single allocation problem. In fact generally the solution to multiple allocation problems has nodes allocated to no more than two or three hubs, making it easy to resolve the allocation through further branching. A similar enumerative approach that uses the shortest path problem as a lower bound for hub/D/SA/•/$\sum_{flow}$ is given by Abdinnour-Helm and Venkataramanan (1998b).

Klincewicz (2000) has shown that the concave flow cost models of Bryan (1998), O'Kelly and Bryan (1998) can be solved efficiently to optimality as classic uncapacitated facility location problems using an enumeration procedure.

12.6.5 Heuristic Algorithms

Hub location problems are difficult to solve exactly. The best methods available so far cannot solve instances with more than 50 nodes unless the number of (potential) hubs is significantly restricted. This has led to a proliferation of heuristics to tackle the many types hub location problems proposed in the literature for various applications.

The large number of different heuristic methods used are not only due to the variety of problem variants, but also an indication that there are many different approaches that can be used to produce satisfactory solutions to hub location problems. Hence rather than being able to recommend a particular approach or being able to list every heuristic method that has been published in this area, we will give a brief overview of the range of methods used.

The largest number of heuristics has been published for the single allocation p-hub median problem. Klincewicz (1991) presents a number of heuristics based on local neighbourhood search and clustering of nodes and also developed a tabu search and GRASP heuristic in Klincewicz (1992). Of these, the tabu search heuristic seems to be marginally better than the others, however the GRASP approach also shows some promise. Klincewicz (2000) applies tabu search and GRASP to the concave flow cost models. Another tabu search algorithm for this problem with better performance has been developed by Skorin-Kapov and Skorin-Kapov (1994). Yet another approach is to use Simulated Annealing Ernst and Krishnamurthy (1996). An interesting use of the heuristic solutions is given by O'Kelly et al. (1995) who use the reference solution to generate lower bounds that can give an estimate of the quality of the solution.

Abdinnour-Helm (1998a, 1999) uses a hybrid heuristic combining genetic algorithms and tabu search to obtain good solutions to the uncapacitated single allocation hub location problem. A rounding heuristic for a variant of this problem with fixed costs on both hubs and arcs is proposed by Yoon and Current (1998) based on a dual ascent solution of the LP relaxation.

Heuristics for capacitated problems include randomized descent for problem hub/D/SA,cap/•/$\sum_{flow} + \sum_{hub}$ in Ernst and Krishnamoorthy (1999), a greedy-interchange heuristic combined with lagrangean lower bounds for problem
p-hub/D/SA,cap/•/$\sum_{flow} + \sum_{hub}$ in Aykin (1994), and constructive heuristics with limited interchange for problem hub/D/MA,cap/•/$\sum_{flow} + \sum_{hub}$ in Ebery et al. (2000). Local search heuristics tend to work less well for capacitated problems as the capacity constraints can make it more difficult to move through the solution space without encountering infeasibilities.

12.7 Conclusions

12.7.1 Key Themes

Hub location problems are an interesting and worthwhile area of study from both an applied and theoretical perspective. Hub networks occur in many different applications, and hub location problems provide great challenges and opportunities for the researcher. Although transportation and telecommunications organizations do not regularly re-design their networks starting from scratch, the study of hub location problems is important in providing answers about the types of networks that should be used and the degree to which any existing networks are sub-optimal.

While much of the early work in hub location focused on the location aspects, attention is now shifting to consideration of network design and hub location together. Several approaches are being considered by research teams around the globe and new developments appear regularly. Attention is also focusing on developing better, more realistic models that capture more of the complexities of real world operations.

One main motivation for researchers has been the intellectual challenge posed by hub location problems. In some sense hub location problems are significantly harder than such well studied problems as the quadratic assignment problem, or the classical facility location problems. As has been shown above, the problem of determining the optimum single allocation of nodes to a given set of hubs is already a multi-processor assignment problem (which is related to the quadratic assignment problem). On the other hand just the choice of hub location even without any consideration to the interaction among nodes and using simply a "closest hub" assignment rule turns the p-hub location problem into a p-median problem. Hence the complete hub location problem combines the challenges of both of these well studied problems.

12.7.2 Future Research Direction

The area of hub location has come a long way since the early work by O'Kelly (1987). The present work shows that there are a large number of variations of the basic hub location problems and several different methods for obtaining solutions. While it is always difficult to predict future directions, a number of different developments are anticipated.

Hub location models have evolved from the original idealized (first generation) models, and newer models are expected to be even more realistic and useful in applied settings. However, in spite of the importance of hub networks to modern transportation and telecommunications systems, only a limited number of publications have reported on real world applications. (Several applied projects involving hub location in large freight networks in the USA are underway.) A study of different applications would serve two purposes. Firstly it would inform the development of better, more realistic models. The other advantage of having more applications published is that it would allow better insights to be obtained by a broader analysis of more general data sets. The current work in this area, and hence the conclusions drawn, are heavily biased towards the CAB data set for passenger airlines in the USA.

Another broad area of future research is in improving the algorithmic techniques for solving hub location problems, so that larger instances can be solved optimally. The tools for improvements could come from a number of sources: work has only just started on the polyhedral properties of hub location problems. Having better insight into the polyhedral structure of these problems should allow more efficient formulations and algorithms to be developed. Better algorithms may also be found by leveraging of developments in more established areas, such as has already been done in adapting dual methods from the facility location area to hub location problems. Furthermore, combinations of different methods for solving hub location may prove useful for solving larger problems. Although future hub location research will require solving very difficult problems, developments in algorithms, software and hardware promise to open new avenues for solving larger and more realistic problems.

References

Abdinnour-Helm, S. (1998). A hybrid heuristic for the uncapacitated hub location problem. *European Journal of Operational Research*, 106, 489–499.

Abdinnour-Helm, S. (1999). Network design in supply chain management. *International Journal of Agile Management Systems*, 1(2), 99–106.

Abdinnour-Helm, S. and Venkataramanan, M. A. (1998). Solution approaches to hub location problems. *Annals of Operations Research*, 78, 31–50.

Aykin, T. (1994). Lagrangian relaxation based approaches to capacitated hub-and-spoke network design problem. *European Journal of Operational Research*, 79, 501–523.

Aykin, T. (1995a). Networking policies for hub-and-spoke systems with application to the air transportation system. *Transportation Science*, 29(3), 201–221.

Aykin, T. (1995b). The hub location and routing problem. *European Journal of Operational Research*, 83, 200–219.

Aykin, T. (1996). On modeling scale economies in hub-and-spoke network design. In *Presented at the Fall Conference of INFORMS, Atlanta*.

Aykin, T. and Brown, G. F. (1992). Interacting new facilities and location-allocation problems. *Transportation Science*, 26(3), 212–222.

Beasley, J. E. (1990). Or-library: distributing test problems by electronic mail. *Journal of the Operational Research Society*, 41, 1069–1072. URL: http://mscmga.ms.ic.ac.uk/info.html.

Boland, N., Krishnamoorthy, M., Ernst, A. T., and Ebery, J. (2000). Preprocessing and cutting for multiple allocation hub location problems. *Submitted to EJOR*.

Bryan, D. L. (1998). Extensions to the hub location problem: Formulations and numerical examples. *Geographical Analysis*, 30(4), 315–330.

Bryan, D. L. and O'Kelly, M. E. (1999). Hub-and-spoke networks in air transportation: An analytical review. *Journal of Regional Science*, 39(2), 275–295.

Campbell, J. F. (1993). Continuous and discrete demand hub location problems. *Transportation Research-B*, 27B(6), 473–482.

Campbell, J. F. (1994a). A survey of network hub location. *Studies in Locational Analysis*, 6, 31–49.

Campbell, J. F. (1994b). Integer programming formulations of discrete hub location problems. *European Journal of Operational Research*, 72, 387–405.

Campbell, J. F. (1998). Hub location and network design. Unpublished.

Campbell, J. F., Ernst, A., and Krishnamoorthy, M. (1999). Locating hub arcs and isolated hubs. *Paper presented at ISOLDE VIII (International Symposium on Locational Decisions), Coimbra, Portugal*.

Campbell, J. F., Ernst, A., and Krishnamoorthy, M. (2000a). Hub arc location problems: Part I – Introduction and results. Working paper.

Campbell, J. F., Ernst, A., and Krishnamoorthy, M. (2000b). Hub arc location problems: Part II – Formulations and optimal algorithms. Working paper.

Chou, Y.-H. (1990). The hierarchical-hub model for airline networks. *Transportation Planning and Technology*, 14, 243–258.

Chung, S.-H., Myung, Y.-S., and Tcha, D.-W. (1992). Optimal design of a distributed network with a two-level hierarchical structure. *European Journal of Operational Research*, 62, 105–115.

Crainic, T. G. (1999). Long-haul freight transportation. *Handbook of Transportation Science, (ed.) R W Hall, Kluwer Academic Publishers, Norwell, MA*, pages 433–491.

Crainic, T. G., Gendreau, M., and Farvolden, J. M. (2000). A simplex-based tabu search method for capacitated network design. *INFORMS Journal on Computing*, 12, 223–236.

Daganzo, C. F. (1999). *Logistics Systems Analysis*. Springer-Verlag, 3rd edition.

Donaldson, H., Johnson, E. L., Ratliff, H. D., and Zhang, M. (1999). Schedule driven cross-docking networks. Working paper.

T. Drezner and Z. Drezner (2001) A Note on Applying the Gravity Rule to the Airline Hub Problem. *Journal of Regional Science*, 41, 67–73.

Ebery, J. (2001). Solving large single allocation p-hub location problems with 2 or 3 hubs. *European Journal of Operational Research*, 128(2), 447–458.

Ebery, J. E., Krishnamoorthy, M., Ernst, A. T., and Boland, N. (2000). The capacitated multiple allocation hub location problem: Formulations and algorithms. *European Journal of Operational Research*, 120(3), 614–631.

Ernst, A. T. and Krishnamoorthy, M. (1996). Efficient algorithms for the uncapacitated single allocation p-hub median problem. *Location Science*, 4(3), 139–154.

Ernst, A. T. and Krishnamoorthy, M. (1998a). An exact solution approach based on shortest-paths for p-hub median problems. *INFORMS Journal on Computing*, 10(2), 149–162.

Ernst, A. T. and Krishnamoorthy, M. (1998b). Exact and heuristic algorithms for the uncapacitated multiple allocation p-hub median problem. *European Journal of Operational Research*, 104, 100–112.

Ernst, A. T. and Krishnamoorthy, M. (1999). Solution algorithms for the capacitated single allocation hub location problem. *Annals of Operations Research*, 86, 141–159.

Fotheringham, A. S. (1983). A new set of spatial-interaction models: the theory of competing destinations. *Environment and Planning A*, 15, 15–36.

Garfinkel, R. S., Sundararaghavan, P. S., Noon, C., and Smith, D. R. (1996). Optimal use of hub facilities: A two-hub model with fixed arc costs. *Trabajos de Investigacion Operativa (TOP)*, 4(2), 331–343.

Gavish, B. (1987). Optimization models for configuring distributed computer systems. *IEEE Transactions on Computers*, C-36(7), 773–793.

Gavish, B. (1992). Topological design of computer communication networks – the overall design problem. *European Journal of Operational Research*, 58(2), 149–172.

Gendron, B., Crainic, T. G., and Frangioni, A. (1999). Multicommodity capacitated network design. *Telecommunications Network Planning, (eds.) B Sanso and P Soriano, Kluwer, Norwell, MA.*, pages 1–19.

Goldman, A. J. (1969). Optimal locations for centers in a network. *Transportation Science*, 3, 352–360.

Grove, P. G. and O'Kelly, M. E. (1986). Hub networks and simulated schedule delay. *Papers of the Regional Science Assoc.*, 59, 103–119.

Hakimi, S. L. (1964). Optimum locations of switching centers and the absolute centers and medians of a graph. *Operations Research*, 12, 450–459.

Hakimi, S. L. (1965). Optimum distribution of switching centers in a communication network and some related graph theoretic problems. *Operations Research*, 13, 462–475.

Hall, R. W. (1989). Configuration of an overnight package air network. *Transportation Research A*, 23A(2), 139–149.

Hamacher, H. W., Labbe, M., Nickel, S., and Sonneborn, T. (2000). Polyhedral properties of the uncapacitated multiple allocation hub location problem. Technical Report 20, Institut für Techno- und Wirtschaftsmathematik (ITWM). Available at http://www.itwm.fhg.de/zentral/berichte/bericht20.html.

Hamacher, H. W. and Nickel, S. (1998). Classification of location models. *Location Science*, 6, 229–242. URL: http://kluedo.ub.uni-kl.de/Mathematik/.

Hunt, G. W. (1998). *Transport Relay Network Design*. PhD thesis, Georgia Institute of Technology.

Iyer, A. V. and Ratliff, H. D. (1990). Accumulation point location on tree networks for guaranteed time distribution. *Management Science*, 36(8), 958–969.

Jaillet, P., Song, G., and Yu, G. (1996). Airline network design and hub location problems. *Location Science*, 4, 195–211.

Kara, B. Y. and Tansel, B. C. (1998). On the allocation phase of the p-hub location problem. Technical report, Department of Industrial Engineering, Bilkent University, Bilkent 06533, Ankara, Turkey.

Kara, B. Y. and Tansel, B. C. (1999a). On the single-assignment p-hub covering problem. Technical report, Department of Industrial Engineering, Bilkent University, Bilkent 06533, Ankara, Turkey.

Kara, B. Y. and Tansel, B. C. (1999b). The latest arrival hub location problem. Technical report, Department of Industrial Engineering, Bilkent University, Bilkent 06533, Ankara, Turkey.

Kara, B. Y. and Tansel, B. C. (2000). On the single-assignment p-hub center problem. *European Journal of Operational Research*, 125(3), 648–655.

Kim, H.-J., Chung, S.-H., and Tcha, D.-W. (1995). Optimal design of the two-level distributed network with dual homing local connections. *IIE Transactions*, 27, 555–563.

Kim, J.-G. and Tcha, D.-W. (1992). Optimal design of a two-level hierarchical network with tree-star configuration. *Computers and Industrial Engineering*, 22(3), 273–281.

Klincewicz, J. G. (1991). Heuristics for the p-hub location problem. *European Journal of Operational Research*, 53(1), 25–37.

Klincewicz, J. G. (1992). Avoiding local optima in the p-hub location problem using tabu search and grasp. *Annals of Operations Research*, 40, 283–302.

Klincewicz, J. G. (1996). A dual algorithm for the uncapacitated hub location problem. *Location Science*, 4, 173–184.

Klincewicz, J. G. (1998). Hub location in backbone/tributary network design: A review. *Location Science*, 6, 307–335.

Klincewicz, J. G. (2000). Enumeration and search procedures for a hub location problem with economies of scale. Technical report.

Laporte, G. (1989). A survey of algorithms for location-routing problems. *Investigacion Operativa*, 1, 93–123.

Lederer, P. J. and Nambimadom, R. S. (1998). Airline network design. *Operations Research*, 46, 785–804.

Lee, C.-H., Ro, H.-B., and Tcha, D.-W. (1993). Topological design of a two-level network with ring-star configuration. *Computers and Operations Research*, 20(6), 625–637.

Lee, Y., Lim, B. H., and Park, J. S. (2000). A hub location problem in designing digital data service networks: Lagrangian relaxation approach. Working paper.

Marianov, V. and Serra, D. (2000). Location models for airline hubs behaving as m/d/c queues. Technical Report 453, Department of Economics IET and GRES, Universitat Pompeu Fabra, Barcelona, Spain.

Marianov, V., Serra, D., and ReVelle, C. (1999). Location of hubs in a competitive environment. *European Journal of Operational Research*, 114, 363–371.

Mayer, G. and Wagner, B. (1998). Hublocater: An exact solution method for the multiple allocation hub location problem. *Computers and Operations Research (to appear)*.

Minoux, M. (1989). Network synthesis and optimum network design-problems – models, solution methods and applications. *Networks*, 19(3), 313–360.

Nagy, G. and Salhi, S. (1998). The many-to-many location-routing problem. *TOP, Socied de Estadistica & Investigacion Operativa*, 6(2), 261–275.

Nickel, S., Schobel, A., and Sonneborn, T. (2000). Hub location problems in urban traffic networks. *Mathematical Methods and Optimisation in Transportation Systems, (eds.) Niittymahi and Pursula, Kluwer Academic Publishers*, pages 1–12.

O'Kelly, M. E. (1986a). Activity levels at hub facilities in interacting networks. *Geographical Analysis*, 18(4), 343–356.

O'Kelly, M. E. (1986b). The location of interacting hub facilities. *Transportation Science*, 20(2), 92–106.

O'Kelly, M. E. (1987). A quadratic integer program for the location of interacting hub facilities. *European Journal of Operational Research*, 32, 393–404.

O'Kelly, M. E. (1992a). A clustering approach to the planar hub location problem. *Annals of Operations Research*, 40, 339–353.

O'Kelly, M. E. (1992b). Hub facility location with fixed costs. *Papers in Regional Science The Journal of the RSAI*, 71(3), 293–306.

O'Kelly, M. E. (1998a). On the allocation of a subset of nodes to a mini-hub in a package delivery network. *Papers in Regional Science. The Journal of the RSAI*, 77(1), 77–99.

O'Kelly, M. E. (1998b). A geographer's analysis of hub-and-spoke networks. *Journal of Transport Geography*, 6(3), 171–186.

O'Kelly, M. E., Bryan, D., Skorin-Kapov, D., and Skorin-Kapov, J. (1996). Hub network design with single and multiple allocation: A computational study. *Location Science*, 4(3), 125–138.

O'Kelly, M. E. and Bryan, D. L. (1998). Hub location with flow economies of scale. *Transportaion Research B*, 32(8), 605–616.

O'Kelly, M. E. and Bryan, D. L. (2000). Interfacility interaction in models of hubs and spoke networks. *Submitted for publication*. Presentations based on this were given at an INFORMS meeting in Cincinnati 1999 and RSAI Meeting in Dublin 1999.

O'Kelly, M. E. and Lao, Y. (1991). Mode choice in a hub-and-spoke network: A zero-one linear programming approach. *Geographical Analysis*, 23(4), 283–297.

O'Kelly, M. E. and Miller, H. J. (1991). Solution strategies for the single facility minimax hub location problem. *Papers in Regional Science The Journal of the RSAI*, 70(4), 367–380.

O'Kelly, M. E. and Miller, H. J. (1994). The hub network design problem. *Journal of Transport Geography*, 2(1), 31–40.

O'Kelly, M. E., Skorin-Kapov, D., and Skorin-Kapov, J. (1995). Lower bounds for the hub location problem. *Management Science*, 41(4), 713–721.

Pirkul, H. and Schilling, D. A. (1998). An efficient procedure for designing single allocation hub and spoke systems. *Management Science*, 44(12), S235–S242.

Podnar, H. and Skorin-Kapov, J. (1999). Genetic algorithm for cost minimization applied to networks with threshold based discounting. Working paper.

Podnar, H., Skorin-Kapov, J., and Skorin-Kapov, D. (1999). Network cost minimization using threshold based discounting. Working paper.

Powell, W. B. (1986). A local improvement heuristic for the design of less-than-truckload motor carrier networks. *Transportation Science*, 20, 246–257.

Powell, W. B. and Sheffi, Y. (1983). The load-planning problem of motor carriers: problem description and proposed solution approach. *Transportation Research A*, 17, 471–480.

Powell, W. B. and Sheffi, Y. (1989). Design and implementation of an interactive optimization system for the network design in the motor carrier industry. *Operations Research*, 37, 12–29.

Sasaki, M., Suzuki, A., and Drezner, Z. (1997). On the selection of relay points in a logistics system. *Asia-Pacific Journal of Operational Research*, 14, 39–54.

Sasaki, M., Suzuki, A., and Drezner, Z. (1999). On the selection of hub airports for an airline hub-and-spoke. *Computers & Operations Research*, 26, 1411–1422.

Skorin-Kapov, D. (1998). Hub network games. *Networks*, 31(4), 293–302.

Skorin-Kapov, D. (2000). On cost allocation in networks with threshold based discounting. *Proceedings, 22nd International Conference on Information Technology Interfaces, Pula, Croatia, June 13-16, 2000.*, pages 409–414.

Skorin-Kapov, D. and Skorin-Kapov, J. (1994). On tabu search for the location of interacting hub facilities. *European Journal of Operational Research*, 73, 502–509.

Skorin-Kapov, D., Skorin-Kapov, J., and O'Kelly, M. E. (1996). Tight linear programming relaxations of uncapacitated p-hub median problems. *European Journal of Operational Research*, 94, 582–593.

Sohn, J. and Park, S. (1997). A linear program for the two-hub location problem. *European Journal of Operational Research*, 100(3), 617–622.

Sohn, J. and Park, S. (1998). Efficient solution procedure and reduced size formulations for p-hub location problems. *European Journal of Operational Research*, 108(1), 118–126.

Sohn, J. and Park, S. (2000). The single allocation problem in the interacting three-hub network. *Networks*, 35, 17–25.

Suzuki, A. and Drezner, Z. (1997). On the airline hub problem: The continuous model. *Journal of the Operations Research Society of Japan*, 40, 62–74.

Taha, T. T. and Taylor, G. D. (1994). An integrated modeling framework for evaluating hub-and-spoke networks in truckload trucking. *Logistics & Transportation Review*, 30(2), 141–166.

Taha, T. T., Taylor, G. D., and Taha, H. A. (1996). A simulation-based software system for evaluation of hub-and-spoke transportation networks. *Simulation Practice & Theory*, 3, 327–346.

Taylor, G. D., Harit, S., English, J. R., and Whicker, G. (1995). Hub and spoke networks in truckload trucking: Configuration, testing and operational concerns. *Logistics & Transportation Review*, 31(3), 209–238.

Taylor, G. D., Meinert, T. S., Killian, R. C., and Whicker, G. L. (1999). Development and analysis of alternative dispatching methods in truckload trucking. *Transportation Research Part E*, 35, 191–205.

Vahrenkamp, R. (1998). Web versus hub structure in the parcel services: A comparison of their respective efficiency. Working paper.

Yoon, M.-G., Back, Y.-H., and Tcha, D.-W. (1998a). On a large-scale distributed communication network design problem with fixed-charge hub facilities. Working paper.

Yoon, M.-G., Baek, Y.-H., and Tcha, D.-W. (1998b). Design of a distributed fiber transport network with hubbing topology. *European Journal of Operational Research*, 104, 510–520.

Yoon, M.-G. and Current, J. (1998). A dual based heuristic for hub network design. Working paper.

Yoon, M.-G. and Tcha, D.-W. (1996). Topological design of a two-level centralized network with service survivability. Working paper.

13 Location and Robotics

Oliver Karch[1], Hartmut Noltemeier[1], and Thomas Wahl[2]

[1] Department of Computer Science I, University of Würzburg, Am Hubland, 97074 Würzburg, Germany. e-mail: karch@informatik.uni-wuerzburg.de; noltemei@informatik.uni-wuerzburg.de
[2] Department of Computer Sciences, University of Texas at Austin, TAY 2.124, Austin, TX 78712-1188, USA. e-mail: wahl@cs.utexas.edu

13.1 Introduction

Nowadays, mobile robots play an important role in more and more fields of application. Mobile robots are used in dangerous or harmful environments, where humans cannot operate or are not willing to operate. Some examples are disarming bombs, maintaining oil platforms, exploring the Mars, see Volpe et al. (1997), tedious cleaning tasks, see Rencken et al. (1999), and many more. Although some of these tasks can also be accomplished by using a telecontrolled robot, the autonomy and the mobility of the robot are often essential for a good solution of the respective problem. Furthermore, there also exist applications of mobile robots, where the capability of performing low-level actions (e.g. *"Follow the wall"*, *"Pass the doorway"*, etc.) and autonomous high-level actions (e.g. *"Bring me to platform 5 at the railway station"*) is absolutely necessary. An example may be an autonomous wheelchair for disabled persons as presented by Praßler et al. (1999).

Since an autonomous mobile robot has to navigate in its environment, it must at least be able to answer the three following questions, summarized by Leonard and Durrant-Whyte (1991): "Where am I?", "Where am I going?", and "How should I get there?" At first, these questions appear to be relatively simply answerable, but in fact finding good solutions is very hard. Concerning the first question, Borenstein et al. (1996) claim that "perhaps the most important result from surveying the vast body of literature on mobile robot positioning is that to date there is no truly elegant solution for the problem."

In the following we will concentrate on the first of the three questions above, "Where am I?", that is, on the problem of *localizing a robot*. Some authors also use the term "self localization" instead of "localization" to express that the task is performed by the robot itself, in contrast to problems where someone else has to localize the robot from an outside point of view. In the following we will only use the shorter term "localization", meaning the robot's task of determining its position and orientation.

After taking a look onto related problems in Sect. 13.2, we give a short overview of the localization problem in Sect. 13.3. That is, we consider the sensors with which the robots are equipped, the different types of localization (relative and absolute), and the corresponding localization methods. Then, we introduce a variant of the localization problem, where some restrictive assumptions about the robot and its environment (e.g., that the robot has a compass and all sensors and the robot's map are *exact*) allow a formulation as a pure geometric problem. This problem can then be solved using methods of the field of "Computational Geometry" and in Sect. 13.5 we show by a detailed analysis, how some of the occuring complexity bounds can be expressed slightly sharper. Due to the idealizing assumptions the solution of the geometric problem is not directly applicable in practice, where the data normally is *noisy*. In Sect. 13.6 we consider these problems and show in the next two sections an approach to avoiding them, which uses *distance functions* to model the resemblance between the noisy sensor data and the structures of the original method (extracted from a possibly inexact map). The results of the following sections represent a summary of the papers of Karch and Wahl (1999) and Karch et al. (1999).

13.2 Related Problems

A robot that has to localize itself first needs a description of its environment, that is, some kind of a map. Such a map can be created by hand from an operator, but generally it is more reasonable to let the robot itself create the map using its on-board sensors. To autonomously perform this map generation process the robot must *explore* its environment. As well as for the localization problem many papers exist about exploration and map generation strategies, see for example the work of Puttkamer et al. (1999), Gonzalez et al. (1994), Pierce (1995), and Hoffmann et al. (1997). The latter work in this list investigates the problem of how to explore an arbitray polygonal environment in a competitive way, which is known as the problem of the *shortest watchman tour*. A related problem is the placement of stationary watchmen, for example, to keep the painting in an art gallery under surveillance. This problem is called the *art gallery problem*; for a comprehensive survey see the work of Urrutia (1997).

13.3 A Short Overview of the Localization Problem

13.3.1 The Sensors

In the following we describe some common types of sensors, with which autonomous mobile robots are equipped:

Odometry is a method to measure the wheel rotation and the steering orientation of the robot. This way we can record the path, which the robot has travelled so far, and provide the robot with an estimate of its position and orientation. Unfortunately, the position and orientation estimate from odometry is corrupted by inevitable drifts if the robot travels long distances without recalibrating.

Gyroscopes are used to measure the rotation much more exactly than using odometry. But as well as odometry also gyroscope data drifts (with time), even if the robot does not move at all. Therefore, also gyroscopes must be periodically recalibrated using an independet reference, see for example Borenstein et al. (1996).

Ultrasonic sensors determine distances to objects in front of the sensor by emitting a ultrasonic signal and measuring the running time until the sensor receives the reflected signal. Since ultrasonic waves are likely to be reflected at smooth surfaces, the distance measurements are not very accurate. Consequently, ultrasonic sensors are often used only to avoid collisions with obstacles.

Laser scanners work similar to ultrasonic sensors, but use a laser beam instead of ultrasonic waves. This results in much more precise distance measurements. The right part of Fig. 13.5 shows a 360° full range scan, which consists of 720 individual range measurements.

Cameras provide digitized pictures of the robot's environment, from which relevant features (e.g., walls, doorways, etc.) can be extracted and used for localization and navigation purposes.

Additionally, the robot may make use of systems that allow an absolute position measurement. This could be, for example, active beacons that usually transmit light or radio signals, whose direction of incidence is measured by the robot. From three or more such measurements the robot can compute its absolute position, provided that the transmitters are located at known sites in the environment. Similarly, artificial landmarks that are placed at known position and which can be recognized by the robot's sensors, can be used. For example, reflecting stripes (for the usage with laser scanners) or signs on the walls with special shapes on them (for vision sensors) could be placed in the environment. Another way of performing an absolute localization is to use the Navstar Global Positioning System (GPS), where 24 satellites orbit, which orbit the earth, transmit radio signals that allow a position measurement with an error of about 3 m; for details see Borenstein et al. (1996). But for indoor-applications (like the ones that we have in mind) GPS is less applicable, since the GPS signals are likely to be absorbed by the walls of the buildings.

In the following we will not consider localization methods using this kind of absolute position measurement, since often it is not possible or too expensive to modify the environment this way. Furthermore, for the long-term objective of an autonomous robot it is not reasonable to rely on modifications to the environment.

Typically, a robot is equipped with odometry and gyroscopes for maintaining a rough position estimate, ultrasonic sensors for collision avoidance, and laser scanners for the localization task. Of course, also the ultrasonic sensors and eventual additional sensors can be used for localization, for example, by integrating their data using Kalman filtering techniques, see Borenstein et al. (1996) and Gutmann (1996).

13.3.2 Relative and Absolute Localization

Generally it must be distinguished between two different types of localization problems: the relative localization problem, where the robot already has an estimate of its position and orientation (e.g., using its odometry), and the absolute localization problem, where the robot has no knowledge about previous configurations. For the former problem many approaches exist, which can be classified into scan-based and feature-based approaches, techniques using artificial landmarks and a few other methods, see for example the work of Praßler and Milios (1995), Edlinger and Weiß (1995), and Cox (1991).

The latter problem can be seen as a kind of wake-up situation (e.g., after a power failure or maintenance works), where the robot is placed somewhere in its environment, powered on, and then "wants" to know where it is located. Since here the search space usually is much greater than for a relative localization query, we cannot expect to solve an absolute localization query as efficient as a relative one. The methods for answering such queries are similiar to the ones described above for the relative localization and can be classified the same way: scan-based and feature-based approaches and methods using artificial landmarks. See Rencken et al. (1999), Kluge (1999), Buck et al. (1999a), and Kluge et al. (2000) for examples of feature-based techniques. In the following we consider an idealization of a scan-based approach as described in the next section.

13.3.3 An Idealized Version of the Localization Problem

In our setting we assume that the robot has a polygonal map of its environment and a range sensor (e.g., a laser radar), which provides the robot with a set of range measurements (usually at equidistant angles). The localization should be performed using only this minimal equipment. In particular, the robot is not allowed to use landmarks (e.g., marks on the walls or on the floor).

The localization process usually consists of two stages. First, the non-moving robot enumerates all hypothetical positions that are *consistent* with its sensor data, i.e., that yield the same (or at least a similiar) visibility polygon. There can very well be several such positions if the map contains identical parts at different places (e.g., buildings with many identical corridors, like hospitals or libraries). All those positions cannot be distinguished by a non-moving robot. Figure 13.1 shows an example: the marked positions at the bottom of the two outermost niches cannot be distinguished using only their visibility polygons.

If there is more than one hypothetical position, the robot eliminates the wrong hypotheses in the second stage and determines exactly where it is by traveling around in its environment. This is a typical on-line problem, because the robot has to consider the new information that arrives while the robot is exploring its environment. Its task is to find a path as efficient (i.e.,

Fig. 13.1. Polygonal map and its decomposition into visibility cells

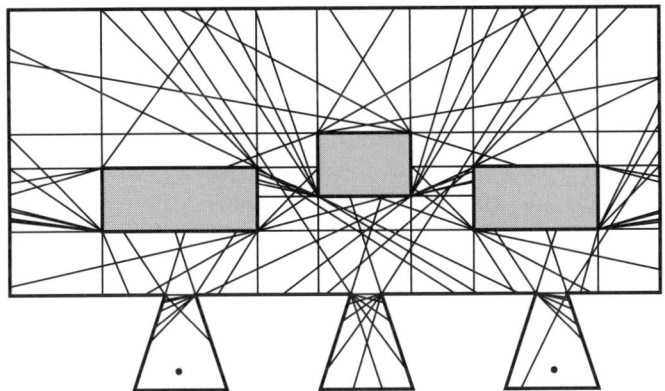

short) as possible for eliminating the wrong hypotheses. Dudek et al. (1998) have already shown that finding an optimal localization strategy is NP-hard, and described a competitive greedy strategy, the running time of which was recently improved by Schuierer (1997).

We will concentrate only on the first stage of the localization process, that is, on generating the possible robot configurations (i.e., positions and orientations), although our current work addresses also the second stage, see Buck et al. (1999b). With the additional assumption that the robot already knows its orientation (i.e., the robot has a *compass*) and all sensors and the map are *exact* (i.e., without any noise), this problem turns into a pure geometric one, stated as follows: For a given map polygon $\mathcal{M}$ and a star-shaped polygon $\mathcal{V}$ (the visibility polygon of the robot), find all points $p \in \mathcal{M}$ that have $\mathcal{V}$ as their visibility polygon.

Guibas et al. (1995) described a scheme for efficiently solving this idealized version of the localization problem and we will briefly sketch their method in the following section.

13.4 Solving the Geometric Problem

In the following we sketch the method of Guibas et al., for which we will give a sharper preprocessing bound in the next section and which also is the basis for our approach described in Sections 13.7 and 13.8. We assume that the robot navigates on a plain surface with mostly vertical walls and obstacles such that the environment can be described by a polygon $\mathcal{M}$, called the *map polygon*. Additionally, we assume that $\mathcal{M}$ has *no holes* (i.e., there are no free-standing obstacles in the environment), although the algorithm remains the same for map polygons with holes; the preprocessing costs, however, may be higher in that case.

The (exact) range sensor generates the star-shaped *visibility polygon* $\mathcal{V}$ of the robot. As the range sensor is only able to measure the distances relative to its own position, we assume the origin of the coordinate system of $\mathcal{V}$ to be the position of the robot.

Using the assumption that we have a compass, the geometric problem is then to find all points $p \in \mathcal{M}$ such that their visibility polygon $\mathcal{V}_p$ is identical with the visibility polygon $\mathcal{V}$ of the robot, that is, the equality $\mathcal{V} + p = \mathcal{V}_p$ holds.

The main idea of Guibas et al. (1995) to solving this problem is to divide the map into finitely many visibility cells such that a certain structure (the visibility skeleton, which is closely related to the visibility polygon) does not change inside a cell.

For a localization query we then do not search for points where the visibility polygon fits into the map, but instead for points where the corresponding skeleton does. That is, the continuous problem[1] of fitting a visibility polygon into the map is discretized in a natural way by decomposing the map into visibility cells.

13.4.1 Decomposing the Map into Cells

At preprocessing time the map $\mathcal{M}$ is divided into convex *visibility cells* by introducing straight lines forming the boundary of the cells such that the following property holds:

> The set of visible map vertices does not change when we travel around within a cell.

As the visibility of a vertex only changes if we cross a straight line induced by that vertex and an occluding *reflex* vertex (i.e., having an internal angle $\geq 180°$), the subdivision into visibility cells can be constructed in the following way: We consider all pairs consisting of a vertex v and a reflex vertex v_r that are visible from each other; for each such pair (v, v_r) we introduce the ray into the map that goes along the line through v and v_r, starts at v_r, and is oriented as to move away from v. An example of such a decomposition is depicted in the left part of Fig. 13.2. The introduced rays are drawn as dashed lines. The points p and q from cell $\mathcal{C}$ see the same set of five map vertices (marked gray in the corresponding visibility polygons in the middle). Figure 13.1 shows a decomposition for a more complex map with three obstacles (gray), generated with our software RoLoPro described in Sect. 13.9.

If the map consists of a total number of n vertices, of which r are reflex, the number of introduced rays is in $\mathcal{O}(nr)$ and therefore the complexity of

---

[1] "Continuous" in the sense that we cannot find an $\varepsilon > 0$ such that the visibility polygon $\mathcal{V}_p$ of a point p moving by at most ε does not change.

Fig. 13.2. Decomposition of a map polygon into visibility cells (left), two visibility polygons (middle), and the corresponding skeleton (right)

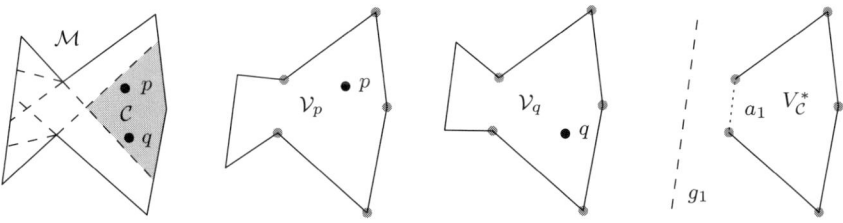

the decomposition is in $\mathcal{O}(n^2 r^2)$. For map polygons without holes it can be shown that this complexity is actually in $\mathcal{O}(n^2 r)$. Moreover, it is easy to give worst-case examples that show that these bounds are tight.

13.4.2 The Visibility Skeleton

When we compare two visibility polygons of points from the same cell (see Fig. 13.2), we see that they are very similar and differ only in certain aspects, namely in the *spurious edges* that are caused by the reflex vertices and are collinear with the viewpoint, and in those map edges that are only *partially visible*. The remaining *full edges* (which are completely visible) are the same in both polygons. This observation can be used to define a structure that does not change inside a visibility cell, the *visibility skeleton*.

For a visibility polygon $\mathcal{V}_p$ with viewpoint p, the corresponding visibility skeleton $\mathcal{V}_p^*$ is constructed by removing the spurious edges, and by substituting the partially visible edges (they lie between two spurious vertices or between a spurious and a full vertex) with an *artificial edge* a_i together with the corresponding line g_i on which the original (partially visible) edge of $\mathcal{V}_p$ lies. Thus, we simply ignore the spurios edges and the spurios vertices, as this information continuously depends on the exact position p.

As the skeleton does not change inside a cell, we can define the *cell skeleton* $\mathcal{V}_\mathcal{C}^*$ as the common skeleton of all visibility polygons of points from cell $\mathcal{C}$. Figure 13.2 shows an example of the common skeleton $\mathcal{V}_\mathcal{C}^*$ of two visibility polygons $\mathcal{V}_p$ and $\mathcal{V}_q$ for points p and q from the same cell $\mathcal{C}$.

13.4.3 Embeddings of a Skeleton

Just as a star-shaped polygon $\mathcal{V}$ can fit into the map $\mathcal{M}$ at several positions, the same holds for the skeleton V^* defined above. A mapping $h(V^*)$, which fits V^* into the map $\mathcal{M}$, is called an *embedding* of the skeleton V^*. It can be shown that for every skeleton only $\mathcal{O}(r)$ different embeddings exist (r being the number of reflex vertices). Therefore, for a localization query the maximum number of possible robot locations is also bounded by $\mathcal{O}(r)$.

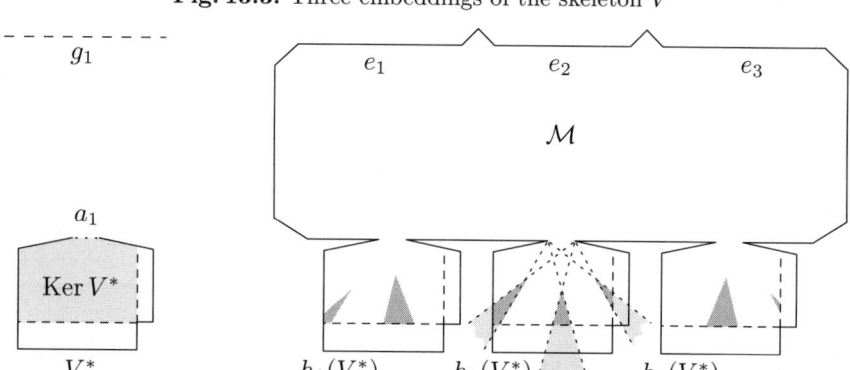

Fig. 13.3. Three embeddings of the skeleton V^*

The connection between a skeleton and its embeddings is illustrated in Fig. 13.3. The skeleton V^*, which has one artificial edge a_1, has three embeddings $h_1, \ldots, h_3$ into the map $\mathcal{M}$. In each embedding, each full edge of V^* matches an edge of $\mathcal{M}$. Furthermore, for the artificial edge a_1 there exists at least one map edge that lies on the corresponding embedded line $h_j(g_1)$, for $1 \leq j \leq 3$. Such edges are called *candidate edges*.

For a fixed skeleton V^* with an artificial edge a_i, which is embedded into the map by h_j, we denote by $C_{i,j}$ the set of all candidate edges for a_i in embedding $h_j(V^*)$ and by $c_{i,j}$ the cardinality of $C_{i,j}$. Note that one edge may serve as a candidate edge in several embeddings of the same skeleton. For example, in Fig. 13.3 the sets of candidate edges for the embeddings of the skeleton V^* are $C_{1,1} = C_{1,2} = C_{1,3} = \{e_1, e_2, e_3\}$. The visibility cells with skeleton V^* are drawn in dark-gray in the figure. Note that from each of these cells exactly one of the candidate edges $e_1, \ldots, e_3$ for the (embedded) artificial edge $h_j(a_1)$ is visible "through" a_1.

When we construct the skeleton V_p^* from the visibility polygon $\mathcal{V}_p$, we "throw away" some information about the spurious edges and the partially visible edges. But this information can be reconstructed using the position of the viewpoint p relative to the skeleton. This relationship between a visibility polygon and the embeddings of its corresponding skeleton is described in the following theorem:

Theorem 1 (Guibas et al.). *Let $\mathcal{V}_p$ be a visibility polygon, V_p^* the corresponding skeleton, $h(V_p^*)$ an embedding of this skeleton into the map, and $h(p)$ the corresponding viewpoint in the embedding. Then, $\mathcal{V}_{h(p)} = \mathcal{V}_p$ if and only if $V_{h(p)}^* = V_p^*$.*

The consequence of this theorem is that in order to determine all points in the map that have $\mathcal{V}_p$ as their visibility polygon it is sufficient to consider all embeddings $h(V_p^*)$ of the skeleton V_p^* and then to check whether the corresponding embedded viewpoint $h(p)$ induces the same skeleton V_p^*, that is, the point $h(p)$ must lie in a cell $\mathcal{C}$ with $V_\mathcal{C}^* = V_p^*$.

That means, the reduction of the visibility polygon to the corresponding skeleton and the decomposition of the map into visibility cells discretizes our problem in a natural way: Instead of testing infinitely many visibility polygons we have to test only a finite number of skeletons to determine all possible robot locations.

13.4.4 Costs of the Localization Query and the Preprocessing

As already stated above, the decomposition into visibility cells has a complexity of $\mathcal{O}(n^2 r)$ for map polygons without holes. At preprocessing time this decomposition is computed and the cells are divided into equivalence classes according to their skeletons, that is, two cells $\mathcal{C}_1$ and $\mathcal{C}_2$ are said to be *equivalent* if $V_{\mathcal{C}_1}^*$ equals $V_{\mathcal{C}_2}^*$ up to a translation. For the resulting set of equivalence classes a search structure (e.g., a multidimensional search tree) is constructed that allows for a given skeleton the retrieval of the corresponding class in an efficient way (i.e., in time logarithmic in the number of classes).

For the localization query, the skeleton V_p^* of the given visibility polygon $\mathcal{V}_p$ is computed and the corresponding equivalence class is determined. As we know the position of the point p relative to the skeleton V_p^* and as we also know the position of each cell $\mathcal{C}$ relative to its cell skeleton $V_\mathcal{C}^*$, we can easily determine for each cell in the equivalence class the embedded point $h(p)$ and check whether it lies in $\mathcal{C}$. If the point $h(p)$ is in $\mathcal{C}$, then $h(p)$ is a valid robot position by Theorem 1. The test whether $h(p)$ is in $\mathcal{C}$ or not can be performed very efficiently by using a point location structure; in fact, only a single point location query is necessary.

This way we get a query time of $\mathcal{O}(m + \log n + A)$ where m is the number of vertices of $\mathcal{V}_p$ and A denotes the size of the output, that is, the number of all reported robot locations.

The total preprocessing time and space is in $\mathcal{O}(n^2 r \cdot |\mathcal{EC}|)$ for map polygons without holes. Here, $|\mathcal{EC}|$ denotes the worst-case complexity of an equivalence class, where for a given skeleton V^* the complexity of the corresponding equivalence class $\mathcal{EC}_{V^*}$ is the total number of vertices and edges of all visibility cells with skeleton V^*. Guibas et al. (1995) have already shown that $|\mathcal{EC}|$ is in $\mathcal{O}(n^2)$.

13.5 A Sharper Bound for $|\mathcal{EC}|$

In this section we establish a tighter upper bound of $\mathcal{O}(n + r^2)$ for the complexity of an equivalence class, such that the dependence on the number of reflex vertices becomes clearer; we also show that this bound is worst-case optimal. Due to lack of space the proofs will only be sketched. For the complete proofs see Karch (1996).

The main idea of the proof is to concentrate on a single skeleton V^* and to determine the complexity of $\mathcal{EC}_{V^*}$ by counting the edges of all visibility cells whose skeleton equals V^*. Therefore, we first study the structure of the visibility cells and classify their bounding edges into two groups. It turns out that only the edges of the second type, which are determined by the candidate edges, are difficult to count, and we will first give an upper bound for their number in a *single* embedding of V^*.

Unfortunately, summing up these numbers for all embeddings of the skeleton V^* does not yield the desired bound. The reason is that one edge may serve as a candidate edge in several embeddings as already stated above. Therefore, we will examine such situations and show how to perform the summation in a more sophisticated way.

13.5.1 The Structure of the Visibility Cells

For a fixed skeleton V^*, we examine the visibility cells whose visibility skeleton equals V^*. Each edge of such a visibility cell is either a *kernel edge* or an *AC edge*:

Kernel edges are edges that lie on the boundary of the embedded *kernel* $h(\mathrm{Ker}\, V^*)$ of the skeleton. Here, the kernel $\mathrm{Ker}\, V^*$ of a skeleton V^* is defined analogously to the kernel of a polygon (see the left part of Fig. 13.3): $\mathrm{Ker}\, V^*$ consists of all points that are visible from all edges of the skeleton (i.e., from all points of all full edges and all artificial edges). For example, all horizontal and vertical edges of the visibility cells of Fig. 13.3 are kernel edges.

AC edges are determined by pairs consisting of an embedded <u>a</u>rtificial edge and a corresponding <u>c</u>andidate edge. An AC edge lies on the ray induced by a reflex vertex of the artificial edge and a vertex of the candidate edge. In Fig. 13.3 these rays are drawn as dotted lines starting at the vertices of the embedded artificial edge $h_2(a_1)$.

The next lemma, the proof of which we omit, gives us an upper bound for the total number of kernel edges. It shows that it suffices to count only the AC edges for establishing the $\mathcal{O}(n + r^2)$ bound.

Lemma 1. *The total number of kernel edges in $\mathcal{EC}_{V^*}$ is in $\mathcal{O}(n + r^2 + t)$, where t is the total number of AC edges.*

In order to count all AC edges we consider each embedding h_j of the skeleton V^* and count the number of AC edges in $h_j(V^*)$. Summing up over all embeddings yields the total number of AC edges in $\mathcal{EC}_{V^*}$. To this end, let k be the number of artificial edges of V^* and let s be the number of all embeddings of V^*. Recall that s (as well as k) is in $\mathcal{O}(r)$, the number of reflex vertices.

The following lemma bounds the number of AC edges in a single embedding $h_j(V^*)$.

Lemma 2. *The number of all AC edges in embedding $h_j(V^*)$ is in*

$$\mathcal{O}\Big(\sum_{i=1}^{k} c_{i,j} + \big(\sum_{i=1}^{k} c_{i,j}\big)^2 - \sum_{i=1}^{k} c_{i,j}^2\Big). \tag{13.1}$$

Proof. This can be shown by considering how the visibility cells are created: Each pair consisting of an embedded artificial edge and a candidate edge induces a *visibility wedge*. This wedge consists of all points that can see the candidate edge "through" the embedded artificial edge. Figure 13.3 shows the visibility wedges (drawn in light-gray) of $h_2(V^*)$ as an example.

It can easily be seen that each AC edge lies on the boundary of one visibility wedge. Therefore, the complexity of the arrangement of the visibility wedges for all artificial and corresponding candidate edges of $h_j(V^*)$ gives us an upper bound for the number of AC edges. When we take into account that each visibility wedge in $h_j(V^*)$ corresponds to a candidate edge in one of the sets $C_{i,j}$, for $1 \leq i \leq k$, this complexity is contained in the class stated in (13.1). □

13.5.2 The Total Number of Possible Candidate Edges

Since the number of candidate edges plays an important role in Lemma 2, we establish an upper bound for the total number of all *possible* candidate edges of cells in $\mathcal{EC}_{V^*}$. To this end, let $\overline{C}$ be the set of all candidate edges for all artificial edges in all embeddings of the skeleton V^*, that is

$$\overline{C} := \bigcup_{\substack{1 \leq i \leq k \\ 1 \leq j \leq s}} C_{i,j}.$$

For the cardinality of $\overline{C}$ the following can be shown:

Lemma 3. *In map polygons without holes or with only convex holes, the cardinality of $\overline{C}$ is in $\mathcal{O}(r)$.*

Proof. Basically, this holds because at least one reflex vertex lies between two consecutive candidate edges. For example, in Fig. 13.3 there are two reflex vertices between the edges e_1 and e_2. □

If we assume that the sets $C_{i,j}$ of candidate edges for a *single* embedding $h_j(V^*)$ are disjoint[2], the sum $\sum_{i=1}^{k} c_{i,j}$ is in $\mathcal{O}(r)$ by Lemma 3 and using Lemma 2 we obviously get an upper bound of $\mathcal{O}(r^2)$ for the number of AC edges in a *single* embedding.

Furthermore, if we could expect that *all* sets $C_{i,j}$ of candidate edges are disjoint, each possible candidate edge of $\overline{C}$ could be assigned to *exactly one*

---
[2] Note that this assumption does *not* hold if the map polygon is allowed to have holes.

visibility wedge and the complexity of the arrangement of the $\mathcal{O}(|\overline{C}|)$ visibility wedges in *all* embeddings of V^* would also be bounded by $\mathcal{O}(|\overline{C}|^2) \subseteq \mathcal{O}(r^2)$. That is, we would have established our desired bound of $\mathcal{O}(r^2)$ for the total number of AC edges. But unfortunately, this assumption is not true, since the sets $C_{i,j}$ need not be disjoint as already noted in Sect. 13.4.3 in the example on page 416.

Therefore, we have to take a closer look at situations where one edge may serve as a candidate edge in more than one embedding of V^*. In this context it is useful to take *pairs* of candidate edges into account.

13.5.3 Pairs of Possible Candidate Edges

Although one possible candidate edge of $\overline{C}$ may be contained in more than one of the sets $C_{i,j}$, the following lemma shows that for one *pair* $(e, f) \in \overline{C} \times \overline{C}$ of possible candidate edges there exist at most one embedding h_j and at most one pair of artificial edges such that e and f are candidate edges for the two artificial edges in embedding $h_j(V^*)$. Informally speaking, each pair of candidate edges can be assigned to at most one pair of visibility wedges.

Lemma 4. *In map polygons without holes, the cardinality of the set*

$$\{(i_1, i_2, j) \mid i_1 \neq i_2 \land (e, f) \in C_{i_1,j} \times C_{i_2,j}\}$$

is at most one, for every pair $(e, f) \in \overline{C} \times \overline{C}$.

Proof. Omitting the details and simplifying the situation, the argument is as follows: Assume that for a skeleton V^* two pairs of visibility wedges exist that are induced by the same pair (e, f) of possible candidate edges. This situation is depicted in Fig. 13.4: The skeleton V^* has two embeddings h_1 and h_2 such that the edge e determines a visibility wedge for the artificial edges $h_1(a_1)$ and $h_2(a_1)$. Analogously, f determines a wedge for $h_1(a_2)$ and $h_2(a_2)$. (Note that the lines g_1 and g_2 and the visibility wedges are omitted in the figure.) As e and f must be visible "through" the embedded artificial edges, there must exist lines of sight (drawn dashed in the figure) from the candidate edges e and f, respectively, to points inside the embeddings $h_1(V^*)$ and $h_2(V^*)$, respectively. These lines intersect and create a circle (drawn in light-gray) that does not intersect any map edge and contains at least one map vertex. Therefore, the map must have at least one hole (drawn in dark-gray), which contradicts our assumption that the map has no holes. □

Using the same idea as in this proof, it can furthermore be shown that for each possible candidate edge e and for each embedding h_j, at most one artificial edge exists such that e is a candidate edge in embedding $h_j(V^*)$, which corresponds to the artificial edge. Informally speaking, in each embedding each candidate edge can be assigned to at most one visibility wedge. This fact is expressed in the following lemma.

Fig. 13.4. Assigning two candidate edges to two pairs of artificial edges

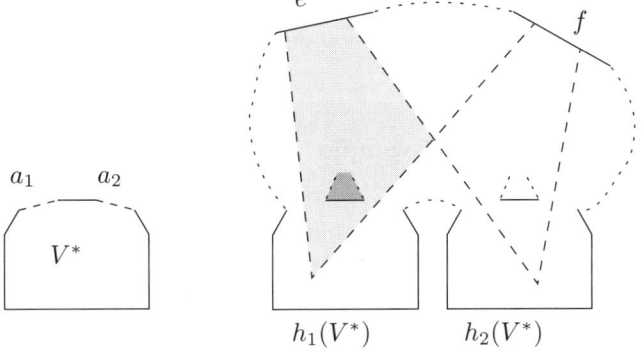

Lemma 5. *In map polygons without holes, the cardinality of the set $\{i \mid e \in C_{i,j}\}$ is at most one for every $e \in \overline{C}$ and for $1 \leq j \leq s$.*

Now we are able to show the $\mathcal{O}(r^2)$ bound for the total number of AC edges. Using (13.1) from Lemma 2, this number is in

$$\mathcal{O}\left(\sum_{j=1}^{s}\sum_{i=1}^{k}c_{i,j} + \sum_{j=1}^{s}\left(\left(\sum_{i=1}^{k}c_{i,j}\right)^2 - \sum_{i=1}^{k}c_{i,j}^2\right)\right).$$

This sum can be rewritten as

$$\sum_{j=1}^{s}\sum_{i=1}^{k}|C_{i,j}| + 2\sum_{j=1}^{s}\sum_{\substack{1\leq i_1,i_2\leq k \\ i_1<i_2}}|C_{i_1,j} \times C_{i_2,j}|,$$

and applying Lemmas 4 and 5 yields

$$\sum_{j=1}^{s}|\overline{C}| + 2\left|\bigcup_{j=1}^{s}\bigcup_{\substack{1\leq i_1,i_2\leq k \\ i_1<i_2}}C_{i_1,j} \times C_{i_2,j}\right|.$$

Using Lemma 3 and the fact that $s \in \mathcal{O}(r)$, this number is in $\mathcal{O}(r|\overline{C}| + |\overline{C}|^2) \subseteq \mathcal{O}(r^2)$.

Summarizing our results, we get the following theorem.

Theorem 2. *The total complexity of any equivalence class for a map polygon without holes is in $\mathcal{O}(n + r^2)$.*

In the case of many reflex vertices (i.e., $r \in \Omega(n)$) this yields the same bound as Guibas et al. (1995). However, the dependence on the number of reflex vertices is now expressed more clearly. For example, if the number of reflex vertices r is bounded by $\mathcal{O}(\sqrt{n})$, the complexity of an equivalence class depends only linearly and not quadratically on the total number n of map vertices.

13.5.4 Effects on the Preprocessing Costs

Using Theorem 2 and the same arguments as Guibas et al. (1995), the time and space bounds for the preprocessing (see Sect. 13.4) can be sharpened from $\mathcal{O}(n^4 r)$ to $\mathcal{O}(n^2 r \cdot (n + r^2))$.

13.5.5 A Worst-Case Example

Consider the map polygon shown in Fig. 13.3, which can be looked at as a corridor with large niches on one side and small niches on the other side. If we insert additional collinear candidate edges e_i and additional large niches $h_j(V^*)$, we get a map polygon with $\Omega(r)$ embeddings of V^*, each with $\Omega(r)$ possible candidate edges for the artificial edge a_1 in embedding $h_j(V^*)$. If we scale the scene in an appropriate way, we can achieve that each edge e_i induces a visibility cell with skeleton V^* in each embedding $h_j(V^*)$. Therefore, the total number of AC edges of all visibility cells equivalent to V^* is in $\Omega(r^2)$. Furthermore, by appropriately inserting additional points to V^* we can achieve that the number of kernel edges is in $\Omega(n)$.

Therefore, we get a total complexity of $\Omega(n+r^2)$ for the equivalence class of the skeleton V^* for this worst-case example.

13.5.6 Map Polygons with Holes

The above results only hold for map polygons *without holes*. That is, no freestanding obstacles are allowed in the robot's environment. For map polygons with holes, the bounds become worse, see Guibas et al. (1995) and Karch (1996). But for the special case of map polygons with l *convex holes*, an upper bound of $\mathcal{O}(n + (l+1)r^2)$ for the worst-case complexity of an equivalence class can be proven.

The idea for establishing this bound is essentially the same as in the case without holes: As in that proof, it can be shown that the cardinality of the two sets defined in Lemma 4 and Lemma 5 is at most l (instead of one). Again, this is done by introducing lines of sight from points inside the embedded skeletons to the candidate edges and counting the number of created holes.

13.6 Problems in Realistic Scenarios

The idealizing assumptions of the method described in Sect. 13.4 prevent us from using it in realistic scenarios, as we encounter several problems:

- Realistic range sensors do not generate a visibility *polygon* $\mathcal{V}$ as assumed for the method, but only a finite sequence $\mathcal{S}$ of *scan points* (usually, measured at equidistant angles). Furthermore, these scan points do not lie exactly on the robot's visibility polygon, but are perturbed due to sensor uncertainties. An example is depicted in Fig. 13.5, which shows in

Fig. 13.5. Exact visibility polygon (left) and approximated visibility polygon (middle) of a noisy scan (right)

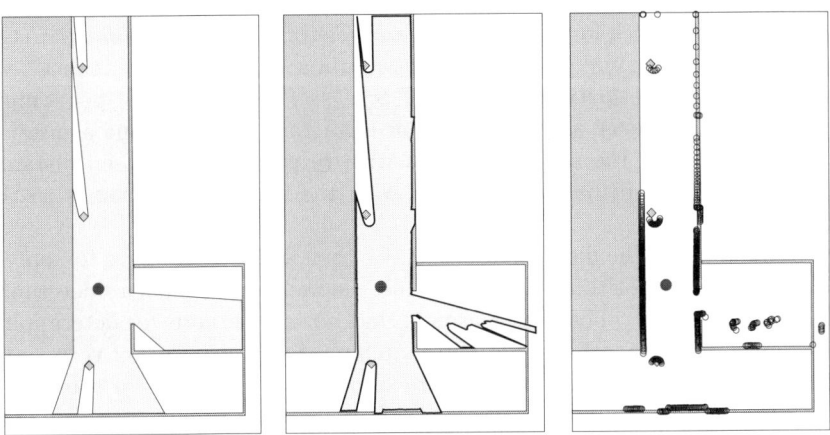

the left part an exact visibility polygon of a robot (based on a map of our department). In the right part of the figure we see a real noisy laser range scan taken at the corresponding position in our department using a SICK LMS 200 laser scanner. Even if we connect the scan points by straight line segments as shown in the middle part of the figure, we only get an approximation $\mathcal{V}_\mathcal{S}$ of the exact visibility polygon $\mathcal{V}$ based on the map.

– For the localization process we assume that we already know the exact orientation of the robot. But in practice this is often not the case, and we only have inexact knowledge or no knowledge at all about the robot's orientation.

– There may be obstacles in the environment that are not considered in the map and which may affect the robot's view. For example, furniture that is too small to be considered for map generation or even dynamic obstacles like people or other robots. Such obstacles can also be recognized in the right part of the example scan of Fig. 13.5.

– Realistic range sensors have a limited sensing range and obstacles that have a greater distance to the robot cannot be detected.

The consequence is that the (approximated) visibility skeleton $V_\mathcal{S}^*$, which the robot computes from its approximated visibility polygon $\mathcal{V}_\mathcal{S}$, usually does not match any of the preprocessed skeletons exactly. That is, the robot is not able to determine the correct equivalence class, and the localization process completely fails.

13.7 Adaptation to Practice

Our approach to tackling these problems is, for a given range scan $\mathcal{S}$ (from the sensor), to search for the preprocessed skeleton that is *most similar* to the scan. For modeling the resemblance between a scan $\mathcal{S}$ and a skeleton V^* we use an appropriate *distance function* $d(\mathcal{S}, V^*)$. Then, instead of performing an *exact match* query as in the original algorithm, we carry out a *nearest-neighbor* query in the set of skeletons with respect to the chosen distance function $d(\mathcal{S}, V^*)$ to find the skeleton with the highest resemblance to the scan.

Depending on the distance function, we then additionally have to apply a local matching algorithm to the scan and the skeleton in order to determine the position of the robot. The reason is that not all methods for determining a distance measure yield an optimal matching (i.e., a translation vector and a rotation angle) as well. Consider, for example, the algorithm for computing the *Arkin metric* for polygons, see Arkin et al. (1991), which, besides the distance measure, only provides the optimal rotation angle and no translation vector. In contrast to this, algorithms for computing the *minimum Hausdorff distance* under rigid motions, see Alt et al. (1995), provide both, the distance measure and the corresponding matching.

13.7.1 Requirements to the Distance Function

In order to be useful in practice, a distance function $d(\mathcal{S}, V^*)$ should at least have the following properties:

Continuity The distance function should be continuous in the sense that small changes in the scan (e.g., caused by noisy sensors) or even in the skeleton (e.g., caused by an inexact map) should only result in small changes of the distance. More precisely: Let $d_\mathcal{S}(\mathcal{S}_1, \mathcal{S}_2)$ and $d_{V^*}(V_1^*, V_2^*)$ be functions that measure the resemblance between two scans $\mathcal{S}_1$ and $\mathcal{S}_2$ and between two skeletons V_1^* and V_2^*, respectively. An appropriate reference distance measure for $d_\mathcal{S}(\mathcal{S}_1, \mathcal{S}_2)$ and $d_{V^*}(V_1^*, V_2^*)$ is, for example, the Hausdorff distance (see Sect. 13.8.1).

The distance $d(\mathcal{S}, V^*)$ is said to be *continuous with respect to scans* if

$$\forall_{\varepsilon>0} \exists_{\delta>0} : d_\mathcal{S}(\mathcal{S}_1, \mathcal{S}_2) < \delta \Rightarrow |d(\mathcal{S}_1, V^*) - d(\mathcal{S}_2, V^*)| < \varepsilon$$

holds, for all scans $\mathcal{S}_1, \mathcal{S}_2$ and all skeletons V^*. Analogously, $d(\mathcal{S}, V^*)$ is said to be *continuous with respect to skeletons* if

$$\forall_{\varepsilon>0} \exists_{\delta>0} : d_{V^*}(V_1^*, V_2^*) < \delta \Rightarrow |d(\mathcal{S}, V_1^*) - d(\mathcal{S}, V_2^*)| < \varepsilon$$

holds, for all skeletons V_1^*, V_2^* and all scans $\mathcal{S}$.

The requirement of continuity is also motivated by the fact that particularly the classification of the edges of the visibility polygon into different types (spurious edges, partially visible edges, etc.) makes the original method susceptible to perturbations: Even a small translation of a vertex can change the type of an edge which yields a skeleton that does not match any equivalence class. In this sense, the exact match query of the original method can also be interpreted as a discrete distance between a visibility polygon and a skeleton, which, however, strongly violates the continuity requirement, because it takes only two values (e.g., 0 – "match" and 1 – "no match").

Similarity Preservation A skeleton V^* that is similar to $\mathcal{S}$ should have a small distance value $d(\mathcal{S}, V^*)$. Otherwise, the distance would not give any advice for finding a well-matching skeleton and therefore would be useless for the localization algorithm. In particular, if we take a scan $\mathcal{S}$ from a point p whose skeleton equals V^*, we want the distance $d(\mathcal{S}, V^*)$ to be zero or at least small, depending on the amount of noise and the resolution of the scan.

Translational Invariance As the robot has no knowledge about the relative position of the coordinate systems of the scan and the skeleton to each other, a translation of the scan or the skeleton in their local coordinate systems must not influence the distance. Rather finding this position is the goal of the localization algorithm.

Rotational Invariance If the robot does not have a compass, the distance must also be invariant under rotations of the scan (or the skeleton, respectively).

Fast Computability As the distance $d(\mathcal{S}, V^*)$ has to be determined several times for a single localization query (for different skeletons, see Sect. 13.7.2), the computation costs should not be too high.

13.7.2 Maintaining the Skeletons

As we do not want to compare a scan with all skeletons to find the skeleton with the highest resemblance (remember that their number can be in $\Omega(n^2 r^2)$, see Sect. 13.4.1), the skeletons should be stored in an appropriate data structure that we can search through efficiently.

For this purpose we can use the *Monotonous Bisector Tree* of Noltemeier et al. (1993), a spatial index that allows to partition the set of skeletons hierarchically with respect to a second distance function $D(V_1^*, V_2^*)$ that models the resemblance between two skeletons V_1^* and V_2^*. The set of skeletons is recursively divided into clusters with monotonously decreasing cluster radii in a preprocessing step. This division then represents the similarities of the skeletons among each other.

The distance function $D(V_1^*, V_2^*)$ should be chosen "compatible" to the function $d(\mathcal{S}, V^*)$, such that in the nearest-neighbor query not all clusters have to be investigated. That is, at least the *triangle inequality*

$$d(\mathcal{S}, V_2^*) \leq d(\mathcal{S}, V_1^*) + D(V_1^*, V_2^*)$$

should be satisfied. This way, we can determine lower bounds for the distance values $d(\mathcal{S}, V^*)$ of complete clusters, when traversing the tree. Such a cluster can then be rejected and does not have to be examined.

13.8 Suitable Distances for $d(\mathcal{S}, V^*)$ and $D(V_1^*, V_2^*)$

It is hard to find distance functions that have all the properties from Section 13.7.1. Particularly, the fifth requirement (fast computability) is contrary to the remaining ones. Moreover, it is often not possible to simply use existing polygon distances, because in our problem we have to cope with scans and skeletons instead of polygons. Therefore, a careful adaptation of the distance functions is almost always necessary. And of course, it is even more difficult to find for a given scan-skeleton distance $d(\mathcal{S}, V^*)$ a compatible distance $D(V_1^*, V_2^*)$, which we need for performing the nearest-neighbor query efficiently.

In the following we investigate two distance functions, the Hausdorff distance and the polar coordinate metric, and illustrate the occurring problems.

13.8.1 The Hausdorff Distance

For two point sets $A, B \subset \mathbb{R}^2$, their *Hausdorff distance* $\delta(A, B)$ is defined as

$$\delta(A, B) := \max\{\vec{\delta}(A, B), \vec{\delta}(B, A)\} ,$$

where

$$\vec{\delta}(A, B) := \sup_{a \in A} \inf_{b \in B} \|a - b\|$$

is the *directed Hausdorff distance* from A to B, and $\|\cdot\|$ is the Euclidean norm. Accordingly, the term $\vec{\delta}(A, B)$ stands for the maximum distance of a point from A to the set B.

Let $\mathcal{T}$ be the set of all Euclidean transformations (i.e., combinations of translations and rotations). The undirected and directed *minimum Hausdorff distances* with respect to these transformations are defined as

$$\delta_{\min}(A, B) := \inf_{t \in \mathcal{T}} \delta(A, t(B)) \quad \text{and} \quad \vec{\delta}_{\min}(A, B) := \inf_{t \in \mathcal{T}} \vec{\delta}(A, t(B)) .$$

It can easily be shown that the minimum Hausdorff distances are continuous and by definition also fulfill the third and forth property of Sect. 13.7.1.

But their computation is very expensive. According to Alt et al. (1995), this can be done in time $\mathcal{O}((ms)^4(m+s)\log(m+s))$ if m is the complexity of the scan and s is the complexity of the skeleton. This is surely too expensive in order to be used in our application.

On the other hand, the computation of the Hausdorff distance without minimization over transformation application is relatively cheap according to Alt et al. (1995), namely in $\mathcal{O}((m+s)\log(m+s))$. The property of continuity is also not affected, but we now have to choose a suitable translation vector and a rotation angle by hand.

An obvious choice for such a translation vector for a scan $\mathcal{S}$ and a skeleton V^* is the vector that moves the scan origin (i.e., the position of the robot) somewhere into the corresponding visibility cell $\mathcal{C}_{V^*}$ (e.g., the center of gravity of $\mathcal{C}_{V^*}$). This is reasonable, because by the definition of the visibility cells, exactly the points in $\mathcal{C}_{V^*}$ induce the skeleton V^*. Of course, the consequence of doing so is that all cells with the same skeleton (e.g., the big cells in the two outermost niches in Fig. 13.1) must be handled separately, because the distance $d(\mathcal{S}, V^*)$ now does not only depend on V^*, but also on the visibility cell itself.[3] Besides, their intersection may be empty and we might not find a common translation vector for all cells. Of course, the bigger the cell is that the scan has to be placed into, the bigger is the error of this approach, compared with the minimum Hausdorff distance.

A compromise for computing a good matching, which does have the advantages of the previous algorithms, is using an *approximate matching strategy*, which yields only a pseudo-optimal solution. This means, the algorithm finds a transformation $t \in \mathcal{T}$ with $\delta(A, t(B)) \leq c \cdot \delta_{\min}(A, B)$, for a constant $c \geq 1$. Alt et al. (1995) showed that for any constant $c > 1$ an approximate matching with respect to Euclidean transformations can be computed in time $\mathcal{O}(ms \log(ms) \log^*(ms))$ using so-called *reference points*. If we only want an approximate matching with respect to translations instead of Euclidean transformations, the time complexity would even be in $\mathcal{O}((m+s)\log(m+s))$.

Another point to consider is that a skeleton (interpreted as a point set) in general is not bounded, because it includes a straight line for each artificial edge. The result is that the directed distances $\vec{\delta}(V^*, \mathcal{S})$ and $\vec{\delta}_{\min}(V^*, \mathcal{S})$ almost always return an infinite value (except for the trivial case when V^* equals the convex map polygon and has no artificial edges). Therefore, we must either modify the skeletons or we can only use the directed distances $\vec{\delta}(\mathcal{S}, V^*)$ and $\vec{\delta}_{\min}(\mathcal{S}, V^*)$. Note that if we pursue the second approach, the distance $\vec{\delta}_{\min}(\mathcal{S}, V^*)$ is also similarity preserving, provided that the resolution of the scan is high enough such that no edge, in particular, no artificial edge, is missed.

---

[3] In this case, the notation $d(\mathcal{S}, V^*)$ is a bit misleading, since there might exist several cells that have the same skeleton V^*. To be correct, we should use the notation $d(\mathcal{S}, \mathcal{C}_{V^*})$, where the dependence of the distance from the cell is expressed more clearly. But we will use the easier-to-understand expression $d(\mathcal{S}, V^*)$.

Fig. 13.6. The polar coordinate function $\mathrm{pcf}_P(\varphi)$ for a star-shaped polygon P

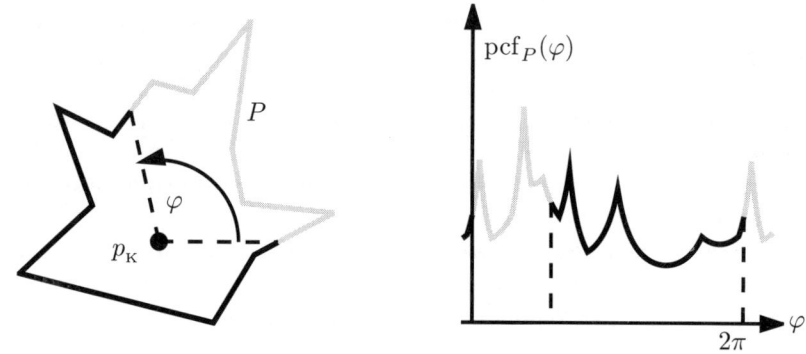

13.8.2 The Polar Coordinate Metric

A more specialized distance for our problem than the Hausdorff distance is the *polar coordinate metric* (PCM for short) investigated by Wahl (1997), which takes a fundamental property of our problem into account: All occurring polygons are *star-shaped* in the following sense, and we even know a kernel point:

- The approximate visibility polygon $\mathcal{V}_S$ (generated from the scan points) is star-shaped by construction with the origin as a kernel point.
- Every skeleton V^* is star-shaped in the sense that from every point in the corresponding visibility cell $\mathcal{C}_{V^*}$ all full edges are completely visible, and for each artificial edge a_i a part of the corresponding straight line g_i is visible.

To define the PCM between two (star-shaped) polygons P and Q with kernel points p_K and q_K we first define the value of the *polar coordinate function* (PCF for short)

$$\mathrm{pcf}_P(\varphi) : \mathbb{R} \to \mathbb{R}_{\geq 0}$$

as the distance from the kernel point p_K to the intersection point of a ray starting at p_K in direction φ with the boundary of P. That is, the function $\mathrm{pcf}_P(\varphi)$ corresponds to a description of the polygon P in polar coordinates (with p_K as the origin) and is periodical with a period of 2π. Figure 13.6 depicts the PCF for a star-shaped polygon as an example. In the same way we define the function $\mathrm{pcf}_Q(\varphi)$ for the polygon Q.

Then, the PCM between the polygons P and Q is the minimum integral norm between the functions pcf_P and pcf_Q in the interval $[0, 2\pi[$ over all horizontal translations between the two graphs (i.e., rotations between the

Fig. 13.7. Computation of the PCM as minimum integral norm

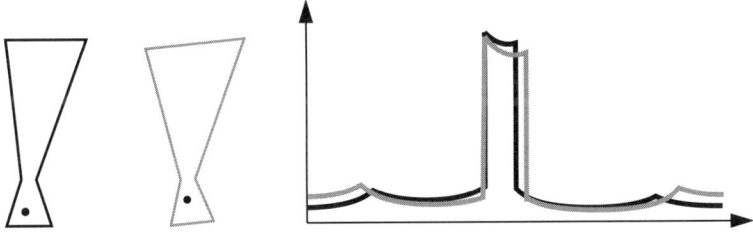

corresponding polygons):

$$\mathrm{pcm}(P,Q) := \min_{t \in [0, 2\pi[} \sqrt{\int_0^{2\pi} \left(\mathrm{pcf}_P(\varphi - t) - \mathrm{pcf}_Q(\varphi)\right)^2 \mathrm{d}\varphi} \quad (13.2)$$

Figure 13.7 shows an example, where the two graphs are already translated in a way such that this integral norm is minimized.

For a fixed kernel point the function pcf_P is continuous in φ except for one special case: When we move a vertex of a polygon edge such that the edge becomes collinear to p_K, the function pcf_P has a discontinuity at the corresponding angle, the height of which represents the length of the collinear edge. Moreover, the PCF is also continuous in the sense of the definitions in Sect. 13.7.1 with respect to translations of the polygon vertices or translations of the kernel point unless this special case occurs. But as pcf_P and pcf_Q may have only finitely many such discontinuities, the integration makes them continuous with respect to *all* translations of polygon vertices and translations of the kernel points, provided that P and Q remain star-shaped with kernel points p_K and q_K.

It can easily be seen that the PCM fulfills the continuity requirement of Sect. 13.7.1, if the kernel points are considered as a part of the polygons (i.e., part of the *input* of the PCM). This means that, given two polygons P and Q and an $\varepsilon > 0$, we can find a $\delta > 0$ such that $|\mathrm{pcm}(P,Q) - \mathrm{pcm}(P',Q)| < \varepsilon$, for all polygons P' that are created from P by moving all vertices *and* the kernel point p_K by at most δ. Moreover, if the kernel points are *not* considered as input of the PCM (that is, they are computed from P and Q by the algorithm that computes $\mathrm{pcm}(P,Q)$), the PCM is continuous as well, provided that the kernel points depend continuously on the polygons. For example, the center of gravity of the kernel of a polygon P depends continuously on P and can be used as a kernel point p_K, whereas, for example, the left-most kernel point does *not* depend continuously on the polygon.

Wahl (1997) also showed that the function $\mathrm{pcm}(P,Q)$ is a polygon *metric*, provided that the kernel points are invariant under Euclidean transformations. That is, if p'_K denotes the kernel point of a polygon $P' = t(P)$ for a transformations $t \in \mathcal{T}$, the equality $t(p_K) = p'_K$ must hold, for all polygons P

and all $t \in \mathcal{T}$. For example, the center of gravity of the kernel of the polygon has this property.

Using the PCM as Distance $d(\mathcal{S}, V^*)$ If we want to use the PCM as a distance function $d(\mathcal{S}, V^*)$ we need corresponding star-shaped polygons for $\mathcal{S}$ and V^* that can be used as polygonal representatives for the scan and the skeletons:

- For the scan, we choose the approximated visibility polygon $\mathcal{V}_\mathcal{S}$, which is star-shaped by construction. Again, the coordinate origin can be used as a kernel point.
- For generating a polygon from a skeleton V^* (with corresponding cell $\mathcal{C}_{V^*}$), we choose a point c inside the cell $\mathcal{C}_{V^*}$ (e.g., the center of gravity) and determine the visibility polygon $\mathcal{V}_c$ of this point. By construction, c is a kernel point of V_c^*.

In the sequel we will only use $\mathcal{V}_\mathcal{S}$ and $\mathcal{V}_c$ for determining the distance measures. Then, our goal is to find the polygon $\mathcal{V}_c$ that is most similar to the approximated visibility polygon $\mathcal{V}_\mathcal{S}$ with respect to pcm$(\mathcal{V}_\mathcal{S}, \mathcal{V}_c)$. With this choice we obtain the following theorem about the polar coordinate metric as a distance function:

Theorem 3. *The distance function $d(\mathcal{S}, V^*) := \mathrm{pcm}(\mathcal{V}_\mathcal{S}, \mathcal{V}_c)$, with $\mathcal{V}_\mathcal{S}$ and $\mathcal{V}_c$ as defined above, fulfills the following requirements from Sect. 13.7.1: continuity and invariance against translations and rotations.*

Note that the PCM is not similarity preserving: If the point c chosen above for computing a corresponding polygon for a visibility cell $\mathcal{C}$ does not equal the robot's position, the two polygons that are compared by the PCM are different and their distance value cannot be zero. But in practice, the visibility cells usually are not too large. That means, if we take a scan at a position $p \in \mathcal{C}$, the distance from p to the corresponding point $c \in \mathcal{C}$ is not too large. Thus, the approximated visibility polygon $\mathcal{V}_\mathcal{S}$ and the visibility polygon $\mathcal{V}_c$ differ not too much, and the value of pcm$(\mathcal{V}_\mathcal{S}, \mathcal{V}_c)$ is small.

Computing the PCM Value Efficiently The exact computation of the minimum in (13.2) seems to be difficult and time consuming, since the polar coordinate functions of a polygon with p edges consists of p pieces of functions of the form $c_i/\sin(\varphi + \alpha_i)$. For one fixed translation t (this corresponds to the case, when the robot already knows its exact orientation), the integral

$$\sqrt{\int_0^{2\pi} \left(\mathrm{pcf}_P(\varphi - t) - \mathrm{pcf}_Q(\varphi) \right)^2 \, \mathrm{d}\varphi}$$

can be computed straightforward in linear time $\mathcal{O}(p+q)$, where p and q stand for the complexities of the two polygons. But the global minimum over all possible values of $t \in [0, 2\pi[$ seems to be much harder to determine.

Fig. 13.8. Two approaches for an approximative PCM value

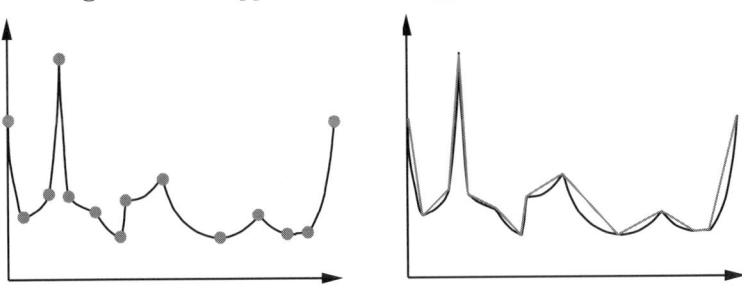

Therefore, we use two different approximative approaches for computing a suitable PCM value. Both approaches use a set of supporting angles/points for each of the two involved polar coordinate functions. For a given polygon its supporting angles are the angles that correspond to a vertex of the polygon plus the angles of the local minima of the inverse sine functions (see the left part of Fig. 13.8). The ideas of the two approximative approaches are then as follows:

1. For the first approach we concentrate on the $\mathcal{O}(p)$ (or $\mathcal{O}(q)$, respectively) supporting angles (see the left part of Fig. 13.8). Namely, we do not minimize over *all rotation angles* of the two polygons, but only over the $\mathcal{O}(pq)$ rotation angles, that place one supporting angle of the first polygon on a supporting angle of the second one. The integral values are then computed exactly in time $\mathcal{O}(p+q)$. Summing up, we need $\mathcal{O}(pq(p+q))$ time to compute this approximated value of the PCM.

2. In the second approach we compute an *exact minimum* over all rotation angles, but use a modified polar coordinate function. Namely, we introduce a linear approximation of the PCM, which also has all metric properties and is sufficient for our applications. This approximation is depicted in the right part of Fig. 13.8: The supporting points (that correspond to a polygon vertex or a local minima of the PCF) are connected by straight line segments to get a modified PCF. The minimum integral norm is then defined like in the non-approximated version of the PCM, see (13.2). Following an idea of Arkin et al. (1991), the actual computation of the minimum integral norm between the two *piecewise linear* functions can now be carried out much faster than the computation of the original PCM: Arkin et al. compute the minimum integral norm between two *piecewise constant* functions in time $\mathcal{O}(pq)$. This idea can be generalized to compute the approximated PCM in time $\mathcal{O}(pq \cdot (p+q))$. Of course, if we do not want to minimize over the rotations, the computation time is again in $\mathcal{O}(p+q)$ like for the non-approximated version.

Using the PCM as Distance $D(V_1^*, V_2^*)$ Since the PCM has all metric properties, it particularly fulfills the triangle inequality. Therefore, we can use it not only for defining the scan-skeleton distance $d(\mathcal{S}, V^*)$, but also for defining a *compatible* skeleton-skeleton distance $D(V_1^*, V_2^*)$.

For this task, we again use for each pair of a skeleton V^* and its corresponding cell $\mathcal{C}_{V^*}$ the polygonal representative $\mathcal{V}_c$ as defined in the last section. Then, the triangle inequality

$$d(\mathcal{S}, V_2^*) \leq d(\mathcal{S}, V_1^*) + D(V_1^*, V_2^*)$$

follows immediately from the triangle inequality of the PCM,

$$\text{pcm}(\mathcal{V}_\mathcal{S}, \mathcal{V}_{c_2}) \leq \text{pcm}(\mathcal{V}_\mathcal{S}, \mathcal{V}_{c_1}) + \text{pcm}(\mathcal{V}_{c_1}, \mathcal{V}_{c_2}) \ ,$$

and we can apply the Monotonous Bisector Tree to the set of skeletons for increasing the performance of the nearest-neighbor query.

The PCM in Realistic Scenarios Since the PCM is continuous and invariant under Euclidean transformations, we can hope that the first two problems mentioned in Sect. 13.6 (the noisy scans and the unknown robot orientation) are solved satisfactorily.

Also the third problem (unknown obstacles in the environment) does not strongly influence the PCM as long as the obstacles take up only a small interval of the whole scanning angle. But the other case, where the obstacles occupy a large part of the robot's view, poses a problem to our localization method. Here, additional (heuristic) algorithms are needed to detect such cases.

In contrast to this, a possibly limited sensing range (the fourth problem addressed in Sect. 13.6) can be easily tackled in a straightforward manner: When we compute the distances $d(\mathcal{S}, V^*)$ and $D(V_1^*, V_2^*)$, we simply cut off all distance values larger than the sensing range, that is, for all occuring polygons we use a modified PCF,

$$\text{pcf}'_P(\varphi) := \min\{\text{pcf}_P(\varphi), \text{sensing range}\} \ .$$

13.9 Our Implementation RoLoPro

Using LEDA, the Library of Efficient Datatypes and Algorithms of Mehlhorn and Näher (1995), we have implemented two versions of the localization algorithm in C++, namely the original method described in Sect. 13.4 for exact sensors as well as the modification for realistic scenarios introduced above. Here, the original algorithm was modified and simplified at some points, since we did not focus our efforts on handling sophisticated but rather complicated data structures and algorithmic ideas that were suggested by Guibas et al. Rather, we wanted to have an instrument to experiment with different inputs for the algorithm that is reasonably stable to be used in real-life environments in the future and that can serve as a basis for own modifications. A consequence is that the program does not keep to all theoretical time and space bounds of Guibas et al. (1995) and Karch (1996), as this would have required a tremendous programming effort. Nevertheless, it is reasonably efficient. Figure 13.9 shows some screen shots of our robot localization program ROLOPRO processing localization queries in real as well as in simulated environments.

As distance function $d(\mathcal{S}, V^*)$ we have implemented the Hausdorff distance and the polar coordinate metric described in Sect. 13.8. Only for the PCM the efficient skeleton management described in Sect. 13.7.2 was implemented, since for the Hausdorff distance we could not find a suitable distance $D(V_1^*, V_2^*)$. Therefore, in this case the scan has to be compared with *all skeletons*, which is much more time consuming than the PCM approach, which uses the Monotonous Bisector Tree.

Furthermore, we have implemented some additional features into ROLO-PRO, described in the following.

Noisy Compass We are able to model different kinds of compasses by using modified *minimization intervals* in (13.2), namely

- an exact compass: Here the minimization interval consists only of a single value that represents the orientation of the robot with respect to the map,
- no compass at all: Since we know nothing about the robot's orientation, we use $[0, 2\pi[$ as the minimization interval,
- a noisy compass: Here we use a smaller interval $[o - \varepsilon, o + \varepsilon[\subset [0, 2\pi[$ that depends on the orientation of the robot and the uncertainty $\varepsilon < \pi$ of the compass.

In most cases we have to carefully adapt the distances $d(\mathcal{S}, V^*)$ and $D(V_1^*, V_2^*)$ in different ways, such that the requirements from Sect. 13.7.1 and the triangle inequality remain fulfilled.

Partial Range Scans By modifying the *integration interval* in (13.2) we also can process partial range scans, where the scanning angle is less than 2π. As well as in the noisy-compass feature the distances have to be carefully modified, in particular if we want to use both features at the same time.

Additional Coarsening Step The complexity of the visibility cell decomposition, which is necessary for the geometrical approach described in Sect. 13.4 and which is the basis for our own approach, is quite high. But in practice many of the visibility cells need not be considered, because their skeletons differ only slightly from the skeletons of neighboring cells and the cells themselves are often very small.

Therefore, we have implemented an additional coarsening step, where neighboring cells with small distances $D(V_1^*, V_2^*)$ are combined into one bigger cell. This way, we need less space for storing the cells as well as less

Fig. 13.9. Screen shots of RoLoPro

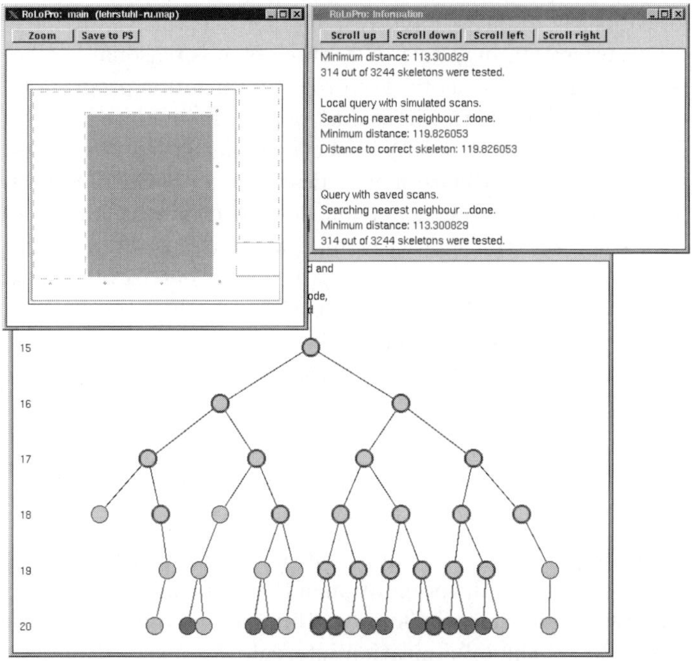

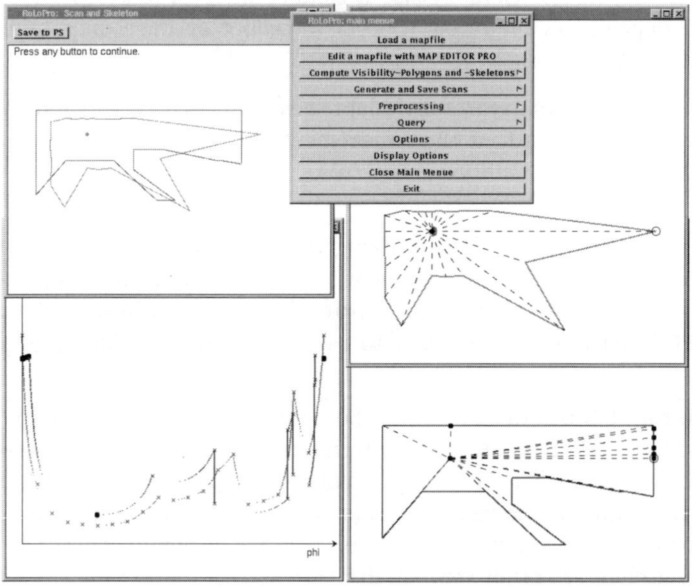

Fig. 13.10. Some localization examples using the PCM

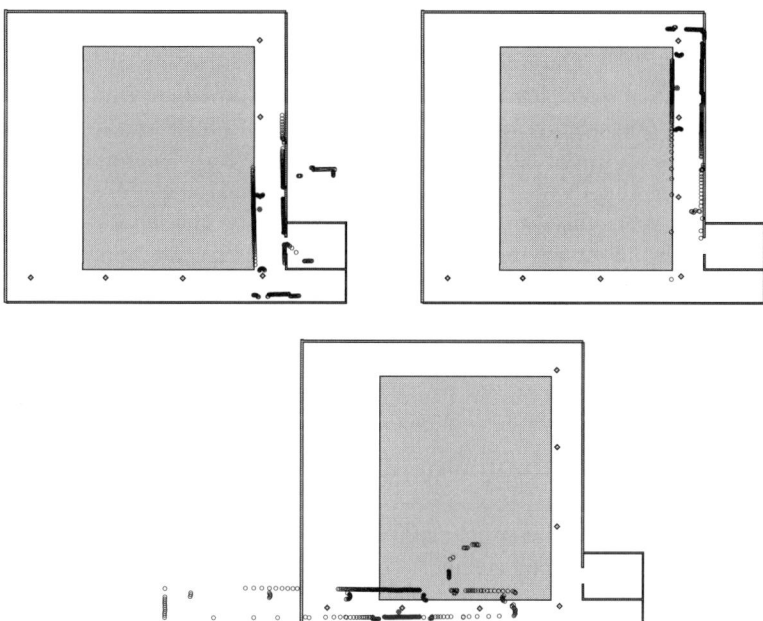

time for constructing the Monotonous Bisector Tree. Of course, this coarsening procedure reduces the quality of the localization, since we obtain both, a fewer number of distance values and larger visibility cells. Thus, the threshold value for the merging step must be carefully chosen by the user.

13.10 Experimental Tests

Tests in small simulated scenes have shown a success rate of approximately 60% for the directed Hausdorff distance, i.e., the scan origins of about 60 out of 100 randomly generated scans were inside that cell with the smallest distance $d(\mathcal{S}, V^*)$ to the scan. In the same scenes the success rate of the polar coordinate metric was about 80-90 %.

We also have evaluated our approach in real environments (using the PCM as distance function), namely in our department and in a "less friendly" supermarket environment, where the scans were very noisy. In both cases the scans were generated by a SICK PLS/LMS 200 laser scanner and we always assumed that the robot has a compass.

Figure 13.10 shows some localization examples for the first environment: We used a partial map of our department, which consists of 76 vertices, 54 of them reflex. This led to a total number of about 4 300 visibility cells, which were finally reduced to about 3 300 cells by our coarsening step (using a suitable threshold). Each of the full range scans consists of 720 range measurements.

Almost all localization queries resulted in solutions like the first two examples in Fig. 13.10. Only about 5–10 % of the localization queries failed. For example, the reason for the bad result of the third query in Fig. 13.10 probably were some scan points at a distance of about 50 m in the right part of the scan (cut off in the figure), which were produced by scanning through a window. Since the PCM "tries" to minimize a kind of *average* quadratic error, the scan was shifted in the opposite direction to the left.

In the supermarket environment our current localization approach completely failed, probably because of the very noisy range scans, which produced effects like the third example of Fig. 13.10. Moreover, in this environment only half range scans were available such that the localization problem gets even harder.

13.11 Possible Enhancements to the Algorithms

Currently, our approach has some limitations, which might be overcome using some of the following enhancements:

- The approach described above has quite large space requirements (caused by the space consuming visibility cell decomposition). To reduce these requirements we want to use a more sophisticated approach than our current coarsening step, for example by considering also the scan resolution and the limited sensing range.
- Currently, the performance of our approach is not good enough, if we have no compass information, because the distance computation for this case takes $\mathcal{O}(pq(p+q))$ time, instead of linear time for the case of an exact compass. Using a suitable scan preprocessing, the distance computation may be sped up significantly.
- For the case of extremely disordered scans we get very bad localization results (like in the supermarket environment described above). To overcome this problem, we try to modify the PCM appropriately and try to do an additional preprocessing for the scans.

The ongoing work at our institute tries to use also local feature matching algorithms for the localization task, see Kluge (1999), and to integrate navigation algorithms into the approach, such that the robot can move around to eliminate ambiguous positions, see Buck et al. (1999b).

References

Alt, H., Behrends, B., and Blömer, J. (1995). Approximate Matching of Polygonal Shapes. *Annals of Mathematics and Artificial Intelligence*, 13:251–265

Arkin, E. M., Chew, L. P., Huttenlocher, D. P., Kedem, K., and Mitchell, J. S. B. (1991). An Efficiently Computable Metric for Comparing Polygonal Shapes. *IEEE Transactions on Pattern Analysis and Machine Intelligence*, 13:209–216

Borenstein, J., Everett, H. R., and Feng, L. (1996). *"Where am I?" – Systems and Methods for Mobile Robot Positioning.* The University of Michigan

Buck, M., Kluge, B., Noltemeier, H., and Schäfer, D. (1999). RoLoPro - Simulationssoftware für die Selbstlokalisation eines autonomen mobilen Roboters. In *Autonome Mobile Systeme 1999 (AMS'99)*, pages 118–127

Buck, M., Noltemeier, H., and Schäfer, D. (1999). Practical Strategies for Hypotheses Elimination on the Self-Localization Problem. Technical Report No. 236, Department of Computer Science I, University of Würzburg

Cox, I. J. (1991). Blanche — An Experiment in Guidance and Navigation of an Autonomous Robot Vehicle. *IEEE Transactions on Robotics and Automation*, 7(2):193–204

Dudek, G., Romanik, K., and Whitesides, S. (1998). Localizing a Robot with Minimum Travel. *SIAM Journal on Computing*, 27(2):583–604

Edlinger, Th., and Weiß, G. (1995). Exploration, Navigation and Self-Localization in an Autonomous Mobile Robot. In Rüdiger Dillmann, Ulrich Rembold, and Tim Lüth (Hrsg.), editors, *Autonome Mobile Systeme 1995 (AMS'95)*, pages 142–151. Springer

Gonzalez, J., Ollero, A., and Reina A. (1994). Map Building for a Mobile Robot Equipped with a 2D Laser Rangefinder. In *Proceedings of the IEEE International Conference on Robotics and Automation*, pages 1905–1909

Guibas, L. J., Motwani, R., and Raghavan, P. (1995). The Robot Localization Problem. In Goldberg, K., Halperin, D., Latombe, J. C., Wilson, R., editors, *Algorithmic Foundations of Robotics*, pages 269–282. A K Peters

Gutmann, J. S. (1996). Vergleich von Algorithmen zur Selbstlokalisierung eines mobilen Roboters. Master's thesis, Fakultät für Informatik, Universität Ulm

Hoffmann, F., Icking, Ch., Klein, R., and Kriegel, K. (1997). An Competitive Strategy for Learning a Polygon. In *Proceedings of the 8th Annual ACM-SIAM Symposium on Discrete Algorithms*, pages 166–174

Karch, O. (1996). A Sharper Complexity Bound for the Robot Localization Problem. Technical Report No. 139, Department of Computer Science I, University of Würzburg, http://www-info1.informatik.uni-wuerzburg.de/publications/karch/tr139.ps.gz.

Karch, O., Noltemeier, H., and Wahl, Th. (1999). Robot Localization Using Polygon Distances. In Christensen, H. T., Bunke, H., Noltemeier, H., editors, *Sensor Based Intelligent Robots*, volume 1724 of *Lecture Notes in Artificial Intelligence*, pages 200–219. Springer

Karch, O., and Wahl, Th. (1999). Relocalization — Theory and Practice. *Discrete Applied Mathematics (Special Issue on Computational Geometry)*, 93:89–108

Kluge, B. (1999). Lokale Features – Erkennung und Matching in Lokalisationsszenarien. Master's thesis, Department of Computer Science I, University of Würzburg

Kluge, B., Noltemeier, H., and Schäfer, D. (2000). Feature-based Localization of an Autonomous Mobile Robot: An Experimetal Case Study. In *Proceedings of the 6th International Conference on Intelligent Autonomous Systems, Venice, Italy*

Leonard, J. J., and Durrant-Whyte, H. F. (1991). Mobile Robot Localization by Tracking Geometric Beacons. *IEEE Transactions on Robotics and Automation*, 7(3):237–256

Mehlhorn, K., and Näher, S. (1995). LEDA – A Platform for Combinatorial and Geometric Computing. *Communications of the ACM*, 38:96–102, http://www.mpi-sb.mpg.de/LEDA/articles/leda.ps.gz.

Noltemeier, H., Verbarg, K., and Zirkelbach, Ch. (1993). A Data Structure for Representing and Efficient Querying Large Scenes of Geometric Objects: MB^*-Trees. In Farin, G., Hagen, H., Noltemeier, H., editors, *Geometric Modelling*, volume 8 of *Computing Supplement*, pages 211–226. Springer

Pierce, D. M. (1995). *Map Learning with Uninterpreted Sensors and Effectors*. PhD thesis, University of Texas at Austin

Praßler, E. A., and Milios, E. E. (1995). Position Estimation Using Equidistance Lines. In *Proceedings of the IEEE International Conference on Robotics and Automation*

Praßler, E. A., Scholz, J., Strobel, M., and Fiorini, P. (1999). MAid: A Robotic Wheelchair Operating in Public Environments. In Henrik I. Christensen, Horst Bunke, and Hartmut Noltemeier, editors, *Sensor Based Intelligent Robots*, volume 1724 of *Lecture Notes in Artificial Intelligence*, pages 68–95. Springer

von Puttkamer, E., Weiss, G., and Edlinger, Th. (1999). Localiztion and On-Line Map Building for an Autonomous Mobile Robot. In Christensen, H. I., Bunke, H., Noltemeier, H., editors, *Sensor Based Intelligent Robots*, volume 1724 of *Lecture Notes in Artificial Intelligence*, pages 36–48. Springer

Rencken, W. D., Feiten, W., and Zöllner, R. (1999). Relocalisation by Partial Map Matching. In Christensen, H. I., Bunke, H., Noltemeier, H., editors, *Sensor Based Intelligent Robots*, volume 1724 of *Lecture Notes in Artificial Intelligence*, pages 21–35. Springer

Schuierer, S. (1997). Efficient Robot Self-Localization in Simple Polygons. In Bolles, R. C., Bunke, H., Noltemeier, H., editors, *Intelligent Robots – Sensing, Modelling, and Planning*, pages 129–146. World Scientific

Urrutia, J. (1997). Art Gallery and Illumination Problems. http://www.csi.uottawa.ca/~jorge/online_papers/Art-Galleries.ps.gz

Volpe, R., Ohm, T., Petras, R., Welch, R., Balaram, J. B., and Ivlev, R. (1997). A Prototype Manipulation System for Mars Rover Science Operations. In *Proceedings of the 10th IEEE/RSJ International Conference on Intelligent Robots and Systems (IROS '97)*, pages 1486–1492

Wahl, Th. (1997). *Distanzfunktionen für Polygone und ihre Anwendung in der Roboterlokalisation*. Master's thesis, University of Würzburg

14 The Quadratic Assignment Problem

Franz Rendl

Universität Klagenfurt, Institut für Mathematik, Universitätsstraße 65-67, A-9020 Klagenfurt, Austria. e-mail: franz.rendl@uni-klu.ac.at

14.1 Introductory Example

Koopmans and Beckmann (1957) introduced the Quadratic Assignment Problem (QAP) as a model for location problems that take into account not only the cost of placing a new facility on a certain site, but also the interaction with other facilities.

To explain this model, let us consider the following campus planning model, which is based on Dickey and Hopkins (1972). A new campus is planned to consist of buildings $B_1, \ldots, B_N$. Each of the buildings B_i has a certain (unique) function, such as library or cafetaria. The university owns a piece of land on which N sites $X_1, \ldots, X_N$ have been identified as possible sites for the buildings B_i. Let c_{ij} denote the cost of erecting the building B_i on site X_j. If we want to assign the buildings to the sites so as to minimize total construction cost, we obtain the **Linear Assignment Problem**.

$$\min\{\sum_i c_{i,\phi(i)} : \; \phi \text{ permutation}\}.$$

Asking for a permutation ϕ insures that each site is used and each building is built on a distinct site.

Suppose now that in addition to the construction costs we have the following data available. The distance between sites X_k and X_l is denoted by $d(X_k, X_l)$. These distances are collected in the matrix $D = (d_{kl}) = (d(X_k, X_l))$. Finally, suppose that f_{ij} denotes the number of people (per day) that typically go from building B_i to B_j.

If B_i is assigned to site $X_{\phi(i)}$, and B_j is assigned to $X_{\phi(j)}$, then

$$f_{ij} d_{\phi(i),\phi(j)}$$

is the total distance walked from B_i to B_j by all users of the campus. Summing over all pairs of buildings, we get

$$\sum_{ij} f_{ij} d_{\phi(i),\phi(j)}$$

as the total distance walked by all campus users (in one day). Minimizing this quantity over all possible assignments ϕ of buildings to sites is certainly a reasonable objective. Problems of this type are called *Quadratic Assignment Problem* (QAP).

Minimizing the total distance walked by all users of the campus will in general be in conflict with the goal of minimizing construction cost. Thus some linear combination of both cost functions might reflect a good compromise to serve as an objective.

Formally, we define the QAP (in Koopmans-Beckmann form) as the following optimization problem.

For given matrices A, B, C of size $n \times n$, minimize

$$\sum_{ij} a_{ij} b_{\phi(i),\phi(j)} + \sum_{i} c_{i,\phi(i)} \tag{14.1}$$

over all permutations ϕ. It is often the case that A and B are symmetric matrices. This is a crucial feature for some of the relaxations of QAP to be described below, therefore we will always explicitly state the symmetry properties of the input matrices.

Before proceeding, we introduce some notation that is used in this chapter.

$$QAP(A, B, C) := \min\{\sum_{ij} a_{ij} b_{\phi(i),\phi(j)} + \sum_{i} c_{i,\phi(i)} : \phi \text{ permutation }\}$$

denotes the optimal value of a QAP given by matrices A, B, C. Similarly

$$LAP(C) := \min\{\sum_{i} c_{i,\phi(i)} : \phi \text{ permutation }\}$$

is the value of the linear assignment problem with cost matrix C. The vector of all ones is denoted by e. Finally, we will use the following sets of square matrices:

$$\mathcal{N} := \{X : x_{ij} \geq 0\} \text{ nonnegative matrices,}$$

$$\mathcal{O} := \{X : X^T X = X X^t = I\} \text{ orthogonal matrices.}$$

The set Π of permutation matrices,

$$\Pi := \{X = (x_{ij}) : x_{i,\phi(i)} = 1 \text{ for a permutation } \phi\}$$

can be expressed as $\Pi = \mathcal{O} \cap \mathcal{N}$. The set of doubly stochastic matrices is

$$\Omega := \{X \in \mathcal{N} : Xe = X^T e = e\}.$$

It is well known that $\text{conv}(\Pi) = \Omega$ and the set of extreme points of Ω is Π.

14.2 Equivalent Formulations of QAP

The combinatorial formulation (14.1) can be rewritten as a mathematical programming problem in Boolean variables by introducing the matrix $X = (x_{ij})$ of decision variables $x_{ij} \in \{0,1\}$ which model permutations ϕ as $x_{ik} = 1$

if and only if $\phi(i) = k$, i.e. $X \in \Pi$. Note that if i is assigned to k and j is assigned to l, in other words $\phi(i) = k, \phi(j) = l, x_{ik} = x_{jl} = 1$, we get

$$\sum_{kl} a_{ij} b_{kl} x_{ik} x_{jl} = a_{ij} b_{\phi(i),\phi(j)}.$$

Therefore we get the following equivalent formulation of QAP as a Boolean problem.

For given matrices A, B, C of size $n \times n$, minimize

$$\sum_{ijkl} a_{ij} b_{kl} x_{ik} x_{jl} + \sum_{ik} c_{ik} x_{ik} \qquad (14.2)$$

over all matrices $X \in \Pi$.

The quadratic terms $x_{ik} x_{jl}$ in this formulation make it clear why this problem is called *quadratic* assignment problem. This formulation also suggests an immediate generalization by introducing a four-dimensional array $Q = (q_{ijkl})$ of numbers instead of $a_{ij} b_{kl}$. In this chapter we always assume that $q_{ijkl} = a_{ij} b_{kl}$, which is the case in many applications.

The formulation (14.2) with four summation variables may seem not very elegant. A more compact formulation of the problem in terms of the matrices A, B, C and X can be obtained using the *trace* of a matrix. We recall that for a square matrix M, the trace of M is given by $\text{tr}(M) = \sum_i m_{ii}$. We now observe that for the ik-entry of AXB^T, we have

$$(AXB^T)_{ik} = \sum_{jl} a_{ij} x_{jl} b_{kl}.$$

Therefore we get

$$\sum_{ijkl} a_{ij} b_{kl} x_{ik} x_{jl} = \sum_{ik} (AXB^T)_{ik} x_{ik} = \text{tr}(AXB^T X^T),$$

which corresponds to the first term in the cost function of (14.2). Thus we get the *trace formulation* of QAP as

$$QAP(A, B, C) = \min\{\text{tr}(AXB^T + C)X^T : X \in \Pi\}. \qquad (14.3)$$

The last formulation of QAP uses instead of the matrix X the vector $x = \text{vec}(X)$ which is obtained from X by stacking its columns into a vector (of order n^2). The following identity is well known in matrix analysis.

$$\text{vec}(AXB) = (B^T \otimes A)\text{vec}(X). \qquad (14.4)$$

Using this, we get

$$\text{tr } AXB^T X^T = x^T \text{vec}(AXB^T) = x^T (B \otimes A) x.$$

Setting $c = vec(C)$, we get the *Kronecker-product form* of QAP.

$$QAP(A,B,C) = \{\min x^T(B \otimes A)x + c^T x : X \in \Pi, x = vec(X)\}. \quad (14.5)$$

Finally, since $x_{ij} = x_{ij}^2$, we get the equivalent formulation

$$\min x^T(B \otimes A + Diag(c))x \text{ over all permutation matrices } X.$$

These last two formulations lead to quadratic problems in dimension n^2.

14.3 Applications

In this section we provide an overview of published applications of QAP. Steinberg (1961) modeled the problem of wiring electronic devices on a board so as to minimize the total wiring length as a QAP. The similarity with the initial example from the introduction is obvious. The buildings take the place of the electronic devices, and the possible locations correspond to potential positions on the board. Steinberg considers not only the Euclidean distance, but also the L_1 distance measure.

Burkard and Offerman (1977) investigate the question of assigning letters to a typewriter keyboard. Here the input matrices A and B take the following interpretation. We are given an alphabet (of some specific language) consisting of n letters $z_1, \ldots, z_n$. Suppose we have a representative sample of text in this language. In this case we can determine experimentally a probability estimate b_{ij} for the appearance of letter z_i followed by letter z_j. Note that in this case there is no reason to assume that B is symmetric. Suppose we also have a typewriter keyboard with n keys, numbered from 1 to n which are arranged in some prescribed fashion. We can also determine experimentally the time that it takes to hit the keys k followed by l, for all possible combinations (k,l) of keys. In this case the QAP with A and B corresponds to an assignment of the letters z_i to the keys i of the keyboard, so that the expected time to type a typical text in the given language is minimized.

Elshafei (1977) models the assignment of rooms in a hospital as a QAP. The interesting point here is that in the hospital setting, it is more interesting to minimize the largest distance rather than the sum of all distances.

As an application from archaeology, we recall that Krarup and Pruzan (1978) use the QAP to rank archaeological data. A similar application in sports is reported by Heffley (1976) who models the ranking of a team in a relay race as a QAP.

From a graph-theoretical point of view, there is a series of graph optimization problems, that can be modeled as QAP with special structure.

1. The Traveling Salesman Problem (TSP) is a QAP where matrix A is the distance matrix of the problem, and B is the adjacency matrix of a cycle.

2. In case of the graph partition problem, matrix A is again the weighted adjacency matrix of a graph, while the matrix B is the adjacency matrix of two disjoint complete graphs. More general partition problems into k partiton blocks of prespecified sizes m_i can also be modeled as QAP. The extreme case $k = \frac{n}{2}$, $m_i = 2$ corresponds to weighted (nonbipartite) Matching, which can of course be solved by other methods.
3. Finally we can reduce other types of graph problems, such as Max-Clique, Bandwidth minimization, graph isomorphism and the largest common subgraph problem to the solution of (one or several) QAP. We refer to Helmberg et al. (1993) for further details on these applications in connection with the QAP.

14.4 Computational Complexity of QAP

The QAP is from a worst case computational complexity point of view one of the most difficult combinatorial optimization problems. It is well-known that QAP is NP-hard, and even finding an ε-approximation of QAP is NP-hard, as shown by Sahni and Gonzales (1976).

To arrive at tractable subclasses, one has to simplify the general problem considerably. The graph-theoretical interpretation of QAP offers such simplifications. Interpreting A and B as weighted adjacency matrices of the complete graph K_n, the QAP asks to find an isomorphism ϕ of K_n onto itsself, that minimizes the "graph product"

$$\sum_{ij} a_{ij} b_{\phi(i)\phi(j)}.$$

By considering this model for trees, it was shown by Christofides and Gerrards (1981) that the resulting QAP can be solved in polynomial time using dynamic programming. If one of the graphs is complete, and the other is a path, the resulting problem is again difficult, as this corresponds to the Traveling Salesman Problem.

A different class of tractable instances of QAP was investigated by Burkard et al. (1998). They consider matrices where the cost elements satisfy certain additional structure, such as being a Monge matrix. Again, only very specialised cases lead to polynomially solvable instances of QAP. The monograph Cela (1998) contains a detailed investigation of the borderline for tractability of QAP with structured cost coefficients.

14.5 Relaxations of QAP

The QAP has attracted interest not only for its wide applicability, but also because it permits a rich variety of ways to get relaxations. We are going to describe the most relevant ones of them.

The by now classical approach to deal with hard combinatorial optimization problems consists in studying the convex hull of its feasible solutions and the use of Linear Programming, see for instance the book by Lawler et al. (1985). This polyhedral approach has been applied successfully to many types of problems, perhaps most prominently the Traveling Salesman Problem. The feasible region of QAP is the assignment polytope Ω, which has the favourable property of possessing only integral vertices. Unfortunately, the QAP has a quadratic cost function. To deal with this computationally, we have to convexify the cost function. There are several ways to achieve this. We can either linearize the cost function, or maintain the quadratic function, but modify it in such a way that it becomes convex. Before looking at linearizations, we recall *minimal scalar products* and how they are used to derive a classical relaxation, proposed by Gilmore and Lawler, and therefore called *Gilmore-Lawler bound (GLB)*.

14.5.1 Minimal Scalar Products and the Gilmore-Lawler Bound

To bound the objective function (14.1), we can study the easier question of minimizing for $x, y \in \mathbb{R}^n$

$$\sum_i x_i y_{\phi(i)}$$

over all permutations ϕ. Note that this is a linear assignmnent problem, but with a cost matrix $C = xy^T$ of rank one. Thus we look at the scalar products of x and y, but under the additional possibility of permuting the elements of y. We denote the optimal value for this assignment problem by

$$LAP(xy^T) = \langle x, y \rangle_- := \min_\phi \sum_i x_i y_{\phi(i)}.$$

It is combinatorial folklore to characterize optimal permutations for this problem. We denote by x^- the vector obtained from x by reordering its components nonincreasingly, and by y^+ the vector y sorted nondecreasingly. One easily verifies the following proposition.

Proposition 1 *Let $x, y \in \mathbb{R}^n$. Then*

$$\langle x, y \rangle_- = \langle x^-, y^+ \rangle.$$

The maximal scalar product $\langle x, y \rangle_+$ is defined similarly, and given by $\langle x, y \rangle_+ = \langle x^+, y^+ \rangle$.

Before applying this concept we rewrite (14.1), by separating the off-diagonal contribution from the main-diagonal contribution.

$$\sum_{ij} a_{ij} b_{\phi(i),\phi(j)} + \sum_i c_{i,\phi(i)} = \sum_i (a_{ii} b_{\phi(i),\phi(i)} + c_{i,\phi(i)}) + \sum_{i \neq j} a_{ij} b_{\phi(i),\phi(j)}.$$

(14.6)

We get for all permutations ϕ:

$$\sum_{j, j \neq i} a_{ij} b_{k,\phi(j)} \geq \langle A_i, B_k \rangle_- =: \gamma_{ik}.$$

Here A_i denotes the vector of elements of row i of A with a_{ii} omitted, and B_k is row k of B without b_{kk}. We substitute this into the double sum in (14.6) and get

$$QAP(A,B,C) \geq \min_\phi \sum_i (a_{ii} b_{\phi(i),\phi(i)} + c_{i,\phi(i)} + \gamma_{i,\phi(i)}) =: GLB(A,B,C). \tag{14.7}$$

This is the bound initially proposed by Gilmore (1962) and Lawler (1963). We note that its computation amounts to first forming the matrix $\Gamma = (\gamma_{ik})$ of scalar products of all rows of A with all rows of B, and then solving one linear assignment problem with cost matrix $C + \Gamma + (a_{ii} b_{kk})$. Even though this lower bound is quite easy to derive, it is nontheless still a very popular bound used in Branch and Bound codes to solve QAP to optimality, see for instance Burkard and Derigs (1980), Roucairol (1987). The derivation of GLB presented above does not immediately suggest that it has any connections with linear relaxations. We will see below however, that indeed GLB can alternatively be formulated through linearization.

14.5.2 Linear Relaxations of QAP

Lawler (1963) suggested the following obvious linearization. For each term $x_{ik} x_{jl}$ a new binary variable y_{ikjl} is introduced to express

$$y_{ikjl} = x_{ik} x_{jl} \quad \forall i,j,k,l. \tag{14.8}$$

He showed that relations (14.8) can be expressed by linear inequalities in binary variables.

Lemma 2 *Let X be a permutation matrix of order n and $y = (y_{ikjl})$ be an array of n^4 numbers. Then (14.8) holds if and only if there exist y_{ikjl} such that*

$$y_{ijkl} \in \{0,1\}, \ x_{ik} + x_{jl} - 2 y_{ikjl} \geq 0 \ \forall i,j,k,l \ \text{and} \ \sum_{ikjl} y_{ikjl} = n^2. \tag{14.9}$$

The proof of this result is easy, and therefore left to the interested reader. It can be used to express QAP as a binary LP.

$$\min \sum_{ikjl} a_{ij} b_{kl} y_{ikjl} + \sum_{ik} c_{ik} x_{ik} \text{ such that } X \in \Pi \text{ and (14.9) holds.}$$

This LP has $O(n^4)$ constraints and binary variables, so it is of only limited practical use.

A linearization with only $O(n^2)$ variables and constraints was proposed by Kaufman and Broecks (1978). They propose the following substitution.

$$w_{ik} = x_{ik} \sum_{jl} a_{ij} b_{kl} x_{jl} \tag{14.10}$$

Let us also introduce

$$\gamma_{ik} := \sum_{jl} a_{ij} b_{kl}.$$

With this notation Kaufman and Broecks show the following equivalent formulation of QAP in case that $A \geq 0$ and $B \geq 0$, which can be assumed without loss of generality.

Theorem 3 *QAP with $A \geq 0$, $B \geq 0$ is equivalent to the following mixed-integer LP.*

$$\min \sum_{ik} w_{ik} \text{ such that}$$

$$X \in \Pi, \ \gamma_{ik} x_{ik} + \sum_{jl} a_{ij} b_{kl} x_{jl} - w_{ik} \leq \gamma_{ik} \text{ and } w_{ik} \geq 0 \ \forall i, k.$$

Again, the proof of this result is easy, and left to the reader. Nonnegativity of A and B is used to show that at the optimum of this LP, the relations (14.10) hold.

A very general linear approach for QAP was initally proposed by Adams and Johnson (1994). They apply the linearization idea not only to the cost function, but also to the assignment constraints, which are multiplied with all decision variables x_{jl}. Thus they get from

$$\sum_i x_{ik} x_{jl} = x_{jl}$$

the constraint

$$\sum_i y_{ikjl} = x_{jl}, \ \forall k, j, l, \tag{14.11}$$

and similarly

$$\sum_k y_{ikjl} = x_{jl}, \ \forall i, j, l, \tag{14.12}$$

The variables y_{ikjl} model pairs of products $x_{ik} x_{jl}$, hence one also has

$$y_{ikjl} = y_{jlik} \ \forall i, k, j, l. \tag{14.13}$$

The key observation underlying the approach of Adams and Johnson consists in the following easy observation.

Lemma 4 *(Adams and Johnson, 1994) Let $X = (x_{ik})$ be a permutation matrix of order n and $y = (y_{ikjl})$ be an array of n^4 numbers. Then*

$$x_{ik}x_{jl} = y_{ikjl} \quad \forall i,j,k,l$$

if and only if $y \geq 0$ satisfies (14.11), (14.12), (14.13).

Proof. The necessity of the conditions on y is obvious. Suppose now that (14.11), (14.12), (14.13) holds for $y \geq 0$. To show that $x_{ik}x_{jl} = y_{ikjl}$ we consider the following cases.

Case 1: $x_{rs} = 0$. Since $\sum_i y_{ikrs} = \sum_k y_{ikrs} = 0$ and $y \geq 0$, we get $x_{rs}x_{ik} = y_{ikrs} = 0 \; \forall i,k$.

Case 2: $x_{rs} = 1$. We first look at $\sum_i y_{isrs} = 1$. Since $x_{is} = 0$ for $i \neq r$, and hence $y_{isrs} = 0$ for $i \neq r$, we conclude that $y_{rsrs} = x_{rs}x_{rs} = 1$. Next let $x_{uv} = 1$ with $u \neq r$, $v \neq s$. The sum $\sum_i y_{ivrs} = 1$ contains only one nonzero term, hence $y_{uvrs} = 1$. ∎

Thus we can solve QAP by solving the mixed integer linear program with n^2 binary variables x_{ik} and $O(n^4)$ continuous nonnegative variables y and $O(n^4)$ linear constraints (14.11), (14.12), (14.13). Adams and Johnson propose to solve its continuous relaxation, i.e. requiring only $x \geq 0$.

$$LPB(A,B,C) := \min\{\sum_{ikjl} a_{ij}b_{kl}y_{ikjl} + \sum_{ik} c_{ik}x_{ik} :$$

$$X \in \Omega, y \geq 0 \text{ satisfies } (14.11), (14.12), (14.13)\}$$

This linear program is still nontrivial to solve, hence they propose to find feasible solutions of the dual to bound the primal problem. They show in particular that the Gilmore-Lawler bound can be obtained from $LPB(A,B,C)$ by ignoring the symmetry constraints (14.13), i.e. by setting the associated dual variables to zero, see Adams and Johnson (1994).

More recently, Karisch et al. (1999) have shown that virtually all the existing linear relaxations of QAP can be derived through a careful choice of the dual variables. Of course, the best lower bound obtainable this way is to solve either the primal or the dual linear relaxation to optimality. It is currently not clear however, how this can be done efficiently. Resende et al. (1995) currently have the most efficient implementation to compute the linear relaxation to optimality. It is based on an interior point algorithm. Unfortunately, the quality of this relaxation in comparison to its computational cost does not make it a very attractive alternative in practical computations to solve QAP to optimality.

14.5.3 Orthogonal Relaxations of QAP

The set of permutation matrices Π can be expressed as the intersection of orthogonal matrices with the set of (elementwise) nonnegative matrices, $\Pi =$

$\mathcal{O} \cap \mathcal{N}$. In the previous section we simplified by omitting orthogonality of X. Now we maintain $X \in \mathcal{O}$ and drop nonnegativity. To explain how this can be achieved, we need to introduce some more notation. We denote by λ_A the vector of eigenvalues of the matrix A (in arbitrary order, but counting multiplicities).

The following theorem from matrix analysis is the basis for the orthogonal relaxation of QAP. It can be traced back to the work of John von Neumann (1963). A more recent proof can be found in Horn and Johnson (1985).

Theorem 5 *Let A and B be symmetric matrices of order n. Then*

$$\min_{X \in \mathcal{O}} \operatorname{tr} AXB^T X^T = \langle \lambda_A, \lambda_B \rangle_-.$$

The minimum is attained for $X = PQ^T$, where $P \in \mathcal{O}$ contains the eigenvectors of A corresponding to the eigenvalues sorted nondecreasingly, and $Q \in \mathcal{O}$ diagonalizes B with the eigenvalues in nonincreasing order.

Proof. The following proof follows (Helmberg et al., 1993). Since A and B are symmetric, there exist orthogonal matrices P, Q such that $A = PLP^T, B = QMQ^T$, where L, M are diagonal matrices with $L = \operatorname{diag}(\lambda_A), M = \operatorname{diag}(\lambda_B)$ and $l_{11} \leq \ldots \leq l_{nn}$ and $m_{nn} \leq \ldots \leq m_{11}$. Setting $X = PYQ^T$, with Y orthogonal leads to

$$\operatorname{tr} AXB^T X^T = \operatorname{tr} LYMY^T = \sum_{ij} l_{ii} m_{jj} y_{ij}^2 \geq$$

$$\geq \min\{\sum_{ij} l_{ii} m_{jj} z_{ij} : Z \in \Omega\} = \sum_i l_{ii} m_{ii} = \langle \lambda_A, \lambda_B \rangle_-.$$

The inequality follows from the fact that (y_{ij}^2) is doubly stochastic. The last two equations follow from the order of the eigenvalues of A and B. Setting $Y = I$, i.e. $X = PQ^T$ gives equality throughout. ∎

As an immediate consequence, we get the following relaxation introduced in Finke et al. (1987).

Theorem 6 *Let $A = A^T, B = B^T$ and C be matrices of the same size. Then*

$$QAP(A, B, C) \geq \langle \lambda_A, \lambda_B \rangle_- + LAP(C).$$

In Finke et al. (1987) several improvement strategies are discussed, which are essentially based on taking the Lagrangian dual of the assignment constraints, with a careful choice of the Lagrange multipliers.

A different approach is taken in Hadley et al. (1989). It is based on the observation that the general solution of the linear equations

$$Xe = X^T e = e$$

can be written down explicitly as

$$X = \frac{1}{n}ee^T + VYV^T, \tag{14.14}$$

where the $n \times (n-1)$ matrix V of rank $n-1$ satisfies $V^T e = 0$. In other words, V is basis of the orthogonal complement of e. This result is easy to prove, and can be found in Hadley et al. (1989). We also want to maintain orthogonality of X, so we assume in addition that $V^T V = I_{n-1}$. In this case we get for matrices X and Y related by (14.14)

$$X^T X = I \text{ if and only if } Y^T Y = I.$$

Substituting the representation (14.14) for X gives

$$\operatorname{tr}\,(AXB + C)X^T = \operatorname{tr}\,\hat{A}Y\hat{B}Y^T + \operatorname{tr}\,DX^T - \frac{1}{n^2}s(A)s(B), \tag{14.15}$$

where $\hat{A} = V^T A V, \hat{B} = V^T B V, D = C + \frac{2}{n} Aee^T B$ and $s(A) = e^T A e, s(B) = e^T B e$. Bounding the quadratic term again by the eigenvalues and solving (independently) the linear assignment problem with matrix D gives the *projection bound* introduced in Hadley et al. (1989).

Theorem 7

$$QAP(A, B, C) \geq PB(A, B, C) := \langle \lambda_{\hat{A}}, \lambda_{\hat{B}} \rangle_- + LAP(D) - \frac{1}{n^2}s(A)s(B).$$

The computational effort to evaluate $PB(A, B, C)$ is dominated by the eigenvalue computation of $\lambda_{\hat{A}}$ and $\lambda_{\hat{B}}$, which is of order $O(n^3)$.

This relaxation is often much stronger than $GLB(A, B, C)$, see Hadley et al. (1989), but it has a serious disadvantage as it bounds the quadratic part of QAP independent of the linear term. We will see in the next section that this relaxation can also be interpreted in the context of semidefinite programming SDP.

14.5.4 Semidefinite Relaxations

If we start from the Kronecker-product form of QAP, (14.5), it is a natural idea to linearize $x^T(B \otimes A)x$ by introducing a new variable Y for xx^T and using

$$x^T(B \otimes A)x = \operatorname{tr}\,(B \otimes A)(xx^T) = \operatorname{tr}\,(B \otimes A)Y.$$

This linearization is equivalent to (14.8), if we think of the entries of Y as being indexed by pairs (i, k) and (j, l). Now the hard part consists in describing matrices Y which are in the convex hull of xx^T where x runs through all permutation matrices, written as vectors in $\mathbb{R}^{n^2}$. We denote this set by

$$P := \operatorname{conv}\{xx^T : x = vec(X), X \in \Pi\}.$$

We are now interested in finding (at least) a partial description of this set by linear equalities, linear inequalities and the fact that $Y \in P$ implies $Y \succeq 0$. We first show that we can in fact strengthen this constraint by exploiting the fact that x is a binary vector. The following proposition is well known and can be considered the basis for semidefinite relaxations of problems in binary variables.

Lemma 8 *Let $Y \in P$ and $y = \text{diag}(Y)$. Then $Y - yy^T \succeq 0$.*

Proof. We only use the fact that P is generated from binary vectors x and consider
$$\text{conv}\{\begin{pmatrix} 1 \\ x \end{pmatrix}\begin{pmatrix} 1 \\ x \end{pmatrix}^T : X \in \Pi, x = vec(X)\}.$$

Elements in this set can be written as $\begin{pmatrix} 1 & u^T \\ u & U \end{pmatrix}$, where $U \in P$ and $\text{diag}(U) = u$ because P is generated by binary vectors. We clearly have
$$\begin{pmatrix} 1 & u^T \\ u & U \end{pmatrix} \succeq 0,$$
which is equivalent to $U - uu^T \succeq 0$. ∎

We next show that we can further tighten this constraint, see Zhao et al. (1998), by exploiting the fact that the row and column sums of X are one. Let us define the $n^2 \times ((n-1)^2 + 1)$ matrix
$$W = (\frac{1}{n} e \otimes e, V \otimes V).$$

Lemma 9 *Let $Y \in P$. Then there exists a symmetric matrix R of order $(n-1)^2 + 1$, indexed from 0 to $(n-1)^2$, such that*
$$r_{00} = 1, \ R \succeq 0 \ and \ Y = WRW^T.$$

Proof. (See also Zhao et al. (1998).) We use (14.14) and (14.4) to get, with $y = vec(Y)$
$$x = vec(X) = \frac{1}{n} e \otimes e + (V \otimes V) y = W \begin{pmatrix} 1 \\ y \end{pmatrix} = Wz.$$

Hence $xx^t = WRW^T$ with the symmetric positive semidefinite matrix $R = zz^T$ having its first element on the main diagonal equal 1. ∎

Thus our first approximation to QAP, based on semidefinite programming is the following problem.
$$SDP1(A, B, C) := \min \text{tr} \ (B \otimes A + \text{diag}(vec(C))) Y$$

such that
$$Y = WRW^T,\ Y - \text{diag}(Y)\text{diag}(Y)^T \succeq 0,\ r_{00} = 1.$$
This semidefinite program is formulated in the variable Y which represents elements from P. In view of Lemma 8 and 9 however, the actual degrees of freedom are given by the matrix R. In a computational setting, Y would be eliminated by the substitution $Y = WRW^T$. The (nonlinear) semidefiniteness constraint $Y - \text{diag}(Y)\text{diag}(Y)^T \succeq 0$ is handled as
$$\begin{pmatrix} 1 & r^T \\ r & WRW^T \end{pmatrix} \succeq 0,\ r = \text{diag}(WRW^T),$$
see also the proof of Lemma 9. The number of variables of this semidefinite program is of the same order as the linear program LPB, but maintaining semidefiniteness is computationally significantly more expensive.

There are further refinements of the bound SDP1, which are investigated in Zhao et al. (1998). Just as in LPB, we can require $Y_{rs} \geq 0$ elementwise. While nonnegativity is trivial to maintain for linear programs, these constraints have to be included explicitly in semidefinite programs. On the other hand, the symmetry conditions (14.13) are satisfied by construction in SDP1, but have to be included explicitly in LPB. The constraints $Y_{rs} = 0$ for $r = (i,k),\ s = (i,l),\ k \neq l$ may also be included. Finally, the orthogonality constraints $X^T X = I,\ XX^T = I$ are quadratic in X, hence linear in Y and may also be included as linear constraints in the relaxation SDP1, yielding an additional refinement of the relaxtion.

We conclude this section with the following theorem of Anstreicher and Wolkowicz (2000), which relates Theorem 5 to semidefinite programming.

Theorem 10 *Let G and H be symmetric matrices of the same order. Then*
$$\min_{X \in \mathcal{O}} tr\ GXH^T X^T = \max\{tr\ S + tr\ T : H \otimes G - I \otimes S - T \otimes I \succeq 0, S = S^T, T = T^T\}.$$

Proof. As in the proof of Theorem 5, we use the diagonalization of G and H, $G = PLP^T, H = QMQ^T$, to parameterize matrices in $\mathcal{O}$ as PXQ^T with $X \in \mathcal{O}$. As before, $L = \text{diag}(\lambda_G), M = \text{diag}(\lambda_H)$. Thus we get
$$\min_{X \in \mathcal{O}} \text{tr}\ G(PXQ^T)H(PXQ^T)^T =$$
$$\min_{X \in \mathcal{O}} \text{tr}\ LXMX^T =$$
$$\min_X \max_{S=S^T,T=T^T} \text{tr}\ (LXMX^T + S(I - XX^T) + T(I - X^T X)) \geq$$
$$\max_{S=S^T,T=T^T} \min_x \text{tr}\ S + \text{tr}\ T + x^T(M \otimes L - I \otimes S - T \otimes I)x =$$
$$\max \text{tr}\ S + \text{tr}\ T \text{ such that } S = S^T, T = T^T, M \otimes L - I \otimes S - T \otimes I \succeq 0 \geq$$
$$\max e^T s + e^T t \text{ such that } l_{ii}m_{jj} - s_i - t_j \geq 0\ \forall i,j =$$
$$\min\{\sum_{ij} m_{ii}l_{jj}z_{ij} \text{ such that } Z = (z_{ij}) \in \Omega\} = \langle \lambda_G, \lambda_H \rangle_-.$$

Hence there is equality throughout, because of Theorem 5. ■

We will see in the next section how this result can be used to establish connections between orthogonal, semidefinite, and also convex quadratic relaxations.

14.5.5 Convex Quadratic Relaxations of QAP

Since the cost function of QAP is quadratic and the feasible region is the assginment polytope, it is tempting to use convex quadratic programming to bound QAP. This means that the cost function has to be "convexified" on the feasible region. An interesting possibility for convexification was investigated by Anstreicher and Brixius (1999). Let us first observe that for any orthogonal X and any symmetric S we have

$$0 = \text{tr } S(I - XX^T) = \text{tr } S - \text{tr } (I \otimes S)xx^T.$$

As usual, we set $x = vec(X)$. This yields the following identity, which is true for any $X \in \mathcal{O}$ and any symmetric S, T:

$$\text{tr } AXB^T X^T = \text{tr } S + \text{tr } T + \text{tr } (B \otimes A - I \otimes S - T \otimes I)xx^T.$$

The more refined approach of Anstreicher and Brixius uses the parametrization $X = \frac{1}{n}ee^t + VYV^T$, where $V^Te = 0, V^TV = I$. We note that similar as before we have for any symmetric matrix $\hat{S}$ of order $n-1$, and any orthogonal X:

$$0 = \text{tr } \hat{S}(I - V^TV) = \text{tr } \hat{S}(I - V^T XX^T V) = \text{tr } \hat{S} - \text{tr } (I \otimes V\hat{S}V^T)xx^t.$$

Hence we get for all symmetric $\hat{S}, \hat{T}$ and all orthogonal X

$$\text{tr } (AXB + C)X^T = \text{tr } \hat{S} + \text{tr } \hat{T} + x^T Qx + c^T x, \tag{14.16}$$

where $x = vec(X), c = vec(C), Q = B \otimes A - I \otimes V\hat{S}V^T - V\hat{T}V^T \otimes I$. Fixing $\hat{S}$ and $\hat{T}$ such that $x^T Qx \geq 0$ for $X \in \Omega$ makes the right hand side a convex quadratic problem in x, hence tractable.

Anstreicher and Brixius (1999) propose to use the optimal dual solutions $\hat{s}$ and $\hat{t}$ from Theorem 10 applied to $\hat{A}, \hat{B}$. These are given by

$$\langle \lambda_{\hat{A}}, \lambda_{\hat{B}} \rangle_- = \sum_i \hat{s}_i + \sum_i \hat{t}_i. \tag{14.17}$$

Then they set $\hat{S} = \text{diag}(\hat{s}), \hat{T} = \text{diag}(\hat{t})$. Dual feasibility of $\hat{s}$ and $\hat{t}$ implies that

$$x^T Qx \geq 0 \text{ for } Xe = X^Te = e.$$

This yields the following bound for QAP, based on convex quadratic programming applied to (14.16).

$$QAP(A,B,C) \geq \sum \hat{s} + \sum \hat{t} + \min\{x^T Q x + c^T x : X \in \Omega\} := ABB(A,B,C). \tag{14.18}$$

It is remarkable, that this bound is at least as good as the projection bound PB(A,B,C), as proved in Anstreicher and Brixius (1999).

Theorem 11 *(Anstreicher and Brixius, 1999)* $PB(A,B,C) \leq ABB(A,B,C)$.

Proof. From (14.16) applied to $\hat{A}$, $\hat{B}$ and $C = 0$ we get for orthogonal Y and any $\hat{S}, \hat{T}$

$$\operatorname{tr} \hat{A} Y \hat{B} Y^T = \operatorname{tr} \hat{S} + \operatorname{tr} \hat{T} + y^T \hat{Q} y,$$

where

$$\hat{Q} := \hat{B} \otimes \hat{A} - I \otimes \hat{S} - \hat{T} \otimes I.$$

Using as before $\hat{S}$ and $\hat{T}$ from (14.17) we conclude that $\hat{Q} \succeq 0$, because of dual feasibility of $\hat{s}$ and $\hat{t}$. Substituting this term into (14.15) we get

$$\operatorname{tr}(AXB^T + C)X^T = \sum \hat{s} + \sum \hat{t} + y^T \hat{Q} y + \operatorname{tr} DX^T - \frac{1}{n^2} s(A) s(B). \tag{14.19}$$

We recall that PB(A,B,C) is obtained by dropping the quadratic term $y^T \hat{Q} y$ and minimizing the remaining part over $X \in \Omega$, i.e. compute LAP(D). Comparing (14.19) with (14.16) we conclude that

$$y^T \hat{Q} y + \operatorname{tr} DX^t - \frac{1}{n^2} s(A) s(B) = x^T Q x + c^T x.$$

Leaving out the term $y^T \hat{Q} y \geq 0$ on the left hand side makes the right hand side for any $X \in \Omega$ at least as big as the left hand side. ∎

In Anstreicher et al. (2000) a Branch and Bound approach was used in conjunction with the bound ABB, which turned out strong enough to find the optimal solution to an outstanding difficult instance for QAP, Nugent $n = 30$. Extrapolated computation times based on the Gilmore-Lawler bound indicated that GLB was simply not strong enough to solve this instance to optimality within years of computation time.

14.6 Heuristics

To get upper bounds on $QAP(A,B,C)$, one of the most natural ideas consists in minimizing over a subset Π' of Π only. Perhaps the simplest strategy here would be to take some permutation ϕ and then try to search in the neighbourhood of ϕ for a better permutation, and repeat until no further improvement can be found. This *local improvement method* needs several ingredients, such as the precise definition of neighbourhood, and how multiple candidates for improvement are dealt with.

A simple neighbourhood structure on Π is given by considering as neighbourhood $N(\phi)$ of permutation ϕ all permutations ϕ' which can be obtained from ϕ by changing only two assignments, i.e. by applying a single transposition to ϕ. It turns out that computing the change of the function value Δ by going from ϕ to ϕ' can be done in $O(n)$ operations.

This opens the door for a variety of heuristics, deterministic and with random moves, allowing decreases as well as increases (simulated annealing and variations) of the objective function, exploring larger (or smaller) neighbourhoods to arrive at some local optimum.

The fine-tuning of these heuristics requires high skills from computer science, and substantial computational experiments to get some practical experience. We refer to Pardalos and Wolkowicz (1994) for a discussion of recent heuristic approaches to QAP and Drezner (2001) for a recent computational study of a hybrid-genetic algorithm.

14.7 Computational Experience

After having seen various ways to obtain lower and upper bounds on QAP (A,B,C) we are now going to look at some computational results. We compare the various relaxations on a set of test problems, that have become standard in the study of QAP, the instances of Nugent et al., which can be found on the internet at http://www.imm.dtu.dk/ sk/qaplib. We compare the following lower bounds. The bound GLB corresponds to the Gilmore-Lawler bound and is the fastest to compute. The currently best bound obtained by linear programming was proposed by Adams and Johnson, see (LPB) above. The implementation of Resende et al. seems to be at present the most efficient way to compute LPB. We quote results from Resende et al. (1995) in the column labeled LPB. We observe that this bound is not much stronger than GLB, but significantly more expensive to compute. Next comes the projection bound PB. Its computational effort is again $O(n^3)$, but takes longer than GLB. This bound seems to be getting stronger as n gets larger. Next we include the improvement of Anstreicher and Brixius of PB, based on convex quadratic programming in the column ABB. Finally, we also include the semidefinite relaxation SDP from Zhao et al. (1998). This bound is based on SDP1 described above, but it includes also the orthogonality constraints $XX^T = X^TX = I$, epxressed as linear constraints in the variable Y. Again, this bound is extremely expensive to compute, but it gives the best approximation.

How do these bounds help to solve QAP to optimality? There are several Branch and Bound based approaches that use GLB or improved variants of this bound. It turns out that problems of moderate size, say $n \approx 20$, are

Table 14.1. Comparison of lower bound for QAP on the Nugent test problems.

n	opt	GLB	LPB	PB	ABB	SDP
12	578	493	523	472	482	486
15	1150	963	1044	973	994	1009
20	2570	2057	2182	2196	2252	2281
30	6124	4539	4805	5266	5362	5413

manageable by this approach. It becomes also clear from Table 14.1, that GLB is too weak for $n = 30$. In contrast, Anstreicher et al. (2000) use convex quadratic programming bounds and manage to solve the Nugen problem with $n = 30$ to optimality.

14.8 Bibliographical Notes

There is still intensive research around QAP. The dissertations of Karisch (1995) and Zhao (1996), see also the journal paper Zhao et al. (1998) deal with semidefinite relaxations for QAP, and in particular show the severe computational limits of this approach. Kaibel (1997) presents the most recent polyhedral investigations related to QAP. Again, it is rather frustrating to see that larger problems are unlikely to be manageable by this approach.

The monograph Cela (1998) provides a recent review on theoretical results related to QAP, and also generalizations.

References

Adams, W. P., and Johnson, T. A. (1994) Improved linear programming based lower bounds for the quadratic assignment problem. In Pardalos, P. M., Wolkowicz, H., editors, *Quadratic Assignment and related problems*, 43–75. American Mathematical Society.

Anstreicher, K., and Brixius, N. (1999) A new bound for the quadratic assignment problem based on convex quadratic programming. Technical report.

Anstreicher, K., Brixius, N., Goux, J. P., and Linderoth, J. (2000) Solving large quadratic assignment problems on computational grids. Technical report.

Anstreicher, K., and Wolkowicz, H. (2000) On Lagrangian relaxation of quadratic matrix constraints. *SIAM Journal on Matrix Analysis and Applications*, 22, 41–55.

Burkard, R. E., Cela, E., Rote, G., and Woeginger, G. J. (1998) The Quadratic Assignment Problem with a monotone anti-Monge and a symmetric Toeplitz matrix: easy and hard cases. *Mathematical Programming*, 82, 125–158.

Burkard, R. E., and Derigs, U. (1980) *Assignment and matching problems: Solution methods with Fortran programs*, volume 184 of *Lecture Notes in Economics and Mathematical Systems*. Springer, Berlin.

Burkard, R. E., and Offerman, J. (1977) Entwurf von Schreibmaschinentastaturen mittels quadratischer Zuordnungsprobleme. *Z. Operations Res.*, 21, B121–B132.

Cela, E. (1998) *The Quadratic Assignment Problem: theory and applications.* Kluwer Academic Publishers, Dordrecht, Boston, London.

Christofides, N., and Gerrard, M. (1981) A graph theoretic analysis of bounds for the quadratic assignment problem. In P. Hansen, editor, *Studies on graphs and discrete programming*, 61–68. North-Holland.

Dicky, J. W., and Hopkins, J. W. (1972) Campus building arrangement using TOPAZ. *Transportation Research*, 6, 59–68.

Drezner, Z. (2001) A new genetic algorithm for the quadratic assignment problem. Submitted for publication.

Elshafei, A. N. (1977) Hospital layout as a quadratic assignment problem. *Operations Research Quarterly*, 28, 167–179.

Finke, G., Burkard, R. E., and Rendl, F. (1987) Quadratic assignment problems. *Annals of Discrete Mathematics*, 31, 61–82.

Gilmore, P. C. (1962) Optimal and suboptimal algorithms for the quadratic assignment problem. *J. SIAM*, 10, 305–313.

Hadley, S. W., Rendl, F., and Wolkowicz, H. (1992) A new lower bound via projection for the quadratic assignment problem. *Mathematics of Operations Research*, 17(3), 727–739.

Heffley, D. R. (1977) Assigning runners to a relay team. In Ladany, S. P., Machol, R. E., editors, *Optimal Strategies in Sports*, 169–171. North-Holland, Amsterdam.

Helmberg, C., Mohar, B., Poljak, S., and Rendl, F. (1995) A spectral approach to bandwidth and separator problems in graphs. *Linear and Multilinear Algebra*, 39, 73–90.

Horn, R., and Johnson, C. (1985) *Matrix Analysis.* Cambridge University Press, New York.

Kaibel, V. (1997) *Polyhedral combinatorics of the quadratic assignment problem.* PhD thesis, Universität zu Köln, Germany.

Karisch, S. E. (1995) *Nonlinear approaches to quadratic assignment problems and graph partition problems.* PhD thesis, Technical University of Graz, Austria.

Karisch, S. E., Cela, E., Clausen, J., and Espersen, T. (1999) A dual framework for lower bounds of the quadratic assignment problem based on linearization. *Computing*, 63, 351–403.

Kaufman, T. C., and Broeckx, F. (1978) An algorithm for the quadratic assignment problem using Bender's decomposition. *European Journal of Operations Research*, 2, 204–211.

Koopmans, T. C., and Beckmann, M. J. (1957) Assignment problems and the location of economic activities. *Econometrica*, 25, 53–76.

Krarup, J., and Pruzan, P. M. (1978) Computer-aided layout design. *Mathematical Programming Study*, 9, 75–94.

Lawler, E. L. (1963) The quadratic assignment problem. *Management Science*, 9, 586–599.

Lawler, E. L., Lenstra, J. K., Rinnooy Kan, A. H. G., and Shmoys, D. B. (1985) *The traveling salesman problem: A guided tour of combinatorial optimization.* John Wiley & Sons.

Pardalos, P. M., and Wolkowicz, H., editors (1994) *Quadratic Assignment and related problems.* American Mathematical Society.

Resende, M. G. C., Ramakrishnan, K. G., and Drezner, Z. (1995) Computing lower bounds for the quadratic assignment problem with an interior point algorithm for linear programming. *Operations Research*, 43, 781–791.

Roucairol, C. (1987) A parallel branch and bound algorithm for the quadratic assignment problem. *Discrete Applied Mathematics*, 15, 211–225.

Sahni, S., and Gonzalez, T. (1976) P-complete approximation problems. *Journal of the Association of Computing Machinery*, 23, 555–565.

Steinberg, L. (1961) The backboard wiring problem: A placement algorithm. *SIAM Review*, 3, 37–50.

von Neumann, J. (1963) Some matrix inequalities and metrization of matrix space. In A.H. Taub, editor, *John von Neumann: Collected works, volume 4*, 205–219.

Zhao, Q. (1996) *Semidefinite programming for assignment and partition problems*. PhD thesis, University of Waterloo, Canada.

Zhao, Q., Karisch, S. E., Rendl, F., and Wolkowicz, H. (1998) Semidefinite programming relaxations for the quadratic assignment problem. *Journal of Combinatorial Optimization*, 2, 71–109.

[Dr. Jan Teichmann, Katrin Gerstenberg]

Why choose between learning and working?
Now you can do both – at SAP!

"Learning by doing" is the foundation of SAP's success. That applies to beginners, whom we immediately integrate into our team, as well as to our experienced staff members. Only people who continue to learn all the time are fit for interesting project tasks and responsibilities. SAP does a lot to keep your knowledge and skills on the cutting edge: from active personnel development all the way to individually tailored training programmes involving courses, seminars and workshops.

Have you successfully obtained your degree – computer science, business computing, business administration, (business) engineering, mathematics or physics – and perhaps already acquired skills in information technology and programming? Then you should definitely get acquainted with one of our departments for Development, Consulting or Service & Support: We rely on our interdisciplinary approach and professional teamwork in a motivating work atmosphere – and not on bureaucracy or hierarchical structures.

SAP AG is a global player in the field of standard business application software: More than 21,000 employees, represented in 50 countries. Would you like to know more?
Look us up at http://www.sap.com/germany/jobs

SAP AG · Postfach 1461 · 69185 Walldorf/Baden

Routing Technology for the Real World

ArcLogistics™ Route software reduces unnecessary and duplicate miles your vehicles travel by using a system based on a real street network. It adds money to your bottom line by

- Recommending the optimal use of vehicles and drivers for the minimum daily cost
- Helping your fleet make more stops in less time
- Allowing your business to grow with existing resources

Powered by geographic information system (GIS) standards from ESRI, ArcLogistics Route is an affordable, complete, and easy-to-use solution for your vehicle routing needs. Interested in saving 10% to 30% on your daily operating costs? Do the math. Then call ESRI to start saving money today.

Quality components from these fine companies

GDT

Seagate Crystal Reports

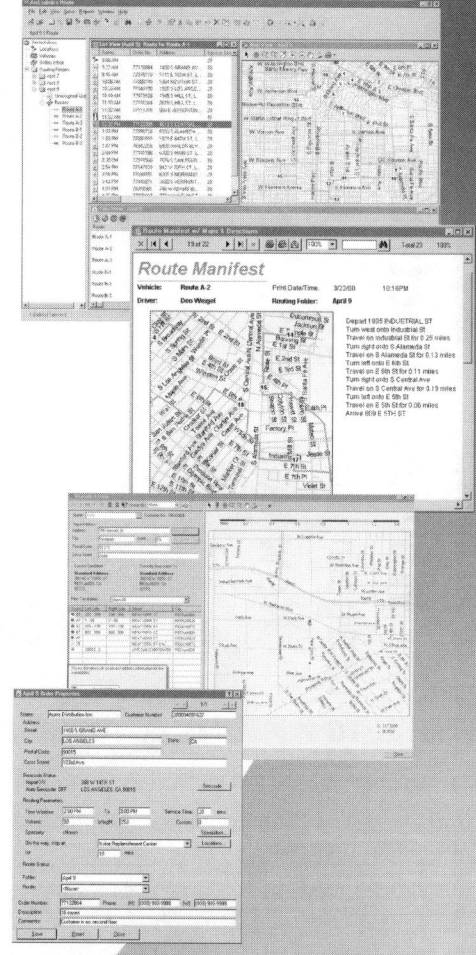

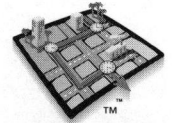

ArcLogistics Route
Desktop Vehicle Routing Software

ArcLogistics Route 2 now available

1-888-288-1386
www.esri.com/arclogisticsroute
E-mail: **info@esri.com**
Join the ArcLogistics Route Reseller Team.

Copyright © 2001 ESRI. All rights reserved. ESRI and the ESRI globe logo are trademarks of ESRI, registered in the United States and certain other countries; registration is pending in the European Community. ArcLogistics and the ArcLogistics Route logo are trademarks and @esri.com and www.esri.com are service marks of ESRI. The GDT logo is a trademark of Geographic Data Technology, Inc. The Seagate Crystal Reports logo is a trademark of Seagate Software. Other companies and products mentioned herein are trademarks or registered trademarks of their respective trademark owners.

Druck: Strauss Offsetdruck, Mörlenbach
Verarbeitung: Schäffer, Grünstadt